Compliments of

Bay Area Fire Protection Forum
and
CIGNA Loss Control Services

D0743150

National
Fire Alarm Code
Handbook

National Fire Alarm Code Handbook

Based on the 1993 edition of NFPA 72,
National Fire Alarm Code

Edited by
Wayne D. Moore

National Fire Protection Association
Quincy, Massachusetts

This first edition of the *National Fire Alarm Code Handbook* provides an essential reference for everyone associated with the design, selection, installation, or testing of fire alarm systems. It explains the requirements found in NFPA 72, *National Fire Alarm Code*. The 1993 edition of the code is the end result of a six-year project combining ten different NFPA standards into one comprehensive document. This project was accomplished by the committees who write NFPA's codes and standards. These committees consist of a broad range of representatives from many related fields, including equipment manufacturers, code enforcers, insurers, safety officers, testing laboratories, and fire protection engineers.

All NFPA codes and standards are processed in accordance with NFPA's *Regulations Governing Committee Projects*. The commentary in this handbook is the opinion of the author(s), recognized experts in the field of fire alarm systems. It is not, however, processed in accordance with the NFPA *Regulations Governing Committee Projects*, and therefore shall not be considered to be, nor relied upon as, a Formal Interpretation of the meaning or intent of any specific provision or provisions of NFPA 72. The language contained in the code, rather than this commentary, represents the official position of the Committees on Signaling Systems and the NFPA.

The handbook contains the complete text of the code, along with explanatory commentary, which is integrated between requirements. Commentary text answers questions frequently asked about the code, and includes intent and interpretations, historical perspective on provisions, application of requirements, cross-references to previous editions, and reasoning behind the code changes. All of the commentary material has been researched, written, and reviewed by experts in the field of fire alarm systems — members of the committees that develop the code.

This handbook is a must for all who design, install, inspect, approve, or maintain signaling systems.

®Registered trademark of the National Fire Protection Association. *National Electrical Code*® and *Life Safety Code*® are registered trademarks of the National Fire Protection Association.

Project Manager: Jennifer Evans

Project Editor: Elizabeth Contarino

Composition: Kathy Barber, Cathy Ray

Illustrations: George Nichols

Cover Design: Chris Jagmin Design

Production: Donald McGonagle and Stephen Dornbusch

Copyright © 1994

National Fire Protection Association

All rights reserved.

NFPA No. 72HB94

ISBN 0-87765-392-5

Library of Congress Catalog Card Number 93-87464

Printed in the United States of America

First Printing, January 1994

96   95   94   3   2   1

Dedication

This handbook is dedicated to the memory of Joseph J. Carney, Jr., who, during his career in the fire alarm systems field, provided the example for all professionals to follow.

Contents

Figure Credits ix

Preface xi

Chapter 1 Fundamentals of Fire Alarm Systems . . . 3

1-1 Scope 4

1-2 Purpose 4

1-3 General 5

1-4 Definitions 6

1-5 Fundamentals 22

1-6 System Interfaces 34

1-7 Documentation 34

Chapter 2 Household Fire Warning Equipment . . 41

2-1 General 42

2-2 Basic Requirements 44

2-3 Power Supplies 47

2-4 Equipment Performance 50

2-5 Installation 54

2-6 Maintenance and Tests 58

2-7 Markings and Instructions 59

Chapter 3 Protected Premises Fire Alarm Systems . . 61

3-1 Scope 62

3-2 General 62

3-3 Applications 64

3-4 Performance of Initiating Device, Notification Appliance, and Signaling Line Circuits 64

3-5 Performance and Capacities of Initiating Device Circuits (IDC) 66

3-6 Performance and Capacities of Signaling Line Circuits (SLC) 69

3-7 Notification Appliance Circuits (NAC) 71

3-8 System Requirements 73

3-9 Fire Safety Control Functions 84

3-10 Suppression System Actuation 87

3-11 Interconnected Fire Alarm Control Units . . 88

3-12 Emergency Voice/Alarm Communications 89

3-13 Special Requirements for Low Power Radio (Wireless) Systems 92

Chapter 4 Supervising Station Fire Alarm Systems . . 95

4-1 Scope 97

4-2 Communication Methods for Off-Premises Fire Alarm Systems 97

4-3 Fire Alarm Systems for Central Station Service 121

4-4 Proprietary Supervising Station Systems . . . 128

4-5 Remote Supervising Station Fire Alarm Systems 134

4-6 Public Fire Alarm Reporting Systems 137

4-7 Auxiliary Fire Alarm Systems 153

Chapter 5   Initiating Devices   159

   5-1   General   161

   5-2   Heat-Sensing Fire Detectors   169

   5-3   Smoke-Sensing Fire Detectors   181

   5-4   Radiant Energy-Sensing Fire Detectors   199

   5-5   Gas-Sensing Fire Detectors   207

   5-6   Other Fire Detectors   210

   5-7   Sprinkler Waterflow Alarm-Initiating
          Devices   211

   5-8   Detection of the Operation of Other Automatic
          Extinguishing Systems   212

   5-9   Manually Actuated Alarm-Initiating
          Devices   213

   5-10   Supervisory Signal-Initiating Devices   216

   5-11   Smoke Detectors for Control of Smoke
          Spread   218

**Chapter 6   Notification Appliances for Fire Alarm
                  Systems**   227

   6-1   Scope   227

   6-2   General   228

   6-3   Audible Characteristics   230

   6-4   Visible Characteristics, Public Mode   232

   6-5   Visible Characteristics, Private Mode   237

   6-6   Supplementary Visible Signaling Method . .   237

   6-7   Coded Appliance Characteristics   238

   6-8   Textual Audible Appliances   238

   6-9   Textual Visible Appliances   239

Chapter 7   Inspection, Testing, and Maintenance . .   241

   7-1   General   242

   7-2   Test Methods   245

   7-3   Inspection and Testing Frequency   258

   7-4   Maintenance   264

   7-5   Records   265

Chapter 8   Referenced Publications   275

Appendix A   Explanatory Material   277

**Appendix B   Engineering Guide for Automatic Fire
                    Detector Spacing**   279

   B-1   Introduction   280

   B-2   Fire Development and Ceiling Height
          Considerations   281

   B-3   Heat Detector Spacing   286

   B-4   Analysis of Existing Heat Detection Systems .   324

   B-5   Smoke Detector Spacing for Flaming Fires . .   365

   B-6   Theoretical Considerations   367

Appendix C   Referenced Publications   371

Cross-References to Previous Editions   375

Index   391

Figure Credits

Chapter 1

1.1	FIREPRO Incorporated, Burlington, MA
1.2	FIREPRO Incorporated, Burlington, MA
1.3	FIREPRO Incorporated, Burlington, MA
1.4	AFA Protective Systems, Inc., North Brunswick, NJ
1.5	Mammoth Fire Alarms, Lowell, MA
1.6	AFA Protective Systems, Inc. (Radionics), North Brunswick, NJ
1.7	AFA Protective Systems, Inc. (Ademco), North Brunswick, NJ
1.8	AFA Protective Systems, Inc. (Ademco AlarmNet), North Brunswick, NJ
1.9	AFA Protective Systems, Inc. (Ademco AlarmNet), North Brunswick, NJ
1.10	Fire Control Instruments, Newton, MA
1.11	Simplex, Gardner, MA
1.12	FIREPRO Incorporated, Burlington, MA
1.13	Mammoth Fire Alarms (Gamewell), Lowell, MA
1.14	Mammoth Fire Alarms (Gamewell), Lowell, MA
1.15	Mammoth Fire Alarms, Lowell, MA
1.16	FIREPRO Incorporated, Burlington, MA
1.17	FIREPRO Incorporated, Burlington, MA
1.18	Simplex, Gardner, MA
1.19	Audiosone Corp., Stratford, CT
1.20	Simplex, Gardner, MA
1.21	Fire Control Instruments, Newton, MA

Chapter 2

2.2(a)	Wheelock, Inc., Long Branch, NJ
2.2(b)	Mammoth Fire Alarms (Gentex), Lowell, MA
2.3	Mammoth Fire Alarms (ESL, Inc.), Lowell, MA
2.4	Mammoth Fire Alarms (System Sensor), Lowell, MA
2.5	Mammoth Fire Alarms (ESL, Inc.), Lowell, MA
2.6	Mammoth Fire Alarms (Edwards Systems Technology), Lowell, MA
2.7	the Fire Protection Alliance, Inc., North Reading, MA
2.8	Radionics, Salinas, CA
2.9	World Electronics, Inc., Coral Springs, FL

Chapter 3

3.1	Audiosone Corp., Stratford, CT
3.2(a)	the Fire Protection Alliance, Inc., North Reading, MA
3.2(b)	the Fire Protection Alliance, Inc., North Reading, MA
3.3	Underwriters Laboratories, Inc., Northbrook, IL
3.4	FIREPRO Incorporated, Burlington, MA
3.5	Mammoth Fire Alarms (System Sensor), Lowell, MA
3.6	Potter Electric Signal Co., St. Louis, MO
3.7	FIREPRO Incorporated, Burlington, MA
3.8	Potter Electric Signal Co., St. Louis, MO
3.9	Potter Electric Signal Co., St. Louis, MO
3.10	FIREPRO Incorporated, Burlington, MA
3.11	FIREPRO Incorporated, Burlington, MA
3.12	Kidde-Fenwal Protection Systems, Ashland, MA
3.13	Simplex, Gardner, MA
3.14	Simplex, Gardner, MA
3.15	World Electronics, Inc., Coral Springs, FL

Chapter 4

4.1–4.6	Robert P. Schifiliti Associates (RPSA) Inc., Reading, MA

Chapter 5

5.1(a)	Mammoth Fire Alarms (ESL, Inc.), Lowell, MA
5.1(b)	Mammoth Fire Alarms (Chemetron), Lowell, MA
5.2	J.M. Cholin Consultants, Inc., Oakland, NJ
5.3(a)	Mammoth Fire Alarms (Kidde-Fenwal), Lowell, MA
5.3(b)	Mammoth Fire Alarms (Fire Control Instruments), Lowell, MA
5.4	Mammoth Fire Alarms (Edwards Systems Technology, Chemetron), Lowell, MA
5.5	J.M. Cholin Consultants, Inc., Oakland, NJ
5.6	Mammoth Fire Alarms (Chemetron), Lowell, MA
5.7(a)	Protectowire Company, Hanover, MA
5.7(b)	Protectowire Company, Hanover, MA
5.8	Kidde-Fenwal Protection Systems, Ashland, MA
5.9	Mammoth Fire Alarms (Thermotech), Lowell, MA
5.10	Mammoth Fire Alarms (Chemetron), Lowell, MA
5.11(a)	Mammoth Fire Alarms (Chemetron), Lowell, MA
5.11(b)	Mammoth Fire Alarms (Edwards Systems Technology), Lowell, MA
5.12	Mammoth Fire Alarms (ESL, Inc.), Lowell, MA

5.13	System Sensor, St. Charles, IL
5.14	System Sensor, St. Charles, IL
5.15	Kidde-Fenwal Protection Systems, Ashland, MA
5.16	Kidde-Fenwal Protection Systems, Ashland, MA
5.17(a)	J.M. Cholin Consultants, Inc., Oakland, NJ
5.17(b)	Simplex, Gardner, MA
5.18	J.M. Cholin Consultants, Inc., Oakland, NJ
5.19(a)	ESL, Inc., Portland, OR
5.19(b)	Kidde-Fenwal Protection Systems, Ashland, MA
5.20	IEI(VESDA), Hingham, MA (NFPA Journal)
5.21(a)	Kidde-Fenwal Protection Systems, Ashland, MA
5.21(b)	Kidde-Fenwal Protection Systems, Ashland, MA
5.21(c)	IEI(VESDA), Hingham, MA
5.21(d)	IEI(VESDA), Hingham, MA
5.22–5.28	J.M. Cholin Consultants, Oakland, NJ
5.29	Kidde-Fenwal Protection Systems, Ashland, MA
5.30	Kidde-Fenwal Protection Systems, Ashland, MA
5.31(a)	Simplex, Gardner, MA
5.31(b)	Simplex, Gardner, MA
5.32	Mammoth Fire Alarms (Gamewell), Lowell, MA
5.33–5.36	Potter Electric Signal Co., St. Louis, MO
5.37	Mammoth Fire Alarms (Edwards Systems Technology), Lowell, MA
5.38	Mammoth Fire Alarms (Edwards Systems Technology), Lowell, MA

Chapter 6

6.1	Edwards Systems Technology, Norwalk, MA
6.2	Wheelock, Inc., Long Branch, NJ
6.3	Gentex Corp., Zeeland, MI
6.4	Gentex Corp., Zeeland, MI
6.5	Audiosone Corp., Stratford, CT
6.6	Wheelock, Inc., Long Branch, NJ
6.7	Radionics, Inc., Salinas, CA

Chapter 7

7.1	Simplex, Gardner, MA
7.2	Simplex, Gardner, MA
7.3	ESL, Inc., Portland, OR

Preface

"We've come full circle. The 'Signaling Standard' (as it was called in 1898) began as a single document, but certainly not as comprehensive as the one we now have." So states Charlie Zimmerman, one of the sages in our industry. Although he is officially retired, Charlie's interest in the development process of the codes and standards has not diminished. He was there when the signaling standard was divided into NFPA 71, 72A, 72B, 72C, and 72D. He was there later when NFPA 72E, under the chairmanship of Ed Reid, also became a stand-alone document.

Throughout the last 60 years, each standard has weathered changes — some controversial, some far-reaching — but always in the interest of a single goal: providing reliable fire alarm systems for the general public. The goal of the 1993 *National Fire Alarm Code* is the same.

The obvious reason for the current recombination effort is to put all of the standards in one location, without the duplication found in previous editions. Along with this recombined information came both the need and opportunity to develop this first edition of the handbook to assist the users of the code.

Fire alarm systems are normally required by building codes, the *Life Safety Code®*, as well as state and local jurisdictions. Many systems are installed to meet the building owner's fire protection goals. In any of these cases, this code provides the **minimum** requirements for the application, location, and device/appliance spacing and location. The design of fire protection systems, including fire alarm systems covered by this code, is an engineering discipline requiring knowledge of special concepts. This knowledge is not normally at the command of engineers or technicians who have not been specifically trained in this field. Neither the code nor this handbook contain every fact and/or concept that may be relevant to the design of a fire alarm system for every specific situation.

I wish to acknowledge the time and effort spent by Dean K. Wilson, John M. Cholin, Joseph A. Drouin, Irving Mande, Robert McPherson, Charles Zimmerman, and Robert P. Schifiliti, the contributors to much of this handbook. Dean Wilson, in particular, provided constant technical advice, research information, and moral support in addition to writing the commentary for Chapters 4, 6, and 7.

Joe Drouin and John Cholin are responsible for the commentary for Chapters 1 and 5 respectively. Irv Mande's assistance and direction with Chapter 3 commentary; Robert McPherson's review of Chapter 1, 3, and 4 commentary; and Bob Schifiliti's review of Chapter 5 and 6 commentary proved invaluable. Charles Zimmerman provided the historical background of the fire alarm standards as well as the insight to the changes made to the standards over the years.

This handbook would not be possible without the efforts of all the individuals, past and present, who have willingly given their time and talent to the NFPA standards-making process, specifically in this case, those involved with the *National Fire Alarm Code*, NFPA 72.

I would be remiss if I did not acknowledge the people "behind the scenes" who help take a project like this handbook from concept to finished product. The NFPA staff members who helped make this possible include Elizabeth Contarino, Mark Earley, Mark Ode, and the project coordinator, Jennifer Evans. I especially thank Jen Evans for her patience and guidance through the publication process.

And finally, without the support and understanding of my wife Maureen and all of our children, this project would never have seen the "light of day." They endured my absence during many evenings and weekends while I "camped" at the computer. I thank them for their love and patience.

WAYNE D. MOORE

About the Editor

Wayne D. Moore is a graduate engineer with over 23 years of practical experience in the fire detection systems field. He is the founder and president of the Fire Protection Alliance, a North Reading, MA, fire protection consulting and engineering firm. He currently serves as chairman of the NICET Fire Alarm Systems Certification Exam Committee and is chairman of both the NFPA 72-1993 Protected Premises Committee and the Fire Detection Institute.

Mr. Moore has written numerous articles and is a contributor to NFPA's *Fire Protection Handbook* and *Fire Alarm Signaling Systems Handbook*. He is an instructor of the NFPA Fire Alarm Systems Seminar and has contributed to its development. He is also co-editor of the national newsletter, the *Moore-Wilson Signaling Report,* and teaches a course on fundamentals of fire alarm systems at Northeastern University.

How to Use this Handbook

The text and illustrations that make up the commentary on the various sections of NFPA 72 are printed in black. The text of the code itself is printed in blue.

The Formal Interpretations included in this handbook were issued as a result of questions raised on specific editions of the code. They apply to all previous and subsequent editions in which the text in question remains substantially unchanged. Formal Interpretations are not part of the code and therefore are printed in black in a shaded blue box.

Paragraphs that begin with the letter "A" are extracted from Appendix A of the code. Appendix A material is not mandatory. It is designed to help users apply the provisions of the code. In this handbook, material from Appendix A is integrated with the text, so that it follows the paragraph it explains. An asterisk (*) following a paragraph number indicates that explanatory material from Appendix A will follow.

A cross-reference to previous editions appears at the end of the handbook.

1

Fundamentals of Fire Alarm Systems

Contents, Chapter 1

1-1 Scope
1-2 Purpose
1-3 General
1-4 Definitions
1-5 Fundamentals
  1-5.1 Common System Fundamentals
    1-5.1.2 Equipment
  1-5.2 Power Supplies
    1-5.2.1 Scope
    1-5.2.2 Code Conformance
    1-5.2.3 Power Sources
    1-5.2.4 Primary Supply
    1-5.2.5 Secondary Supply Capacity and Sources
    1-5.2.6 Continuity of Power Supplies
      1-5.2.6.1 Uninterruptible Power System Bypass
    1-5.2.7 Power Supply for Remotely Located Control
        Equipment
    1-5.2.8 Light and Power Service
      1-5.2.8.3 Overcurrent Protection
    1-5.2.9 Storage Batteries
      1-5.2.9.1 Location
      1-5.2.9.2 Battery Charging
      1-5.2.9.3 Overcurrent Protection
      1-5.2.9.4 Metering
      1-5.2.9.5 Under-Voltage Detection
    1-5.2.10 Engine-Driven Generator
      1-5.2.10.3 Capacity
      1-5.2.10.4 Fuel
    1-5.2.11 Primary Batteries
      1-5.2.11.1 Location
      1-5.2.11.2 Separation of Cells
      1-5.2.11.3 Capacity
  1-5.3 Compatibility
  1-5.4 System Functions
    1-5.4.1 Local Fire Safety Functions
    1-5.4.2 Alarm Signals
      1-5.4.2.1 Coded Alarm Signal
    1-5.4.3 Supervisory Signals
      1-5.4.3.1 Coded Supervisory Signal
      1-5.4.3.2 Combined Coded Alarm and Supervisory
          Signal Circuits

1-5.4.6 Trouble Signal
  1-5.4.6.1 General
  1-5.4.6.4 Audible Trouble Signal Silencing Switch
1-5.4.7 Distinctive Signals
1-5.4.8 Alarm Signal Deactivation
1-5.4.9 Supervisory Signal Silencing
1-5.4.10 Presignal Feature
1-5.5 Performance and Limitations
  1-5.5.1 Voltage, Temperature, and Humidity Variation
  1-5.5.2 Installation and Design
  1-5.5.4 Wiring
  1-5.5.5 Grounding
  1-5.5.6 Initiating Devices
1-5.6 Protection of Control Equipment
1-5.7 Visible Indication (Annunciation)
  1-5.7.1 Visible Zone Alarm Indication
  1-5.7.1.2 Zone of Origin
1-5.8 Monitoring Integrity of Installation Conductors and
    Other Signaling Channels
  1-5.8.5 Monitoring Integrity of Emergency Voice/Alarm
    Communication Systems
  1-5.8.5.1 Monitoring Integrity of Speaker Amplifier
    and Tone-Generating Equipment
  1-5.8.6 Monitoring Integrity of Power Supplies
1-6 System Interfaces
1-7 Documentation
  1-7.1 Approval and Acceptance
  1-7.2 Certificate of Completion
  1-7.2.3 Central Station Fire Alarm Systems
  1-7.3 Records

1-1 Scope.

This code deals with the application, installation, performance, and maintenance of fire alarm systems and their components.

This code provides the minimum test, maintenance, and performance requirements for fire alarm systems. The application, location, and limitations of fire alarm components such as manual fire alarm boxes, detectors, and sprinkler devices is also provided.

1-2 Purpose.

1-2.1* The purpose of this code is to define the means of signal initiation, transmission, notification, and annunciation; the levels of performance; and the reliability of the various types of fire alarm systems. This code defines the features associated with these systems, and also provides the information necessary to modify or upgrade an existing system to meet the requirements of a particular system classification. It is the intent of this code to establish the required levels of performance, extent of redundancy, and quality of installation, but not the methods by which these requirements are to be achieved.

This code describes the various types of initiating devices, supervisory devices, visible and audible notification appliances, and how and where they should be used. Types of systems and methods of system transmission acceptable for each type of system are described, as are system reliability and performance requirements.

A-1-2.1 In determining the performance criteria of circuits, consult the performance and capacity tables in Chapters 3 and 4. On modifying an existing system, the system should be tested to determine the style of each circuit for the proper description and understanding of the system.

Chapters 3 and 4 provide system loading capacity tables for protected premises fire alarm systems and for each type of available off-premises system transmission method. Any time a system is modified, it is very important to have a thorough understanding of the existing equipment and its capabilities, and the system's wiring style, type, and configuration. In many cases the existing equipment is older technology and may not interface easily with the planned additional equipment, or may not be able to be modified to conform to current code requirements.

1-2.2 Any reference or implied reference to a particular type of hardware is for the purpose of clarity and shall not be interpreted as an endorsement.

NFPA does not manufacture, distribute, endorse, approve, or list products or components.

1-3 General.

1-3.1 This code classifies fire alarm systems as follows:

(a) Household fire warning systems

Household fire warning systems are installed in households to warn the occupants of a fire emergency so they may immediately evacuate the building. The system may consist of a complete fire alarm system with initiating devices and notification appliances, or of single or multiple station smoke detectors located in family living units. Single and multiple station smoke detectors are also used in hotel rooms and living units of apartment buildings. Household fire warning systems are incorporated in Chapter 2 in this code.

(b) Protected premises fire alarm systems

The primary purpose of a protected premises fire alarm system is to warn building occupants to evacuate the premises; its secondary purpose is to activate the building fire protection features for property protection. Chapter 3 of this code covers protected premises fire alarm systems.

1. Local fire alarm systems

The local fire alarm system provides the evacuation notification and system functions only at the protected premises and is the most common form of a protected premises fire alarm system.

(c) Off-premises fire alarm systems

Off-premises fire alarm systems, described in Chapter 4, provide the means to communicate between the protected premises and a location called a *supervising station*. The types of off-premises fire alarm systems are listed below.

1. Auxiliary fire alarm systems

These systems provide a direct means of communicating between the protected premises and the fire department using fire-department-maintained transmission lines.

(i) Local energy type

Contacts in the fire alarm control unit activate a city fire alarm box mounted on the inside or outside of the building, initiating an alarm at the fire department. Power to activate this system is provided from the fire alarm control unit.

The fire alarm control unit circuit is isolated from the city circuit.

(ii) Parallel telephone type

In one version, telephone lines are used to communicate between each protected premises fire alarm box and the fire department. In another version, the leased telephone circuits interconnect between the municipal communication center and a supervised circuit of the fire alarm system.

(iii) Shunt type

A closed contact at the protected premises is electrically connected to and is an integral part of the city circuit. Shunt circuits are normally connected to waterflow initiating devices and a manual fire alarm box. An open in the reporting circuit will send an alarm to the fire department. A shunt type system has very specific requirements and is not allowed to be interconnected to a protected premises system unless the city circuits entering the protected premises are installed in rigid conduit.

2. Remote station fire alarm systems

These systems provide a means of transmitting alarm and trouble signals from the protected premises to a remote supervising station. These systems are normally used in remote locations where no municipal system is available, but where the local fire department has installed a remote supervising station to receive fire alarm signals.

3. Proprietary fire alarm systems

These systems require trained, competent personnel in constant attendance at a supervising station that monitors the protected premises and takes appropriate action when necessary. The supervising station is located at the protected property and is owned and operated by the property owner. The property may consist of a single building, such as a high-rise building, or several buildings, such as a college campus where the dormitories and other buildings report to a supervising station on campus.

The property may be contiguous or noncontiguous. If noncontiguous, it consists of protected property at remote locations, such as across town.

4. Central station fire alarm systems

These are privately owned systems or groups of systems that provide competent operators, who, upon receipt of a signal from the protected premises, take appropriate action. Central station service is controlled and operated by an individual or a group whose business is the furnishing, maintaining, and monitoring of supervised fire alarm systems.

5. Municipal fire alarm systems.

These are systems used to transmit alarms to the fire department from street locations, using street fire alarm boxes, or from a building, using master fire alarm boxes.

1-3.2 A device or system having materials or forms different from those detailed in this code shall be permitted to be examined and tested according to the intent of the code and, if found equivalent, shall be approved.

A device or system that does not meet the specific requirements of this code may be submitted to a testing laboratory for listing by the laboratory to determine if the device or system meets the intent of the code. The authority having jurisdiction has the responsibility to determine whether or not a product or device is suitable once it is in the field.

1-3.3 The intent and meaning of the terms used in this code are, unless otherwise defined herein, the same as those of NFPA 70, *National Electrical Code.®*

1-4 Definitions.

For the purposes of this code, the following terms have the meanings shown below:

Active Multiplex System. A multiplexing system in which transponders are employed to transmit status signals of each initiating device or initiating device circuit within a prescribed time interval.

Active Signaling Element. A component within a circuit interface such as a transistor, silicon controlled rectifier, or relay whose function is to impress a signal on the multiplexed signaling line circuit.

Addressable Device. A fire alarm system component with discreet identification that can have its status individually identified or that is used to individually control other functions.

Addressable devices may be either initiating or control devices. An initiating device will provide the control panel with its status and actual location. The control device will operate specific circuits or devices.

Adverse Condition. Any occurrence to a communications or transmission channel that interferes with the proper transmission and/or interpretation of status change signals at the supervising station. (*See also Trouble Signal.*)

An open, short circuit or electrical interference on the transmission line that prevents the normal operation of the system are adverse conditions.

Air Sampling-Type Detector. A detector that consists of a piping or tubing distribution network from the detector to the area(s) to be protected. An aspiration fan in the detector housing draws air from the protected area back to the detector through air sampling ports, piping, or tubing. At the detector, the air is analyzed for fire products.

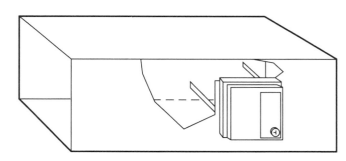

Figure 1.1 Duct smoke detector.

Alarm. A warning of fire danger.

Only the word "alarm" is used to indicate a fire alarm condition. The phrases "trouble alarm" or "supervisory alarm" are not acceptable references.

Alarm Service. The service required following the receipt of an alarm signal.

Alarm service provides immediate retransmission of the alarm to the fire department, dispatching of a runner to the protected premises for either an alarm or trouble signal, notification of the subscriber, and notification of the authority having jurisdiction (if the signal received was not an alarm), when necessary.

Alarm Signal. A signal indicating an emergency requiring immediate action, such as a signal indicative of fire.

Alarm Verification Feature. A feature of automatic fire detection and alarm systems to reduce unwanted alarms wherein smoke detectors must report alarm conditions for a minimum period of time, or confirm alarm conditions within a given time period, after being reset to be accepted as a valid alarm initiation signal.

This feature has been used in facilities with a history of unwanted alarms. It is available as a listed component of most fire alarm control units. The timing sequence and operation is defined by the listing laboratory.

Alert Tone. An attention-getting signal to alert occupants of the pending transmission of a voice message.

The alert tone is used in voice communication systems to warn the occupants that a voice announcement is about to be made. The alert tone is not considered an alarm.

Analog Initiating Device (Sensor). An initiating device that transmits a signal indicating varying degrees of condition as contrasted with a conventional initiating device, which can only indicate an on/off condition.

Examples of analog devices are devices that can measure and transmit smoke density, temperature variation, water level, or water pressure, etc., to a fire alarm control unit. New smoke detector technology uses the analog feature to provide a warning signal to the owner when the detector is dirty or when the detector drifts outside of its listed sensitivity values.

Annunciator. A unit containing two or more indicator lamps, alpha-numeric displays, or other equivalent means in which each indication provides status information about a circuit, condition, or location.

Approved. Acceptable to the "authority having jurisdiction."

NOTE: The National Fire Protection Association does not approve, inspect or certify any installations, procedures, equipment, or materials nor does it approve or evaluate testing laboratories. In determining the acceptability of installations or procedures, equipment or materials, the authority having jurisdiction may base acceptance on compliance with NFPA or other appropriate standards. In the absence of such standards, said authority may require evidence of proper installation, procedure or use. The authority having jurisdiction may also refer to the listings or labeling practices of an organization concerned with product evaluations which is in a position to determine compliance with appropriate standards for the current production of listed items.

The authority having jurisdiction may require a product to be listed or labeled, but the listing or label alone does not constitute approval. The authority having jurisdiction has the right to approve products or systems that are not labeled or listed.

Authority Having Jurisdiction. The "authority having jurisdiction" is the organization, office or individual responsible for "approving" equipment, an installation or a procedure.

NOTE: The phrase "authority having jurisdiction" is used in NFPA documents in a broad manner since jurisdictions and "approval" agencies vary as do their responsibilities. Where public safety is primary, the "authority having jurisdiction" may be a federal, state, local or other regional department or individual such as a fire chief, fire marshal, chief of a fire prevention bureau, labor department, health department, building official, electrical inspector, or others having statutory authority. For insurance purposes, an insurance inspection department, rating bureau, or other insurance company representative may be the "authority having jurisdiction." In many circumstances the property owner or his designated agent assumes the role of the "authority having jurisdiction"; at government installations, the commanding officer or departmental official may be the "authority having jurisdiction."

Automatic Extinguishing System Operation Detector. A device that detects the operation of an extinguishing system by means appropriate to the system employed.

Examples of alarm initiating devices provided to initiate an alarm are agent flow switches, agent pressure switches, or other appropriate means.

Automatic Extinguishing System Supervision. Devices that respond to abnormal conditions that could affect the proper operation of an automatic sprinkler system or other fire extinguishing system, including but not limited to control valves; pressure levels; liquid agent levels and temperatures; pump power and running, engine temperature and overspeed; and room temperature.

When an abnormal condition is detected, a supervisory signal different from the alarm or trouble signal is activated to warn the owner/attendant that the extinguishing system needs attention.

Automatic Fire Detectors. Fire is a phenomenon that occurs when a substance reaches a critical temperature and reacts chemically with oxygen (for example) to produce heat, flame, light, smoke, water vapor, carbon monoxide, carbon dioxide, or other products and effects.

An automatic fire detector is a device designed to detect the presence of fire and initiate action. For the purpose of this code, automatic fire detectors are classified as listed below.

Fire-Gas Detector. A device that detects gases produced by a fire.

Heat Detector. A device that detects abnormally high temperature or rate-of-temperature rise.

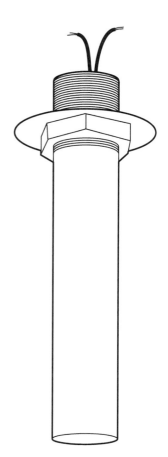

Figure 1.2 *Rate-compensated heat detector.*

Other Fire Detectors. Devices that detect a phenomenon other than heat, smoke, flame, or gases produced by a fire.

These devices are designed to detect the abnormal concentration of combustion effects that occur during a fire, such as water vapor, ionized molecules, hydrogen chloride, or other phenomena.

Radiant Energy Sensing Fire Detector. A device that detects radiant energy (such as ultraviolet, visible, or infrared) that is emitted as a product of combustion reaction and obeys the laws of optics.

Smoke Detector. A device that detects visible or invisible particles of combustion.

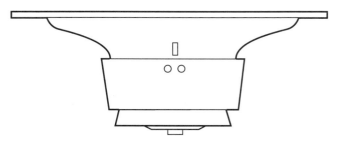

Figure 1.3 *Typical spot-type smoke detector.*

Auxiliary Box. A fire alarm box that can be operated from one or more remote actuating devices.

In an auxiliary fire alarm system, the auxiliary fire alarm box is normally located on the outside of a building. The protected premises fire alarm system is connected to this box and will automatically activate the box when an alarm occurs. The auxiliary box may also be manually activated.

Auxiliary Fire Alarm System. A system connected to a municipal fire alarm system for transmitting an alarm of fire to the public fire service communication center. Fire alarms from an auxiliary fire alarm system are received at the public fire service communication center on the same equipment and by the same methods as alarms transmitted manually from municipal fire alarm boxes located on streets.

(a) *Local Energy Type.* An auxiliary system that employs a locally complete arrangement of parts, initiating devices, relays, power supply, and associated components to automatically trip a municipal transmitter or master box over electric circuits that are electrically isolated from the municipal system circuits.

(b) *Parallel Telephone Type.* An auxiliary system connected by a municipally controlled individual circuit to the protected property to interconnect the initiating devices at the protected premises and the municipal fire alarm switchboard.

(c) *Shunt Auxiliary Type.* An auxiliary system electrically connected to an integral part of the municipal alarm system extending the municipal circuit into the protected premises to interconnect the initiating devices, which, when operated, open the municipal circuit shunted around the trip coil of the municipal transmitter or master box, which is thereupon energized to start transmission without any assistance whatsoever from a local source of power.

Box Battery. The battery supplying power for an individual fire alarm box where radio signals are used for the transmission of box alarms.

Bridging Point. The location where the distribution of signaling line circuits to trunk facilities or leg facilities, or both, occurs.

Carrier. High frequency energy that can be modulated by voice or signaling impulses.

Carrier System. A means of conveying a number of channels over a single path by modulating each channel on a different carrier frequency and demodulating at the receiving point to restore the signals to their original form.

Ceiling. The upper surface of a space, regardless of height. Areas with a suspended ceiling would have two ceilings, one visible from the floor and one above the suspended ceiling.

Ceiling Height. The height from the continuous floor of a room to the continuous ceiling of a room or space.

Ceiling Surfaces. Ceiling surfaces referred to in conjunction with the locations of initiating devices are as follows:

(a) *Beam Construction.* Ceilings having solid structural or solid nonstructural members projecting down from the ceiling surface more than 4 in. (100 mm) and spaced more than 3 ft (0.9 m), center to center.

(b) *Girders.* Girders support beams or joists and run at right angles to the beams or joists. When the top of girders are within 4 in. (100 mm) of the ceiling, they are a factor in determining the number of detectors and are to be considered as beams. When the top of the girder is more than 4 in. (100 mm) from the ceiling, it is not a factor in detector location.

Central Station. A supervising station that is listed for central station service.

The central station is the attended location where the central station fire alarm system is located, where alarm signals are received, and which provides runner service.

Central Station Fire Alarm System. A system or group of systems in which the operations of circuits and devices are transmitted automatically to, recorded in, maintained by, and supervised from a listed central station having competent and experienced servers and operators who, upon receipt of a signal, take such action as required by this code. Such service is to be controlled and operated by a person, firm, or corporation whose business is the furnishing, maintaining, or monitoring of supervised fire alarm systems.

Figure 1.4   Central station fire alarm system.

Central Station Service. The use of a system or a group of systems in which the operations of circuits and devices at a protected property are signaled to, recorded in, and supervised from a listed central station having competent and experienced operators who, upon receipt of a signal, take such action as required by this code. Related activities at the protected property such as equipment installation, inspection, testing, maintenance, and runner service are the responsibility of the central station or a listed fire alarm service - local company. Central station service is controlled and operated by a person, firm, or corporation whose business is the furnishing of such contracted services or whose properties are the protected premises.

Certificate of Completion. A document that acknowledges the features of installation, operation (performance), service, and equipment with representation by the property owner, system installer, system supplier, service organization, and the authority having jurisdiction.

The certificate of completion is a very important document used to confirm that a system has been installed properly and operates in conformance with this code. Many problems with fire alarm systems occur where the authority having jurisdiction does not require the certificate of completion and allows systems that are not properly installed.

Certification. A systematic program using randomly selected follow-up inspections of the certified systems installed under the program, which allows the listing organization to verify that a fire alarm system complies with all the requirements of this code. A system installed under such a program is identified by the issuance of a certificate and is designated as a certificated system.

Certification of Personnel. A formal program of related instruction and testing as provided by a recognized organization or the authority having jurisdiction.

NOTE: This definition applies only to municipal fire alarm systems.

Channel. A path for voice or signal transmission utilizing modulation of light or alternating current within a frequency band.

Circuit Interface. A circuit component that interfaces initiating devices and/or control circuits, indicating appliances and/or circuits, system control outputs, and other signaling line circuits to a signaling line circuit.

Combination Detector. A device that either responds to more than one of the fire phenomenon or employs more than one operating principle to sense one of these phenomenon. Typical examples are a combination of a heat detector with a smoke detector or a combination rate-of-rise and fixed-temperature heat detector.

Combination Fire Alarm and Guard's Tour Box. Manually operated box for separately transmitting a fire alarm signal and a distinctive guard patrol tour supervisory signal.

Combination System. A local fire alarm system for fire alarm, supervisory, or guard's tour supervisory service, or a household fire warning system whose components may be used in whole or in part in common with a nonfire signaling system, such as a paging system, a burglar alarm system, or a process

Figure 1.5 *Combination rate-of-rise and fixed temperature heat detector.*

monitoring supervisory system, without degradation of or hazard to the fire alarm system.

Communication Channel. A circuit or path connecting subsidiary station(s) to supervising station(s) over which signals are carried.

Compatibility Listed. A specific listing process that applies only to two-wire devices (such as smoke detectors) designed to operate with certain control equipment.

It is important that two-wire, circuit-powered detectors be listed for compatibility with a fire alarm control unit. Detectors that are not listed for this purpose may appear to operate properly under normal operating conditions but may not work under adverse conditions such as low voltage.

Compatible (Equipment). Equipment that interfaces mechanically or electrically together as manufactured without field modification.

Control Unit. A system component that monitors inputs and controls outputs through various types of circuits.

Delinquency Signal. A signal indicating the need of action in connection with the supervision of guards or system attendants.

This definition applies to guard's tour systems. If the guard fails to activate a tour station or fails to make a tour in a prescribed amount of time, the supervising station is notified.

Derived Channel. A signaling line circuit that uses the local leg of the public switched network as an active multiplex channel, while simultaneously allowing that leg's use for normal telephone communications.

Digital Alarm Communicator Receiver (DACR). A system component that will accept and display signals from digital alarm communicator transmitters (DACTs) sent over the public switched telephone network.

Figure 1.6   Digital alarm communicator receiver (DACR).

Digital Alarm Communicator System (DACS). A system in which signals are transmitted from a digital alarm communicator transmitter (DACT) located at the protected premises through the public switched telephone network to a digital alarm communicator receiver (DACR).

Digital Alarm Communicator Transmitter (DACT). A system component at the protected premises to which initiating devices or groups of devices are connected. The DACT will seize the connected telephone line, dial a preselected number to connect to a DACR, and transmit signals indicating a status change of the initiating device.

Digital Alarm Radio Receiver (DARR). A system component composed of two subcomponents: one that receives and decodes radio signals, the other that annunciates the decoded data. These two subcomponents can be coresident at the central station or separated by means of a data transmission channel.

Digital Alarm Radio System (DARS). A system in which signals are transmitted from a digital alarm radio transmitter (DART) located at a protected premises through a radio channel to a digital alarm radio receiver (DARR).

See Figure 1.8.

Figure 1.7   Digital alarm communicator transmitter (DACT).

Digital Alarm Radio Transmitter (DART). A system component connected to or an integral part of a DACT that is used to provide an alternate radio transmission channel.

See Figure 1.9.

Display. The visual representation of output data other than printed copy.

Dual Control. The use of two primary trunk facilities over separate routes or different methods to control one communication channel.

Evacuation. The withdrawal of occupants from a building.

NOTE: Evacuation does not include relocation of occupants within a building.

Evacuation Signal. Distinctive signal intended to be recognized by the occupants as requiring evacuation of the building.

In the appendix of the previous editions of NFPA 72, the code-three temporal pattern was recommended as the national evacuation signal. This signal meets the requirements of the *American National Standard Audible Emergency Evacuation Signal*, ANSI S3.41-1990. This code will make the ANSI evacuation signal a requirement for household fire warning systems and for all protected premises fire alarm systems effective July 1, 1996.

Exit Plan. Plan for the emergency evacuation of the premises.

Figure 1.8 Digital alarm radio system (DARS).

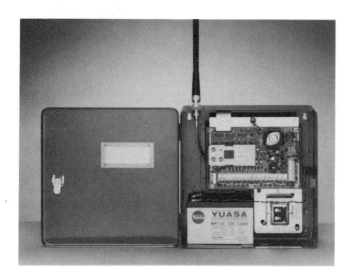

Figure 1.9 Digital alarm radio transmitter (DART).

Family Living Unit. That structure, area, room, or combination of rooms in which a family (or individual) lives. This is meant to cover living area only and not common usage areas in multifamily buildings such as corridors, lobbies, basements, etc.

Fire Alarm Control Unit (Panel). A system component that receives inputs from automatic and manual fire alarm devices and may supply power to detection devices and transponder(s) or off-premises transmitter(s). The control unit may also provide transfer of power to the notification appliances and transfer of condition to relays or devices connected to the control unit. The fire alarm control unit can be a local fire alarm control unit or master control unit.

Fire Alarm/Evacuation Signal Tone Generator. A device that, upon command, produces a fire alarm/evacuation tone.

Figure 1.10 Typical fire alarm control unit.

Fire Alarm Signal. A signal initiated by a fire alarm initiating device such as a manual fire alarm box, automatic fire detector, waterflow switch, or other device whose activation is indicative of the presence of a fire or fire signature.

Fire Alarm System. A system or portion of a combination system consisting of components and circuits arranged to monitor and annunciate the status of fire alarm or supervisory signal initiating devices and to initiate appropriate response to those signals.

Fire Command Center. The principal manned or unmanned location where the status of the detection, alarm communications, and control systems is displayed and from which the system(s) can be manually controlled.

Fire Rating. The classification indicating in time (hours) the ability of a structure or component to withstand fire conditions.

Fire Safety Function Control Device. The fire alarm system component that directly interfaces with the control system that controls the fire safety function.

These devices are used to control safety functions that increase the level of life safety and property protection. They may be operated automatically by the fire alarm control unit or operated manually. Examples of functions that these devices control are door unlocking, suppression system activation, and HVAC (heating, ventilating, and air conditioning) systems.

Figure 1.11 A fire command center.

Fire Safety Functions. Building and fire control functions that are intended to increase the level of life safety for occupants or to control the spread of harmful effects of fire.

Fire Warden. Building staff or tenant trained to perform assigned duties in the event of a fire emergency.

Guard Signal. A supervisory signal monitoring the performance of guard patrols.

Guard's Tour Supervision. Devices that are manually or automatically initiated to indicate the route being followed and the timing of a guard's tour.

Household. The family living unit in single-family detached dwellings, single-family attached dwellings, multifamily buildings, and mobile homes.

Household Fire Alarm System. A system of devices that produces an alarm signal in the household for the purpose of notifying the occupants of the presence of a fire so they may evacuate the premises.

Hunt Group. A group of associated telephone lines within which an incoming call is automatically routed to an idle (not busy) telephone line for completion.

Initiating Device. A system component that originates transmission of a change of state condition, such as a smoke detector, manual fire alarm box, supervisory switch, etc.

Initiating Device Circuit. A circuit to which automatic or manual initiating devices are connected where the signal received does not identify the individual device operated.

Integrated System. A computer-based control system, listed for use as a fire alarm system, in which certain components are common to nonfire monitoring and control functions.

An example of an integrated system is a proprietary system that uses a computer-based control system in a manufacturing facility to monitor and control fire alarm, security, guard's tour, and process monitoring functions.

Intermediate Fire Alarm or Fire Supervisory Control Unit. A control unit used to provide area fire alarm or area fire supervisory service that, when connected to the proprietary fire alarm system, becomes a part of that system.

A fire alarm control unit at a protected premises connected to a proprietary system would be part of the proprietary system and designated as an intermediate fire alarm control unit.

Labeled. Equipment or materials to which has been attached a label, symbol or other identifying mark of an organization acceptable to the "authority having jurisdiction" and concerned with product evaluation, that maintains periodic inspection of production of labeled equipment or materials and by whose labeling the manufacturer indicates compliance with appropriate standards or performance in a specified manner.

Leg Facility. That portion of a communication channel that connects not more than one protected premises to a primary or secondary trunk facility. The leg facility includes the portion of the signal transmission circuit from its point of connection with a trunk facility to the point where it is terminated within the protected premises at one or more transponders.

Level Ceilings. Those ceilings that are actually level or have a slope of $1\frac{1}{2}$ in. or less per ft (41.7 mm per m).

Line-Type Detector. A device in which detection is continuous along a path. Typical examples are rate-of-rise pneumatic tubing detectors, projected beam smoke detectors, and heat-sensitive cable.

Listed. Equipment or materials included in a list published by an organization acceptable to the "authority having jurisdiction" and concerned with product evaluation, that maintains periodic inspection of production of listed equipment or materials and whose listing states either that the equipment or material meets appropriate standards or has been tested and found suitable for use in a specified manner.

NOTE: The means for identifying listed equipment may vary for each organization concerned with product evaluation, some of which do not recognize equipment as listed unless it is also labeled. The "authority having jurisdiction" should utilize the system employed by the listing organization to identify a listed product.

Loading Capacity. The maximum number of discrete elements of fire alarm systems permitted to be used in a particular configuration.

The loading capacity of a system is dependent on the type of system, and the style and class of circuit being used. Chapters 3 and 4 provide the loading capacities for various styles and classes of circuits.

Local Control Unit (Panel). A control unit that serves the protected premises or a portion of the protected premises and indicates the alarm via notification appliances inside the protected premises.

Local Fire Alarm System.   A local system sounding an alarm at the protected premises as the result of the manual operation of a fire alarm box or the operation of protection equipment or systems, such as water flowing in a sprinkler system, the discharge of carbon dioxide, the detection of smoke, or the detection of heat.

Local Supervisory System.   A local system arranged to supervise the performance of guard's tours, or the operative condition of automatic sprinkler systems or other systems for the protection of life and property against a fire hazard.

Local System.   A system that produces a signal at the premises protected.

Loss of Power.   The reduction of available voltage at the load below the point at which equipment will function as designed.

Low Power Radio Transmitter.   Any device that communicates with associated control/receiving equipment by some kind of low power radio signals.

Maintenance.   Repair service, including periodically recurrent inspections and tests, required to keep the fire alarm system and its component parts in an operative condition at all times, together with replacement of the system or of its components, when for any reason they become undependable or inoperable.

Manual Fire Alarm Box.   A manually operated device used to initiate an alarm signal.

Master Box.   A municipal fire alarm box that may also be operated by remote means.

Figure 1.12   Manual fire alarm box.

Figure 1.13   Master box.

Master Control Unit (Panel).   A control unit that serves the protected premises or portion of the protected premises as a local control unit and accepts inputs from other fire alarm control units.

Where more than one fire alarm control unit has been installed in a facility, one of the control units must act as the master control unit to monitor alarm, trouble, and supervisory conditions from the other panels.

Multiple Station Alarm Device.   Two or more single-station alarm devices that may be interconnected so that actuation of one causes all integral or separate audible alarms to operate. It may also consist of one single-station alarm device having connections for other detectors or manual fire alarm box.

Multiplexing.   A signaling method characterized by simultaneous or sequential transmission, or both, and reception of multiple signals on a signaling line circuit or a communication channel including means for positively identifying each signal.

Municipal Fire Alarm Box (Street Box).   An enclosure housing a manually operated transmitter used to send an alarm to the public fire service communication center.

See Figure 1.14.

Municipal Fire Alarm System.   A system of alarm initiating devices, receiving equipment, and connecting circuits (other than a public telephone network) used to transmit alarms from street locations to the public fire service communication center.

Figure 1.14 Municipal fire alarm box (street box).

Figure 1.15 Nonrestorable fixed-temperature heat detector.

ment to access the public switched network and automatically provide an off-hook condition prior to transmission.

On-Hook.   To disconnect from the public switched telephone network.

Municipal Transmitter.   A transmitter that can only be tripped remotely, used to send an alarm to the public fire service communication center.

Nonrestorable Initiating Device.   A device whose sensing element is designed to be destroyed in the process of operation.

A fixed-temperature heat detector using a fusible element that melts when subjected to heat is an example of a nonrestorable initiating device.

Notification Appliance.   A fire alarm system component such as a bell, horn, speaker, strobe, printer, etc., that provides an audible or visible output, or both.

See Figures 1.16 and 1.17.

Notification Appliance Circuit.   A circuit or path directly connected to a notification appliance(s).

Off-Hook.   To make connection with the public switched telephone network in preparation to dial a telephone number.

When a telephone is raised from the receiver, it is considered off-hook. Some transmission technologies use equip-

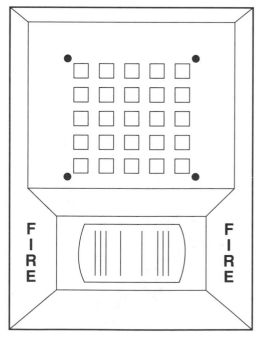

Figure 1.16 Speaker/strobe.

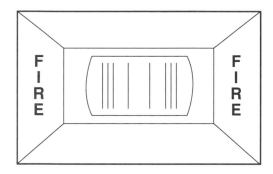

Figure 1.17 Strobe.

When a telephone is returned to the receiver, it is considered on-hook. Some transmission technologies accomplish this function automatically.

Ownership. Any property, building, contents, etc., under legal control by occupant, by contract, or by holding of title or deed.

Paging System. A system intended to page one or more persons such as by means of voice over loudspeaker, by means of coded audible signals or visible signals, or by means of lamp annunciators.

See Figures 1.18 an 1.19.

Parallel Telephone System. A telephone system in which an individually wired circuit is used for each fire alarm box.

Permanent Visual Record (Recording). Immediately readable, not easily alterable, print, slash, punch, etc., listing all occurrences of status change.

Plant. One or more buildings under the same ownership or control on a single property.

Figure 1.18 Fire fighter using paging system.

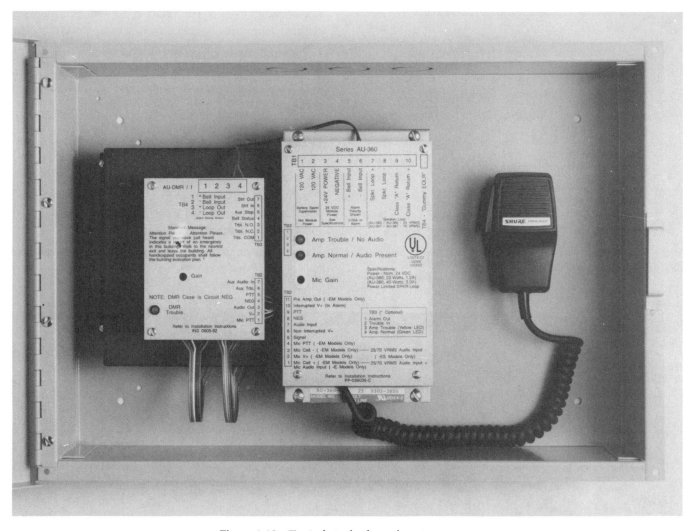

Figure 1.19 *Typical single channel paging system.*

Positive Alarm Sequence. An automatic sequence that results in an alarm signal, even if manually delayed for investigation, unless the system is reset.

Power Supply. A source of electrical operating power including the circuits and terminations connecting it to the dependent system components.

Primary Battery (Dry Cell). A nonrechargeable battery requiring periodic replacement.

Primary Trunk Facility. That part of a transmission channel connecting all leg facilities to a supervising or subsidiary station.

Prime Contractor. The one company contractually responsible for providing central station services to a subscriber as required by this code. This may be either a listed central station or a listed fire alarm service — local company.

Private Radio Signaling. A radio system under control of the proprietary supervising station.

Proprietary Fire Alarm System. An installation of fire alarm systems that serve contiguous and noncontiguous properties under one ownership from a proprietary supervising station located at the protected property, where trained, competent personnel are in constant attendance. This includes the proprietary

Figure 1.20 *Proprietary supervising station.*

supervising station; power supplies; signal initiating devices; initiating device circuits; signal notification appliances; equipment for the automatic, permanent visual recording of signals; and equipment for initiating the operation of emergency building control services.

Proprietary Supervising Station. A location to which alarm or supervisory signaling devices on proprietary fire alarm systems are connected and where personnel are in attendance at all times to supervise operation and investigate signals.

Protected Premises. The physical location protected by a fire alarm system.

Public Fire Service Communication Center. The building or portion of the building used to house the central operating part of the fire alarm system; usually the place where the necessary testing, switching, receiving, transmitting, and power supply devices are located.

In a municipal system, the public fire service communication center is normally at the main fire station.

Public Switched Telephone Network. An assembly of communications facilities and central office equipment operated jointly by authorized common carriers that provides the general public with the ability to establish communications channels via discrete dialing codes.

Radio Alarm Central Station Receiver (RACSR). A system component that receives data and annunciates that data at the central station.

Radio Alarm Satellite Station Receiver (RASSR). A system component that receives radio signals. This component is resident at a satellite station, located at a remote receiving location.

Radio Alarm System (RAS). A system in which signals are transmitted from a radio alarm transmitter (RAT) located at a protected premises through a radio channel to two or more radio alarm satellite station receivers (RASSR), and that are annunciated by a radio alarm central station receiver (RACSR) located at the central station.

Radio Alarm Transmitter (RAT). A system component at the protected premises to which initiating devices or groups of devices are connected. The RAT transmits signals indicating a status change of the initiating devices.

Radio Channel. A band of frequencies of a width sufficient to permit its use for radio communications.

NOTE: The width of the channel depends on the type of transmissions and the tolerance for the frequency of emission. Normally allocated for radio transmission in a specified type for service by a specified transmitter.

Record Drawings. Drawings (as-built) that document the location of all devices, appliances, wiring sequences, wiring methods, and connections of the components of the fire alarm system as installed.

Relocation. The movement of occupants from a fire zone to a safe area within the same building.

In high-rise buildings or large facilities where it is impossible to evacuate all occupants, occupants in the fire zone are relocated to a safe area.

Remote Station Fire Alarm System. A system installed in accordance with this code to transmit alarm, supervisory, and trouble signals from one or more protected premises to a remote location at which appropriate action is taken.

Repeater Facility. Equipment needed to relay signals between supervisory stations, subsidiary stations, and protected premises.

Repeater Station. The location of the equipment needed for a repeater facility.

Restorable Initiating Device. A device whose sensing element is not ordinarily destroyed in the process of operation. Restoration may be manual or automatic.

Runner. A person other than the required number of operators on duty at central, supervising, or runner stations (or otherwise in contact with these stations) available for prompt dispatching, when necessary, to the protected premises.

Runner Service. The service provided by a runner at the protected premises, including resetting and silencing of all equipment transmitting fire alarm or supervisory signals to the off-premise location.

Runner service is normally provided as part of the central station service. A runner is sent to the protected premises from which the signal was transmitted and takes appropriate action.

Satellite Trunk. A circuit or path connecting a satellite to its central or proprietary supervising station.

Scanner. Equipment located at the telephone company wire center that monitors each local leg and relays status changes to the alarm center. Processors and associated equipment may also be included.

Secondary Trunk Facility. That part of a transmission channel connecting two or more, but less than all, leg facilities to a primary trunk facility.

Separate Sleeping Area. The area or areas of the family living unit in which the bedrooms (or sleeping rooms) are located. For the purpose of this code, bedrooms (or sleeping rooms) separated by other use areas, such as kitchens or living rooms (but not bathrooms), shall be considered as separate sleeping areas.

Shall. Indicates a mandatory requirement.

Shapes of Ceilings. The shapes of ceilings are classified as follows:

Sloping Ceilings. Those having a slope of more than 1½ in. per ft (41.7 mm per m). Sloping ceilings are further classified as follows:

(a) *Sloping-Peaked Type.* Those in which the ceiling slopes in two directions from the highest point. Curved or domed ceilings may be considered peaked with the slope figured as the slope of the chord from highest to lowest point. (*See Figure A-5-2.7.4.1.*)

(b) *Sloping-Shed Type.* Those in which the high point is at one side with the slope extending toward the opposite side. (*See Figure A-5-2.7.4.2.*)

Smooth Ceiling. A surface uninterrupted by continuous projections, such as solid joists, beams, or ducts, extending more than 4 in. (100 mm) below the ceiling surface.

NOTE: Open truss constructions are not considered to impede the flow of fire products unless the upper member in continuous contact with the ceiling projects below the ceiling more than 4 in. (100 mm).

Should. Indicates a recommendation or that which is advised but not required.

Signal. A status indication communicated by electrical or other means.

Signal Transmission Sequence. A DACT that obtains dial tone, dials the number(s) of the DACR, obtains verification that the DACR is ready to receive signals, transmits the signals, and receives acknowledgment that the DACR has accepted that signal before disconnecting (going on-hook).

Signaling Line Circuit. A circuit or path between any combination of circuit interfaces, control units, or transmitters over which multiple system input signals or output signals, or both, are carried.

Signaling Line Circuit Interface. A system component that connects a signaling line circuit to any combination of initiating devices, initiating device circuits, notification appliances, notification appliance circuits, system control outputs, and other signaling line circuits.

Single Station Alarm Device. An assembly incorporating the detector, control equipment, and the alarm-sounding device in one unit operated from a power supply either in the unit or obtained at the point of installation.

Solid Joist Construction. Ceilings having solid structural or solid nonstructural members projecting down from the ceiling surface a distance of more than 4 in. (100 mm) and spaced at intervals 3 ft (0.9 m) or less, center to center.

Spacing. A horizontally measured dimension relating to the allowable coverage of fire detectors.

The distance allowed between each initiating device (according to the device's listing) or between each notification appliance that is necessary to meet code requirements.

Spot-Type Detector. A device whose detecting element is concentrated at a particular location. Typical examples are bimetallic detectors, fusible alloy detectors, certain pneumatic rate-of-rise detectors, certain smoke detectors, and thermoelectric detectors.

Story. The portion of a building included between the upper surface of a floor and upper surface of the floor or roof next above.

Subscriber. The recipient of contractual supervising station signal service(s). In case of multiple, noncontiguous properties having single ownership, the term "subscriber" refers to each protected premises or its local management.

Subsidiary Station. A subsidiary station is a normally unattended location, remote from the supervising station and linked by communication channel(s) to the supervising station. Interconnection of signal-receiving equipment or communication channel(s) from protected buildings with channel(s) to the supervising station is accomplished at this location.

Supervising Station. A facility that receives signals and where personnel are in attendance at all times to respond to these signals.

Supervisory Service. The service required to monitor performance of guard tours and the operative condition of fixed suppression systems or other systems for the protection of life and property.

Supervisory Signal. A signal indicating the need of action in connection with the supervision of guard tours, fire suppression systems or equipment, or with the maintenance features of related systems.

Supplementary. As used in this code, supplementary refers to equipment or operations not required by this code and designated as such by the authority having jurisdiction.

In order for anything to be considered supplementary, the authority having jurisdiction, as defined in this section, must designate it as supplementary. This additional requirement is to avoid having the installer or supplier simply label a device or operation as supplementary.

Switched Telephone Network. An assembly of communications facilities and central office equipment operated jointly by authorized service providers, which provide the general public with the ability to establish transmission channels via discrete dialing.

System Unit. The active subassemblies at the central station utilized for signal receiving, processing, display, or recording of status change signals; a failure of one of these subassemblies would cause the loss of a number of alarm signals by that unit.

Transmission Channel. A circuit or path connecting transmitters to supervising stations or subsidiary stations on which signals are carried.

Transmitter. A system component that provides an interface between signaling line circuits, initiating device circuits, or control units and the transmission channel.

Transponder. A multiplex alarm transmission system functional assembly located at the protected premises.

Trouble Signal. A signal initiated by the fire alarm system, indicative of a fault in a monitored circuit or component.

Trunk Facility. That part of a transmission channel connecting two or more leg facilities to the central supervising station or subsidiary station.

Trunk Primary Facility. That part of a transmission channel connecting all leg facilities to a central or proprietary supervising station or subsidiary station.

Trunk Secondary Facility. That part of a transmission channel connecting two or more, but less than all, leg facilities to a primary trunk facility.

WATS (Wide Area Telephone Service). Telephone company service allowing reduced costs for certain telephone call arrangements; may be in-WATS or 800-number service where calls can be placed from anywhere in the continental U.S. to the called party at no cost to the calling party, or out-WATS, a service whereby, for a flat-rate charge, dependent on the total duration of all such calls, a subscriber may make an unlimited number of calls within a prescribed area from a particular telephone terminal without the registration of individual call charges.

Zone. A defined area within the protected premises. A zone may define an area from which a signal can be received, an area to which a signal can be sent, or an area in which a form of control can be executed.

1-5 Fundamentals.

1-5.1 Common System Fundamentals. The provisions of this chapter shall apply to Chapters 3 through 7.

The basic requirements for all fire alarm systems except household fire warning systems are contained in Chapter 1. Chapter 2, Household Fire Warning Equipment, is a self-contained chapter with specific requirements for detection in residential family occupancies.

1-5.1.1 The provisions of this chapter cover the basic functions of a complete fire alarm system. These systems are primarily intended to provide notification of fire alarm, supervisory, and trouble conditions, alert the occupants, summon appropriate aid, and control fire safety functions.

1-5.1.2 Equipment. Equipment constructed and installed in conformity with this code shall be listed for the purpose for which it is used.

It is not only required that fire alarm products be listed, but they must be listed for fire alarm application. Since fire alarm systems are used for life safety, the listing requirements are more stringent than for those products listed for electrical safety only.

1-5.2 Power Supplies.

1-5.2.1 Scope. The provisions of this section apply to power supplies used for fire alarm systems.

1-5.2.2 Code Conformance. All power supplies shall be installed in conformity with the requirements of NFPA 70, *National Electrical Code*, for such equipment, and with the requirements indicated in this section.

1-5.2.3 Power Sources. Fire alarm systems shall be provided with at least two independent and reliable power supplies, one primary and one secondary (standby), each of which shall be of adequate capacity for the application.

Previous editions of NFPA 72 required three sources of power: primary, secondary (standby), and trouble. The requirement for trouble power supply had an exception allowing the secondary power source to be used as the trouble power source. The majority of fire alarm systems followed this exception; therefore, the requirement for a separate trouble power source was eliminated.

Exception No. 1: Where the primary power is supplied by a dedicated branch circuit of an emergency system in accordance with NFPA 70, National Electrical Code, Article 700, or a legally required standby system in accordance with NFPA 70, National Electrical Code, Article 701, a secondary supply is not required.

Exception No. 2: Where the primary power is supplied by a dedicated branch circuit of an optional standby system in accordance with NFPA 70, National Electrical Code, Article 702, which also

---

Formal Interpretation 85-13
Reference: 1-5.2.3

Background: In reading the 1993 code under NFPA 72, 1-5.2.3, Exception No. 1 stipulates that fire alarm system power supply can be fed from the emergency generator of an emergency system or a legally required standby system.

Question 1: When Exception No. 1 is complied with, are the requirements of 1-5.2.10.5 also required?
Answer: Yes, for generator option.

Question 2: When Exception No. 1 is complied with do the requirements for a 2-hour fuel supply stated in NFPA 70, Section 700-12(b)(2) apply?
Answer: No, 1-5.2.10.5 is more stringent.

Issue Edition: NFPA 72A-1985
Reference: 2-3.2.1, 2-3.4.1
Issue Date: November 1986 ∎

---

meets the performance requirements of Article 700 or Article 701, a secondary supply is not required.

NOTE to Exceptions No. 1 and No. 2: A trouble signal is not required where operating power is being supplied by either of the two sources of power indicated in Exceptions No. 1 and No. 2 above, if they are capable of providing the hours of operation required by 1-5.2.5 and loss of primary power is otherwise indicated (e.g., by loss of building lighting).

This is a very important note, since the normal hours of operation for an emergency system meeting the requirements of Article 701 is 1½ hours, and the optional standby system meeting the requirements of Article 702 does not specify an operating time.

Where dc voltages are employed they shall be limited to no more than 350 volts above earth ground.

1-5.2.4 **Primary Supply.** The primary supply shall have a high degree of reliability, shall have adequate capacity for the intended service, and shall consist of one of the following:

(a) Light and power service arranged in accordance with 1-5.2.8,

(b) Engine-driven generator or equivalent arranged in accordance with 1-5.2.10.

1-5.2.5 **Secondary Supply Capacity and Sources.** The secondary supply shall automatically supply the energy to the system within 30 seconds and without loss of signals, wherever the primary supply is incapable of providing the minimum voltage required for proper operation. The secondary (standby) power supply shall supply energy to the system in the event of total failure of the primary (main) power supply or when the primary voltage drops to a level insufficient to maintain functionality of the control equipment and system components. Under maximum normal load, the secondary supply shall have sufficient capacity to operate a local, central station or proprietary system for 24 hours, or an auxiliary or remote station system for 60 hours; and then, at the end of that period, operate all alarm notification appliances used for evacuation or to direct aid to the location of an emergency for 5 minutes. The secondary power supply for emergency voice/alarm communications service shall be capable of operating the system under maximum normal load for 24 hours and then be capable of operating the system during a fire or other emergency condition for a period of 2 hours. Fifteen minutes of evacuation alarm operation at maximum connected load shall be considered the equivalent of 2 hours of emergency operation.

The proper amount of battery standby can be calculated. These calculations should include the standby supervisory time as well as the 5-minute alarm time.

The emergency voice/alarm communications system must be capable of operating for 2 hours during the emergency condition because communication to the relocated occupants is necessary.

The secondary supply shall consist of one of the following:

(a) A storage battery arranged in accordance with 1-5.2.9.
(b) An automatic starting engine-driven generator arranged in accordance with 1-5.2.10 and storage batteries with 4 hours capacity arranged in accordance with 1-5.2.9.
(c) Multiple engine-driven generators, one of which is arranged for automatic starting, arranged in accordance with 1-5.2.10, capable of supplying the energy required herein with the largest generator out of service. It shall be permitted for the second generator to be pushbutton start.

Operation on secondary power shall not affect the required performance of a fire alarm system. The system shall produce the same alarm, supervisory, and trouble signals and indications (excluding the ac power indicator) when operating from the standby power source as produced when the unit is operating from the primary power source.

Manufacturers have supplied systems in the past that, in order to save battery power, eliminated annunciation and some supplementary functions when in the standby power mode. This new code paragraph eliminates that practice by requiring the system to operate with all of the same features as when powered by the primary power source.

1-5.2.6   Continuity of Power Supplies.

(a) Where signals could be lost on transfer of power between the primary and secondary sources, rechargeable batteries of sufficient capacity to operate the system under maximum normal load for 15 minutes shall assume the load in such a manner that no signals are lost if either of the following conditions exists:

1. Secondary power is supplied in accordance with 1-5.2.5(a) or 1-5.2.5(b), and the transfer is made manually; or

2. Secondary power is supplied in accordance with 1-5.2.5(c).

(b) Where signals will not be lost due to transfer of power between the primary and secondary sources, one of the following arrangements shall be made:

1. The transfer shall be automatic.
2. Special provisions shall be made to allow manual transfer within 30 seconds of loss of power.
3. The transfer shall be arranged in accordance with 1-5.2.5(a).

(c)* Where a computer system of any kind or size is used to receive or process signals, an uninterruptible power supply (UPS) with sufficient capacity to operate the system for at least 15 minutes, or until the secondary supply is capable of supplying the UPS input power requirements, shall be required if either of the following conditions apply:

1. Status of signals previously received will be lost upon loss of power.

2. The computer system cannot be restored to full operation within 30 seconds of loss of power.

A-1-5.2.6(c)   An engine-driven generator without standby battery supplement is not assumed to be capable of reliable power transfer within 30 seconds of a primary power loss.

1-5.2.6.1*   **Uninterruptible Power System Bypass.** A positive means for disconnecting the input and output of the UPS system while maintaining continuity of power supply to the load shall be provided.

A-1-5.2.6.1   UPS equipment often contains an internal bypass arrangement to supply the load directly from the line. These internal bypass arrangements are a potential source of failure. UPS equipment also requires periodic maintenance. It is therefore necessary to provide a means of promptly and safely bypassing and isolating the UPS equipment from all power sources while maintaining continuity of power supply to the equipment normally supplied by the UPS.

1-5.2.7   **Power Supply for Remotely Located Control Equipment.** Additional power supplies, where provided for control units, circuit interfaces, or other equipment essential to system operation, located remote from the main control unit, shall be comprised of a primary and secondary power supply that shall meet the same requirements as for 1-5.2.1 through 1-5.2.8

1-5.2.7.1   Power supervisory devices shall be arranged so as not to impair the receipt of fire alarm or supervisory signals.

1-5.2.8   **Light and Power Service.**

1-5.2.8.1   A light and power service employed to operate the system under normal conditions shall have a high degree of reliability and capacity for the intended service. This service shall consist of one of the following:

(a) *Two-Wire Supplies.* A two-wire supply circuit may be used for either the primary operating power supply or the trouble signal power supply of the signaling system.

(b) *Three-Wire Supplies.* A three-wire ac or dc supply circuit having a continuous unfused neutral conductor, or a polyphase ac supply circuit having a continuous unfused neutral conductor where interruption of one phase does not prevent operation of the other phase, may be used with one side or phase for the primary operating power supply and the other side or phase for the trouble signal power supply of the fire alarm system.

1-5.2.8.2 Connections to the light and power service shall be on a dedicated branch circuit. The circuit and connections shall be mechanically protected. The circuit disconnecting means shall have a red marking, be accessible only to authorized personnel, and be identified as "FIRE ALARM CIRCUIT CONTROL." The location of the circuit disconnecting means shall be permanently identified at the fire alarm control unit.

1-5.2.8.3 Overcurrent Protection. An overcurrent protective device of suitable current-carrying capacity and capable of interrupting the maximum short-circuit current to which it may be subject shall be provided in each ungrounded conductor. The overcurrent protective device shall be enclosed in a locked or sealed cabinet located immediately adjacent to the point of connection to the light and power conductors.

1-5.2.8.4 Circuit breakers or engine stops shall not be installed in such a manner as to cut off the power for lighting or for operating elevators.

1-5.2.9* Storage Batteries.

A-1-5.2.9 Rechargeable (Storage-)Type Batteries. The following newer types of rechargeable batteries are normally used in protected premises applications:

(a) *Vented Lead-Acid, Gelled, or Starved Electrolyte Battery.* This rechargeable-type battery is generally used in place of primary batteries in applications having a relatively high current drain or requiring extended standby capability of much lower currents. Nominal voltage of a single cell is 2 volts, and the battery is available in multiples of 2 volts (2, 4, 6, 12, etc.). Batteries should be stored according to manufacturer's recommendations.

(b) *Nickel-Cadmium Battery.* The sealed-type nickel-cadmium battery generally used in applications where the battery current drain during a power outage is low to moderate (typically up to a few hundred milliamperes) and is fairly constant. Nickel-cadmium batteries are also available in much larger capacities for other applications. The nominal voltage per cell is 1.42 volts (12.78, 25.56, etc.). Batteries in storage can be stored in any state of charge for indefinite periods. However, a battery in storage will lose capacity (will self-discharge) according to storage time and temperature. Typically, batteries stored more than 1 month will require an 8- to 14-hour charge period to restore capacity.

In service, the battery should receive a continuous constant charging current sufficient to keep it fully charged (typically, the charge rate equals $\frac{1}{10}$ to $\frac{1}{20}$ of the ampere-hour rating of the battery). Because batteries are made up of individual cells connected in series, the possibility exists that during deep discharge one or more cells that may be low in capacity will reach complete discharge prior to other cells. The cells with remaining life tend to charge the depleted cells, causing a polarity reversal resulting in permanent battery damage. This condition can be determined by measuring the open cell voltage of a fully charged battery (voltage should be a minimum of 1.28 volts per cell multiplied by the number of cells). Voltage depression effect is a minor change in discharge voltage level caused by constant current charging below the system discharge rate.

In some applications of nickel-cadmium batteries (for example, battery-powered shavers) a memory characteristic also exists. Specifically, if the battery is discharged for 1 minute a day, day after day, followed by a recharge, an attempt to have it operate for 5 minutes will not result in obtaining the rated ampere-hour output. The reason for this is that the battery has developed a 1-minute discharge memory.

(c) *Sealed Lead-Acid Battery.* In a sealed lead-acid battery, the electrolyte is totally absorbed by the separators, and no venting normally occurs. Gas evolved during recharge is internally recombined, resulting in minimal loss of capacity life. A high-pressure vent, however, is provided to avoid damage under abnormal conditions. Other battery characteristics are comparable to those described under A-1-5.2.11(a).

1-5.2.9.1 Location. Storage batteries shall be so located that the fire alarm equipment, including overcurrent devices, are not adversely affected by battery gases and shall conform to the requirements of NFPA 70, *National Electrical Code*, Article 480. Cells shall be suitably insulated against grounds and crosses and shall be substantially mounted in such a manner as not to be subject to mechanical injury. Racks shall be suitably protected against deterioration.

1-5.2.9.2 Battery Charging.

(a) Adequate facilities shall be provided to automatically maintain the battery fully charged under all conditions of normal operation and, in addition, to recharge batteries within 48 hours after fully charged batteries have been subject to a single discharge cycle as specified in 1-5.2.5. Upon attaining a fully charged condition, the charge rate shall not be so excessive as to result in battery damage.

(b) A reliable source of power shall be provided for charging the batteries.

(c) Central stations shall maintain spare parts or units available and employed to restore failed charging capacity prior to the consumption of one-half of the capacity of the batteries for the central station equipment.

(d)* Batteries shall be either trickle or float charged.

A-1-5.2.9.2(d) Batteries are trickle charged if they are off-line and waiting to be put under load in the event of a loss of power.

Float-charge batteries are fully charged and connected across the output of the rectifiers to smooth the output and serve as a standby source of power in the event of a loss of line power.

(e) A rectifier employed as a battery charging source of supply shall be of adequate capacity. A rectifier employed as a charging means shall be energized by an isolating transformer.

1-5.2.9.3 Overcurrent Protection. The batteries shall be protected against excessive load current by overcurrent devices having a rating not less than 150 percent and not more than 250 percent of the maximum operating load in the alarm condition. The batteries shall be protected from excessive charging current by overcurrent devices or by automatic current-limiting design of the charging source.

1-5.2.9.4 Metering. The charging equipment shall provide either integral meters or readily accessible terminal facilities for the connection of portable meters by which the battery voltage and charging current can be determined.

1-5.2.9.5 Under-Voltage Detection. An under-voltage detection device shall be provided to detect a failure of the charging source and initiate a trouble signal.

This requirement was part of the metering requirement of the 1989 edition of NFPA 71, *Standard for the Installation, Maintenance, and Use of Signaling Systems for Central Station Service.* The battery charging circuits of all systems are now required to be monitored and produce a trouble signal upon failure.

1-5.2.10 Engine-Driven Generator.

1-5.2.10.1 An engine-driven generator shall be used only where a person specifically trained in its operation is on duty at all times.

Exception: Where acceptable to the authority having jurisdiction and where the requirements of 1-5.2.5(b) and (c) are met, a person specifically trained in the operation of a generator dedicated to the fire alarm system shall not be required to be on duty at all times.

1-5.2.10.2 The installation of such units shall conform to the provisions of NFPA 110, *Standard for Emergency and Standby Power Systems,* except as restricted by the provisions of this section.

1-5.2.10.3 Capacity. The unit shall be of a capacity sufficient to operate the system under the maximum normal load conditions in addition to all other demands placed upon the unit, such as those of emergency lighting.

1-5.2.10.4 Fuel. Fuel shall be stored in outside underground tanks whenever possible, and gravity feed shall not be used. Gasoline deteriorates with age. Where gasoline-driven generators are used, fuel shall be supplied from a frequently replenished "working" tank, or other means provided, to ensure that gasoline will always be fresh.

1-5.2.10.5 Sufficient fuel shall be available in storage for 6 months of testing plus the capacity specified in 1-5.2.5. For public fire alarm reporting systems, refer to 4-6.7.3.4.

Exception No. 1: If a reliable source of supply is available at any time on 2-hour notice, sufficient fuel shall be in storage for 12 hours of operation at full load.

Exception No. 2: Fuel systems using natural or manufactured gas supplied through reliable utility mains shall not be required to have fuel storage tanks unless located in seismic risk zone 3 or greater as defined in ANSI A-58.1, Building Code Requirements for Minimum Design Loads in Buildings and Other Structures.

1-5.2.10.6 A separate storage battery and separate automatic charger shall be provided for starting the engine-driven generator and shall not be used for any other purpose.

1-5.2.11* **Primary Batteries.**

A-1-5.2.11 **Maximum Load.** The maximum normal load of a No. 6 primary battery should not be more than 2 amperes per cell. No. 6 batteries should be replaced under the following conditions:

(a) An individual primary battery cell rated 1 volt should be replaced when a test load of 1 ohm reduces the potential below 1 volt.

(b) A unit assembly of primary battery cells rated 6 volts should be replaced when a test load of 4 ohms reduces the potential of the unit below 4 volts.

1-5.2.11.1 **Location.** Primary batteries shall be located in a clean, dry place accessible for servicing and where the ambient air temperature will not be less than 40°F (4.4°C) and not more than 100°F (37.8°C).

1-5.2.11.2 **Separation of Cells.** Primary batteries shall be housed in a locked, substantial enclosure or otherwise suitably protected against movement, injury, and moisture. Reliable separation between cells shall be provided to prevent contact between terminals of adjacent cells and between battery terminals and other metal parts, which may result in depletion of the battery or other deterioration. Battery cells having containers constructed of other than suitable electrical insulating material shall be located on insulating supports.

1-5.2.11.3 **Capacity.** A primary battery shall have sufficient capacity to supply 125 percent of the maximum normal load for not less than one year.

1-5.3 **Compatibility.**

1-5.3.1 All initiating devices, notification appliances, and control equipment constructed and installed in conformity with this code shall be listed for the purpose for which they are intended.

The fact that a component is listed is not sufficient for fire alarm application. The component must be tested and listed specifically for use in fire alarm systems.

1-5.3.2 All fire detection devices that receive their power from the initiating device circuit or signaling line circuit of a fire alarm control unit shall be listed for use with the control unit.

A two-wire smoke detector obtains its power from the control unit initiating device circuit; therefore, it is mandatory that the smoke detector be listed for use with the control unit and its associated initiating device circuit.

The listing organizations have developed specific requirements for this listing process and should be consulted if there is any doubt as to the detector's compatibility with a specific control unit.

1-5.4 **System Functions.**

1-5.4.1 **Local Fire Safety Functions.** Fire safety functions shall be permitted to be performed automatically. The performance of automatic fire safety functions shall not interfere with power for lighting or for operating elevators. This does not preclude the combination of fire alarm services with other services requiring monitoring of operations.

1-5.4.2 **Alarm Signals.**

1-5.4.2.1* **Coded Alarm Signal.** A coded alarm signal shall consist of not less than three complete rounds of the number transmitted, and each round shall consist of not less than three impulses.

A-1-5.4.2.1 **Coded Alarm Signal Designations.** The following suggested coded signal assignment for buildings having four floors and multiple basements is provided as a guide:

Location	Coded Signal
4th floor	2-4
3rd Floor	2-3
2nd Floor	2-2
1st Floor	2-1
Basement	3-1
Sub-Basement	3-2

1-5.4.3 **Supervisory Signals.**

1-5.4.3.1 **Coded Supervisory Signal.** A coded supervisory signal shall be permitted to consist of two rounds of the number transmitted to indicate a supervisory off-normal condition, and one round of the number transmitted to indicate the restoration of the supervisory condition to normal.

1-5.4.3.2 Combined Coded Alarm and Supervisory Signal Circuits. Where both coded sprinkler supervisory signals and coded fire or waterflow alarm signals are transmitted over the same signaling line circuit, provision shall be made either to obtain alarm signal precedence or sufficient repetition of the alarm signal to prevent the loss of an alarm signal.

1-5.4.4 Fire alarms, supervisory signals, and trouble signals shall be distinctively and descriptively annunciated.

1-5.4.5 Where status indicators are required to be provided for emergency equipment or fire safety functions, they shall be arranged to reflect accurately the actual status of the associated equipment or function.

1-5.4.6 Trouble Signal.

1-5.4.6.1 General. Trouble signals and their restoration to normal shall be indicated within 200 seconds at the locations identified in 1-5.4.6.2 or 1-5.4.6.3. Trouble signals required to indicate at the protected premises shall be indicated by distinctive audible signals. These audible trouble signals shall be distinctive from alarm signals. If an intermittent signal is used, it shall sound at least once every 10 seconds with a minimum time duration of one-half second. An audible trouble signal may be common to several supervised circuits. The trouble signal(s) shall be located in an area where it is likely to be heard.

1-5.4.6.2 Visible and audible trouble signals and visible indication of their restoration to normal shall be indicated at the following locations:

(a) Control unit (central equipment) for local fire alarm systems

(b) Building fire command center for emergency voice/alarm communication systems

(c) Central station or remote station location for systems installed in compliance with Chapter 4.

1-5.4.6.3 Trouble signals and their restoration to normal shall be visibly and audibly indicated at the proprietary supervising station for systems installed in compliance with Chapter 4.

1-5.4.6.4 Audible Trouble Signal Silencing Switch.

1-5.4.6.4.1 A switch for silencing the trouble notification appliance(s) shall be permitted only if it transfers the trouble indication to a lamp or other acceptable visible indicator adjacent to the switch. The visible indication shall persist until the trouble has been corrected. The audible trouble signal shall sound if the switch is in its silence position and no trouble exists.

1-5.4.6.4.2 Where an audible trouble notification appliance is also used to indicate a supervisory condition, as permitted in 1-5.4.7(b), a trouble signal silencing switch shall not prevent subsequent sounding of supervisory signals.

1-5.4.7 Distinctive Signals. Audible alarm notification appliances for a fire alarm system shall produce signals that are distinctive from other similar appliances used for other purposes in the same area. The distinction among signals shall be as follows:

(a) Fire alarm signals shall be distinctive in sound from other signals and this sound shall not be used for any other purpose. (See 3-7.2.)

(b)* Supervisory signals shall be distinctive in sound from other signals. This sound shall not be used for any other purpose except that it may be employed to indicate a trouble condition. Where the same sound is used for both supervisory signals and trouble signals, distinction between signals shall be by other appropriate means such as visible annunciation.

A-1-5.4.7(b) A tamper switch, low pressure switch, or other device intended to cause a supervisory signal when actuated should not be connected in series with the end-of-line supervisory device of initiating device circuits unless a distinctive signal, different from a trouble signal, is indicated.

(c) Fire alarm, supervisory, and trouble signals shall take precedence over all other signals.

Exception: Signals from hold-up alarms or other life threatening signals shall be permitted to take precedence over supervisory and trouble signals if acceptable to the authority having jurisdiction.

1-5.4.8 Alarm Signal Deactivation. A means for turning off the alarm notification appliances shall be permitted only if it is key-operated, located within a locked cabinet, or arranged to provide equivalent protection against unauthorized use. Such a means shall be permitted only if a visible zone alarm indication or equivalent has been provided as specified in 1-5.7.1 and subsequent alarms on other initiating device circuits will cause the notifica-

<table>
<tr><td>

Formal Interpretation
Reference: 1-5.4.7

Statement: Is it proper, within the meaning of the code, to interconnect the gate valve signal with the trouble signal of the fire alarm system?

Question: Is it proper, within the meaning of the code, to utilize a common audible device to indicate a closed gate valve on a sprinkler system supervisory circuit, as well as to indicate trouble on a separate waterflow alarm circuit, with the understanding that there will be visual means for identifying the specific circuit involved?
Answer: Yes.

Question: Is it proper, within the meaning of the code, to interconnect the gate valve switch(es) on the waterflow alarm circuit so that a closed gate valve will be indicated as a trouble on the waterflow alarm circuit?
Answer: No.

Issue Edition: NFPA 72A-1972
Reference: 3610
Issue Date: June 1974 ∎

</td><td>

Formal Interpretation 85-9
Reference: 1-5.4.7, 3-8.7.3, 3-8.7.5

Question 1: Is it the intent of the Committee to prohibit the use of a dedicated closed loop circuit to which only normally closed supervisory switches (one or more) are connected where a break in the line or an off normal supervisory switch produce the same signal at the control unit?
Answer: Yes.

Question 2: Is it the intent of the Committee to permit the use of the same audible signal for both a supervisory signal and a trouble signal?
Answer: Yes.

Question 3: Is it the intent of the Committee to permit silencing an audible supervisory signal?
Answer: Yes.

Question 4: Is it the intent of the Committee to permit an arrangement where silencing an audible trouble signal would prevent the receipt of the first (in case there are several supervisory circuits and the answer to Question 3 is "yes") audible supervisory signal?
Answer: No.

Question 5: Is it the intent of the Committee to prohibit the use of a common trouble signal silencing switch to silence both trouble and supervisory audible signals when operation of the switch to silence the audible signal caused by a trouble condition will prevent the receipt of an audible signal associated with a supervisory signal?
Answer: Yes.

Issue Edition: NFPA 72-1985
Reference: 2-5.5, 3-5.4.2, et al.
Issue Date: October 1985 ∎

</td></tr>
</table>

tion appliances to reactivate. A means that is left in the "off" position when there is no alarm shall operate an audible trouble signal until the means is restored to normal. Where automatically turning off the alarm notification appliances is permitted by the authority having jurisdiction, the alarm shall not be turned off in less than 5 minutes unless otherwise permitted by the authority having jurisdiction.

1-5.4.9 Supervisory Signal Silencing. A switch for silencing the supervisory signal sounding appliance(s) shall be permitted only if it is key-operated, located within a locked cabinet, or arranged to provide equivalent protection against unauthorized use. Such a switch shall be permitted only if it transfers the supervisory indication to a lamp or other visible indicator and subsequent supervisory signals from other zones will cause the supervisory signal indicating appliances to resound. A switch left in the "silence" position where there is no supervisory off-normal signal shall operate a visible signal silence indicator and cause the trouble signal to sound until the switch is restored to normal.

See Figure 1.21.

1-5.4.10 Presignal Feature. Where permitted by the authority having jurisdiction, systems shall be permitted to have a feature where initial fire alarm signals will sound only in department offices, control rooms, fire brigade stations, or other constantly attended central locations and where human action is subsequently required to activate a general alarm, or a feature where

Formal Interpretation 87-3
Reference: 1-5.4.7

Background: An initiating device circuit has a waterflow device and a valve supervisory device connected to it and by using current limiting techniques provides the distinctive signals (i.e., separate alarm, trouble and supervisory signals) required by NFPA 72, 1-5.4.7.

Question 1: Does this meet the intent of NFPA 72-1993, 1-5.4.7?
Answer: Yes.

Issue Edition: NFPA 72-1987
Reference: 2-8.5
Issue Date: April 30, 1990 ∎

Formal Interpretation 85-3
Reference: 1-5.4.9

Background: Previous interpretations by this Committee and actions by the membership at the fall of 1984 meeting still leave unclear the Committee's intent on the questions of permitted means to connect supervisory devices to fire alarm control units.

Question 1: If a control unit is arranged to sound the same audible signal for trouble indication as it does for a supervisory signal, is it the intent of the Committee that a supervisory device be permitted to be connected in such a manner that it is not possible to differentiate between an actuated supervisory device or an open circuit trouble condition on the same circuit?
Answer: No.

Question 2: If the answer to Question 1 is "no," would the answer be "yes" if the circuit involved was individually annunciated in some manner?
Answer: No.

Question 3: In a control unit arranged as in Question 1, is it the intent of the Committee to permit an audible trouble signal silencing switch to prevent subsequent sounding of audible supervisory signals?
Answer: No.

Question 4: In a control unit arranged as in Question 1, is it the intent of the Committee to permit an audible supervisory signal silencing switch to prevent subsequent sounding of audible trouble signals or supervisory signals?
Answer: Yes.

Issue Edition: NFPA 72A-1985
Reference: 2-5.5, 3-5.4.2, et al.
Issue Date: June 1985 ∎

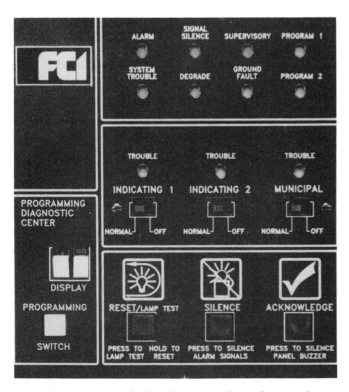

Figure 1.21 *Interior of a fire alarm control unit showing alarm, supervisory, and trouble signal indications.*

the control equipment delays general alarm by more than one minute after the start of the alarm processing. Where there is a connection to a remote location, it shall activate upon initial alarm signal.

NOTE: A system provided with an alarm verification feature as permitted by 3-8.2.3 is not considered a presignal system since the delay in signal produced is 60 seconds or less and requires no human intervention.

Presignal systems rely on human action, which can be unreliable. The *Life Safety Code*®, NFPA *101*®, provides guidance for using this feature. Caution is recommended when delaying alarm signals.

1-5.5 Performance and Limitations.

1-5.5.1 Voltage, Temperature, and Humidity Variation. Unless otherwise listed, equipment shall be installed in locations where conditions do not exceed the following:

(a)* Eighty-five percent and at 110 percent of the nameplate primary (main) and secondary (standby) input voltage(s)

A-1-5.5.1(a) This requirement does not preclude transfer to secondary supply at less than 85 percent of nominal primary voltage as long as the requirements of 1-5.2.6 are met.

(b) Ambient temperatures of 32°F (0°C) and 120°F (49°C) for a minimum duration at each extreme of 3 hours

(c) Relative humidity of 85 percent ± 5 percent and an ambient temperature of 86°F ± 3°F (30°C ± 2°C) for a duration of at least 24 hours.

1-5.5.2 Installation and Design.

1-5.5.2.1 All systems shall be installed in accordance with the specifications and standards approved by the authority having jurisdiction.

1-5.5.2.2 Devices and appliances shall be so located and mounted that accidental operation or failure will not be caused by vibration or jarring.

1-5.5.2.3 All apparatus requiring rewinding or resetting to maintain normal operation shall be restored to normal as promptly as possible after each alarm and kept in normal condition for operation.

1-5.5.3 To reduce the possibility of damage by induced transients, circuits and equipment shall be properly protected in accordance with requirements as set forth in NFPA 70, *National Electrical Code*, Article 800.

1-5.5.4* Wiring. The installation of all wiring, cable, and equipment shall be in accordance with NFPA 70, *National Electrical Code*, and specifically with Article 760, *Fire Protective Signal-*

ing Systems; Article 770, *Optical Fiber Cables*; and Article 800, *Communication Circuits, National Electrical Code*, where applicable. Optical fiber cables shall be protected against mechanical injury in accordance with Article 760.

A-1-5.5.4 Wiring and Equipment. The installation of all fire alarm system wiring should take into account the fire alarm system manufacturer's published installation instructions and the limitations of the applicable product listings or approvals.

1-5.5.5 Grounding. All systems shall test free of grounds.

Exception: Parts of circuits or equipment that are intentionally and permanently grounded to provide ground-fault detection, noise suppression, emergency ground signaling, and circuit protection grounding.

1-5.5.6 Initiating Devices.

1-5.5.6.1 Initiating devices of both the manual or automatic type shall be selected and installed as to minimize false alarms.

1-5.5.6.2 Fire alarm boxes of the manually operated type shall comply with 3-8.1.

1-5.6 Protection of Control Equipment. In areas that are not continuously occupied, automatic smoke detection shall be provided at each control unit(s) location to provide notification of fire at that location.

Exception: Should ambient conditions prohibit installation of automatic smoke detection, automatic heat detection shall be permitted.

1-5.7 Visible Indication (Annunciation).

1-5.7.1 Visible Zone Alarm Indication. Where required, the location of an operated initiating device shall be visibly indicated by building, floor, fire zone, or other approved subdivision by annunciation, printout, or other approved means. The visible indication shall not be canceled by the operation of an audible alarm silencing means.

1-5.7.1.1 The primary purpose of fire alarm system annunciation is to enable responding personnel to quickly and accurately identify the location of a fire, and to indicate the status of emergency equipment or fire safety functions that might affect the safety of occupants in a fire situation. All required annunciation means shall be readily accessible to responding personnel and shall be located as required by the authority having jurisdiction to facilitate an efficient response to the fire situation.

1-5.7.1.2 Zone of Origin. Fire alarm systems serving two or more zones shall identify the zone of origin of the alarm initiation by annunciation or coded signal.

1-5.7.2 Alarm annunciation at the fire command center shall be by means of audible and visible indicators.

1-5.7.3 For the purpose of alarm annunciation, each floor of the building shall be considered as a separate zone.

1-5.7.4 A system supervisory signal shall be annunciated at the fire command center by means of an audible and visible indicator.

1-5.7.5 A system trouble signal shall be annunciated at the fire command center by means of an audible and visible indicator.

1-5.7.6 Where the system serves more than one building, each building shall be indicated separately.

1-5.8 Monitoring Integrity of Installation Conductors and Other Signaling Channels.

1-5.8.1 All means of interconnecting equipment, devices, and appliances and wiring connections shall be monitored for the integrity of the interconnecting conductors or equivalent path so that the occurrence of a single open or a single ground fault condition in the installation conductors or other signaling channels and their restoration to normal shall be automatically indicated within 200 seconds.

Connections to devices and appliances must be made so that the opening of any installer's connection to the device or appliance will cause a trouble condition. Many installers "loop" the conductor around the terminal without cutting the conductor and making the necessary two connections. This practice is a violation of the code. Where a listed device is furnished with pigtail connections, the installer must use separate "in/out" wires for each circuit passing into or through the device in order to prevent pigtail connections in the installation wiring.

NOTE: The provisions of a double loop or other multiple path conductor or circuit to avoid electrical monitoring is not acceptable.

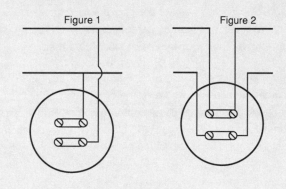

Formal Interpretation 75-5
Reference: 1-5.8.1

Question: Is it the intent of 1-5.8.1, with reference to initiating device circuits, that all wires installed by the installer be supervised?

Answer: Yes.   a) In Figure 1 the two field installed wires to the screw terminals of the normally open device are not supervised and therefore unacceptable. b) In Figure 2 all four of the installed wires connected to the normally open device are supervised and therefore acceptable.

Figure 1   Figure 2

Issue Edition: NFPA 72A-1975
Reference: 2411
Issue Date: August 1977 ∎

Exception No. 1: Styles of initiating device circuits, signaling line circuits, and notification appliance circuits tabulated in Tables 3-5.1, 3-6.1, and 3-7.1 that do not have an "X" under "Trouble" for the abnormal condition indicated.

Exception No. 2: Shorts between conductors, except as required by 1-5.8.3, 1-5.8.4, 1-5.8.5.2, Tables 3-5.1, 3-6.1, and 3-7.1, are not covered by this code.

Exception No. 3: A noninterfering shunt circuit, provided that a fault circuit condition on the shunt circuit wiring results only in the loss of the noninterfering feature of operation.

Exception No. 4: Connections to and between supplementary system components, provided that single open, ground, or short circuit conditions of the supplementary equipment and/or interconnecting means does not affect the required operation of the fire alarm system.

Exception No. 5: The circuit of an alarm notification appliance installed in the same room with the central control equipment, provided that the notification appliance circuit conductors are installed in conduit or equivalently protected against mechanical injury.

Exception No. 6: A trouble signal circuit.

Exception No. 7: Interconnection between equipment within a common enclosure.

NOTE: This code does not have jurisdiction over monitoring integrity of conductors within equipment, devices, or appliances.

The requirement for monitoring applies only to installation conductors. The wiring within equipment, devices, or appliances is not required to be monitored for integrity.

Exception No. 8: Interconnection between enclosures containing control equipment located within 20 ft (6 m) when the conductors are installed in conduit or equivalently protected against mechanical injury.

Exception No. 9: Conductors for ground detection, where a single ground does not prevent the required normal operation of the system.

Exception No. 10: Central station circuits serving notification appliances within a central station.

Exception No. 11: Pneumatic rate-of-rise systems of the continuous line type in which the wiring terminals of such devices are connected in multiple across electrically supervised circuits.

1-5.8.2 Interconnection means shall be arranged so that a single break or single ground fault will not cause an alarm signal.

1-5.8.3 An open, ground, or short circuit fault on the installation conductors of one alarm notification appliance circuit shall not affect the operation of any other alarm notification circuit.

1-5.8.4 The occurrence of a wire-to-wire short circuit fault on any alarm notification appliance circuit shall result in a trouble signal at the protected premises.

Exception No. 1: A circuit employed to produce a supplementary local alarm signal, provided that the occurrence of a short circuit on the circuit in no way affects the required operation of the fire alarm system.

Exception No. 2: The circuit of an alarm notification appliance installed in the same room with the central control equipment, provided that the notification appliance circuit conductors are installed in conduit or equivalently protected against mechanical injury.

Exception No. 3: Central station circuits serving notification appliances within a central station.

1-5.8.5 Monitoring Integrity of Emergency Voice/Alarm Communication Systems.

1-5.8.5.1* Monitoring Integrity of Speaker Amplifier and Tone-Generating Equipment. Where speakers are used to produce audible fire alarm signals, the following shall apply:

(a) Failure of any audio amplifier shall result in an audible trouble signal.

(b) Failure of any tone-generating equipment shall result in an audible trouble signal.

Exception: Tone-generating and amplifying equipment enclosed as integral parts and serving only a single listed loudspeaker need not be monitored.

A-1-5.8.5.1 Backup amplifying and evacuation signal-generating equipment is recommended with automatic transfer upon primary equipment failure to ensure prompt restoration of service in the event of equipment failure.

1-5.8.5.2 Where a two-way telephone communication circuit is provided, its installation wires shall be monitored for a short circuit fault that would make the telephone communication circuit inoperative.

1-5.8.6 Monitoring Integrity of Power Supplies.

1-5.8.6.1 All primary and secondary power supplies shall be monitored for the presence of voltage at the point of connection to the system.

Exception No. 1: A power supply for supplementary equipment.

Exception No. 2: The neutral of a three-, four-, or five-wire ac or dc supply source.

Exception No. 3: In a central station, the main power supply, if the fault condition is otherwise so indicated as to be obvious to the operator on duty.

Exception No. 4: The output of an engine-driven generator that is part of the secondary power supply, if the generator is tested weekly per Chapter 7.

1-5.8.6.2 Power supply sources and electrical supervision for digital alarm communications systems shall be in accordance with 1-5.2 and 1-5.8.1.

NOTE: Since digital alarm communicator systems establish communications channels between the protected premises and the central station via the public switched telephone network, the requirement to supervise circuits between the protected premises and the central station (*see 1-5.8.1*) is considered met when the communications channel is periodically tested in accordance with 4-2.3.2.1.10.

1-5.8.6.3 The primary power failure trouble signal for the DACT shall not be transmitted until the actual battery capacity is depleted at least 25 percent, but not more than 50 percent.

1-6 System Interfaces.

The requirements by which fire alarm systems interface with other fire protective systems and fire safety functions can be found in Chapter 3.

1-7 Documentation.

1-7.1 Approval and Acceptance.

1-7.1.1 The authority having jurisdiction shall be notified prior to installation or alteration of equipment or wiring. At its request, complete information regarding the system or system alterations, including specifications, wiring diagrams, battery calculation, and floor plans shall be submitted for approval.

1-7.1.2 Before requesting final approval of the installation, where required by the authority having jurisdiction the installing contractor shall furnish a written statement to the effect that the system

has been installed in accordance with approved plans and tested in accordance with the manufacturer's specifications and the appropriate NFPA requirements.

1-7.2 Certificate of Completion.

1-7.2.1* A certificate (*see Figure 1-7.2.1*) shall be prepared for each system. Parts 1, 2, and 4 through 10 shall be completed after the system is installed and the installation wiring has been checked. Part 3 shall be completed after the operational acceptance tests have been completed. A preliminary copy of the certificate shall be given to the system owner and, when requested, to other authorities having jurisdiction after completion of the installation wiring tests, and a final copy after completion of the operational acceptance tests.

A-1-7.2.1 The requirements of Chapter 7 should be used to perform the installation wiring and operational acceptance tests required when completing the certificate of compliance.

1-7.2.2 Every system shall include the following documentation, which shall be delivered to the owner or the owner's representative upon final acceptance of the system.

(a)* An owner's manual and installation instructions covering all system equipment, and
(b) Record drawings.

A-1-7.2.2(a) The owner's manual and installation instructions should include the following:

(a) A detailed narrative description of the system inputs, evacuation signaling, ancillary functions, annunciation, intended sequence of operations, expansion capability, application considerations, and limitations.
(b) Operator instructions for basic system operations, including alarm acknowledgment, system reset, interpreting system output (LEDs, CRT display, and printout), operation of manual evacuation signaling and ancillary function controls, changing printer paper, etc.
(c) A detailed description of routine maintenance and testing as required and recommended and as would be provided under a maintenance contract, including testing and maintenance instructions for each type of device installed. This information should include the following:

1. A listing of the individual system components that require periodic testing and maintenance

Certificate of Completion

Name of Protected Property: _____

Address: _____

Rep. of Protected Prop. (name/phone): _____

Authority Having Jurisdiction: _____

Address/Phone Number: _____

1.   Type(s) of System or Service:

_____ NFPA 72, Chapter 3 — Local
         If alarm is transmitted to location(s) off premise, list where received:

         _____

_____ NFPA 72, Chapter 3 — Emergency Voice/Alarm Service
         Quantity of voice/alarm channels: _____   Single: _____   Multiple: _____
         Quantity of speakers installed: _____   Quantity of speaker zones: _____
         Quantity of telephones or telephone jacks included insystem: _____

_____ NFPA 72, Chapter 4 — Auxiliary
         Indicate type of connection:
         Local energy, _____   Shunt, _____   Parallel telephone
         Location and telephone number for receipt of signals:

         _____

_____ NFPA 72, Chapter 4 — Remote Station
         Alarm: _____

         Supervisory: _____

_____ NFPA 72, Chapter 4 — Proprietary
         If alarms are retransmitted to public fire service communications center or others, indicate location and telephone
         number of the organization receiving alarm:

         _____

         Indicate how alarm is retransmitted:

         _____

_____ NFPA 72, Chapter 4 — Central Station
         The Prime Contractor:

         _____

         Central Station Location:

         _____

         Means of transmission of signals from the protected premise to the central station:
         _____ McCulloh   _____ Multiplex   _____ One-Way Radio
         _____ Digital Alarm Communicator _____ Two-Way Radio   _____ Others

Figure 1-7.2.1   *Certificate of Completion.*

Means of transmission of alarms to the public fire service communications center:

1. _____

2. _____

System Location: _____

	Organization Name/Phone	Representative Name/Phone
Installer	_____	_____
Supplier	_____	_____
Service Organization	_____	_____

Location of Record (As-Built) Drawings:

_____

Location of Owners Manuals:

_____

Location of Test Reports:

_____

A contract, dated _____ , for test and inspection in accordance with NFPA standard(s) No.(s) _____ , dated _____ , is in effect.

2. Certification of System Installation
(Fill out after installation is complete and wiring checked for opens, shorts, ground faults, and improper branching, but prior to conducting operational acceptance tests.)

This system has been installed in accordance with the NFPA standards as listed below, was inspected by _____ on _____ , includes the devices listed below and has been in service since _____ .

_____ NFPA 72, Chapters 1 3 4 5 6 7 (circle all that apply)
_____ NFPA 70, *National Electrical Code*, Article 760
_____ Manufacturer's Instructions
_____ Other (specify): _____

Signed: _____ Date: _____

Organization: _____

3. Certification of System Operation
All operational features and functions of this system were tested by _____ on _____ and found to be operating properly in accordance with the requirements of:

_____ NFPA 72, Chapters 1 3 4 5 6 7 (circle all that apply)
_____ NFPA 70, *National Electrical Code*, Article 760
_____ Manufacturer's Instructions
_____ Other (specify): _____

Signed: _____ Date: _____

Organization: _____

Figure 1-7.2.1 Certificate of Completion. (cont.)

4. Alarm Initiating Devices and Circuits (Use blanks to indicate quantity of devices.)

MANUAL

a) _____ Manual Stations _____ Noncoded, Activating _____ Transmitters _____ Coded

b) _____ Combination Manual Fire Alarm and Guard's Tour Coded Stations

AUTOMATIC

Coverage: Complete: _____ Partial: _____

a) _____ Smoke Detectors _____ Ion _____ Photo

b) _____ Duct Detectors _____ Ion _____ Photo

c) _____ Heat Detectors _____ FT _____ RR _____ FT/RR _____ RC

d) _____ Sprinkler Water Flow Switches: _____ Noncoded, activating _____ Transmitters _____ Coded

e) _____ Other (list): _____

5. Supervisory Signal Initiating Devices and Circuits (Use blanks to indicate quantity of devices.)

GUARD'S TOUR

a) _____ Coded Stations

b) _____ Noncoded Stations Activating _____ Transmitters

c) _____ Compulsory Guard Tour System Comprised of _____ Transmitter Stations and _____ Intermediate Stations

Note: Combination devices recorded under 4(b) and 5(a).

SPRINKLER SYSTEM

a) _____ Coded Valve Supervisory Signaling Attachments
_____ Valve Supervisory Switches Activating _____ Transmitters

b) _____ Building Temperature Points

c) _____ Site Water Temperature Points

d) _____ Site Water Supply Level Points

Electric Fire Pump:

e) _____ Fire Pump Power

f) _____ Fire Pump Running

g) _____ Phase Reversal

Engine-Driven Fire Pump:

h) _____ Selector in Auto Position

i) _____ Engine or Control Panel Trouble

j) _____ Fire Pump Running

Engine-Driven Generator:

k) _____ Selector in Auto Position

l) _____ Control Panel Trouble

m) _____ Transfer Switches

n) _____ Engine Running

Figure 1-7.2.1 *Certificate of Completion. (cont.)*

Other Supervisory Function(s) (specify): _____

_____

_____

_____

_____

6. Alarm Notification Appliances and Circuits

Quantity of indicating appliance circuits connected to the system: _____

Types and quantities of alarm indicating appliances installed:

a) _____ Bells   _____ Inch

_____ Speakers

b) _____ Horns

c) _____ Chimes

d) _____ Other: _____

e) _____ Visual Signals   Type: _____

_____ with audible   _____ w/o audible

f) _____ Local Annunciator

7. Signaling Line Circuits:

Quantity and Style (See NFPA 72, Table 3-6.1) of signaling line circuits connected to system:

Quantity: _____   Style: _____

8. System Power Supplies

a) Primary (Main):   Nominal Voltage: _____   Current Rating: _____

Overcurrent Protection: Type: _____   Current Rating: _____

Location: _____

b) Secondary (Standby):

_____ Storage Battery: Amp-Hour Rating _____

Calculated capacity to drive system, in hours: _____ 24 _____ 60

_____ Engine-driven generator dedicated to fire alarm system:

Location of fuel storage: _____

c) Emergency or Standby System used as backup to Primary Power Supply, instead of using a Secondary Power Supply:

_____ Emergency System described in NFPA 70, Article 700

_____ Legally Required Standby System described in NFPA 70, Article 701

_____ Optional Standby System described in NFPA 70, Article 702, which also meets the performance requirements of Article 700 or 701

9. System Software

a) Operating System Software Revision Level(s): _____

b) Application Software Revision Level(s): _____

c) Revision Completed by: _____

                          (name)   (firm)

10. Comments:

_____

(signed) for Central Station or Alarm Service Company   (title)   (date)

Figure 1-7.2.1   *Certificate of Completion. (cont.)*

Frequency of routine tests and inspections, if other than in accordance with the referenced NFPA standards(s):

_____

_____

_____

System deviations from the referenced NFPA standard(s) are: _____

_____

_____

_____

_____

(signed) for Central Station or Alarm Service Company   (title)   (date)

Upon completion of the system(s) satisfactory test(s) witnessed (if required by the authority having jurisdiction):

_____

(signed) representative of the authority having jurisdiction   (title)   (date)

Figure 1-7.2.1 *Certificate of Completion. (cont.)*

2. Step-by-step instructions detailing the requisite testing and maintenance procedures and the intervals at which those procedures shall be performed, for each type of device installed

3. A schedule that correlates the testing and maintenance procedures required by paragraph (2) above with the listing required by paragraph (1) above.

(d) Detailed troubleshooting instructions for each trouble condition generated from the monitored field wiring, including opens, grounds, loop failures, etc. These instructions should include a list of all trouble signals annunciated by the system, a description of the condition(s) that will cause those trouble signals, and step-by-step instructions describing how to isolate those problems and correct them (or call for service, as appropriate).

(e) A service directory, including a list of names and telephone numbers for those who should be called to obtain service on the system.

1-7.2.3 **Central Station Fire Alarm Systems.** It shall be conspicuously indicated by the prime contractor (*see Chapter 4*) that the fire alarm system providing service at a protected premises complies with all applicable requirements of this code by providing a means of verification as specified in either 1-7.2.3.1 or 1-7.2.3.2.

1-7.2.3.1 The installation shall be certificated.

1-7.2.3.1.1 Central station fire alarm systems providing service that complies with all requirements of this code shall be certificated by the organization that has listed the prime contractor, and a document attesting to this certification shall be located on or near the fire alarm system control unit or, if no control unit exists, on or near a fire alarm system component.

> **Formal Interpretation 89-3**
> Reference: 1-7.2.3.1.1
>
> *Question:* Is it the intent of the Committee that this section provide a means for the authority having jurisdiction to require all central station fire alarm systems to be certified to verify compliance with this code, where the central station is listed and provides a certification service in accordance with its listing?
> *Answer:* Yes.
>
> *Issue Edition:* NFPA 71-1989
> *Reference:* 1-2.3.1
> *Issue Date:* March 15, 1993 ■

1-7.2.3.1.2 A central repository of issued certification documents, accessible to the authority having jurisdiction, shall be maintained by the organization that has listed the central station.

1-7.2.3.2 The installation shall be placarded.

Formal Interpretation 89-1
Reference: 1-7.2.3.2

Question: Is placarding (which is not defined as is certification) intended to be an independent method of verification by the central station with no third party agency being involved as assurance?
Answer: No.

Issue Edition: NFPA 71-1989
Reference: 1-2.3.1
Issue Date: June 22, 1992 ■

1-7.2.3.2.1 Central station fire alarm systems providing service that complies with all requirements of this code shall be conspicuously marked by the prime contractor to indicate com-pliance. The marking shall be by one or more securely affixed placards.

1-7.2.3.2.2 The placard(s) shall be 20 sq in. (130 cm²) or larger, shall be located on or near the fire alarm system control unit or, if no control unit exists, on or near a fire alarm system component, and shall identify the central station and, if applicable, the prime contractor by name and telephone number.

1-7.3 Records. A complete unalterable record of the tests and operations of each system shall be kept for at least 2 years. The record shall be available for examination and, where required, reported to the authority having jurisdiction. Archiving of records by any means shall be permitted if hard copies of the records can be provided promptly when requested.

Exception: Where off-premises monitoring is provided, records of all signals, tests, and operations recorded at the supervising station shall be maintained for not less than one year.

2

Household Fire Warning Equipment*

Contents, Chapter 2

2-1 General
    2-1.1 Scope
    2-1.2 General Provisions
    2-1.3 Approval
        2-1.3.3 Equivalency
2-2 Basic Requirements
    2-2.1 Required Protection
    2-2.2 Alarm Notification Appliances
        2-2.2.2 Standard Signal
    2-2.3 Alarm Notification Appliances for the Hearing
        Impaired
2-3 Power Supplies
    2-3.1 General
    2-3.2 Primary Power Supply — AC
    2-3.3 Primary Power Supply — Monitored Battery
    2-3.4 Secondary (Standby) Power Supply
    2-3.5 Primary Power — Nonelectrical
2-4 Equipment Performance
    2-4.1 General
    2-4.2 Smoke Detectors
    2-4.3 Heat Detectors
    2-4.4 Alarm Signaling Intensity
    2-4.5 Control Equipment
    2-4.6 Monitoring Integrity of Installation Conductors
    2-4.7 Combination System
    2-4.8 Low Power Wireless Systems
    2-4.9 Digital Alarm Communicators
2-5 Installation
    2-5.1 General
        2-5.1.1 General Provisions
        2-5.1.2 Multiple-Station Detector Interconnection
    2-5.2 Detector Location and Spacing
        2-5.2.1 Smoke Detectors
        2-5.2.2 Heat Detectors
    2-5.3 Wiring and Equipment
2-6 Maintenance and Tests
    2-6.1 Maintenance
    2-6.2 Tests
        2-6.2.1 Single- and Multiple-Station Smoke Detectors
        2-6.2.2 Fire Alarm Systems
2-7 Markings and Instructions

This chapter was previously a stand-alone document, NFPA 74-1989, entitled *Standard for the Installation, Maintenance, and Use of Household Fire Warning Equipment.*

A-2 Household Fire Warning Protection.

(a) *Fire Danger in the Home.* Fire is the third leading cause of accidental death. Residential occupancies account for most fire fatalities, and most of these deaths occur at night during the sleeping hours.

Most fire injuries also occur in the home. Of the 300,000 Americans who are injured by fire every year, nearly 50,000 lie in hospitals for a period ranging from 6 weeks to 2 years. Many never resume normal lives.

The chances are that the average family will experience one serious fire every generation.

(b) *Fire Safety in the Home.* This code is intended to provide reasonable fire safety for persons in family living units. Reasonable fire safety can be produced through a three-point program:

1. Minimizing fire hazards
2. Providing a fire warning system
3. Having and practicing an escape plan.

(c) *Minimizing Life Safety Hazards.* This code cannot protect all persons at all times. For instance, the application of this code may not protect against the three traditional fire killers:

1. Smoking in bed
2. Leaving children home alone
3. Cleaning with flammable liquids such as gasoline.

But Chapter 2 can lead to reasonable safety from fire when the three items under A-2(b) are observed.

(d) *Fire Warning System.* There are two extremes of fire to which household fire warning equipment must respond. One is the rapidly developing, high heat fire. The other is the slow, smoldering fire. Either can produce smoke and toxic gases.

Household fires are especially dangerous at night when the occupants are asleep. Fires produce smoke and deadly gases that can overcome occupants while they are asleep. Further, dense smoke reduces visibility. Most fire casualties are victims of smoke and gas inhalation rather than burns. To warn against a fire, Chapter 2 requires smoke detectors in accordance with 2-2.1.1.1 and recommends heat or smoke detectors in all other major areas. (*See 2-2.1.1.1.*)

(e) *Family Escape Plan.* There often may be very little time between detection of a fire and the time it becomes deadly. This interval may be as little as 1 or 2 minutes. Thus, this code requires detection means to give a family some advance warning of the development of conditions that will become dangerous to life within a short period of time. Such warning, however, may be wasted unless the family has planned in advance for rapid exit from their residence. Therefore, in addition to the fire warning system, this code requires exit plan information to be furnished.

Planning and practicing for fire conditions with focus on rapid exit from the residence are important. Drills should be held so that all family members know what to do. Each person should plan for the possibility that exit out of the bedroom window may be necessary. An exit out of the residence without requiring the opening of a bedroom door is essential.

(f) *Special Provisions for the Disabled.* For special circumstances where life safety of some occupant(s) depends upon prompt rescue by others, the fire warning system should include means of prompt, automatic notification to those who are to be depended upon for rescue.

2-1 General.

2-1.1* Scope. This chapter contains minimum requirements for the selection, installation, operation, and maintenance of fire warning equipment for use within family living units. The requirements of the other chapters do not apply except as specifically indicated.

Chapter 2 deals strictly with household fire detection equipment and is a stand-alone chapter.

A-2-1.1 Chapter 2 does not attempt to cover all equipment, methods, and requirements that may be necessary or advantageous for the protection of lives and property from fire.

This is what is known as a "minimum code" and it provides a number of requirements related to household fire warning equipment that are deemed to be the practical and necessary minimum for average conditions at the present state-of-the-art.

2-1.2 General Provisions.

2-1.2.1 This code is primarily concerned with life safety, not with protection of property. It presumes that the family has an exit plan.

Family living units lead all other occupancies in fire deaths in the nation. The creation of an exit plan is of paramount importance. Smoke detectors required by this chapter and other means of detection recommended by this chapter provide warning but do not extinguish the fire. After the detector does its job, it is the responsibility of the occupants to exit as quickly as possible and meet at a specified location to ensure everyone is accounted for. Having a designated meeting place also assists the responding fire department, allowing them to focus on their suppression efforts rather than search for someone that may have already made a successful exit. Home fire statistics, recommendations for fire safety and life safety, fire warning system capabilities, a more detailed explanation of an escape plan, and special provisions for the disabled are listed in Section A-2.

2-1.2.2 A control and associated equipment, multiple- or single-station alarm device(s), or any combination thereof shall be permitted to be used as a household fire warning system, provided the requirements of 2-1.3.1 are met.

The minimum **requirement** is for a smoke detector on each level and in each bedroom (new construction only) of a home, which can be met using multiple-station alarm devices. See Figure 2.1(a). The code also allows the use of a complete household fire alarm system that contains system-type (i.e., connected to a control panel) smoke detectors along with other devices, such as heat detectors, that actually exceed the minimum requirements of the code. See Figure 2.1(b). Obviously, the complete system would be used in place of, and not in addition to, the interconnected 120-vac multiple-station smoke detectors. (*See 2-2.2.1.*)

2-1.2.3 Detection and alarm systems for use within the protected household are covered by this chapter.

The equipment necessary to perform other functions such as off-premises transmission are covered by other chapters in this code.

2-1.2.4 Supplementary functions, including the extension of an alarm beyond the household, shall be permitted and shall not interfere with the performance requirements of this chapter.

These supplementary functions include the connection to a remote station, a central station, or to another remote monitoring location. See Chapter 4 for information regarding off-premises connection requirements.

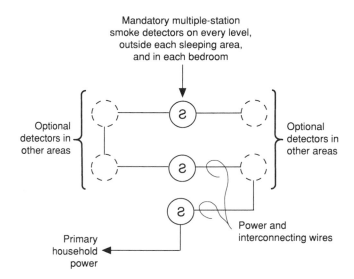

Figure 2.1(a) *Typical household multiple-station smoke detector system.*

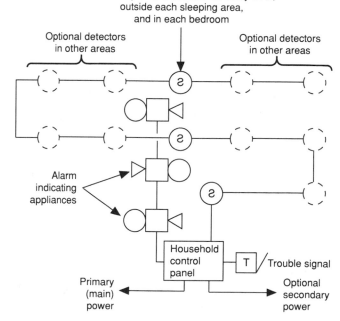

Figure 2.1(b) *Typical household fire alarm system with separate control panel.*

2-1.2.5 Where the authority having jurisdiction requires a household fire warning system to comply with the requirements of Chapter 4 or any other chapters of this code, the requirements of Section 2-2 shall still apply.

Some jurisdictions require a household fire alarm system (utilizing a control panel) to be connected to an off-premises monitoring location. The primary objective of a household fire warning system is life safety. The basic requirements of the occupant warning features in this chapter must still be followed regardless of any off-premises connection.

2-1.2.6 Definitions of Chapter 1 shall apply.

2-1.2.7 This chapter does not exclude the use of fire alarm systems complying with other chapters of this code in household applications, provided all of the requirements of this chapter are met or exceeded.

A listed commercial fire alarm system, installed in accordance with Chapter 3 of this code, meets the basic detection and warning requirements of this chapter.

2-1.3 Approval.

2-1.3.1 All devices, combination of devices, and equipment to be installed in conformity with this chapter shall be approved or listed for the purposes for which they are intended.

The approval or listing is normally from a testing laboratory. It is the responsibility of the authority having jurisdiction to either accept or reject the laboratory's approval or listing.

2-1.3.2 A device or system of devices having materials or forms different from those detailed in the chapter may be examined and tested according to the intent of the chapter and, if found equivalent, may be approved.

The authority having jurisdiction is the individual or group responsible for approval or disapproval based on the information presented for the equipment and may request tests of that equipment to determine equivalency.

2-1.3.3 Equivalency. Nothing in this code is intended to prevent the use of systems, methods, or devices of equivalent or superior quality, strength, fire resistance, effectiveness, durability, and safety over those prescribed by this code, provided technical documentation is submitted to the authority having jurisdiction to demonstrate equivalency and the system, method, or device is approved for the intended purpose.

The authority having jurisdiction has the sole responsibility for accepting a proposed system, method, or device as equivalent.

2-2 Basic Requirements.

2-2.1 Required Protection.

2-2.1.1* This code requires the following detectors within the family living unit.

A-2-2.1.1 Experience has shown that all hostile fires in family living units generate smoke to a greater or lesser degree. The same statement can be made with respect to heat buildup from fires. But the results of full-scale experiments conducted over the past several years in the U.S., using typical fires in family living units, indicate that detectable quantities of smoke precede detectable levels of heat in nearly all cases. In addition, slowly developing, smoldering fires may produce smoke and toxic gases without a significant increase in the room's temperature. Again, the results of experiments indicate that detectable quantities of smoke precede the development of hazardous atmospheres in nearly all cases.

For the above reasons, the required protection in this code utilizes smoke detectors as the primary life safety equipment that provides a reasonable level of protection against fire.

Of course, it is possible to install a lesser number of detectors than required in this code. It may be argued that the installation of only one fire detector, be it a smoke or heat detector, offers some life-saving potential. While this is true, it is the opinion of the committee that developed Chapter 2 that the smoke detector requirements as stated in 2-2.1.1 are the minimum that should be considered.

The installation of additional detectors of either the smoke or heat type should result in a higher degree of protection. Adding detectors to rooms that are normally closed off from the required detectors will increase the escape time because the fire need not build to a higher level needed to force smoke out of the closed room to the required detector. As a consequence, it is recommended that the householder consider the installation of additional fire protection devices. But it should be understood that Chapter 2 does not require additional detectors over and above those called for in 2-2.1.1.

2-2.1.1.1 Smoke detectors shall be installed outside of each separate sleeping area in the immediate vicinity of the bedrooms and on each additional story of the family living unit, including basements and excluding crawl spaces and unfinished attics. In new construction a smoke detector also shall be installed in each sleeping room.

In order to gain the full benefit of smoke detection, it is imperative that detectors be located as outlined in this chapter. However, smoke detectors must also be located away from areas that might cause nuisance alarms. The areas most likely to cause nuisance alarm problems are kitchens, garages, attics, and basements with dirt floors or moisture problems. Reliability of detection and credibility of the smoke detector must be maintained in order for the household members to continually respond properly to an alarm condition.

2-2.1.1.2* For family living units with one or more split levels (i.e., adjacent levels with less than one full story separation between levels), a smoke detector required by 2-2.1.1.1 shall suffice for an adjacent lower level, including basements. (*See Figure A-2-2.1.1.2.*)

Exception: Where there is an intervening door between one level and the adjacent lower level, a smoke detector shall be installed on the lower level.

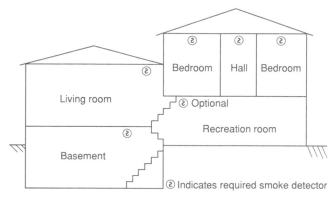

Figure A-2-2.1.1.2 Split level arrangement. Smoke detectors are required where shown. Smoke detectors are optional if door is not provided between living and recreation rooms.

2-2.1.1.3 Automatic sprinkler systems provided in accordance with NFPA 13D, *Standard for the Installation of Sprinkler Systems in One- and Two-Family Dwellings and Mobile Homes*, or NFPA 13R, *Standard for the Installation of Sprinkler Systems in Residential Occupancies Up to and Including Four Stories in Height*, shall be interconnected to sound alarm notification appliances throughout the dwelling when a fire warning system is provided.

The interconnection referenced here should be approved by the authority having jurisdiction. There are multiple-station smoke detectors that have a listed connection that will activate the notification appliances in the interconnected smoke detectors when the connection senses a closure such as a waterflow switch in alarm. A listed control panel would have a waterflow switch connected to a separate zone in the control.

2-2.2* **Alarm Notification Appliances.** Each detection device shall cause the operation of an alarm that shall be clearly audible in all bedrooms over background noise levels with all intervening doors closed. The tests of audibility level shall be conducted with all household equipment that may be in operation at night in full operation.

Examples of such equipment are window air conditioners and room humidifiers. (*See A-2-2.2 for additional information.*)

See Figure 2.2.

A-2-2.2 At times, depending upon conditions, the audibility of detection devices may be seriously impaired to occupants within the bedroom area. For instance, there may be a noisy window air conditioner or room humidifier that may generate an ambient noise level of 55 dBA or higher. The detection devices' alarms must be able to penetrate through the closed doors and be heard over the bedroom's noise levels with sufficient intensity to awaken sleeping occupants therein. Test data indicate that detection devices having sound pressure ratings of 85 dBA of 10 ft (3 m) and installed outside the bedrooms can produce about 15 dBA over ambient noise levels of 55 dBA in the bedrooms. This should be sufficient to awaken the average sleeping person.

The measurements required here can be made using a sound level meter as described in Chapter 6.

Detectors located remote from the bedroom area may not be loud enough to awaken the average person. In such cases, it is recommended that detectors be interconnected in such a way that the operation of the remote detector will cause an alarm of sufficient intensity to penetrate the bedrooms. The interconnection may be accomplished by the installation of a fire detection

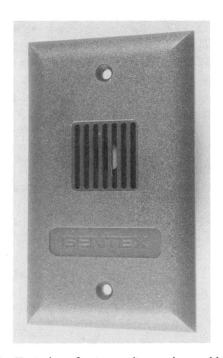

Figure 2.2 *Typical notification appliances that could be used in a household system. Top: Bell. Bottom: Mini-horn.*

system, by the wiring together of multiple-station alarm devices, or by the use of line carrier or radio frequency transmitters/ receivers.

Additional notification appliances connected to the dry contacts of a single-/multiple-station smoke detector relay may also be used to comply with this requirement.

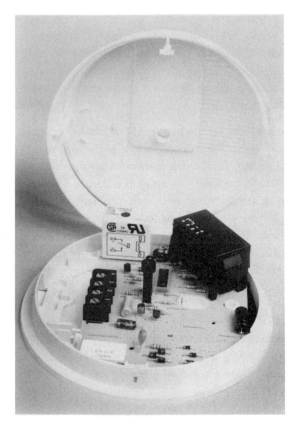

Figure 2.3 *Single-/multiple-station smoke detector with relay contacts for remote notification appliance.*

2-2.2.1 In new construction, where more than one smoke detector is required by 2-2.1, they shall be so arranged that operation of any smoke detector shall cause the alarm in all smoke detectors within the dwelling to sound.

Exception: Configurations that provide equivalent distribution of the alarm signal.

The requirement here is to use listed multiple-station smoke detectors, as a minimum, in all new construction. Equivalent distribution of the alarm signal can be achieved by notification appliances connected to a household control panel. Wireless transmitters are another means that can be used to activate notification appliances distributed throughout the home.

2-2.2.2* **Standard Signal.** Alarm notification appliances used with a household fire warning system and single- and multiple-station smoke detectors shall produce the audible emergency evacuation signal described in ANSI S3.41, *Audible Emergency Evacuation Signals*. This requirement shall become effective on July 1, 1996.

A-2-2.2.2 The use of the distinctive three-pulse temporal pattern fire alarm evacuation signal required by 3-7.2(a) had previously been recommended for this purpose by this code since 1979. It has since been adopted as both an American National Standard (ANSI S3.41, *Audible Emergency Evacuation Signal*) and an International Standard (ISO 8201, *Audible Emergency Evacuation Signal*).

Copies of both of these standards are available from the Standards Secretariat, Acoustical Society of America, 335 East 45th Street, New York, NY 10017-3483. Telephone 212-661-9404 ext. 562.

The standard fire alarm evacuation signal is a three-pulse temporal pattern using any appropriate sound. The pattern consists of an "on" phase (a) lasting 0.5 second ± 10 percent followed by an "off" phase (b) lasting 0.5 second ± 10 percent, for three successive "on" periods, which is then followed by an "off" phase (c) lasting 1.5 seconds ± 10 percent. [*See Figures A-2-2.2.2(a) and (b)*.] The signal should be repeated for a period appropriate for the purposes of evacuation of the building, but for not less than 180 seconds. A single-stroke bell or chime sounded at "on" intervals lasting 1 second ± 10 percent, with a 2-second ± 10 percent "off" interval after each third "on" stroke, is acceptable. [*See Figure A-2-2.2.2(c)*.]

The minimum repetition time may be manually interrupted.

2-2.3 **Alarm Notification Appliances for the Hearing Impaired.** In a household occupied by one or more hearing impaired persons, each initiating device shall cause the operation of visible alarm signal(s) in accordance with 2-4.4.2. Since hearing deficits are often not apparent, the responsibility to advise appropriate persons shall rest with the hearing impaired party. The responsibility for compliance shall rest with the occupants of the family living unit.

Exception: A listed tactile signal shall be permitted to be employed.

Underwriters Laboratories has investigated the effectiveness of strobe lights used for notification of the hearing impaired. This information may be found in Chapter 6 of this code.

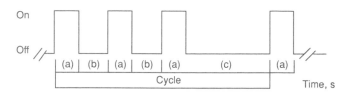

Key:
Phase (a) signal is "on" for 0.5 s ± 10%
Phase (b) signal is "off" for 0.5 s ± 10%
Phase (c) signal is "off" for 1.5 s ± 10% [(c) = (a) + 2(b)]
Total cycle lasts for 4 s ± 10%

Figure A-2-2.2.2(a) *Temporal pattern parameters.*

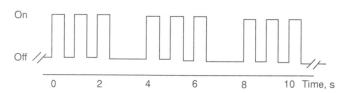

Figure A-2-2.2.2(b) *Temporal pattern imposed on signaling appliances that emit a continuous signal while energized.*

Figure A-2-2.2.2(c) *Temporal pattern imposed on a single stroke bell or chime.*

2-3 Power Supplies.

2-3.1 General.

2-3.1.1 All power supplies shall have sufficient capacity to operate the alarm signal(s) for at least 4 continuous minutes.

This requirement applies to battery-operated smoke detectors as well as smoke detectors powered by 120 vac or by a control panel power supply.

2-3.1.2 For electrically powered detectors, an ac primary power source shall be utilized in all new construction. In existing households, ac primary power is preferred; however, where such is not practical, a monitorized battery primary power source is permitted.

2-3.2 Primary Power Supply — AC.

2-3.2.1 An ac primary (main) power source shall be a dependable commercial light and power supply source. A visible "power-on" indicator shall be provided.

The "power-on" indicator is required on both smoke detectors and control panels used in household systems.

Formal Interpretation 78-2
Reference: 2-3.2.1

Background: Paragraph 2-3.2.1 states that "A visible 'power-on' indicator shall be provided." It is our understanding that this "power-on" indicator requirement is to alert the homeowner that the unit was being powered and no unintentional interruption of the power to the detector has taken place.

The problem at hand involves a 120-vac residential smoke detector also equipped with an alternative power supply in the form of a battery. The battery is fully monitored, in accordance with 2-3.3.1 with or without ac power being applied.

Question: With the above background and under the above conditions, is it the Committee's intent that such a smoke detector as the one described above be required to have a "power-on" indicator?
Answer: Yes.

Issue Edition: NFPA 74-1978
Reference: 2-1.2.1
Issue Date: January 1980
Reissued: January 1986 ■

2-3.2.2 All electrical systems designed to be installed by other than a qualified electrician shall be powered from a source not in excess of 30 volts that meets the requirements for power limited fire alarm circuits as defined in NFPA 70, *National Electrical Code*, Article 760.

Most jurisdictions require licensed electricians to install all 120-vac outlets and connections; however, in some jurisdictions a licensed fire alarm technician may be allowed to install any 120-vac connection that is associated with the fire alarm system. It is imperative that the authority having jurisdiction be consulted as to the requirements of the jurisdiction where the fire alarm system is being installed.

2-3.2.3 A restraining means shall be used at the plug-in of any cord connected installation.

2-3.2.4 AC primary (main) power shall be supplied either from a dedicated branch circuit or the unswitched portion of a branch circuit also used for power and lighting. Operation of a switch (other than a circuit breaker) or a ground fault circuit interrupter shall not cause loss of primary (main) power.

Exception No. 1: Detectors with a supervised rechargeable standby battery that provides at least 4 months' operation with a fully charged battery.

Exception No. 2: Where a ground-fault circuit interrupter serves all electrical circuits within the household.

When installing single- or multiple-station smoke detectors, it is good practice to connect the detector's power to a branch circuit serving electrical outlets in a habitable area such as the living room or family room. This is done to ensure that if for any reason the circuit breaker is in the "off" position, the chance the circuit will remain off for any length of time is small. The power connection to a household control panel can be connected in the same way. When connecting to a power circuit that serves other appliances, one must make sure that the circuit is not overloaded to the point that the power to the control panel is interrupted due to constant circuit breaker operation.

2-3.2.5 Neither loss nor restoration of primary (main) power shall cause an alarm signal in excess of 2 seconds within nor any alarm signal outside the living unit.

Generally, a loss or restoration of power will not cause any alarm signal. However, some 120-vac single- and multiple-station smoke detectors may sound briefly (2 seconds or less) in this situation.

2-3.2.6 Where a secondary (standby) battery is provided, the primary (main) power supply shall be of sufficient capacity to operate the system under all conditions of loading with any secondary (standby) battery disconnected or fully discharged.

2-3.3 Primary Power Supply — Monitored Battery.

2-3.3.1 Household fire warning equipment shall be permitted to be powered by a battery, provided that the battery is monitored to ensure that the following conditions are met:

(a) All power requirements are met for at least 1 year's life, including monthly testing.

Formal Interpretation 78-1
Reference: 2-3.3.1

Question 1: Is it the intent of 2-3.3.1(a) for all power requirements to be met for at least one year's life at 90°F including weekly testing?
Answer: No.

Question 2: Is it the intent of 2-3.3.1 that inferior batteries (batteries with poor shelf life or incapable of meeting the power requirements over the normal environmental conditions of a household) can be used provided they meet the requirements of 2-3.3.1(a) for at least six months?
Answer: No.

Question 3: Is it the intent of 2-3.3.1 that the requirements of 2-3.3.1(a) do not have to be met provided the requirements of 2-3.3.1(d) and 2-3.3.1(e) are met?
Answer: No.

Issue Edition: NFPA 74-1978
Reference: 2-1.3.1, 2-1.3.1(a)
Issue Date: November 1978
Reissued: January 1986 ■

(b) A distinctive audible trouble signal is given before the battery is incapable of operating (from aging, terminal corrosion, etc.) the device(s) for alarm purposes.

(c) For a unit employing a lock-in alarm feature, automatic transfer is provided from alarm to a trouble condition.

(d) The unit is capable of producing an alarm signal for at least 4 minutes at the battery voltage at which a trouble signal is normally obtained, followed by not less than 7 days of trouble signal operation.

(e) The audible trouble signal is produced at least once every minute for 7 consecutive days.

(f) Acceptable replacement batteries are clearly identified by manufacturer's(s') name and model number(s) on the unit near the battery compartment.

(g) A readily noticeable visible indication shall be displayed when a primary battery is removed from the unit.

Figure 2.4 *Battery operated smoke detector.*

(h) Any unit that uses a nonrechargeable battery as a primary power supply that is capable of a 10-year or greater service life, including testing, and meets the requirements of (b) through (e) above, shall not be required to have a replaceable battery.

2-3.4 Secondary (Standby) Power Supply.

2-3.4.1 Removal or disconnection of a battery used as a secondary (standby) power source shall cause a distinctive audible or visible trouble signal.

2-3.4.2 Acceptable replacement batteries shall be clearly identified by manufacturer's(s') name and model number(s) on the unit near the battery compartment.

2-3.4.3 If required by law for disposal reasons, rechargeable batteries shall be removable.

2-3.4.4 Automatic recharging shall be provided where a rechargeable battery is used as the secondary (standby) supply. The supply shall be capable of operating the system for at least 24 hours in the normal condition, followed by not less than 4 minutes of alarm. Loss of the secondary (standby) source shall sound an audible trouble signal at least once every minute.

Where standby is required by the authority having jurisdiction or is supplied for other reasons, it must meet the requirements as stated above. When standby is to be provided, the authority having jurisdiction should require the submission of battery calculations used to determine the time required for standby and alarm time.

2-3.4.4.1 The battery shall be recharged within 4 hours if power is provided from a circuit that can be switched on or off other than by a circuit breaker, or within 48 hours when power is provided from a circuit that cannot be switched on or off other than by a circuit breaker.

2-3.4.5 Where automatic recharging is not provided, the battery shall be monitored to ensure that the following conditions are met:

(a) All power requirements are met for at least 1 year's life.
(b) A distinctive audible trouble signal is given before the battery capacity has been depleted below the level required to produce an alarm signal for 4 minutes.

2-3.5 **Primary Power — Nonelectrical.** A suitable spring-wound mechanism shall provide power for the nonelectrical portion of a listed single station detector. A visible indication shall be provided to show that sufficient operating power is not available.

2-4 Equipment Performance.

2-4.1 **General.** The failure of any nonreliable or short-life component that renders the detector inoperable shall be readily apparent to the occupant of the living unit without the need for test.

This subsection requires the supervision of the circuitry in the smoke detector and some form of audible or visible indication of detector component failure. One acceptable means of indication could be a fail-safe feature, where the detector fails in the alarm mode.

2-4.2 **Smoke Detectors.**

2-4.2.1 Each smoke detector shall detect abnormal quantities of smoke that may occur in a dwelling, shall properly operate in the normal environmental conditions of a household, and shall

be in compliance with ANSI/UL 268, *Smoke Detectors for Fire Protective Signaling Systems*, or ANSI/UL 217, *Single and Multiple Station Smoke Detectors*.

ANSI/UL 268 is the standard for "system" smoke detectors. These detectors are connected to a control panel and may also have integral notification appliances, depending on the model and manufacturer. ANSI/UL 217 is the standard that all ac-powered, battery-powered, or combination ac-/battery-powered single- and multiple-station smoke detectors must comply with in order to be listed. These detectors are not normally connected to a control panel unless they have been specifically listed for that purpose. Both the ionization type and the photoelectric type smoke detectors are available under ANSI/UL 268 or ANSI/UL 217. See Chapter 5.

Formal Interpretation 89-1
Reference: 2-4.2.1

Question: Is it the intent of the Committee that only devices which comply with the definition of a smoke detector in 1-4 meet the intent of 2-4.2.1?

Answer: No. Paragraph 2-1.3.2 allows for any device or system that can demonstrate equivalent performance to be "approved." Data currently exists that could be used by any interested party to demonstrate such equivalency for a device sensing carbon monoxide. If an analysis of this data shows an equivalent performance, carbon monoxide sensing devices could be listed or approved as meeting the requirements of NFPA 72.

Issue Edition: NFPA 74-1989
Reference: 1-4, 4-2.1
Issue Date: January 2, 1990 ■

2-4.3* **Heat Detectors.** Each heat detector, including a heat detector integrally mounted on a smoke detector, shall detect abnormally high temperature or rate-of-temperature rise, and all such detectors shall be listed or approved for not less than 50-ft (15-m) spacing.

A-2-4.3 The linear space rating is the maximum allowable distance between heat detectors. The linear space rating is also a measure of their response time to a standard test fire when tested at the same distance. The higher the rating, the faster the response time. This code recognizes only those heat detectors with ratings of 50 ft (15 m) or more.

The heat detector types that are allowed by this requirement are either fixed-temperature or rate-of-rise detectors.

Figure 2.5   Combination smoke detector and heat detector.

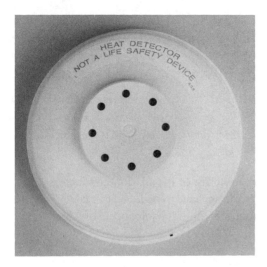

Figure 2.6   Low profile rate-of-rise heat detector.

2-4.3.1*   Fixed-temperature detectors shall have a temperature rating at least 25°F (14°C) above the normal ambient temper-

ature and shall not exceed 50°F (28°C) higher than the maximum anticipated ambient temperature in the room or space where installed.

A-2-4.3.1   A heat detector with a temperature rating somewhat in excess of the highest normally expected ambient temperature is specified in order to avoid the possibility of premature operation of the heat detector to nonfire conditions.

Some areas or rooms of the family living unit can experience ambient temperatures considerably higher than in the normally occupied living spaces. Examples are unfinished attics, the space near hot air registers, and some furnace rooms. This fact should be considered in the selection of the appropriate temperature rating for fixed temperature heat detectors to be installed in these areas or rooms.

For similar reasons, in applying rate-of-rise heat detectors, which respond to rapid temperature rises, one should consider the environment in which these detectors are to be installed. Areas near dishwashers, hot air vents, and ovens are examples of areas to be avoided.

2-4.4   Alarm Signaling Intensity.

2-4.4.1   All alarm-sounding appliances shall have a minimum rating of 85 dBA at 10 ft (3 m).

The requirement of 85 dBA at 10 ft (3 m) is measured under specific conditions and marked on the notification appliance. The minimum rating of a notification appliance is not measured in the field. The sound level needed to awaken someone varies; however, the recommended sound level for household occupancies is 15 dBA above ambient conditions. For more information regarding notification appliances, refer to Chapter 6.

Exception: An additional sounding appliance intended for use in the same room as the user, such as a bedroom, may have a sound pressure level as low as 75 dBA at 10 ft (3 m).

2-4.4.2   Visible notification appliances used in rooms where hearing impaired person(s) sleep shall have a minimum rating of 177 candela for a maximum room size of 14 ft by 16 ft (4.27 m by 4.88 m). For larger rooms, the visible notification appliance shall be located within 16 ft (4.88 m) of the pillow. Visible notification appliances in other areas shall have a minimum rating of 15 candela.

Exception: Where a visible notification appliance in a sleeping room is mounted more than 24 in. below the ceiling, a minimum rating of 110 candela shall be permitted.

There are federal laws and regulations that affect the placement and intensity of visible signals. Underwriters Laboratories has conducted research that has been accepted by the committee for Chapter 2 and the engineering community as sufficient for use in household fire alarm systems. The intensity levels required by this subsection are the results of that research.

2-4.5 Control Equipment.

2-4.5.1 The control equipment shall be automatically restoring on restoration of electrical power.

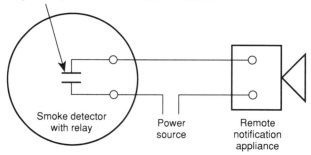

Normally open contacts close when single-/multiple-station smoke detector alarms.

Smoke detector with relay

Power source

Remote notification appliance

Figure 2.7   Single-station smoke detector with remote notification appliance.

2-4.5.2   The control equipment shall be of a type that "locks in" on an alarm condition. Smoke detection circuits need not lock in.

The control panel must have the "lock-in" feature. The circuits powering single- or multiple-station smoke detectors need not lock in. However, system-connected smoke detectors utilizing a control panel are required by ANSI/UL 268 to provide a lamp or equivalent on a spot-type detector head or base to identify it as the unit from which the alarm was initiated. ANSI/UL 268 requires that "the means incorporated to identify the initiation of an alarm shall remain activated after the smoke has dissipated from within the detector." The lock-in feature is, therefore, required on all spot-type smoke detectors connected to a control panel.

2-4.5.3   If a reset switch is provided, it shall be a self-restoring type.

2-4.5.4   An alarm-silencing switch or an audible trouble-silencing switch shall not be provided unless its silenced position is indicated by a readily apparent signal.

2-4.5.5   Each electrical fire warning system and each single station smoke detector shall have an integral test means to permit the householder to check the system and sensitivity of the detector(s).

This sensitivity test is generally a test of either the outside limits of detector operation, as allowed by the testing laboratories, or the detector's sensitivity "window" of operation. If a more exact sensitivity reading is deemed necessary, see Chapter 5.

2-4.6 Monitoring Integrity of Installation Conductors.

2-4.6.1   All means of interconnecting initiating devices or notification appliances shall be monitored for the integrity of the interconnecting pathways up to the connections to the device or appliance so that the occurrence of a single open or single ground fault, which prevents normal operation of the system, will be indicated by a distinctive trouble signal.

Exception No. 1: Conductors connecting multiple-station detectors, provided a single fault on the wiring will not prevent single-station operation of any of the interconnected detectors.

Exception No. 2: Circuits extending from single- or multiple-station detectors to required remote notification appliances provided operation of the test feature on any detector will cause all connected appliances to activate.

The monitoring of installation conductors applies only to devices and appliances connected to a control panel installed in accordance with the requirements of Chapter 3.

2-4.7 Combination System.

2-4.7.1   Where common wiring is employed for a combination system, the equipment for other than the fire warning signaling system shall be connected to the common wiring of the system so that short circuits, open circuits, grounds, or any fault in this equipment or interconnection between this equipment and the fire warning system wiring shall not interfere with the supervision of the fire warning system, or prevent alarm or trouble signal operation.

In order to comply with this requirement, it may be necessary to provide additional detection beyond the basic requirements of 2-2.1.1.1. With some combination fire/burglar alarm control panels, a fault on the burglar alarm system wiring may affect the operation of the entire control panel. If this situation could exist, protection of the wiring and control panel accessories should be considered. This protection could take the form of additional detectors (heat or smoke detectors, depending on the location) or physical protection of the wiring using metal raceway, for example. See Figure 2.8.

2-4.7.2 In a fire/burglar system, the operation shall be as follows:

(a) A fire alarm signal shall take precedence or be clearly recognizable over any other signal even when the nonfire alarm signal is initiated first.

(b) Distinctive alarm signals shall be obtained between fire alarms and other functions such as burglar alarms. The use of a common sounding appliance for fire and burglar alarms is acceptable if distinctive signals are obtained. (*See 2-2.2.2.*)

The distinctive alarm signals called for in 2-4.7.2(b) can be coded signals, where a steady tone for the burglar alarm and a three-pulse temporal pattern for the fire alarm can be used, with one appliance supplying both tones. See 2-2.2.2.

2-4.8 **Low Power Wireless Systems.** Household fire warning systems utilizing low power wireless transmission of signals within the protected household shall comply with the requirements of Section 3-13, except for 3-13.4.5.

See Figure 2.9.

Figure 2.8 *Combination fire/burglar alarm control unit.*

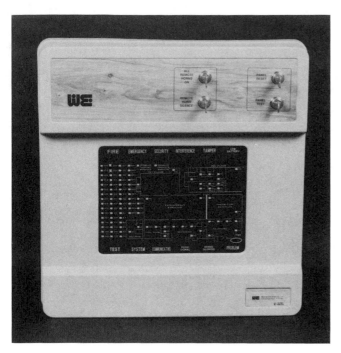

Figure 2.9 Low power wireless combination system control unit.

2-4.9 Digital Alarm Communicators.

2-4.9.1 Household fire warning systems that employ off-premises transmission of signals via digital alarm communicators shall comply with the provisions of section 4-2.3.2 with the following exceptions:

(a) For 4-2.3.2.1.6 only one telephone line shall be required for one- and two-family residences.

(b) For 4-2.3.2.1.8 each DACT need only be programmed to call a single DACR number.

(c) For 4-2.3.2.1.10 each DACT serving a one- or two-family residence shall transmit a test signal to its associated receiver at least once a month.

The exceptions allow the homeowner to utilize a DACT without the additional restrictions imposed on commercial systems. Obviously, should the homeowner desire the additional protection required for commercial systems, the code would allow it.

2-5 Installation.

2-5.1 General.

2-5.1.1 General Provisions.

2-5.1.1.1 All equipment shall be installed in a workmanlike manner.

"Workmanlike manner" is a phrase that most people intuitively assume to mean the installation is neat, safe, easily maintained, and complies with all appropriate codes and standards.

2-5.1.1.2 All devices shall be so located and mounted that accidental operation will not be caused by jarring or vibration.

2-5.1.1.3 All installed household fire warning equipment shall be mounted so as to be supported independently of its attachment to wires.

2-5.1.1.4 All equipment shall be restored to normal as promptly as possible after each alarm or test.

2-5.1.1.5 The supplier or installing contractor shall provide the owner with:

(a) An instruction booklet illustrating typical installation layouts

(b) Instruction charts describing the operation, method and frequency of testing, and proper maintenance of household fire warning equipment

(c) Printed information for establishing a household emergency evacuation plan

(d) Printed information to inform owners where they may obtain repair or replacement service, and where and how parts requiring regular replacement (such as batteries or bulbs) may be obtained within two weeks.

2-5.1.2 Multiple-Station Detector Interconnection.

(a) Where the interconnected wiring is unsupervised, no more than 18 detectors shall be interconnected in a multiple station configuration.

(b) Where the interconnecting wiring is supervised, the number of interconnected detectors shall be limited to 64.

The intent here is to recognize that the requirements for large residences should differ from the requirements for smaller residences. It is important to remember that the configuration allowed in 2-5.1.2(a) incorporates heat detectors that may be connected to multiple-station smoke detectors or to their unsupervised circuits. The number of multiple-station smoke detectors used to protect a residence is limited to 12 (*see 2-5.1.2.2*). The cautions regarding misapplication of devices mentioned earlier in the chapter should be reviewed. This requirement applies only to one- and two-family residences.

2-5.1.2.1* Interconnection that causes other detectors to sound shall be limited to an individual family living unit. Remote annunciation from single- or multiple-station detectors shall be permitted.

A-2-5.1.2.1 One of the common problems associated with residential smoke detectors is the unwanted alarms that are usually triggered by products of combustion from cooking, smoking, or other household particulates. While an alarm for such a condition would be anticipated and tolerated by the occupant of a family living unit through routine living experience, the alarm would not be acceptable if it also sounded alarms in other family living units or in common use spaces. Unwanted alarms from cooking are a very common occurrence, and inspection authorities should be aware of the ramifications that could result if the coverage is extended beyond the limits of the family living unit.

2-5.1.2.2 No more than 12 smoke detectors may be interconnected in a multiple-station connection.

2-5.2* **Detector Location and Spacing.**

A-2-5.2 One of the most critical factors of any fire alarm system is the location of the fire detecting devices. This appendix is not a technical study. It is an attempt to state some fundamentals on detector location. For simplicity, only those types of detectors recognized by Chapter 2, i.e., smoke and heat detectors, will be discussed. In addition, special problems requiring engineering judgment, such as locations in attics and in rooms with high ceilings, will not be covered.

For more information on special problems, see Chapter 5.

2-5.2.1* Smoke Detectors.

A-2-5.2.1 Smoke Detection.

(a) *Where to Locate the Required Smoke Detectors in Existing Construction.* The major threat from fire in a family living unit is at night when everyone is asleep. The principal threat to persons in sleeping areas comes from fires in the remainder of the unit; therefore, smoke detector(s) are best located between the bedroom areas and the rest of the unit. In units with only one bedroom area on one floor, the smoke detector should be located as shown in Figure A-2-5.2.1(a).

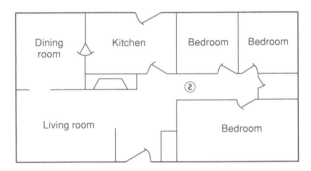

Figure A-2-5.2.1(a) *A smoke detector should be located between the sleeping area and the rest of the family living unit.*

In family living units with more than one bedroom area or with bedrooms on more than one floor, more than one smoke detector will be needed, as shown in Figure A-2.5.2.1(b).

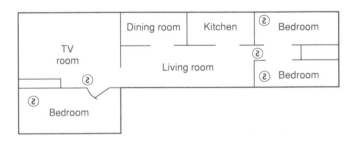

Figure A-2-5.2.1(b) *In family living units with more than one sleeping area, a smoke detector should be provided to protect each sleeping area in addition to detectors required in bedrooms.*

In addition to smoke detectors outside of the sleeping areas, Chapter 2 requires the installation of a smoke detector on each additional story of the family living unit, including the basement.

These installations are shown in Figure A-2-5.2.1(c). The living area smoke detector should be installed in the living room and/or near the stairway to the upper level. The basement smoke detector should be installed in close proximity to the stairway leading to the floor above. If installed on an open-joisted ceiling, the detector should be placed on the bottom of the joists. The detector should be positioned relative to the stairway so as to intercept smoke coming from a fire in the basement before the smoke enters the stairway.

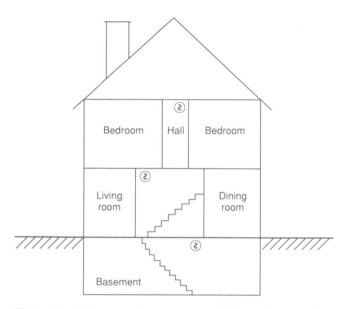

Figure A-2-5.2.1(c) *A smoke detector should be located on each story.*

(b) *Where to Locate the Required Smoke Detectors in New Construction.* All of the smoke detectors specified in (a) for existing construction are required, and, in addition, a smoke detector is required in each bedroom.

(c) *Are More Smoke Detectors Desirable?* The required number of smoke detectors may not provide reliable early warning protection for those areas separated by a door from the areas protected by the required smoke detectors. For this reason, it is recommended that the householder consider the use of additional smoke detectors for those areas for increased protection. The additional areas include: basement, bedrooms, dining room, furnace room, utility room, and hallways not protected by the required smoke detectors. The installation of smoke detectors in kitchens, attics (finished or unfinished), or in garages is not normally recommended, as these locations occasionally experience conditions that may result in improper operation.

2-5.2.1.1 Smoke detectors in rooms with ceiling slopes greater than 1 ft rise per 8 ft (1 m rise per 8 m) horizontally shall be located at the high side of the room.

2-5.2.1.2 A smoke detector installed in a stairwell shall be so located as to ensure that smoke rising in the stairwell cannot be prevented from reaching the detector by an intervening door or obstruction.

2-5.2.1.3 A smoke detector installed to detect a fire in the basement shall be located in close proximity to the stairway leading to the floor above.

2-5.2.1.4 A smoke detector installed to protect a sleeping area in accordance with 2-2.1.1.1 shall be located outside of the bedrooms but in the immediate vicinity of the sleeping area.

2-5.2.1.5 The smoke detector installed to comply with 2-2.1.1.1 on a story without a separate sleeping area shall be located in close proximity to the stairway leading to the floor above.

2-5.2.1.6* Smoke detectors shall be mounted on the ceiling at least 4 in. (102 mm) from a wall or on a wall with the top of the detector not less than 4 in. (102 mm) nor more than 12 in. (305 mm) below the ceiling.

Exception: Where the mounting surface might become considerably warmer or cooler than the room, such as a poorly insulated ceiling below an unfinished attic or an exterior wall, the detectors shall be mounted on an inside wall.

A-2-5.2.1.6 **Smoke Detector Mounting — Dead Air Space.** The smoke from a fire generally rises to the ceiling, spreads out across the ceiling surface, and begins to bank down from the ceiling. The corner where the ceiling and wall meet is an air space into which the smoke may have difficulty penetrating. In most fires, this dead air space measures about 4 in. (0.1 m) along the ceiling from the corner and about 4 in. (0.1 m) down the wall as shown in Figure A-2-5.2.2(b). Detectors should not be placed in this dead air space.

Smoke and heat detectors should be installed in those locations recommended by the manufacturer, except in those cases where the space above the ceiling is open to the outside and little or no insulation is present over the ceiling. Such cases result in the ceiling being excessively cold in the winter or excessively hot in the summer. Where the ceiling is significantly different in temperature from the air space below, smoke and heat has difficulty reaching the ceiling and a detector that may be placed there. In this situation, placement of the detector on a side wall, with the top 4 in. to 12 in. (0.1 m to 0.3 m) from the ceiling, is preferred.

The situation described above for uninsulated or poorly insulated ceilings may also exist, but to a lesser extent, with outside walls. While the recommendation is to place the smoke detector on a side wall, if the side wall is an exterior wall with little or no insulation, then an interior wall should be selected. It should be recognized that the condition of inadequately insulated ceilings and walls can exist in multifamily housing (apartments), single-family housing, and mobile homes.

In those family living units employing radiant heating in the ceiling, the wall location is the preferred location. Radiant heating in the ceiling can create a hot-air, boundary layer along the ceiling surface, which can seriously restrict the movement of smoke and heat to a ceiling-mounted detector.

2-5.2.1.7 Smoke detectors shall not be located within kitchens or garages, or in other spaces where temperatures can fall below 32°F (0°C) or exceed 100°F (38°C). Smoke detectors shall not be located closer than 3 ft (0.9 m) from:

(a) The door to a kitchen or a bathroom containing a tub or shower

(b) Supply registers of a forced air heating or cooling system.

Exception: Detectors specifically listed for the application.

2-5.2.2* Heat Detectors.

A-2-5.2.2 Heat Detection.

(a) *General.* While Chapter 2 does not require heat detectors as part of the basic protection scheme, it is recommended that the householder consider the use of additional heat detectors for the same reasons presented under A-2-5.2.1(c). The additional areas lending themselves to protection with heat detectors are: kitchen, dining room, attic (finished or unfinished), furnace room, utility room, basement, and integral or attached garage. For bedrooms, the installation of a smoke detector is preferable to the installation of a heat detector for protection of the occupants from fires in their bedrooms.

(b) *Heat Detector Mounting — Dead Air Space.* Heat from a fire rises to the ceiling, spreads out across the ceiling surface, and begins to bank down from the ceiling. The corner where the ceiling and the wall meet is an air space into which heat has difficulty in penetrating. In most fires, this dead air space measures about 4 in. (0.1 m) along the ceiling from the corner and 4 in. (0.1 m) down the wall as shown in Figure A-2-5.2.2(b). Heat detectors should not be placed in this dead air space.

The placement of the detector is critical if maximum speed of fire detection is desired. Thus, a logical location for a detector is the center of the ceiling. At this location, the detector is closest to all areas of the room.

If the detector cannot be located in the center of the ceiling, an off-center location may be used on the ceiling.

The next logical location for mounting detectors is on the side wall. Any detector mounted on the side wall should be located as near as possible to the ceiling. A detector mounted on the side wall should have the top of the detector between 4 in. and 12 in. (0.1 m and 0.3 m) from the ceiling.

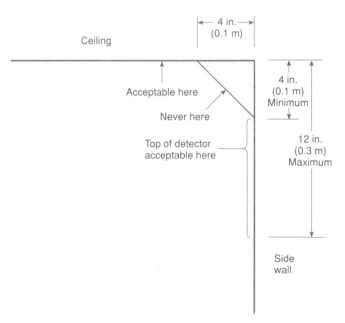

NOTE: Measurements shown are to the closest edge of the detector.
Figure A-2-5.2.2(b) Example of proper mounting for detectors.

(c) *The Spacing of Detectors.* In a room too large for protection by a single detector, several detectors should be used. It is important that they be properly located so all parts of the room are covered. For further information on the spacing of detectors see Chapter 5.

(d) *When the Distance Between Detectors Should Be Further Reduced.* The distance between detectors is based on data obtained from the spread of heat across a smooth ceiling. If the ceiling is not smooth, then the placement of the detector will have to be tailored to the situation.

For instance, with open wood joists heat travels freely down the joist channels so that the maximum distance between detectors [50 ft (15 m)] can be used. Heat, however, has trouble spreading across the joists, so the distance in this direction should be one-half the distance allowed between detectors, as shown in Figure A-2-5.2.2(d), and the distance to the wall is reduced to 12½ ft (3.8 m). Since ½ × 50 ft (15 m) is 25 ft (7.6 m), the distance between detectors across open wood joists should not exceed 25 ft (7.6 m), as shown in Figure A-2-5.2.2(d), and the distance to the wall is reduced [½ × 25 ft (7.6 m)] to 12.5 ft (3.8 m). Paragraph 2-5.2.2.4 requires that detectors be mounted on the bottom of the joists and not up in joist channels.

Walls, partitions, doorways, ceiling beams, and open joists interrupt the normal flow of heat, thus creating new areas to be protected.

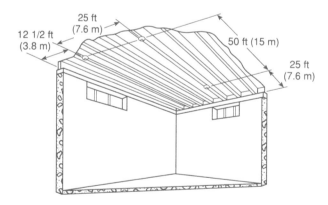

Figure A-2-5.2.2(d) *Open joists, attics, and extra high ceilings are some of the areas that require special knowledge for installation.*

2-5.2.2.1 On smooth ceilings, heat detectors shall be installed within the strict limitations of their listed spacing.

2-5.2.2.2 For sloped ceilings having a rise greater than 1 ft in 8 ft (1 m in 8 m) horizontally, the detector shall be located on or near the ceiling at or within 3 ft (0.9 m) of the peak. The spacing of additional detectors, if any, shall be based on a horizon-

tal distance measurement, not on a measurement along the slope of the ceiling.

2-5.2.2.3* Heat detectors shall be mounted on the ceiling at least 4 in. (102 mm) from a wall or on a wall with the top of the detector not less than 4 in. (102 mm) nor more than 12 in. (305 mm) below the ceiling.

Exception: Where the mounting surface might become considerably warmer or cooler than the room, such as a poorly insulated ceiling below an unfinished attic or an exterior wall, the detectors shall be mounted on an inside wall.

A-2-5.2.2.3 The same comments apply here as under A-2-5.2.1.6.

2-5.2.2.4 In rooms with open joists or beams, all ceiling-mounted detectors shall be located on the bottom of such joists or beams.

2-5.2.2.5* Detectors installed on an open-joisted ceiling shall have their smooth ceiling spacing reduced where this spacing is measured at right angles to solid joists; in the case of heat detectors, this spacing shall not exceed one-half of the listed spacing.

A-2-5.2.2.5 In addition to the special requirements for heat detectors installed on ceilings with exposed joists, reduced spacing may also be required due to other structural characteristics of the protected area, possible drafts, or other conditions that may affect detector operation.

2-5.3 **Wiring and Equipment.** The installation of wiring and equipment shall be in accordance with the requirements of NFPA 70, *National Electrical Code*, Article 760.

The installation of all fire alarm system wiring should take into account the fire alarm system manufacturer's published installation instructions and the limitations of the applicable product listings or approvals.

2-6   Maintenance and Tests.

2-6.1* **Maintenance.** If batteries are used as a source of energy, they shall be replaced in accordance with the recommendations of the alarm equipment manufacturer.

A-2-6.1 Good fire protection requires that the equipment be periodically maintained. If the householder is unable to perform the required maintenance, a maintenance agreement should be considered.

2-6.2* Tests.

A-2-6.2 It is a good practice to establish a specific schedule for these tests.

2-6.2.1 Single- and Multiple-Station Smoke Detectors. Homeowners shall inspect and test smoke detectors and all connected appliances in accordance with the manufacturer's instructions at least once a month.

2-6.2.2 Fire Alarm Systems. Homeowners shall test systems in accordance with the manufacturer's instructions and shall have every residential fire alarm system tested by a qualified service technician at least every 3 years. This test shall be conducted according to the methods of Chapter 7.

2-7 Markings and Instructions.

All household fire warning equipment or systems shall be plainly marked with the following information on the unit:

(a) Manufacturer's or listee's name, address, and model number

(b) A mark or certification that the unit has been approved or listed by a testing laboratory

(c) Electrical rating (if applicable)

(d) Temperature rating (if applicable)

(e) Spacing rating (if applicable)

(f) Operating instructions

(g) Test instructions

(h) Maintenance instructions

(i) Replacement and service instructions.

Exception: When space limitations prohibit inclusion of 2-7.1(g), 2-7.1(h), and 2-7.1(i), a permanent label or plaque suitable for permanent attachment within the living unit shall be provided with the equipment and referenced on the equipment. In the case of a household fire warning system, the required information shall be prominently displayed at the control panel.

3

Protected Premises Fire Alarm Systems

Contents, Chapter 3

3-1 Scope
3-2 General
3-3 Applications
3-4 Performance of Initiating Device, Notification Appliance, and Signaling Line Circuits
   3-4.1 Circuit Designations
      3-4.1.1 Class
      3-4.1.2 Style
3-5 Performance and Capacities of Initiating Device Circuits (IDC)
3-6 Performance and Capacities of Signaling Line Circuits (SLC)
3-7 Notification Appliance Circuits (NAC)
   3-7.1 Performance
   3-7.2 Distinctive Evacuation Signal
3-8 System Requirements
   3-8.1 Manual Fire Alarm Signal Initiation
   3-8.2 Automatic Fire Alarm Signal Initiation
   3-8.3 Positive Alarm Sequence
   3-8.4 Concealed Detectors
   3-8.5 Automatic Drift Compensation
   3-8.6 Waterflow Alarm Signal Initiation
   3-8.7 Supervisory Signal Initiation
      3-8.7.1 General
      3-8.7.7 Pressure Supervision
      3-8.7.8 Water Temperature Supervision
   3-8.8 Signal Annunciation
   3-8.9 Signal Initiation from Automatic Fire Suppression System Other Than Waterflow
   3-8.10 Pump Supervision
   3-8.11 Tampering
   3-8.12 Guard's Tour Supervisory Service
   3-8.13 Suppressed (Exception Reporting) Signal System
   3-8.14 Combination Systems
   3-8.15 Elevator Recall for Fire Fighters' Service
   3-8.16 Elevator Shutdown
3-9 Fire Safety Control Functions
   3-9.1 Scope
   3-9.2 General
   3-9.3 Heating, Ventilation, and Air Conditioning (HVAC) Systems
   3-9.4 Door Release Service

3-9.5 Door Unlocking Devices

3-10 Suppression System Actuation

3-11 Interconnected Fire Alarm Control Units

3-12 Emergency Voice/Alarm Communications

   3-12.1 Application

   3-12.3 Survivability

   3-12.4 Voice/Alarm Signaling Service

      3-12.4.1 General

      3-12.4.2 Multichannel Capability

      3-12.4.3 Functional Sequence

      3-12.4.4 Voice and Tone Devices

      3-12.4.5 Fire Command Station

      3-12.4.6 Loudspeakers

   3-12.5 Evacuation Signal Zoning

   3-12.6 Two-Way Telephone Communications Service

3-13 Special Requirements for Low Power Radio (Wireless) Systems

   3-13.2 Power Supplies

   3-13.3 Alarm Signals

   3-13.4 Supervision

3-1 Scope.

This chapter provides requirements for the application, installation, and performance of fire alarm systems, including fire alarm and supervisory signals, within protected premises.

3-2 General.

The systems covered in this chapter are intended to be used for the protection of life by automatically indicating the necessity for evacuation of the building or fire area, and for the protection of property through the automatic notification of responsible persons and for the automatic activation of fire safety functions. The requirements of the other chapters shall also apply except where they conflict with the requirements of this chapter.

The wording in the first sentence of Section 3-2 has changed from "primarily for the protection of life and secondarily for the protection of property," which appeared in earlier editions of NFPA 72. This change is to emphasize that the protection of both life and property are given equal and full consideration.

Exception: For household fire warning equipment protecting a single living unit, see Chapter 2.

Fire alarm systems as discussed in Chapter 3 can also be used in single living units or as household systems. However, fire alarm systems covered under Chapter 2 are usually more economical and easier to operate for those applications.

3-2.1 Systems requiring transmission of signals to continually manned locations providing supervising station service (e.g., central station, proprietary, remote supervising station) shall also comply with the applicable requirements of Chapter 4.

Reference to Chapter 4 applies to signals transmitted to off-premises locations, utilizing central supervising stations, proprietary supervising stations, remote supervising station systems, and auxiliary systems. The Chapter 4 requirements apply to the transmitter located at the protected premises and the transmission channel between the protected premises and the remotely located supervising station.

There are essentially two choices that can be made where a single multibuilding contiguous property has its proprietary supervising station in one of the on-site buildings. Each building can have its own protected premises system and be connected to its on-site supervisory station through a transmitter and transmission channel that meet the requirements of Section 4-2. Or, as an alternative, the individual building systems can be directly connected to the supervising station using signaling line circuits (SLCs) and must comply with the requirements of Section 3-11 for interconnected fire alarm control units. For a single-building property, the initiating devices and notification appliances are either directly connected or connected through zone or floor subcontrol units to the supervising station using SLC circuits Where subcontrol units are used, the requirements of Section 3-11 apply. Regardless of the method of connection, for both single- and multiple-building properties, the supervising station facilities must comply with the requirements of Section 4-4.

3-2.2 All protected premises fire alarm systems shall be maintained and tested in accordance with Chapter 7.

Many jurisdictions feel the need to develop and enforce their own fire alarm system testing requirements. The purpose of 3-2.2 is to provide the authorities having jurisdiction with the mandatory requirement to test all fire alarm systems in accordance with this code.

3-2.3 Fire alarm systems provided for evacuation of occupants shall have one or more notification appliances listed for the purpose on each floor of the building, so located that they shall have the characteristics for public mode described in Chapter 6.

The phrase "listed for the purpose" is used when all listed equipment is not necessarily acceptable for a specific application. For example, there are listed notification appliances that do not meet all the requirements for fire alarm use and are intended for background music or other applications. Only those notification appliances specifically listed for fire alarm use are acceptable to meet the "listed for the purpose" requirement of 3-2.3.

3-2.4* The system shall be so designed and installed that attack by fire:

(a) In an evacuation zone, causing loss of communications to this evacuation zone, shall not result in loss of communications to any other evacuation zone.

"Evacuation zone" is not defined in the code. An evacuation zone can be an area of a floor, an entire floor, or several floors that are always intended to be evacuated simultaneously. See also 3-12.5.

(b) Causing failure of equipment or a fault on one or more installation wiring conductors of one communications path shall not result in total loss of communications to any evacuation zone.

This paragraph essentially requires the use of separate communications paths, circuits, or some other arrangement to avoid the total loss of communications when the fire attacks the installation wiring conductors of one communications path. For example, if a fire attacked a communications path riser that connected multiple floors, it is possible that there would be a total loss of communications to all the floors connected to that riser. However, if two risers, each protected by a 2-hour fire-rated enclosure, were each connected to one half of the notification appliances on the floors being served, then a fire attack on one communications path riser would not result in a total loss of communications to all the connected floors. See Exception No. 6 to 3-2.4 and A-3-2.4. See Figure 3.1.

Exception No. 1 to (a) and (b): Systems that, on alarm, automatically sound evacuation signals throughout the protected premises.

Figure 3.1 *Supervised speaker loop splitter can be used to separate speaker circuits to paging zones.*

Exception No. 2 to (a) and (b): Where there is a separate means acceptable to the authority having jurisdiction for voice communications to each floor or evacuation zone.

Exception No. 3 to (b): The fire command station and the central control equipment.

Exception No. 4 to (b): Where the installation wiring is enclosed in a 2-hour rated enclosure, other than a stairwell.

Exception No. 5 to (b): Where the installation wiring is enclosed within a 2-hour rated stairwell in a fully sprinklered building in accordance with NFPA 13, Standard for the Installation of Sprinkler Systems.

Exception No. 6 to (b): When the evacuation zone is directly attacked by fire within the zone.

A-3-2.4 This requirement is intended to limit damage to a fire alarm system, resulting from a fire, to the area in which the fire occurs. The concern is maintaining the operability of the system in areas beyond, but threatened by, the fire.

Conformance to this requirement may entail that:

(a) Where common risers or trunk circuits are used:

1. Separately routed, redundant risers or trunk circuits be provided, arranged so that one or more circuit faults on one riser or trunk circuit causes the system to automatically switch over to its associated, alternate circuit without loss of function. This capability should permit full system operation with a damaged or severed riser or trunk circuit.

2. Primary and alternate conductors for redundant circuits be separated by 2-hour fire resistive construction.

(b) Where multiple individual circuits are routed in a common riser, conduit, raceway, cable, bundle of conductors, or other arrangement resulting in close physical proximity and resultant susceptibility to common misfortune, such circuits be Class A, capable of full operation over a single open or single ground fault.

(c) Where Class A circuits are required, that they be installed so that the supply and return conductors are routed separately. Supply and return risers should be separated by at least 2-hour rated fire construction.

The survivability requirements previously applied only to emergency voice/alarm communication systems. They now have been extended to all types of systems installed in buildings where the fire response plan permits either partial evacuation or relocation of the building occupants to a "safe" area during a fire emergency. The survivability requirements have been relocated to Section 3-2 because the intent now is to have them apply to nonvoice systems used to evacuate the building occupants by floor or zone. As long as occupants are allowed or required to remain in the building, it is essential that the fire alarm system remain operational so that additional floors or zones can be evacuated as needed.

3-3 Applications.

Protected premises fire alarm systems include one or more of the following features:

(a) Manual alarm signal initiation
(b) Automatic alarm signal initiation
(c) Monitoring of abnormal conditions in fire suppression systems
(d) Activation of fire suppression systems
(e) Activation of fire safety functions
(f) Activation of alarm notification appliances
(g) Emergency voice/alarm communications
(h) Guard's tour supervisory service
(i) Process monitoring supervisory systems
(j) Activation of off-premises signals
(k) Combination systems
(l) Integrated systems.

3-4 Performance of Initiating Device, Notification Appliance, and Signaling Line Circuits.

This section covers the requirements for the circuits that interconnect initiating devices, notification appliances, and control devices to the control unit of a protected premises fire alarm system.

3-4.1* **Circuit Designations.** Initiating device, notification appliance, and signaling line circuits shall be designated by class or style, or both, depending on the circuits' capability of being able to continue to operate during specified fault conditions.

The circuit designation, if not mandated in a local or state code or ordinance, is generally the responsibility of the system designer. This code does not require the use of one class or style of circuit over another. The system designer obviously should choose a class or style circuit that matches his/her or the owner's reliability and performance requirements.

A-3-4.1 Class A and Class B circuit designations have been added to this edition of the code because they are still preferred by some specifiers and authorities having jurisdiction to the style designations introduced into the code in the late 1970s. The committee had discontinued the use of the Class A and Class B designations because, with the introduction of signaling line circuits, they were no longer adequate for describing the required performance of new technology systems under all fault conditions.

Class A circuits are considered more reliable than Class B circuits because they remain fully operational during the occurrence of a single open or a single ground fault, while Class B circuits only remain operational up to the location of an open fault. However, neither Class A nor Class B circuits remain operational during a wire-to-wire short.

For both Class A and Class B initiating device circuits, a wire-to-wire short was permitted to cause an alarm on the system on the rationale that a wire-to-wire short was the result of a double fault (e.g., both circuit conductors have to become grounded), while the code only considered the consequences of single faults. For many applications, an alarm caused by a wire-to-wire short is unacceptable and being limited to a simple Class A designation was not adequate. Introducing the style designation made it possible to specify the exact performance required during a variety of possible fault conditions.

A more serious problem existed for signaling line circuits. Though a Class A signaling line circuit remains fully operational during the occurrence of a single open or single ground fault, a wire-to-wire short disables the entire circuit. The risk of such a catastrophic failure was not acceptable to many system designers, users, and authorities having jurisdiction. Here again, introducing the style designation made it possible to specify either full system operation during a wire-to-wire short (Style 7), or performance in between that of a Style 7 and a minimum function Class A circuit (Style 2).

As revised, the specifier now can simply specify a circuit as either Class A or Class B where system performance during wire-to-wire shorts is of no concern, or by the appropriate style designation where the system performance during a wire-to-wire short and other multiple fault conditions is of concern.

3-4.1.1 **Class.** Initiating device, notification appliance, and signaling line circuits shall be permitted to be designated as either Class A or Class B, depending on the capability of the circuit to transmit alarm and trouble signals during nonsimultaneous single circuit fault conditions as specified by the following:

(a) Circuits capable of transmitting an alarm signal during a single open or a nonsimultaneous single ground fault on a circuit conductor shall be designated as Class A.

(b) Circuits not capable of transmitting an alarm beyond the location of the fault conditions specified in (a) above shall be designated as Class B.

Faults on both Class A and Class B circuits shall result in a trouble condition on the system in accordance with the requirements of 1-5.8.

3-4.1.2 **Style.** Initiating device, notification appliance, and signaling line circuits shall be permitted to also be designated by style depending on the capability of the circuit to transmit alarm and trouble signals during specified simultaneous multiple circuit fault conditions in addition to the single circuit fault conditions considered in the designation of the circuits by class.

(a) An initiating device circuit shall be permitted to be designated as either Style A, B, C, D, or E, depending on its ability to meet the alarm and trouble performance requirements shown in Table 3-5.1, during a single open, single ground, wire-to-wire short, or loss of carrier fault condition.

(b) A notification appliance circuit shall be permitted to be designated as either Style W, X, Y, or Z, depending on its ability to meet the alarm and trouble performance requirements shown in Table 3-7.1, during a single open, single ground, or wire-to-wire short fault condition.

(c) A signaling line circuit shall be permitted to be designated as either Style 0.5, 1, 2, 3, 3.5, 4, 4.5, 5, 6, or 7, depending on its ability to meet the alarm and trouble performance requirements shown in Table 3-6.1, during a single open, single ground, wire-to-wire short, simultaneous wire-to-wire short and open, simultaneous wire-to-wire short and ground, simultaneous open and ground, and loss of carrier fault conditions.

It is the designer's responsibility to choose the style or class of circuit that meets the desired performance requirements, unless the style or class is already required by another code, project specifications, or the authority having jurisdiction.

3-4.2* All styles of Class A circuits using physical conductors (metallic, optical fiber) shall be installed such that the outgoing and return conductors, exiting from and returning to the control unit respectively, are routed separately. The outgoing and return (redundant) circuit conductors shall not be run in the same cable assembly (multiconductor cable), enclosure, or raceway.

This requirement ensures that the Class A circuit's operation is not defeated by the cutting of the cable or the failure of a section of cable due to attack by fire in a single location. See A-3-4.2.

Exception No. 1: For a distance not to exceed 10 ft (3 m) where the outgoing and return conductors enter or exit initiating device, notification appliance, or control unit enclosures; or

Exception No. 2: Where the vertically run conductors are enclosed (installed) in a 2-hour rated enclosure other than a stairwell; or

Exception No. 3: Where the vertically run conductors are enclosed (installed) in a 2-hour rated stairwell in a building fully sprinklered in accordance with NFPA 13, Standard for the Installation of Sprinkler Systems.

Exception No. 4: Where looped conduit/raceway systems are provided, single conduit/raceway drops to individual devices or appliances shall be permitted.

Exception No. 5: Where looped conduit/raceway systems are provided, single conduit/raceway drops to multiple devices or appliances installed within a single room not exceeding 1000 sq ft (92.9 m²) in area shall be permitted.

A-3-4.2 Where installed within the protected premises, the integrity and reliability of the interconnecting signaling paths (circuits) are influenced by the following:

(a) The transmission media utilized

(b) The length of the circuit conductors

(c) The total building area covered by and the quantity of initiating devices and notification appliances connected to a single circuit

(d) The nature of the hazard present within the protected premises

(e) The functional requirements of the system necessary to provide the level of protection desired by the system.

3-5 Performance and Capacities of Initiating Device Circuits (IDC).

3-5.1* The assignment of class designations, style designations, or both to initiating device circuits shall be based on their performance capabilities under abnormal (fault) conditions in accordance with the requirements of Table 3-5.1.

It is important to evaluate the type of circuit (IDC or SLC) that is to be used based on the planned installation technique. If "T" tapping is used, the type of circuit originally chosen may not be valid with the "T" taps in the circuit.

A-3-5.1 Using Tables 3-5.1 and 3-6.1:

(a) Determine whether the initiating devices are:

1. Directly connected to the initiating device circuit

2. Directly connected to a signaling line circuit interface on a signaling line circuit

3. Directly connected to an initiating device circuit, which in turn is connected to a signaling line circuit interface on a signaling line circuit.

(b) Determine the style of signaling performance required. The columns marked A through Eα in Table 3-5.1, and 0.5 through 7α in Table 3-6.1 are arranged in ascending order of performance and capacities.

(c) Upon determining the style of the system, the charts singularly or together will specify the maximum number of devices, equipment, premises, and buildings allowed to be incorporated into an actual protected premises installation.

(d) In contrast, where the number of devices, equipment, premises, and buildings (in addition to signaling ability) in an installation is known, a required system style can be determined.

(e) The prime purpose of the tables is to enable identification of minimum performance for styles of initiating device circuits and signaling line circuits. It is not the intention that the styles be construed as grades. That is, a Style 3 system is not better than a Style 2, or vice versa. In fact, a particular style may better provide adequate and reliable signaling for an installation than a more complex style number. The quantities tabulated under each style do, unfortunately, tend to imply that one style is better than the one to its left. The increased quantities for the higher style numbers are based on the ability to signal an alarm during an abnormal condition in addition to signaling the same abnormal condition.

(f) The tables allow users, designers, manufacturers, and the authority having jurisdiction to identify minimum performance of present and future systems by determining the trouble and alarm signals received at the control unit for the specified abnormal conditions.

Table 3-5.1 Performance and Capacities of Initiating Device Circuits (IDC)

G = Systems with ground detection shall indicate systems trouble with a single ground.
R = Required capability.
X = Indication required at protected premises and as required by Chapter 4.
α = Style exceeds minimum requirements for Class A.
* = See A-3-5.1.

Class	B			B			B			A			A		
Style	A			B			C			D			Eα		
	Alarm	Trouble	Alarm Receipt Capability During Abnormal Condition	Alarm	Trouble	Alarm Receipt Capability During Abnormal Condition	Alarm	Trouble	Alarm Receipt Capability During Abnormal Condition	Alarm	Trouble	Alarm Receipt Capability During Abnormal Condition	Alarm	Trouble	Alarm Receipt Capability During Abnormal Condition
Abnormal Condition	1	2	3	4	5	6	7	8	9	10	11	12	13	14	15
A. Single Open		X			X			X			X	X		X	X
B. Single Ground		R			G	R		G	R		G	R		G	R
C. Wire-to-Wire Short	X			X				X		X				X	
D. Loss of Carrier (If Used)/Channel Interface								X						X	

Note: The following sections apply only where signals are transmitted to a proprietary supervising station in accordance with Section 4-4.

	Style A	Style B	Style C	Style D	Style Eα
E. Maximum Quantity per Initiating Device Circuit					
1. Fire Alarm					
(a) Manual Fire Alarm Boxes	2	5	5	25	25
(b) Water Flow Alarm Devices	1	2	2	5	5
(c) Discharge Alarm from Other Fire Suppression Systems	1	2	2	5	5
(d) Automatic Fire Detectors	*	*	*	*	*
2. Fire Supervisory					
(a) Sprinkler Supervisory Devices	2	4	4	20	20
(b) Other Fire Suppression Supervisory Devices	2	4	4	20	20
3. Guard's Tour	1	1	1	1	1
4. Process, Security, and Other Devices in Combination with 1, 2, and 3 Above	0	0	0	0	0
5. Process, Security, and Other Devices Not Combined with 1, 2, and 3 Above	5	10	10	20	20
6. Buildings	1	1	1	1	1
7. Intermediate Fire Alarm or Fire Supervisory Control Unit	1	1	1	1	1
F. Maximum Quantity of Initiating Device Circuits per Circuit Interface Between IDC & SLC					
1. Per Limits of E above	10	10	10	10	10
2. With Following Limitations Fulfilled	10	20	20	50	50
(a) One Water Flow per IDC					
(b) Maximum of Four Sprinkler Supervisory Devices					
(c) Maximum of Five Process, Security, and Other Devices on a Separate IDC					
(d) Maximum of One Intermediate Fire Alarm or Fire Supervisory Control Unit per IDC					

(g) The overall system reliability is considered to be equal from style to style when the capacities are at the maximum allowed.

(h) Upon determining the style of the system, the tables indicate the maximum number of devices, equipment, protected buildings, etc., allowed to be incorporated into an actual installation for a protected premises fire alarm system.

(i) The number of automatic fire detectors connected to an initiating device circuit is limited by good engineering practice. If a large number of detectors are connected to one initiating device circuit covering a widespread area, pinpointing the source of alarm becomes difficult and time consuming.

On certain types of detectors, a trouble signal results from faults in the detector. Where this occurs with a large number of detectors on an initiating device circuit, locating the faulty detector also becomes difficult and time consuming.

3-5.2 The loading of initiating device circuits on systems connected to a proprietary supervising station shall not exceed the capacities listed in Table 3-5.1 for their assigned style designations. The loading of initiating device circuits designated only as Class A or Class B (without a style designation) shall not exceed the capacities for the style with the lowest capacities in their class (Style A for Class B circuits, and Styles D or E for Class A circuits).

NOTE: Though Styles D and E have been assigned the same capacities, the choice between the two styles depends on the desired system performance. Style D circuits transmit an alarm signal, while Style E circuits only transmit a trouble signal on the occurrence of a wire-to-wire short on the circuit. A similar distinction exists between Class B, Styles B and C, which have also been assigned the same capacities.

3-5.3 Numbered initiating device groups listed in Table 3-5.1, Section E, shall not be combined on the same initiating device circuit.

This subsection applies only where signals are transmitted to a proprietary supervising station.

Exception No. 1: When implementing 3-8.1.2, manual means and automatic means shall be permitted to be combined on the same initiating device circuit.

Exception No. 2: Where only one fire alarm box is required, it shall be permitted to be connected to the waterflow initiating device circuit.

Formal Interpretation 79-8
Reference: 3-5, 3-6

Background: This is a request for a formal interpretation in regards to the use of addressable initiating devices. Since the control unit or the central supervising station is in two-way communication with these devices, then:

Question 1: Is it the intent to categorize the description of performance of the circuit these devices are on as a "Signaling Line Circuit," rather than an "Initiating Device Circuit"?

Answer: Yes.

Question 2: If the style of this addressable communication circuit has a different performance (style number) than the remaining portions of the multiplex pathways, would these different circuit performance levels have to be individually specified in order to adequately describe the system?

Answer: Yes.

Issue Edition: NFPA 72D-1979
Reference: 3-9, 3-10
Issue Date: June 1985 ■

Formal Interpretation 87-1
Reference: Tables 3-5.1, 3-6.1

Question 1: Is it the intent of the Committee to categorize the description of performance of the circuit these devices are on as a "Signaling Line Circuit" illustrated in Table 3-6.1 rather than an Initiating Device Circuit as described in Table 3-5.1?

Answer: Yes.

Question 2: If the style of this communication circuit has a different performance (style number as illustrated in Table 3-6.1) than the remaining portions of the multiplex pathways, would these different circuit performance levels have to be individually specified in order to adequately describe the system?

Answer: Yes.

Issue Edition: NFPA 72D-1987
Reference: Tables 2-12.1, 2-13.1
Issue Date: June 1987
Reissued: January 1989 ■

Formal Interpretation 86-3
Reference: Tables 3-5.1, 3-6.1

Background: Table 3-6.1 permits between 250 to 1000 initiating devices to be served by a single circuit depending on the circuit style. In Table 3-5.1, initiating devices that may be connected to a single initiating circuit are limited to as low as 1 to a high of 25, depending on type of device and style of the circuit.

Question 1: In Table 3-5.1, is it the intent of the Committee to limit the number of alarm and supervisory inputs that can be indicated at a control unit with only a single common identity?
Answer: Yes.

Question 2: Considering that up to 250 indicating devices minimum are permitted to be connected to a single signaling line circuit: Is it the intent of the Committee that the loadings shown in Table 3-5.1, Section E, not be exceeded for a Style D circuit when all of the following conditions exist:

A. Each input device is separately indicated at the main control for change of state.
B. All functions of Style D circuits are met.
C. No form of polling, time sharing, digital, or addressable multiplexing techniques are used.
D. All signaling is done with conventional direct current circuit techniques.
E. All initiating circuits are connected to the main control.
F. No signal interference would exist even if all initiating devices were initiated simultaneously.

Answer: No. Table 3-5.1 does not contemplate individual identification of the initiating devices connected to a circuit.

Question 3: Considering that Table 3-6.1 permits all types of initiating devices to be served by a single signaling line circuit: Is it the intent of the Committee to prohibit combining initiating devices listed in Table 3-5.1, Section E, on the same initiating circuit when all
continued

Formal Interpretation 86-3, continued
of the conditions listed in Question 2, A through E, exist? (Note: A indicates each device is separately indicated at the main control.)
Answer: No. Table 3-5.1 does not contemplate individual identification of the initiating devices connected to a circuit.

Issue Edition: NFPA 72D-1986
Reference: Tables 3-9.1, 3-10.1
Issue Date: November 1986 ■

3-6 Performance and Capacities of Signaling Line Circuits (SLC).

3-6.1* The assignment of class designations or style designations, or both, to signaling line circuits shall be based on their performance capabilities under abnormal (fault) conditions in accordance with the requirements of Table 3-6.1.

Formal Interpretation 86-1
Reference: Table 3-6.1

Question: For systems where the main (primary) and standby (secondary) power are transmitted over the same circuit separate from the signaling circuit, is it the intent of 72 to require redundancy (i.e., two such circuits) of the power circuit, when a Style 7 signaling line circuit is required?
Answer: Yes.

Issue Edition: NFPA 72D-1986
Reference: Entire Standard
Issue Date: August 1986
Replaces F.I. 86-1 issued February 1986 ■

A-3-6.1 Using Tables 3-5.1 and 3-6.1:

(a) Determine whether the initiating devices are:

1. Directly connected to the initiating device circuit
2. Directly connected to a signaling line circuit interface on a signaling line circuit

Table 3-6.1 Performance and Capacities of Signaling Line Circuits (SLC)

G = Systems with ground detection shall indicate system trouble with a single ground.
M = May be capable of alarm with wire-to-wire short.
R = Required capability.
X = Indication required at protected premises and as required by Chapter 4.
α = Style exceeds minimum requirements for Class A.

Each Style below has three sub-columns: **A** = Alarm, **T** = Trouble, **ARC** = Alarm Receipt Capability During Abnormal Condition.

Class	B			B			A			B			B			B			B			A			A			A		
Style	0.5			1			2α			3			3.5			4			4.5			5α			6α			7α		
Sub-col	A	T	ARC	A	T	ARC	A	T	ARC	A	T	ARC	A	T	ARC	A	T	ARC	A	T	ARC	A	T	ARC	A	T	ARC	A	T	ARC
Abnormal Condition	1	2	3	4	5	6	7	8	9	10	11	12	13	14	15	16	17	18	19	20	21	22	23	24	25	26	27	28	29	30
A. Single Open		X		X			X	R		X			X			X			X	R		X	R		X	R		X	R	
B. Single Ground		X		G	R		G	R		G	R		X			G	R		X			G	R		G	R		G	R	
C. Wire-to-Wire Short								M		X			X			X			X			X			X			X		R
D. Wire-to-Wire Short & Open								M		X			X			X			X			X			X			X		
E. Wire-to-Wire Short & Ground							G	M		X			X			X			X			X			X			X		
F. Open and Ground							X	R		X			X			X			X			X			X	X		X		R
G. Loss of Carrier (If Used)/ Channel Interface													X			X			X			X			X			X		

Note: The following sections apply only where signals are transmitted to a proprietary supervising station in accordance with Section 4-4.

	0.5	1	2α	3	3.5	4	4.5	5α	6α	7α
H. Maximum quanitity per Signaling Line Circuit β										
1. Initiating Devices (All Types)	250	250	250	300	300	500	500	750	1000	unlimited
2. Buildings	25	25	25	50	50	75	75	75	100	100
I. Maximum Quantity per Proprietary Supervising Station (PSS)										
1. Initiating Device Circuits	500	500	500	1000	1000	1000	1000	1500	2000	2000
2. IDCs with Redundant PSS Control Equipment[1]	1000	1000	1000	2000	2000	2000	2000	3000	unlimited	unlimited
3. Buildings	25	25	25	25	25	50	50	75	400	400
4. Buildings with Redundant PSS Control Equipment[1]	25	25	25	50	50	100	100	150	unlimited	unlimited

Note 1: When the supervisory station multiplex control unit is duplicated and a switchover can be accomplished in not more than 90 seconds with no loss of signals during this period, the capacity of the system is unlimited.

β = See the exception to 3-6.2.

3. Directly connected to an initiating device circuit, which in turn is connected to a signaling line circuit interface on a signaling line circuit.

(b) Determine the style of signaling performance required. The columns marked A through Eα in Table 3-5.1, and 0.5 through 7α in Table 3-6.1 are arranged in ascending order of performance and capacities.

(c) Upon determining the style of the system, the charts singularly or together will specify the maximum number of devices, equipment, premises, and buildings allowed to be incorporated into an actual protected premises installation.

(d) In contrast, where the number of devices, equipment, premises, and buildings (in addition to signaling ability) in an installation is known, a required system style can be determined.

(e) The prime purpose of the tables is to enable identification of minimum performance for styles of initiating device circuits and signaling line circuits. It is not the intention that the styles be construed as grades. That is, a Style 3 system is not better than a Style 2, or vice versa. In fact, a particular style may better provide adequate and reliable signaling for an installation than a more complex style number. The quantities tabulated under each style do, unfortunately, tend to imply that one style is better than the one to its left. The increased quantities for the higher style numbers are based on the ability to signal an alarm during an abnormal condition in addition to signaling the same abnormal condition.

(f) The tables allow users, designers, manufacturers, and the authority having jurisdiction to identify minimum performance of present and future systems by determining the trouble and alarm signals received at the control unit for the specified abnormal conditions.

(g) The overall system reliability is considered to be equal from style to style when the capacities are at the maximum allowed.

(h) Upon determining the style of the system, the tables indicate the maximum number of devices, equipment, protected buildings, etc., allowed to be incorporated into an actual installation for a protected premises fire alarm system.

(i) The number of automatic fire detectors connected to an initiating device circuit is limited by good engineering practice. If a large number of detectors are connected to one initiating device circuit covering a widespread area, pinpointing the source of alarm becomes difficult and time consuming.

On certain types of detectors, a trouble signal results from faults in the detector. Where this occurs with a large number of detec-

tors on an initiating device circuit, locating the faulty detector also becomes difficult and time consuming.

3-6.2 The loading of signaling line circuits shall not exceed the capacities listed in Table 3-6.1 for their assigned style designation. The loading of signaling line circuits designated only as Class A or Class B (without a style designation) shall not exceed the capacities for the style with the lowest capacities in their class (Styles 0.5 or 1 for Class B circuits, and Style 2 for Class A circuits).

NOTE: Sections H and I of Table 3-6.1 provide information regarding capacities where protected premises fire alarm equipment, as covered in this chapter, is used in a proprietary fire alarm system. For information regarding proprietary supervising stations, see Section 4-4.

Exception: Where a Class A signaling line circuit is so arranged that a single open or ground fault, or short circuit between wires of the same signaling line circuit does not cause the loss of fire alarm signals from more than a single zone as defined in 1-5.7.3 and 1-5.7.6, the maximum number of initiating devices per signaling line circuit shall be unlimited.

3-7 Notification Appliance Circuits (NAC).

3-7.1 Performance. The assignment of class designations or style designations, or both, to notification appliance circuits shall be based on their performance capabilities under abnormal (fault) conditions in accordance with the requirements of Table 3-7.1.

3-7.2 Distinctive Evacuation Signal.

(a)* Section 1-5.4.7 requires that fire alarm signals be distinctive in sound from other signals and that this sound not be used for any other purpose. To meet this requirement, the fire alarm signal used to notify building occupants of the need to evacuate (leave the building) shall be ANSI S3.41, *American National Standard Audible Emergency Evacuation Signal*. This requirement shall become effective July 1, 1996.

A-3-7.2(a) The use of the distinctive three-pulse temporal pattern fire alarm evacuation signal required by 3-7.2(a) had previously been recommended for this purpose by this code since 1979. It has since been adopted as both an American National

Table 3-7.1   Notification Appliance Circuits (NAC)

	Class	B	B	B	A
	Style	W	X	Y	Z
G = Systems with ground detection shall indicate system trouble with a single ground. X = Indication required at protected premises.		Trouble Indication at Protected Premises / Alarm Capability During Abnormal Conditions	Trouble Indication at Protected Premises / Alarm Capability During Abnormal Conditions	Trouble Indication at Protected Premises / Alarm Capability During Abnormal Conditions	Trouble Indication at Protected Premises / Alarm Capability During Abnormal Condition

Abnormal Condition	1	2	3	4	5	6	7	8
Single Open	X		X	X	X		X	X
Single Ground	X		X		G	X	G	X
Wire-to-Wire Short	X		X		X		X	

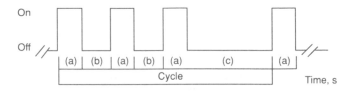

Key:
Phase (a) signal is "on" for 0.5 s ± 10%
Phase (b) signal is "off" for 0.5 s ± 10%
Phase (c) signal is "off" for 1.5 s ± 10% [(c) = (a) + 2(b)]
Total cycle lasts for 4 s ± 10%

Figure A-3-7.2(a)(1)   Temporal pattern parameters.

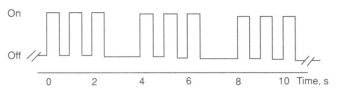

Figure A-3-7.2(a)(2)   Temporal pattern imposed on signaling appliances that emit a continuous signal while energized.

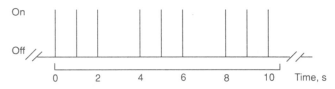

Figure A-3-7.2(a)(3)   Temporal pattern imposed on a single stroke bell or chime.

Standard (ANSI S3.41, *Audible Emergency Evacuation Signal*) and an International Standard (ISO 8201, *Audible Emergency Evacuation Signal*).

Copies of both of these standards are available from the Standards Secretariat, Acoustical Society of America, 335 East 45th Street, New York, NY 10017-3483. Telephone 212-661-9404 ext. 562.

The standard fire alarm evacuation signal is a three-pulse temporal pattern using any appropriate sound. The pattern consists of an "on" phase (a) lasting 0.5 second ± 10 percent followed by an "off" phase (b) lasting 0.5 second ± 10 percent, for three successive "on" periods, which is then followed by an "off" phase (c) lasting 1.5 seconds ± 10 percent. [*See Figures A-3-7.2(a)(1) and (2).*] The signal should be repeated for a period appropriate for the purposes of evacuation of the building, but for not less than 180 seconds. A single-stroke bell or chime sounded at "on" intervals lasting 1 second ± 10 percent, with a 2-second ± 10 percent "off" interval after each third "on" stroke, is acceptable. [*See Figure A-3-7.2(a)(3).*]

The minimum repetition time may be manually interrupted.

(b) The use of the American National Standard Audible Emergency Evacuation Signal shall be restricted to situations where it is desired to have all occupants hearing the signal evacuate the building immediately. It shall not be used where, with the approval of the authority having jurisdiction, the planned action during a fire emergency is not evacuation, but relocation of the occupants from the affected area to a safe area within the building, or their protection in place (e.g., high rise buildings, health care facilities, penal institutions, etc.).

It is important to note that the American National Standard Audible Emergency Evacuation Signal is required only where immediate evacuation of a building or a zone of a building is desired. The goal is to have anyone hearing the signal in any building in the United States (or in other countries that adopt ISO standards requiring the same evacuation signal) immediately recognize the signal as a fire alarm evacuation signal. The use of the American National Standard Audible Emergency Evacuation Signal is also required for household fire alarm systems. See Chapter 2.

3-8 System Requirements.

(See also Section 5-9.)

3-8.1 Manual Fire Alarm Signal Initiation.

3-8.1.1 Fire alarm boxes shall be listed for the intended application, installed in accordance with Chapter 5, and tested in accordance with Chapter 7.

In the previous edition of NFPA 72, the installation requirements of manual fire alarm boxes were covered in Chapter 3. The installation requirements of all automatic or manual initiating devices are now located in Chapter 5.

3-8.1.2 For fire alarm systems employing automatic fire detectors or waterflow detection devices, at least one fire alarm box shall be provided to initiate a fire alarm signal. This fire alarm box shall be located where required by the authority having jurisdiction.

One reason that at least one fire alarm box is required is to allow an alarm to be transmitted if the automatic fire detectors or sprinkler system are out of service during repairs or during a test. Another reason the fire alarm box is required is to permit a building occupant to initiate an alarm signal prior to the actuation of an automatic initiating device, providing early warning of a fire emergency. Where there is more than one fire alarm box on a zone circuit, and where it is practical, placing one at the end of the circuit allows for both testing of the fire alarm box and ensuring that the entire zone circuit is intact.

Exception: Fire alarm systems dedicated to elevator recall control and supervisory as permitted in 3-8.15.1.

3-8.1.3 Where signals from fire alarm boxes and other fire alarm initiating devices within a building are transmitted over the same signaling line circuit, there shall be no interference with fire alarm

box signals when both types of initiating devices are operated at or near the same time. Provision of the shunt noninterfering method of operation shall be acceptable for this performance.

Note that this requirement applies only to systems that have devices reporting to the control panel using a signaling line circuit arrangement. This does not apply to systems utilizing initiating device circuits, since these circuits do not distinguish which initiating device initiated the alarm. The requirement was first introduced to apply to spring-wound coded devices that could only transmit a fixed number of rounds of code. If the first device was interfered with by the simultaneous alarm transmission of another device on the same circuit, the first transmission could be lost or garbled.

3-8.2 Automatic Fire Alarm Signal Initiation.

3-8.2.1 Automatic alarm initiating devices shall be listed for the intended application and installed in accordance with Chapter 5.

The phrases "listed for the purpose" and "listed for the intended application" have similar meanings. "Listed for the intended application" means, for example, that if a device is to be used in a low-temperature environment or a wet location, the device should be listed for that environment. Also, if a device is to be used for releasing service, it should be listed for that application as well.

3-8.2.2 Automatic alarm initiating devices having integral trouble contacts shall be wired on the initiating device circuit so that a trouble condition within a device does not impair the alarm transmission from any other initiating device.

It is possible to disable part or all of an initiating device circuit beyond the initiating device in trouble if the integral trouble contacts are connected improperly. Most automatic alarm initiating devices do not have integral trouble contacts; therefore, this requirement does not apply.

At one time, photoelectric smoke detectors used a tungsten filament lamp as a light source, and the integrity of the filament was required to be monitored. A common method used for reporting an open filament was to open a normally closed trouble contact within the smoke detector, which was wired in series with the initiating circuit.

To properly comply with this requirement, the trouble contacts must be wired in series, with the end-of-line resistor at the end of the circuit after the last initiating device. See Figures 3.2(a) and 3.2(b).

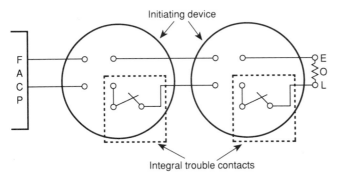

Figure 3.2(a)   *Incorrect method of connection of integral trouble contacts.*

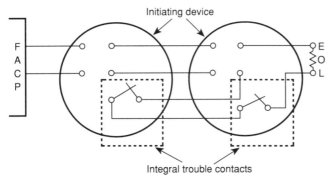

Figure 3.2(b)   *Correct method of connection of integral trouble contacts.*

NOTE: Though a trouble signal is required when a plug-in initiating device is removed from its base, it is not considered as a trouble condition within the device and the requirement of 3-8.2.2 does not apply.

3-8.2.3*   Systems equipped with alarm verification features shall be permitted, provided:

(a) A smoke detector continuously subjected to a smoke concentration above alarm threshold magnitude initiates a system alarm within 1 minute.

(b) Actuation of an alarm initiating device other than a smoke detector shall cause a system alarm signal within 15 seconds.

A-3-8.2.3   The alarm verification feature should not be used as a substitute for proper detector location/applications or regular system maintenance. Alarm verification features are intended to reduce the frequency of false alarms caused by transient conditions. They are not intended to compensate for design errors or lack of maintenance.

Alarm verification can be very useful in reducing false alarms from transient conditions, but not from conditions that remain relatively constant, such as a wet environment or areas subject to insect infestation. The alarm verification feature can reduce both malicious and accidental false alarms caused by such acts as the spraying of aerosols into a smoke detector or a gust of wind blowing into the detector. The feature should not be installed/programmed in a system until a thorough investigation of the causes of false alarms has been accomplished. Alarm verification refers to specific timing sequences of detector/system operation. See Figure 3.3.

3-8.2.4   Where individual alarm initiating devices are used to control the operation of equipment as permitted by 1-5.4.1, this control capability shall remain operable even if all of the initiating devices connected to the same circuit are in an alarm state.

This requirement does not allow the use of multiple, two-wire, circuit-powered smoke detectors with an auxiliary relay to be installed on the same initiating device circuit. In addition, this type of smoke detector (with an auxiliary relay) should not be installed on the same initiating device circuit with other devices, such as manual fire alarm boxes or heat detectors. The reason for this requirement is that with a device such as a heat detector in alarm on the initiating device circuit, power is removed from the remaining smoke detector(s) with the relay. Without power, smoke detectors with control relay cannot operate. This problem does not exist with detectors installed on signaling line circuits because each device reports separately and is not affected by the operation of other devices on the circuit.

3-8.2.5   Systems that require the operation of two automatic detection devices to initiate the alarm response shall be permitted, provided:

(a) They are not prohibited by the authority having jurisdiction.

(b) There are at least two automatic detection devices in each protected space.

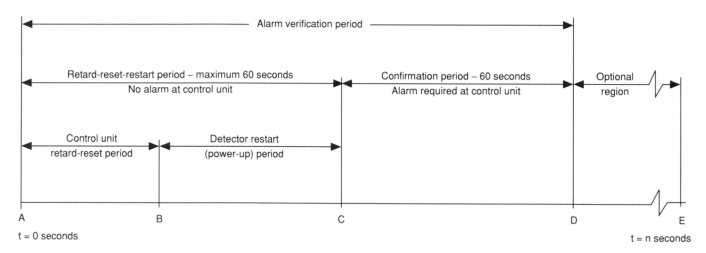

A – Smoke detector goes into alarm.

AB – Retard-reset period (control unit) – Control unit senses detector in alarm and retards (delays) alarm signal, usually by de-energizing power to the detector. Length of time varies with design.

BC – Restart period (detector power-up time) – Power to the detector is reapplied and time is allowed for detector to become operational for alarm. Time varies with detector design.

AC – Retard-reset-restart period – No alarm obtained from control unit. Maximum permissible time is 60 seconds.

CD – Confirmation period – Detector is operational for alarm at point C. If detector is still in alarm at point C, control unit will alarm. If detector is not in alarm, system returns to standby. If the detector realarms at any time during the confirmation period, the control unit will alarm.

DE – Optional region – Either an alarm can occur at control unit or restart of the alarm verification cycle can occur.

AD – Alarm verification period – Consists of the retard-reset-restart and confirmation periods.

Figure 3.3   Alarm verification timing diagram.

(c) Automatic detection device area spacing is no more than one-half that determined by the application of Chapter 5.

(d) The alarm verification feature is not used.

The common names used for this type of configuration are "cross zoning" and "priority matrix zoning." The most common use of this configuration is with special hazard extinguishing systems. In this configuration, false alarms are minimized because more than one detector in alarm is required to activate the special hazard system.

3-8.3   Positive Alarm Sequence.

3-8.3.1   Systems having positive alarm features complying with the following shall be permitted where approved by the authority having jurisdiction.

Positive alarm sequence is simply a delayed alarm under specific controlled conditions. Before the advent of alarm verification, this method was used to minimize false alarms.

The positive alarm sequence feature should be used only with approval of the authority having jurisdiction.

3-8.3.1.1   The signal from an automatic fire detection device selected for positive alarm sequence operation shall be acknowledged at the control unit by trained personnel within 15 seconds of annunciation in order to initiate the alarm investigation phase. If the signal is not acknowledged within 15 seconds, all building and remote signals shall be activated immediately and automatically.

3-8.3.1.2   Trained personnel shall have up to 180 seconds during the alarm investigation phase to evaluate the fire condition and reset the system. If the system is not reset during this investigation phase, all building and remote signals shall be activated immediately and automatically.

3-8.3.2   If a second automatic fire detector selected for positive alarm sequence is actuated during the alarm investigation phase, all normal building and remote signals shall be activated immediately and automatically.

3-8.3.3 If any other initiating device is actuated, all building and remote signals shall be activated immediately and automatically.

Subsections 3-8.3.1.1 through 3-8.3.3 help eliminate the human unreliability factor from the use of the positive alarm sequence feature by requiring time limitations and automatic activation when additional initiating devices are activated.

3-8.3.4* The system shall provide means to bypass the positive alarm sequence.

A-3-8.3.4 The bypass means is intended to enable automatic or manual day/night/weekend operation.

3-8.4* **Concealed Detectors.** Where a remote alarm indicator is provided for an automatic fire detector in a concealed location, the location of the detector and the area protected by the detector shall be prominently indicated either at the remote alarm indicator by a permanently attached placard or by other approved means.

A remote alarm indicator, usually a red light emitting diode (LED) mounted on a single gang plate, is the most common method of indicating an alarm from a concealed detector. It is important to locate and mark the remote alarm indicator so that the detector in alarm can be found easily. Engraved phenolic plates permanently attached to the remote alarm indicator are generally considered the most appropriate way of complying with this requirement. See Figures 3.4 and 3.5.

A-3-8.4 Embossed plastic tape, pencil, ink, crayon, etc., should not be considered a permanently attached placard.

3-8.5 **Automatic Drift Compensation.** Where automatic drift compensation of sensitivity for a fire detector is provided, the control unit shall give an indication identifying the affected detector when the limit of compensation is reached.

Automatic drift compensation of sensitivity is used to eliminate false alarms from smoke detectors that experience dust or dirt build-up within the detection chamber or that respond to minor changes in the environment. The feature allows the detector to maintain its original sensitivity by compensating for effects caused by outside sources. If the

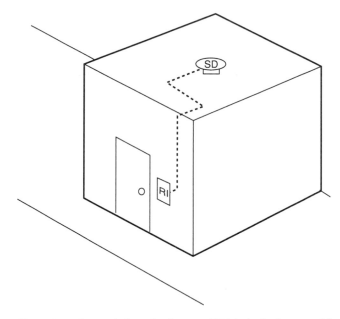

Figure 3.4 *Concealed smoke detector (SD) in locked room with remote indicator (RI).*

Figure 3.5 *Remote indicator used for concealed detectors.*

compensated value places the detector's sensitivity outside its listed window of sensitivity, the control unit indicates that maintenance is needed.

3-8.6 Waterflow Alarm Signal Initiation.

3-8.6.1 The provisions of 3-8.6 apply to sprinkler system signaling attachments that initiate an alarm indicating a flow of water in the system. Waterflow initiating devices shall be listed for the intended application and installed in accordance with Chapter 5.

Waterflow devices are also required by NFPA 13, *Standard for the Installation of Sprinkler Systems*, and additional information regarding their placement and use may be found in NFPA 13. See Figures 3.6 and 3.7.

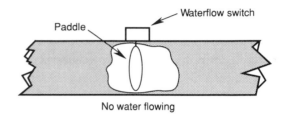

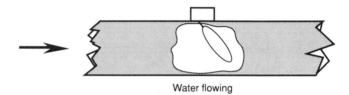

Figure 3.7   Typical waterflow switch operation.

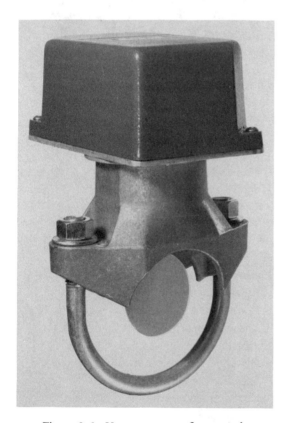

Figure 3.6   Vane-type waterflow switch.

3-8.6.2 A dry-pipe or preaction sprinkler system that is supplied with water by a connection beyond the alarm initiating device of a wet-pipe system shall be equipped with a separate waterflow alarm initiating pressure switch or other approved means to initiate a waterflow alarm.

3-8.6.3 The number of waterflow switches permitted to be connected to a single initiating device circuit shall not exceed five.

The reasoning here is to avoid the potential loss of all waterflow signals in a large sprinklered facility and to avoid confusion as to the location of the activated sprinkler area. In a large protected area, consideration should be given to separately zoning each main waterflow switch.

3-8.7 Supervisory Signal Initiation.

3-8.7.1 General. The provisions of this section apply to the monitoring of sprinkler and other fire protection systems for the initiation of a supervisory signal indicating an off-normal condition that may adversely affect the performance of the system.

3-8.7.1.1 Supervisory devices shall be listed for the intended application and installed in accordance with Chapter 5.

3-8.7.1.2 The number of supervisory devices permitted to be connected to a single initiating device circuit shall not exceed 20.

The code allows up to 20 supervisory devices on a single initiating device circuit. However, design considerations for a given installation might limit the number even further. Locating a supervisory device that is in an off-normal position may prove to be difficult and time consuming when 20 devices are on one circuit. The code does not differentiate between types of supervisory devices (e.g., gate valve versus special hazard control unit), and this also may add to the difficulty of locating the device off-normal. Care should be taken when determining the number and type of supervisory devices installed on an initiating device circuit.

3-8.7.2* Provisions shall be made for supervising the conditions that are essential for the proper operation of sprinkler and other fire suppression systems.

Exception: Those conditions related to water mains, tanks, cisterns, reservoirs, and other water supplies controlled by a municipality or a public utility.

A-3-8.7.2 Supervisory systems are not intended to provide indication of design, installation, or functional defects in the supervised systems or system components and are not a substitute for regular testing of those systems in accordance with the applicable standard.

Supervised conditions should include but not be limited to:

(a) Control valves 1½ in. (38.1 mm) or larger

See Figure 3.8.

Figure 3.8 Control valve supervisory switch.

(b) Pressure:
Dry pipe system air
Pressure tank air
Preaction system supervisory air
Steam for flooding systems
Public water

See Figure 3.9.

Figure 3.9 High/low pressure switch.

(c) Water tanks:
Level
Temperature
(d) Building temperature (including valve closet, fire pump house, etc.)
(e) Fire pumps:
Electric:
Running (alarm or supervisory)
Power failure
Phase reversal
Engine-driven:
Running (alarm or supervisory)
Failure to start
Controller off "automatic"
Trouble (low oil, high temperature, overspeed, etc.)
Steam turbine:
Running (alarm or supervisory)
Steam pressure
Steam control valves
(f) Fire suppression systems appropriate to the system employed.

3-8.7.3 Signals shall distinctively indicate the particular function (such as valve position, temperature, pressure, etc.) of the system that is off-normal and also indicate its restoration to normal.

NOTE: Cancellation of the off-normal signal is acceptable as a restoration signal except where separate recording of all changes of state is a specific requirement. (*See Chapter 4.*)

3-8.7.4 A dry-pipe sprinkler system equipped for waterflow alarm signaling shall be supervised for off-normal system air pressure.

The intent of 3-8.7.4 is to require the monitoring of the air pressure of the dry-pipe sprinkler system. This will result in the receipt of a supervisory signal prior to loss of total air pressure and the subsequent filling of the piping network with water.

3-8.7.5 A control valve shall be supervised to initiate a distinctive signal indicating movement of the valve from its normal position. The off-normal signal shall remain until the valve is restored to its normal position. The off-normal signal shall be obtained during the first two revolutions of the hand wheel or during one-fifth of the travel distance of the valve control apparatus from its normal position.

3-8.7.6 An initiating device for supervising the position of a control valve shall not interfere with the operation of the valve, obstruct the view of its indicator, or prevent access for valve maintenance.

3-8.7.7 **Pressure Supervision.** Pressure sources shall be supervised to obtain two separate and distinct signals, one indicating that the required pressure has been increased or decreased, and the other indicating restoration of the pressure to its required value.

(a) A pressure supervisory signal initiating device for a pressure tank shall indicate both high and low pressure conditions. A signal shall be obtained where the required pressure is increased or decreased 10 psi (70 kPa) from the required pressure value.

(b) A pressure supervisory signal initiating device for a dry-pipe sprinkler system shall indicate both high and low pressure conditions. A signal shall be obtained when the required pressure is increased or decreased 10 psi (70 kPa) from the required pressure value.

(c) A steam pressure supervisory initiating device shall indicate a low pressure condition. A signal shall be obtained where the pressure is reduced to a value that is 110 percent of the minimum operating pressure of the steam operated equipment supplied.

(d) An initiating device for supervising the pressure of sources other than those specified above shall be provided as required by the authority having jurisdiction.

3-8.7.8 **Water Temperature Supervision.** Exposed water storage containers shall be supervised to obtain two separate and distinct signals, one indicating that the temperature of the water has been lowered to 40°F (4.4°C), and the other indicating restoration to a temperature above 40°F (4.4°C).

3-8.8 **Signal Annunciation.** Protected premises fire alarm systems shall be arranged to annunciate alarm, supervisory, and trouble signals in accordance with 1-5.7.

The requirement for three separate and distinct signals was included in previous editions of the code, but was clarified in this edition. All protected premises systems must indicate all three signals in a distinct fashion.

The system trouble LED and audible notification appliance cannot also be used to indicate a supervisory condition (*see 1-5.4.7*). However, the system trouble audible notification appliance can be used in conjunction with a separate supervisory LED indication to comply with 3-8.8.

3-8.9 **Signal Initiation from Automatic Fire Suppression System Other Than Waterflow.**

3-8.9.1 The operation of an automatic fire suppression system installed within the protected premises shall be indicated on the protected premises fire alarm system.

This subsection requires that an alarm condition at the automatic suppression system will cause an alarm indication at the protected premises fire alarm system. The automatic suppression system should be connected to the protected premises control unit as a separate zone or point. The protected premises system would operate as intended and activate all the notification appliances, etc.

3-8.9.2 A supervisory signal shall indicate the off-normal condition and its restoration to normal appropriate to the system employed.

In addition to indicating alarm conditions to the protected premises fire alarm system, the off-normal condition of the automatic suppression system must also be indicated. This supervisory signal must be able to be differentiated from a broken wire in the interconnection wiring between the protected premises fire alarm system and the automatic suppression system. See 3-8.9.3.

3-8.9.3 The integrity of each fire suppression system actuating device and its circuit shall be supervised in accordance with 1-5.8.1 and with other applicable NFPA standards.

The word "supervised" as used here means the same as "monitored for integrity" as used elsewhere in this code.

Because fire protection systems are not exercised on a daily basis, the integrity of the systems is electrically monitored for proper operation regardless of the type of system (protected premises fire alarm or special hazard).

3-8.10 Pump Supervision. Automatic fire pumps and special service pumps shall be supervised in accordance with NFPA 20, *Standard for the Installation of Centrifugal Fire Pumps,* and the authority having jurisdiction.

3-8.10.1 Supervision of electric power supplying the pump shall be made on the line side of the motor starter. All phases and phase reversal shall be supervised.

The word "supervision" is used in 3-8.10 and 3-8.10.1 for the same reason indicated in the commentary following 3-8.9.3.

3-8.10.2 Where both sprinkler supervisory signals and pump running signals are transmitted over the same signaling circuits, provisions shall be made to obtain pump running signal preference unless the circuit is so arranged that no signals will be lost.

3-8.11 Tampering.

3-8.11.1 Automatic fire suppression system alarm and supervisory signal initiating devices and their circuits shall be so designed and installed that they cannot be readily tampered with, opened, or removed without initiating a signal. This provision specifically includes junction boxes installed outside of buildings to facilitate access to the initiating device circuit.

Junction boxes installed outside of buildings must be specifically either tamper-proof or provided with a device to

initiate a signal. Generally this is a supervisory signal, not an alarm signal; however, the authority having jurisdiction should be consulted prior to this type of connection.

3-8.11.2* If a valve is installed in the connection between a signal attachment and the fire suppression system to which it is attached, such a valve shall be supervised in accordance with the requirements of Chapter 5.

Supervision of the valve means providing an indication of an off-normal condition at the valve being supervised at the protected premises fire alarm system. As previously noted, "supervised" as used in this code has the same meaning as "monitored for integrity."

A-3-8.11.2 Sealing or locking such a valve in the open position or removing the handle from the valve does not meet the intent of this requirement.

3-8.12 Guard's Tour Supervisory Service.

Guard's tour supervisory service is used to provide fire protection surveillance during the hours when occupants are not in a building; to facilitate and control the movement of persons into, out of, and within a building; and to carry out procedures for the orderly conduct of specific operations in a building or on the surrounding property.

Guard's tour supervisory services designed to continually report the performance of a guard are often found in connection with protected premises fire alarm systems utilizing off-premises reporting through central or proprietary supervising stations.

3-8.12.1 Guard's tour reporting stations shall be listed for the application.

3-8.12.2 The number of guard's tour reporting stations, their locations, and the route to be followed by the guard for operating the stations shall be approved for the particular installation in accordance with NFPA 601, *Standard on Guard Service in Fire Loss Prevention.*

3-8.12.3 A permanent record indicating every time each signal-transmitting station is operated shall be made at the main control unit. Where intermediate stations that do not transmit a signal are employed in conjunction with signal-transmitting stations, distinctive signals shall be transmitted at the beginning and end of each tour of a guard, and a signal-transmitting station shall be provided at intervals not exceeding ten stations. Intermediate stations that do not transmit a signal shall be capable of operation only in a fixed sequence.

3-8.13 Suppressed (Exception Reporting) Signal System.

This tour arrangement is somewhat less flexible than supervised tours but has the advantages of the absence of interconnected wires between the preliminary stations and the reduction of signal traffic. The usual arrangement is to have the guard transmit only start and finish signals that must be received at the central point at programmed reception times.

3-8.13.1 The system shall comply with the provisions of 3-8.12.2.

3-8.13.2 The system shall transmit a start signal to the signal-receiving location and shall be initiated by the guard at the start of continuous tour rounds.

3-8.13.3 The system shall automatically transmit a delinquency signal within 15 minutes after the predetermined actuation time if the guard fails to actuate a tour station as scheduled.

3-8.13.4 A finish signal shall be transmitted within a predetermined interval after the guard completes each tour of the premises.

3-8.13.5 For periods of over 24 hours, during which tours are continuously conducted, a start signal shall be transmitted at least every 24 hours.

3-8.13.6 The start, delinquency, and finish signals shall be recorded at the signal-receiving location.

3-8.14 Combination Systems.

3-8.14.1* Fire alarm systems shall be permitted to share components, equipment, circuitry, and installation wiring with nonfire alarm systems.

A-3-8.14.1 The provisions of this section apply to the types of equipment used in common for fire alarm systems (such as fire alarm, sprinkler supervisory, or guard's tour service) and for other systems (such as burglar alarm or coded paging systems) and to methods of circuit wiring common to both types of systems.

3-8.14.2 Where common wiring is employed for combination systems, the equipment for other than fire alarm systems shall be permitted to be connected to the common wiring of the system. Short circuits, open circuits, or grounds in this equipment or between this equipment and the fire alarm system wiring shall not interfere with the supervision of the fire alarm system or prevent alarm or supervisory signal transmissions.

Common wiring between fire alarm and nonfire alarm systems means receiving signal inputs or providing outputs for either system on the same wires. This common wiring could be detector power, initiating device, or notification appliance circuits. In any case, a short, a ground, or an open circuit in the common wiring caused by the nonfire equipment should not prevent the receipt of an alarm, a trouble, or a supervisory signal.

3-8.14.3 To maintain the integrity of fire alarm system functions, the removal, replacement, failure, or maintenance procedure on any hardware, software, or circuit not required to perform any of the fire alarm system functions shall not cause loss of any of these functions.

Exception: Where the hardware, software, and circuits are listed for fire alarm use.

This subsection addresses the problem of field maintenance or equipment failure of nonfire alarm system components. Equipment not required for the operation of the fire alarm system that is modified, removed, or is malfunctioning in any way must not impair the operation of the fire alarm system. Subsection 3-8.14 permits fire alarm systems to be installed using some components not specifically listed for fire alarm use. Most applications of this permitted use involve interconnection of the listed fire alarm system with paging systems, burglar alarm systems, and process monitoring systems. Users have also connected fire alarm systems to supplementary equipment such as business computers and monitors not listed for fire alarm use, in order to display system conditions to operators in more detail than may be available from the fire alarm systems alone.

However, authorities having jurisdiction have found that some installations do not satisfy the intent of the code sections covering combination systems. In some installations, program changes or other repairs to the nonfire alarm equipment delayed fire alarm signals or prevented their display altogether. In general, these incorrectly applied systems prevented one or more of the fire alarm system functions from operating as intended. To guard against such failures, it is usually necessary to retest the complete fire alarm system or extensive portions of the system after all changes and repairs have been made to the nonfire alarm components of the system.

Where a nonfire alarm system component is listed for fire alarm use, the listing agency has investigated the timing aspects of the signals, as well as temperature characteristics and other extensive fire alarm safety factors, and found the product suitable for the purpose. Consequently, the exception to 3-8.14.3 exempts this specifically listed equipment from the requirements of 3-8.14.3.

3-8.14.4   Speakers used as alarm notification appliances on fire alarm systems shall not be used for nonemergency purposes.

Exception: Where the fire command station is constantly attended by a trained operator, selective paging shall be permitted.

The fire alarm notification appliances cannot be used for general paging functions unless the exception applies. Conversely, standard background music or paging speakers cannot be used as fire alarm notification appliances. The speakers must be listed for fire alarm use.

3-8.14.5   In combination systems, fire alarm signals shall be distinctive, clearly recognizable, and take precedence over any other signal even when a nonfire alarm signal is initiated first.

This requirement does not mean that two separate notification appliances must be used. A single appliance may be used if it can supply two different, distinctive signals; the fire alarm signal always takes precedence. However, the notification appliance circuit must comply with 3-8.14.2.

3-8.15   Elevator Recall for Fire Fighters' Service.

3-8.15.1*   System type smoke detectors located in elevator lobbies, elevator hoistways, and elevator machine rooms, which are used to initiate fire fighters' service recall, shall be connected to the building fire alarm system. In facilities without a building fire alarm system, these smoke detectors shall be connected to a dedicated fire alarm system control unit that shall be designated "Elevator Recall Control and Supervisory Panel" on the record drawings. Unless otherwise required by the authority having jurisdiction, only the elevator lobby, elevator hoistway, and the elevator machine room smoke detectors shall be used to recall elevators for fire fighters' service.

Elevator hoistway smoke detectors were added in this edition of the code to correlate with ANSI/ASME A17.1-1990, *Safety Code for Elevators and Escalators*, which requires recall of elevators to the designated level when these detectors are activated.

The new type of control unit, designated the "Elevator Recall Control and Supervisory Panel," will now be required for use in buildings where smoke detectors for elevator recall are installed but are not required to have (and do not have) a fire alarm system.

A-3-8.15.1   Dedicated fire alarm system control units are required for elevator recall by 3-8.15.1 in order that the elevator recall systems be monitored for integrity and have primary and secondary power meeting the requirements of this code.

The control unit used for this purpose should be located in an area that is normally occupied and should have audible and visible indicators to annunciate supervisory (elevator recall) and trouble conditions; however, no form of general occupant notification or evacuation signal is required or intended by 3-8.15.1.

The new elevator control unit should obviously be placed in an area that is constantly attended for monitoring, especially when it is installed as a stand-alone control. See Figure 3.10.

3-8.15.2   Each elevator lobby, elevator hoistway, and elevator machine room smoke detector shall be capable of initiating elevator recall when all other devices on the same initiating device circuit have been manually or automatically placed in the alarm condition.

This requirement is simply restating 3-8.2.4. Generally speaking, unless the required smoke detectors are installed

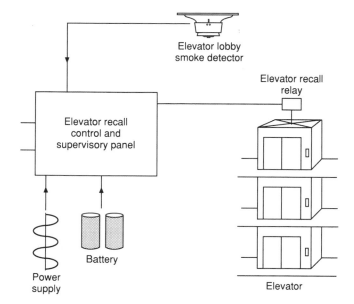

Figure 3.10 *Elevator recall system.*

on individual fire alarm initiating device circuits without any other fire alarm devices installed on the circuits, the smoke detectors should be powered separately from the control unit. Smoke detectors installed on signaling line circuits will not be affected.

3-8.15.3 When actuated, each elevator lobby, elevator hoistway, and elevator machine room smoke detector shall initiate an alarm condition on the building fire alarm system and shall visibly indicate, at the control unit and required remote annunciators, the alarm initiation circuit or zone from which the alarm originated.

The requirements in this subsection apply only where there is a fire alarm system in the building.

Exception: Where approved by the authority having jurisdiction, the elevator hoistway and machine room detectors shall be permitted to initiate a supervisory signal.

This option is provided to minimize the unnecessary recall of elevators due to false alarms. The option should be used only where trained personnel are constantly in attendance and can immediately respond to the supervisory signal and investigate the cause of the signal. Means should be provided for initiating the fire alarm signal if the investigation of the supervisory signal's cause indicates building evacuation and elevator recall is necessary. In addition to hav-

ing trained personnel constantly in attendance, it is recommended that if the supervisory signal is not acknowledged within a given period of time (3–10 minutes), the fire alarm system will automatically and immediately initiate an alarm.

3-8.15.4* For each group of elevators within a building, two elevator control circuits shall be terminated at the designated elevator controller within the group's elevator machine room(s). The operation of the elevators shall be in accordance with ANSI/ASME A17.1, *Safety Code for Elevators and Escalators*, Rules 211.3 through 211.8. The smoke detectors shall be connected to the two elevator control circuits as follows:

(a) The smoke detector located in the designated elevator recall lobby shall be connected to the first elevator control circuit.

(b) The smoke detectors in the remaining elevator lobbies, elevator hoistways, and the elevator machine room shall be connected to the second elevator control circuit except that when the elevator machine room is located at the designated landing, then that elevator machine room smoke detector shall be connected to the first elevator control circuit. In addition, where the elevator is equipped with front and rear doors, then the smoke detectors in both lobbies at the designated level shall be connected to the first elevator control circuit.

Two elevator control circuits are needed to prevent the recalling of the elevators and the discharging of passengers to the lobby when the lobby is the fire location. The two-circuit configuration provides for an alternate recall location (determined by the authority having jurisdiction) when the lobby is reporting a fire condition.

A-3-8.15.4 It is recommended that the installation be in accordance with the following figures. Use Figure A-3-8.15.4(a) when the elevator is installed at the same time as the building fire alarm system. Use Figure A-3-8.15.4(b) when the elevator is installed after the building fire alarm system.

3-8.16 Elevator Shutdown.

This new section is the result of additional sprinkler protection requirements of other codes for elevator machine rooms and hoistways. The purpose of elevator shutdown prior to sprinkler operation is to avoid the hazard of a wet

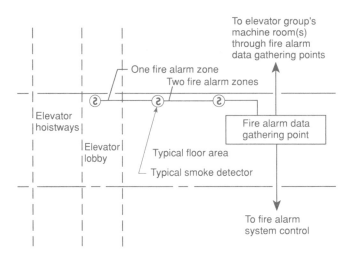

Figure A-3-8.15.4(a) *Elevator zone — elevator and fire alarm system installed at same time.*

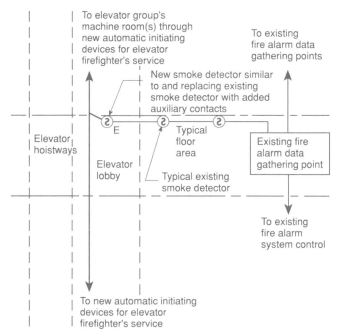

Figure A-3-8.15.4(b) *Elevator zone — elevator installed after fire alarm system.*

elevator braking system. If the elevator brakes are wet, there is the danger of the elevator rising to the top of the shaft or, depending on the load in the car, descending uncontrollably.

3-8.16.1* Where heat detectors are used to shut down elevator power prior to sprinkler operation, the detector shall have both a lower temperature rating and a higher sensitivity [often characterized by a lower Response Time Index (RTI)] when compared to the sprinkler.

A-3-8.16.1 A lower response time index is intended to provide detector response prior to the sprinkler, since a lower temperature rating alone may not provide earlier response. The listed spacing rating of the heat detector should be 25 ft (7.6 m) or greater.

This requirement is extremely important. Often, it is not understood that a 135°F heat detector may not respond prior to a 165°F sprinkler head despite the obvious differences in temperature sensitivity. The response time is based on the RTI of both devices and must be known prior to design and installation of heat detectors for elevator shutdown.

3-8.16.2 Where heat detectors are used for elevator power shutdown prior to sprinkler operation, they shall be placed within 2 feet of each sprinkler head and be installed in accordance with the requirements of Chapter 5. Alternatively, engineering methods (such as in Appendix B) shall be permitted to be used to select and place heat detectors to ensure response prior to any sprinkler head under a variety of fire growth rate scenarios.

3-8.16.3* Where pressure or waterflow switches are used to shut down elevator power immediately upon or prior to the discharge of water from sprinklers, the use of devices with time delays shall not be permitted.

A-3-8.16.3 Care should be taken to ensure that elevator power will not be interrupted due to water pressure surges in the sprinkler system.

3-9 Fire Safety Control Functions.

The control of preprogrammed fire safety control functions is automatically initiated by the fire alarm system in response to fire alarm signals. Depending on the system configuration, either all of the fire safety functions are initiated by any alarm signal, or selected fire safety functions may be initiated in response to a specific initiating device or zone in alarm (e.g., doors released or fans shut down on the floor of fire origin only.)

3-9.1 Scope. The provisions of this section apply to the minimum requirements for the interconnection of fire safety control functions (e.g., fan control, door control, etc.) to the fire alarm system. These fire safety functions are not intended to provide notification of alarm, supervisory, or trouble conditions; alert or control occupants; or summon aid.

3-9.2 General.

3-9.2.1 An auxiliary relay connected to the fire alarm system used to initiate control of fire safety functions shall be located within 3 ft (1 m) of the controlled circuit or device. The auxiliary relay shall function within the voltage and current limitations of the control unit. The installation wiring between the fire alarm system control unit and the auxiliary relay shall be monitored for integrity.

As an example, the fan control for a fan located on the fourth floor may be located in the basement. Subsection 3-9.2.1 requires the monitoring for integrity of the wiring from the fire alarm system control unit to the fire alarm system fan control relay, which actuates the fan control in the basement, not the fan located on the fourth floor. The distance between the fire alarm system fan control relay and the fan control should not exceed 3 ft (1 m).

Exception: Control devices that operate on loss of power or on loss of power to the auxiliary relay shall be considered self-monitoring for integrity.

This subsection and its exception are based on Chapter 7 (7-6.5.5) of NFPA *101, Life Safety Code.* This addition has been made to ensure that when auxiliary relays are used to cause the operation of smoke dampers, fire dampers, fan controls, smoke doors, and fire doors, and when they are connected to the building fire alarm system, the requirements of monitoring the interconnecting wiring for integrity apply.

3-9.2.2 Fire safety functions shall not interfere with other operations of the fire alarm control system.

This requirement is similar to the requirements for combination systems. One way to ensure that the fire safety functions do not interfere with any operations of the fire alarm system is to use auxiliary relays to isolate the fire safety function from the control unit. Fire safety functions connected directly to a notification appliance circuit, for instance, may disable the notification appliance circuit if a fault condition occurs in the fire safety function control equipment.

3-9.2.3 Transfer of data over listed serial communication ports shall be an acceptable means of interfacing between the fire alarm control unit and fire safety function control devices.

3-9.2.4 The fire safety function control devices shall be listed as compatible with the fire alarm control unit, so as not to interfere with the control unit's operation.

Generally, the fire safety control function devices are auxiliary relays. These relays should be listed specifically to operate with the fire alarm control unit and not be "off the shelf" items from an electronics supply store.

3-9.2.5 The interfaced systems shall be acceptance tested together in the presence of the authority having jurisdiction to ensure proper operation of the fire alarm system and the interfaced system(s).

The testing of the interfaced systems is extremely important. The proper interface and subsequent proper operation of the fire safety function is often dependent on more than one contractor and design engineer. See Figure 3.11.

To ensure that the individuals involved communicate and work together to interface the fire safety function properly, all parties involved should be present when the authority having jurisdiction is present for the acceptance test.

3-9.2.6 Where manual controls for emergency control functions are required to be provided, they shall provide visible indication of the status of the associated control circuits.

As used here, "emergency control functions" are the same as "fire safety control functions" mentioned previously in this code. The required visible status indication can be by

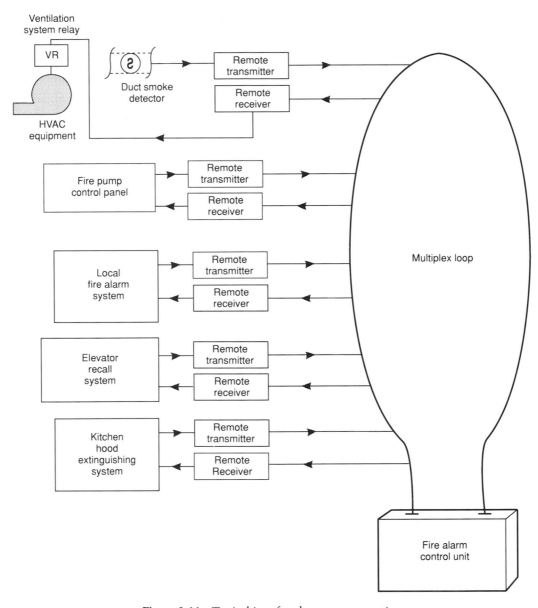

Figure 3.11   Typical interfaced systems connections.

a labeled LED annunciator (or equivalent means) or by the labeled position of a toggle or rotary switch.

3-9.3 Heating, Ventilation, and Air Conditioning (HVAC) Systems.

3-9.3.1 The provisions of 3-9.3 apply to the basic method by which a fire alarm system interfaces with the HVAC systems.

3-9.3.2 All detection devices used to cause the operation of smoke dampers, fire dampers, fan control, smoke doors, and fire doors shall be monitored for integrity in accordance with 1-5.8 where connected to the fire alarm system serving the protected premises.

This requirement precludes the use of stand-alone detection devices (such as duct-type smoke detectors) unless there is no fire alarm system in the building. In most instances, the detection devices referred to in 3-9.3.2 will be the automatic and manual alarm initiating devices used to detect and report a fire emergency. The wiring to these devices is required to be monitored for integrity by subsection 1-5.8.

3-9.3.3 Connections between fire alarm systems and the HVAC system for the purpose of monitoring and control shall operate and be monitored in accordance with applicable NFPA standards.

3-9.4 Door Release Service.

3-9.4.1 This section applies to the methods of connection of door hold release devices and to integral door hold release, closer, and smoke detection devices.

3-9.4.2 All detection devices used for door hold release service, whether integral or stand alone, shall be monitored for integrity in accordance with 1-5.8 where connected to the fire alarm system serving the protected premises.

3-9.4.3 All door hold release and integral door release and closure devices used for release service shall be monitored for integrity in accordance with 3-9.2.

Generally, door hold release magnetic devices are failsafe in that they release on loss of power. If these types of devices are connected in such a fashion to the fire alarm system, the wiring to the control relay or circuit does not need to be monitored for integrity. (*See the exception to 3-9.2.1.*)

3-9.5 Door Unlocking Devices.

3-9.5.1 Any device or system intended to effect the locking/unlocking of emergency exits shall be connected to the fire alarm system serving the protected premises.

3-9.5.2 All emergency exits connected in accordance with 3-9.5.1 shall unlock upon receipt of any fire alarm signal by the fire alarm system serving the protected premises.

3-9.5.3 All emergency exits connected in accordance with 3-9.5.1 shall unlock upon loss of the primary power to the fire alarm system serving the protected premises. The secondary power supply shall not be utilized to maintain these doors in the locked condition.

This subsection confirms by the description of the required operation that the wiring from the interface to the unlocking device need not be monitored for integrity.

3-10 Suppression System Actuation.

3-10.1 Fire alarm systems listed for releasing service shall be permitted to provide automatic or manual actuation of fire suppression systems.

Figure 3.12 Wet chemical kitchen hood and duct cylinder with control head.

3-10.2 The integrity of each releasing device (e.g., solenoid, relay, etc.) shall be supervised in accordance with applicable NFPA standards.

In 3-10.2, the word "supervised" means the same as "monitored for integrity" as used elsewhere in this code. See commentary following 3-8.9.3.

3-10.3 The integrity of the installation wiring shall be monitored in accordance with the requirements of Chapter 1.

3-10.4 Fire alarm systems used for fire suppression releasing service shall be provided with a disconnect switch to permit system testing without activating the fire suppression systems. Operation of the disconnect switch shall cause a trouble signal at the fire alarm control unit.

3-10.5 Sequence of operation shall be consistent with the applicable suppression system standards.

3-10.6* Each space protected by an automatic fire suppression system actuated by the fire alarm system shall contain one or more automatic fire detectors installed in accordance with Chapter 5.

A-3-10.6 Automatic fire suppression systems referred to in 3-10.6 include, but are not limited to, preaction and deluge sprinkler systems, carbon dioxide systems, halon systems, and dry chemical systems.

3-11* Interconnected Fire Alarm Control Units.

Fire alarm systems shall be permitted to be either integrated systems combining all detection, notification, and auxiliary functions in a single system, or a combination of component subsystems. Fire alarm system components shall be permitted to share control equipment or be able to operate as stand alone subsystems, but shall in any case be arranged to function as a single system. All component subsystems shall be capable of simultaneous, full load operation without degradation of the required, overall system performance.

This new section applies both where the interconnected control units are the products of a single manufacturer and where they are the products of two or more manufacturers, regardless of when installed. The section covers the requirements for the interconnection, monitoring, and compatibility of the control units.

A-3-11 This code contemplates field installations interconnecting two or more listed control units, possibly from different manufacturers, which together fulfill the requirements of this code.

Such an arrangement should preserve the reliability, adequacy, and integrity of all alarm, supervisory, and trouble signals and interconnecting circuits intended to be in accordance with the provisions of this code.

Where interconnected control units are in separate buildings, consideration should be given to protecting the interconnecting wiring from electrical and radio frequency interference.

There are many reasons for interconnecting control units. In some cases a building may need additional power supplies and notification appliances to conform to a new law, such as the Americans with Disabilities Act (ADA), and the original fire alarm system may not be able to accommodate the necessary changes. It also may not be feasible to attempt to modify or expand the existing control due to lack of parts, economics, or the system configuration. Some newer systems can consist of two or more subsystems (control units) connected to a single or multiple "master" control unit(s). In other cases, a new wing is added to a building, and the contractor plans to use a new, separate, and different manufacturer's system for the new wing. The problem is that the new wing's fire alarm system must operate as if it were part of the original system installed many years before.

Until this edition of the code, there had been no guidance regarding the correct procedures to follow when interconnecting the control panels or fire alarm systems as described in these examples. The goal of 3-11 and A-3-11 is to advise the fire alarm system designer, the installer, and the authority having jurisdiction that consideration must be given to what will be the appropriate, most reliable, and acceptable method of interconnection with these minimum guidelines as an outline.

3-11.1 The method of interconnection of control units shall be by the following recognized means:

(a) Properly rated electrical contacts

(b) Compatible digital data interfaces

(c) Other listed methods

and shall meet the monitoring requirements of 1-5.8 and the requirements of NFPA 70, *National Electrical Code*, Article 760.

3-11.2 Where approved by the authority having jurisdiction, interconnected control units providing localized detection, evacuation signaling, and auxiliary functions shall be permitted to be monitored by a fire alarm system as initiating devices.

3-11.2.1 Each interconnected control unit shall be separately monitored for alarm, trouble, and supervisory conditions.

3-11.2.2 Interconnected control unit alarm signals shall be permitted to be monitored by zone or combined as common signals as appropriate.

Subsection 3-11.2 covers such installations where, for example, multiple buildings on a contiguous property, under single ownership, and with individual protected premises fire alarm systems may be monitored by a single fire alarm control unit. This control unit may also serve the building it is located in as part of the protected premises fire alarm system. The purpose of this arrangement may be to have only one off-premises connection. In any case, the building with the "master" control unit that is providing the monitoring of the other systems should be identified at the building, at the remote annunciator, and at the supervising station receiving the fire alarm signal.

3-12 Emergency Voice/Alarm Communications.

3-12.1 **Application.** This section describes the requirements for emergency voice/alarm communications. The primary purpose is to provide dedicated manual and automatic facilities for the origination, control, and transmission of information and instructions pertaining to a fire alarm emergency to the occupants (including fire department personnel) of the building. It is the intent of this section to establish the minimum requirements for emergency voice/alarm communications.

3-12.2 Monitoring the integrity of speaker amplifiers, tone-generating equipment, and two-way telephone communications circuits shall be in accordance with 1-5.8.5.

3-12.3 **Survivability.**

3-12.3.1 The fire command station and the central control unit shall be located within a minimum 1-hour rated fire-resistive area and shall have a minimum 3-ft (1-m) clearance about the face of the fire command station control equipment.

Exception: Where approved by the authority having jurisdiction, the fire command station control equipment shall be permitted to be located in a lobby or other approved space.

Emergency voice/alarm communications service is generally used where relocation or partial evacuation (by floor or zone) of the building occupants is part of the fire response plan. As long as occupants are allowed to remain in the building, system survivability must be considered. It is essential that the fire alarm system remain operational during the fire emergency so that additional floors or zones can be evacuated as needed. See 3-2.4 for additional survivability requirements.

3-12.3.2 Where the fire command station control equipment is remote from the central control equipment, the wiring between the two shall be installed in conduit or other metal raceway that is routed through areas whose characteristics are at least equal to the limited combustible characteristics as defined in NFPA 90A, *Standard for the Installation of Air Conditioning and Ventilating Systems*. The maximum run of conduit or raceway shall not exceed 100 ft (30 m) or shall be enclosed in a 2-hour fire rated enclosure.

3-12.3.3 The primary power supply installation wiring between the central control equipment and the main service entrance shall also be routed through areas whose characteristics are at least equal to the limited combustible characteristics as defined in NFPA 90A, *Standard for the Installation of Air Conditioning and Ventilating Systems*.

3-12.3.4 The secondary (standby) power supply shall be provided in accordance with 1-5.2.5.

3-12.4 **Voice/Alarm Signaling Service.**

3-12.4.1* **General.** The purpose of the voice/alarm signaling service is to provide an automatic response to the receipt of a signal indicative of a fire emergency. Subsequent manual control capability of the transmission and audible reproduction of evacuation tone signals, alert tone signals, and voice directions on a selective and all-call basis, as determined by the authority having jurisdiction, is also required from the fire commandstation.

Exception: Where the fire command station or remote monitoring location is constantly attended by trained operators, and operator acknowledgment of receipt of a fire alarm signal is received within 30 seconds, automatic response is not required.

A-3-12.4.1   It is not the intention that emergency voice/alarm communications service be limited to English-speaking populations. Emergency messages should be provided in the language of the predominant building population. Where there is a possibility of isolated groups that do not speak the predominant language, multilingual messages should be provided. It is expected that small groups of transients unfamiliar with the predominant language will be picked up in the traffic flow in the time of emergency, and are not likely to be in an isolated situation.

3-12.4.2   **Multichannel Capability.**   When required by the authority having jurisdiction, the system shall allow the application of an evacuation signal to one or more zones and, at the same time, shall permit voice paging to the other zones selectively or in any combination.

3-12.4.3   **Functional Sequence.**

3-12.4.3.1   In response to an initiating signal indicative of a fire emergency, the system shall automatically transmit, either immediately or after a delay acceptable to the authority having jurisdiction, the following:

(a) An alert tone of 3 to 10 seconds' duration followed by a message (or messages when multichannel capability is provided) shall be repeated at least three times to direct the occupants of the alarm signal initiation zone and other zones in accordance with the building's fire evacuation plan; or

(b) An evacuation signal to the alarm signal initiation zone and other zones in accordance with the building's fire evacuation plan.

3-12.4.3.2   Failure of the message described by 3-12.4.3.1(a), where used, shall sound the evacuation signal automatically. Provisions for manual initiation of voice instructions or evacuation signal generation shall be provided.

Exception: Different functional sequences shall be permitted where approved by the authority having jurisdiction.

3-12.4.3.3   Live voice instructions shall override all previously initiated signals on that channel.

3-12.4.3.4   Where provided, manual controls for emergency voice/alarm communications shall be arranged to provide visible indication of the on/off status for their associated evacuation zones.

3-12.4.4   **Voice and Tone Devices.**   The alert tone preceding any message shall be permitted to be a part of the voice message or to be transmitted automatically from a separate tone generator.

3-12.4.5   **Fire Command Station.**

3-12.4.5.1   A fire command station shall be provided near a building entrance or other location approved by the authority having jurisdiction. The fire command station shall provide a communications center for the arriving fire department and shall provide for control and display of the status of detection, alarm, and communications systems. The fire command station shall be permitted to be physically combined with other building operations and security centers as permitted by the authority having jurisdiction. Operating controls for use by the fire department shall be clearly marked.

3-12.4.5.2   The fire command station shall control the emergency voice/alarm communications signaling service and, where provided, the two-way telephone communications service.

All of the requirements in 3-12.4.5 are in addition to the obvious coordination necessary with the responding fire department. Voice communication systems can be very effective both by calming occupants in areas remote from the fire and by directing others toward safety. All of the guidance and requirements in this section must be complemented with adequate training of the fire service personnel that will be responsible for using the equipment.

3-12.4.6   **Loudspeakers.**

3-12.4.6.1   Loudspeakers and their enclosures shall be listed for voice/alarm signaling service and installed in accordance with Chapter 6.

Loudspeakers used for background music, etc., are not acceptable unless they have been specifically listed for fire alarm system use.

3-12.4.6.2* There shall be at least two loudspeakers in each paging zone of the building, so located that signals can be clearly heard regardless of the maximum noise level produced by machinery or other equipment under normal conditions of occupancy. (See Section 6-3.)

"Paging zone" is not defined in the code. A common definition of a paging zone is a zone where all speakers are selected by the same switch or switching circuit. It can be inferred from 3-12.4.6.2 that a paging zone can be different than an evacuation zone. While this may be true in many cases, some emergency voice/alarm communications systems paging and evacuation zones are served by the same speakers and control equipment. The two-loudspeaker requirement is in consideration for survivability as discussed in subsection 3-2.4.

Figure 3.13 *Paging zone showing typical speaker location in a corridor.*

A-3-12.4.6.2 Placement of loudspeakers should give consideration to interference with normal use of emergency telephones and microphones in the area.

This problem is often overlooked. Loudspeakers located too near a microphone at a command center or a fire fighters' telephone can interfere with critical communications during a fire emergency.

3-12.4.6.3 Each elevator car shall be equipped with a single loudspeaker connected to the paging zone serving the elevator group in which the elevator car is located.

3-12.5 **Evacuation Signal Zoning.**

3-12.5.1 Where two or more evacuation signaling zones are provided, such zones shall be arranged consistent with the fire or smoke barriers within the protected premises. Undivided fire areas shall not be divided into multiple evacuation signaling zones.

NOTE: This section does not prohibit provision of multiple notification appliance circuits within a single evacuation signaling zone (i.e., separate circuits for audible and visible signals, redundant circuits provided to enhance survivability, or multiple circuits necessary to provide sufficient power/capacity).

"Evacuation signaling zones" are not defined in this code. As stated in 3-2.4, an evacuation zone can be an area of a floor, an entire floor, or several floors that are always intended to be evacuated simultaneously. Subsection 3-12.5.1 requires that the evacuation signaling zones (also called "evacuation zones" elsewhere in the code) be consistent with the fire or smoke barriers within the protected premises.

Exception: Stairwells not exceeding two stories in height.

3-12.5.2 Where multiple notification appliance circuits are provided within an single evacuation signaling zone, all of the notification appliances within the zone shall be arranged to activate simultaneously, either automatically or by actuation of a common, manual control.

Exception: Where the different notification appliance circuits within an evacuation signaling zone perform separate functions (i.e., presignal and general alarm signals, predischarge and discharge signals, etc.).

3-12.6 **Two-Way Telephone Communications Service.**

3-12.6.1 Two-way telephone communications equipment shall be listed for two-way telephone communications service and installed in accordance with 3-12.6.

3-12.6.2 Two-way telephone communications service, where provided, shall be available for use by the fire service. Additional uses, where specifically permitted by the authority having jurisdiction, shall be permitted to include signaling and communications for a building fire warden organization, signaling and communications for reporting a fire and other emergencies, (i.e., voice call box service, signaling, and communications for guard's tour service) and other uses. Variation of equipment and system operation provided to facilitate additional use of the two-way telephone communications service shall not adversely affect performance when used by the fire service.

Two-way telephone communications service is provided normally because fire department hand-held radios may be ineffective in buildings with a great deal of structural steel. The authority having jurisdiction can waive this requirement if the hand-held radios work effectively in the specific building in question.

Figure 3.14 *Two-way telephone communications service in use.*

3-12.6.3* Two-way telephone communications service shall be capable of permitting the simultaneous operation of any five telephone stations in a common talk mode.

A-3-12.6.3 Consideration should be given to the type of fire fighters' telephone handset used in areas where high ambient noise levels exist or areas where high noise levels may exist during a fire condition. Push-to-talk handsets, handsets containing directional microphones, or handsets containing other suitable noise-canceling features may be used.

3-12.6.4 A notification signal at the fire command station, distinctive from any other alarm or trouble signal, shall indicate the off-hook condition of a calling telephone circuit. Where a selective talk telephone communications service is supplied, a distinctive visible indicator shall be furnished for each selectable circuit so that all circuits with telephones off-hook are continuously and visibly indicated.

3-12.6.5 A switch for silencing the audible call-in signal sounding appliance shall be permitted only if it is key operated, in a locked cabinet, or given equivalent protection from use by unauthorized persons. Such a switch shall be permitted only if it operates a visible indicator and sounds a trouble signal whenever the switch is in the silence position where there are no telephone circuits in an off-hook condition. Where a selective talk telephone system is used, such a switch shall be permitted only if subsequent telephone circuits going off-hook will operate the distinctive off-hook audible signal sounding appliance.

3-12.6.6 The minimum requirement for fire service use shall be common talk, i.e., a conference or party line circuit. The minimum requirement for fire warden use, where provided, shall be a selective talking system controlled at the fire command station. Either system shall be capable of operation with five telephone stations connected together. There shall be at least one fire service telephone station or jack per floor and at least one per exit stairway. Where provided, there shall be at least one fire warden station or jack to serve each fire paging zone.

The paging zone is normally the same as the evacuation zone. See comment for subsection 3-12.4.6.2. In most designs, a fire fighters' telephone station or jack is located in each stairwell and elevator lobby.

3-12.6.7 Where the control equipment provided does not indicate the location of the caller (common talk systems), each telephone station or phone jack shall be clearly and permanently labeled to allow the caller to readily identify his location to the fire command station by voice.

3-12.6.8 Where telephone jacks are provided, a sufficient quantity of portable handsets, as determined by the authority having jurisdiction, shall be stored at the fire command station for distribution during an incident to responding personnel.

The required quantity of portable handsets is often a percentage of the telephone jacks in the building.

3-13* Special Requirements for Low Power Radio (Wireless) Systems.

A-3-13 Special Requirements for Low Power Radio (Wireless) Systems.

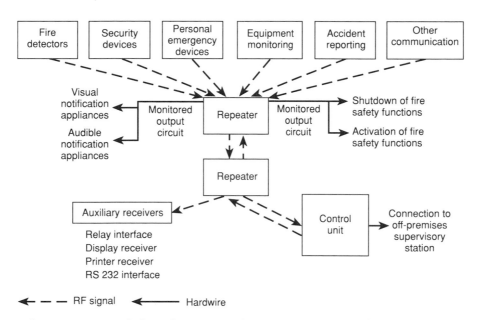

Figure 3.15 *Combination system including a low power radio (wireless) fire alarm system with other nonfire equipment.*

(a) The term "wireless" has been replaced with "low power radio" to eliminate potential confusion with other transmission media such as optical fiber cables.

(b) Low power radio devices are required to comply with the applicable low power requirements of Title 47, *Code of Federal Regulations*, Part 15.

3-13.1 Compliance with this section shall require the use of low power radio equipment specifically listed for the purpose.

NOTE: Equipment solely listed for household use does not comply with this requirement.

3-13.2 **Power Supplies.** A primary battery (dry cell) shall be permitted to be used as the sole power source of a low power radio transmitter when all of the following conditions are met:

(a) Each transmitter shall serve only one device and shall be individually identified at the receiver/control unit.

(b) The battery shall be capable of operating the low power radio transmitter for not less than one year before the battery depletion threshold is reached.

(c) A battery depletion signal shall be transmitted before the battery has depleted to a level insufficient to support alarm transmission after 7 additional days of normal operation. This signal shall be distinctive from alarm, supervisory, tamper, and trouble signals; shall visibly identify the affected low power radio trans-

mitter; and, if silenced, shall automatically resound at least once every 4 hours.

(d) Catastrophic (open or short) battery failure shall cause a trouble signal identifying the affected low power radio transmitter at its receiver/control unit. If silenced, the trouble signal shall automatically resound at least once every 4 hours.

(e) Any mode of failure of a primary battery in a low power radio transmitter shall not affect any other low power radio transmitter.

3-13.3 **Alarm Signals.**

3-13.3.1 When actuated, each low power radio transmitter shall automatically transmit an alarm signal.

NOTE: This requirement is not intended to preclude verification and local test intervals prior to alarm transmission.

3-13.3.2 Each low power radio transmitter shall automatically repeat alarm transmission at intervals not exceeding 60 seconds until the initiating device is returned to its normal condition.

3-13.3.3 Fire alarm signals shall have priority over all other signals.

3-13.3.4 The maximum allowable response delay from activation of an initiating device to receipt and display by the receiver/control unit shall be 90 seconds.

3-13.3.5 An alarm signal from a low power radio transmitter shall latch at its receiver/control unit until manually reset and shall identify the particular initiating device in alarm.

3-13.4 Supervision.

3-13.4.1 The low power radio transmitter shall be specifically listed as using a transmission method that shall be highly resistant to misinterpretation of simultaneous transmissions and to interference (e.g., impulse noise and adjacent channel interference).

3-13.4.2 The occurrence of any single fault that disables transmission between any low power radio transmitter and the receiver/control unit shall cause a latching trouble signal within 200 seconds.

Exception: Where Federal Communications Commission (FCC) regulations prevent meeting the 200-second requirement, the time period for a low power radio transmitter with only a single alarm initiating device connected shall be permitted to be increased to four times the minimum time interval permitted for a one-second transmission up to:

(a) Four hours maximum for a transmitter serving a single initiating device

(b) Four hours maximum for a re-transmission device (repeater) if disabling of the repeater or its transmission does not prevent the receipt of signals at the receiver/control unit from any initiating device transmitter.

3-13.4.3 A single fault on the signaling channel shall not cause an alarm signal.

3-13.4.4 The normal periodic transmission from a low power radio transmitter shall provide assurance of successful alarm transmission capability.

3-13.4.5 Removal of a low power radio transmitter from its installed location shall cause immediate transmission of a distinctive supervisory signal that indicates its removal and individually identifies the affected device. Household fire warning systems do not need to comply with this requirement.

3-13.4.6 Reception of any unwanted (interfering) transmission by a retransmission device (repeater) or by the main receiver/control unit, for a continuous period of 20 seconds or more, shall cause an audible and visible trouble indication at the main receiver/control unit. This indication shall identify the specific trouble condition present as an interfering signal.

4

Supervising Station
Fire Alarm Systems

Contents, Chapter 4

4-1 Scope

4-2 Communication Methods for Off-Premises Fire Alarm
Systems

  4-2.1 Scope

  4-2.2 General

    4-2.2.1 Applicable Requirements

    4-2.2.2 Equipment

    4-2.2.3 Adverse Conditions

    4-2.2.4 Dual Control

  4-2.3 Communication Methods

    4-2.3.1 Active Multiplex Transmission Systems

      4-2.3.1.3 System Classification

      4-2.3.1.4 System Loading Capacities

      4-2.3.1.5 Exceptions to Loading Capacities Listed in
Table 4-2.3.1.4

    4-2.3.2 Digital Alarm Communicator Systems

      4-2.3.2.1 Digital Alarm Communicator Transmitter
(DACT)

      4-2.3.2.2 Digital Alarm Communicator Receiver
(DACR)

        4-2.3.2.2.1 Equipment

        4-2.3.2.2.2 Transmission Channel

      4-2.3.2.3 Digital Alarm Radio System (DARS)

      4-2.3.2.4 Digital Alarm Radio Transmitter (DART)

      4-2.3.2.5 Digital Alarm Radio Receiver (DARR)

        4-2.3.2.5.1 Equipment

      4-2.3.2.6 Derived Local Channel

    4-2.3.3 McCulloh Systems

      4-2.3.3.1 Transmitters

      4-2.3.3.2 Transmission Channels

      4-2.3.3.3 Loading Capacity of McCulloh Circuits

    4-2.3.4 Two-Way RF Multiplex Systems

      4-2.3.4.3 Transmission Channel

      4-2.3.4.5 Loading Capacities

        4-2.3.4.5.2 Exceptions to Loading Capacities Listed
in Table 4-2.3.4.5.1

    4-2.3.5 One-Way Private Radio Alarm Systems

      4-2.3.5.3 Supervision

        4-2.3.5.3.2 Protected Premises

      4-2.3.5.4 Transmission Channel

      4-2.3.5.7 Exceptions to Loading Capacities Listed in
Table 4-2.3.5.6

4-2.3.6 Directly-Connected Noncoded Systems

    4-2.3.6.5 Loading Capacity of Circuits

4-2.3.7 Private Microwave Radio Systems

4-2.4 Display and Recording

4-2.5 Testing and Maintenance

4-3 Fire Alarm Systems for Central Station Service

4-3.1 Scope

4-3.2 General

4-3.3 Supervising Station Facilities

4-3.4 Equipment

4-3.5 Personnel

4-3.6 Operations

    4-3.6.1 Disposition of Signals

    4-3.6.2 Record Keeping and Reporting

4-3.7 Testing and Maintenance

4-4 Proprietary Supervising Station Systems

4-4.1 Scope

4-4.2 General

4-4.3 Supervising Station Facilities

4-4.4 Equipment

4-4.5 Personnel

4-4.6 Operations

    4-4.6.7 Disposition of Signals

        4-4.6.7.1 Alarms

        4-4.6.7.2 Guard's Tour Delinquency

        4-4.6.7.3 Supervisory Signals

        4-4.6.7.4 Trouble Signals

    4-4.6.8 Record Keeping and Reporting

4-4.7 Testing and Maintenance

4-5 Remote Supervising Station Fire Alarm Systems

4-5.1 Scope

4-5.2 General

4-5.3 Supervising Station Facilities

4-5.4 Equipment

4-5.5 Personnel

4-5.6 Operations

4-5.7 Testing and Maintenance

4-6 Public Fire Alarm Reporting Systems

4-6.1 Scope

4-6.2 General Fundamentals

4-6.3 Management and Maintenance

4-6.4 Equipment and Installation

4-6.5 Design of Boxes

4-6.6 Location of Boxes

4-6.7 Power Supply

4-6.7.1 General

    4-6.7.1.7 Form 2

    4-6.7.1.8 Form 3

    4-6.7.1.9 Form 4

4-6.7.2 Rectifiers, Converters, Inverters, and Motor-Generators

4-6.7.3 Engine-Driven Generator Sets

4-6.7.4 Float-Charged Batteries

4-6.8 Requirements for Metallic Systems and Metallic Interconnections

4-6.8.1 Circuit Conductors

4-6.8.2 Cables

    4-6.8.2.1 General

    4-6.8.2.2 Underground Cables

    4-6.8.2.3 Aerial Construction

    4-6.8.2.4 Leads Down Poles

    4-6.8.2.5 Wiring Inside Buildings

4-6.9 Facilities for Signal Transmission

4-6.9.1 Circuits

    4-6.9.1.1 General

    4-6.9.1.2 Box Circuits

    4-6.9.1.3 Tie Circuits

    4-6.9.1.4 Circuit Protection

        4-6.9.1.4.1 General

        4-6.9.1.4.2 Communication Center

        4-6.9.1.4.3 Protection on Aerial Construction

4-6.10 Power

4-6.10.1 Requirements for Constant-Current Systems

4-6.11 Receiving Equipment — Facilities for Receipt of Box Alarms

4-6.11.1 General

4-6.11.2 Visual Recording Devices

4-6.12 Supervision

4-6.13 Coded Wired Reporting Systems

4-6.14 Coded Radio Reporting Systems

4-6.14.1 Radio Box Channel (Frequency)

4-6.14.2 Metallic Interconnections

4-6.14.3 Receiving Equipment — Facilities for Receipt of Box Alarms

    4-6.14.3.1 Type A System

    4-6.14.3.2 Type B Systems

4-6.14.4 Power

4-6.14.5 Testing

4-6.14.6 Supervision

4-6.15 Telephone (Series) Reporting Systems

4-6.16 Telephone (Parallel) Reporting Systems
    4-6.16.1 Box Circuits
    4-6.16.2 Receiving Equipment — Facilities for Receipt of Box Alarms
4-7 Auxiliary Fire Alarm Systems
    4-7.1 Scope
    4-7.2 General
    4-7.3 Communication Center Facilities
    4-7.4 Equipment
        4-7.4.1 Types of Systems
        4-7.4.4 Location of Transmitting Devices
    4-7.5 Personnel
    4-7.6 Operations
    4-7.7 Testing and Maintenance

This chapter first presents the requirements for the various transmission technologies. It then presents the requirements for each of the supervising station services: central station, proprietary, remote station, public fire reporting, and auxiliary.

4-1 Scope.

This chapter covers the requirements for the proper performance, installation, and operation of fire alarm systems between the protected premises and the continuously attended supervising station facility.

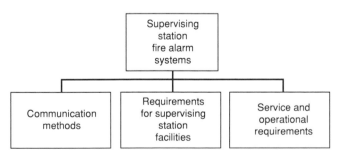

Figure 4.1 Supervising station fire alarm systems.

This chapter of the code covers the requirements for a protected premises fire alarm system connected to and monitored by a continuously attended supervising station. This supervising station may be either a central station, propri-

etary supervising station, remote supervising station, or, in the case of an auxiliary connection to a public fire reporting system, a public fire service communication center.

4-2 Communication Methods for Off-Premises Fire Alarm Systems.

NOTE: The requirements of Chapters 1, 3, 5, 6, and 7 shall apply to off-premises fire alarm systems unless they conflict with requirements of this section.

4-2.1 Scope. This section describes the requirements for the methods of communication between the protected premises and the supervising station. This includes the transmitter, transmission channel, and the signal receiving, processing, display, and recording equipment at the supervising station.

An important feature of the 1993 code is that it makes a full range of transmission technologies available to all of the supervising station services. This gives designers maximum flexibility in choosing the transmission technology most appropriate for the particular application. Available technologies include active multiplex and derived local channel, digital alarm communicator systems, digital alarm radio systems, McCulloh systems, two-way RF multiplex systems, one-way radio alarm systems, directly-connected noncoded systems, and private microwave radio systems. However, the use of each transmission technology by a particular supervising station service is limited by the specific requirements, if any, of that service. Of the available supervising station services, the most limited is the public fire reporting system. In fact, the requirements for this service are so limited that the available transmission technologies are fully defined within Section 4-6.

4-2.2 General.

4-2.2.1 Applicable Requirements. The requirements of Sections 4-1, 4-3, 4-4, 4-5, 4-6, and 4-7 shall apply to active multiplex, including systems utilizing derived channels; digital alarm communicator systems, including digital alarm radio systems; McCulloh systems; two-way RF multiplex systems; and one-way radio alarm systems, except where they conflict with the requirements of this section.

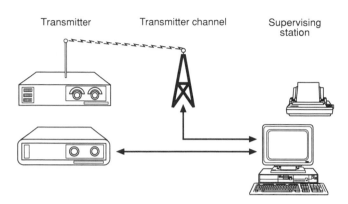

Figure 4.2 *Communications methods for supervising station fire alarm systems.*

Because new transmission technologies added over the years were most often compared to the performance capability of the McCulloh system — since it was the first transmission technology employed — certain operational requirements have been intermixed among the technologies. This subsection becomes a caveat to make certain that no critical operational requirement that appears in only one section is ignored when applying a particular transmission technology.

4-2.2.2 Equipment.

4-2.2.2.1 Wiring, power supplies, and overcurrent protection shall comply with the requirements of 1-5.5.4 and 1-5.8.6.

4-2.2.2.2 Exclusive of the transmission channel, grounding of fire alarm equipment shall be permitted.

4-2.2.2.3 Fire alarm system equipment and installations shall comply with Federal Communication Commission (FCC) rules and regulations, as applicable, concerning electromagnetic radiation; use of radio frequencies; and connection to the public switched telephone network of telephone equipment, systems, and protection apparatus.

This subsection recognizes the jurisdiction that the FCC has over the installation requirements for certain communication equipment used to transmit signals from a protected premises to a supervising station.

4-2.2.2.4 Equipment shall be installed in compliance with NFPA 70, *National Electrical Code*, Article 810.

When a particular transmission technology uses television or radio equipment, that equipment must be installed in compliance with the appropriate article of NFPA 70, *National Electrical Code®*.

4-2.2.2.5 All external antennas shall be protected in order to minimize the possibility of damage by static discharge or lightning.

4-2.2.3 Adverse Conditions.

4-2.2.3.1 For active and two-way RF multiplex systems, the occurrence of an adverse condition on the transmission channel between a protected premises and the supervising station that will prevent the transmission of any status change signal shall be automatically indicated and recorded at the supervising station. This indication and record shall identify the affected portions of the system so that the supervising station operator can determine the location of the adverse condition by trunk or leg facility, or both.

The integrity of active multiplex transmission technology is monitored by interrogation and response transmission back and forth along the communication path. The satisfactory exchange of data ensures that all trunks and legs remain operational. If an interrogation and response sequence is not successfully completed, indicating possible failure of a trunk or a leg, then this section provides for the detailed notification of the supervising station.

4-2.2.3.2 For a one-way radio alarm system, the system shall be supervised to ensure that at least two independent radio alarm repeater station receivers (RARSRs) are receiving signals for each radio alarm transmitter (RAT) during each 24-hour period. The occurrence of a failure to receive a signal by either RARSR shall be automatically indicated and recorded at the supervising station. The indication shall identify which RARSR has failed to receive such supervisory signals. It is not necessary for correctly received test signals to be indicated at the supervising station.

The integrity of one-way radio transmission technology is monitored by the satisfactory receipt of at least one transmission every 24 hours by at least two RARSRs. If such a signal is not received by both receivers, then this section provides for the detailed notification of the supervising station.

4-2.2.3.3 For active and two-way RF multiplex systems, restoration of normal service to the affected portions of the system shall be automatically recorded. When normal service is restored, the first status change of any initiating device circuit, or any initiating device directly connected to a signaling line circuit, or any combination that occurred at any of the affected premises during the service interruption shall also be recorded.

Exception: This requirement does not apply to proprietary systems on contiguous properties.

For services other than a proprietary system that serves a contiguous property, the restoration of interrupted service must be automatically recorded at the supervising station, and the first status change on any initiating device circuit must be reported. This means that the equipment at the protected premises must be able to retain and later report the first status change on each initiating device circuit that occurs during the transmission interruption.

4-2.2.4 Dual Control.

4-2.2.4.1 Dual control, where required, shall provide for redundancy in the form of a standby circuit or similar alternate means of transmitting signals over the primary trunk portion of a transmission channel. The same method of signal transmission shall be permitted to be used over separate routes, or different methods of signal transmission shall be permitted to be utilized. Public switched telephone network facilities shall be used only as the alternate method of transmitting signals.

While dual control does not provide full redundancy for every trunk and leg, it does offer an option that a system designer can choose to employ when signals must be received during interruptions to the primary trunk. Most often, this option is a technology offered by the public telephone utility called DataPhone Select-A-Station (DSAS). When the primary trunk fails, this service allows the supervising station to either automatically or manually dial into the public switched telephone network and establish an alternate path for the signals that would normally be transmitted over the primary trunk. Telephone technicians sometimes refer to this arrangement as "dial up, make good."

4-2.2.4.2 Where utilizing facilities leased from a telephone company, that portion of the primary trunk facility between the super-

vising station and its serving wire center shall be permitted to be excepted from the separate routing requirement of the primary trunk facility. Dual control, where used, requires supervision as follows:

(a) Dedicated facilities, which are available full time and whose use is limited to signaling purposes as defined in this code, shall be exercised at least once every hour.

(b) Public switched telephone network facilities shall be exercised at least once every 24 hours.

To ensure that the alternate path is usable by the dual control arrangement, it must be exercised. If the alternate path is one dedicated solely to this purpose, then it must be exercised once every hour. If it is part of the public switched telephone network — and this is the prevalent arrangement — then it must be exercised once every 24 hours.

4-2.3 Communication Methods.

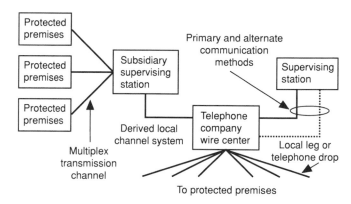

Figure 4.3   Communications methods — active multiplex transmission systems.

4-2.3.1 Active Multiplex Transmission Systems.

4-2.3.1.1 The multiplex transmission channel terminates in a transmitter at the protected premises and in a system unit at the supervising station. The derived channel terminates in a transmitter at the protected premises and in derived channel equipment at a subsidiary station location or a telephone company wire center. The derived channel equipment at the subsidiary station location or a telephone company wire center selects or establishes the communication with the supervising station.

One of three forms of active multiplex ownership may be involved. In the first, the equipment and all transmission facilities are owned by the protected facility, or owned by the fire alarm service provider and leased to the facility. In the second, the equipment is owned by the protected facility, or owned by the fire alarm service provider and leased to the facility; the transmission facilities are leased from the public telephone utility. In the third form, some of the equipment is owned by the protected facility, or owned by the fire alarm service provider and leased to the facility; some of the equipment is owned by the public telephone utility; and transmission facilities are leased from the public telephone company utility. This third form is commonly called "derived local channel."

4-2.3.1.2* Operation of the transmission channel shall conform to the requirements of this code whether channels are private facilities, such as microwave, or leased facilities furnished by a communication utility company. Where private signal transmission facilities are utilized, the equipment necessary to transmit signals shall also comply with the requirements for duplicate equipment or replacement of critical components, as described in 4-2.4.2. The trunk transmission channels shall be dedicated facilities for the main channel. For Type 1 multiplex systems, the public switched telephone network facilities shall be permitted to be used for the alternate channel.

Exception: Derived channel scanners with no more than 32 legs shall be permitted to use the public switched telephone network for the main channel.

One manufacturer of derived local channel equipment offers an option of connecting the scanner in the telephone company wire center to the supervising station by means of a dial-up modem that essentially performs as if it were a DACT. The exception to this subsection limits the loading of such a scanner to no more than 32 legs (protected premises).

A-4-2.3.1.2 Where derived channels are used, normal operating conditions of the telephone equipment will not inhibit or impair the successful transmission of signals. These normal conditions include, but are not limited to, the following:

(a) Intraoffice calls with a transponder on the originating end

(b) Intraoffice calls with a transponder on the terminating end

(c) Intraoffice calls with transponders on both ends

(d) Receipt and origination of long-distance calls

(e) Calls to announcement circuits

(f) Permanent signal receiver off-hook tone

(g) Ringing with no answer, with transponder on either the originating or the receiving end

(h) Calls to tone circuits, i.e., service tone, test tone, busy, and/or reorder

(i) Simultaneous with voice source

(j) Simultaneous with data source

(k) Tip and ring reversal

(l) Cable identification equipment.

4-2.3.1.2.1 Derived channel signals shall be permitted to be transmitted over the leg facility, which shall be permitted to be shared by the telephone equipment under all normal on-hook and off-hook operating conditions.

This subsection embodies the whole concept of derived local channel: it shares the leg facility used by the "plain, old telephone service" (POTS) for a particular protected premises. The derived local channel system permits the transmission of alarm and supervisory signals, even though the leg may be used for normal telephone communication. The operational integrity of the leg is monitored by means of interrogation and response sequence similar to that used by other active multiplex systems.

4-2.3.1.2.2 Where used, the public switched telephone network shall be in compliance with the requirements of 4-2.3.2.

4-2.3.1.2.3 The maximum end-to-end operating time parameters allowed for an active multiplex system are as follows:

(a) The maximum allowable time lapse from the initiation of a single fire alarm signal until it is recorded at the supervising station shall not exceed 90 seconds. When any number of subsequent fire alarm signals occur at any rate, they shall be recorded at a rate no slower than one every 10 additional seconds.

Paragraph (a) effectively ensures that an interrogation and response sequence will be completed at least every 90 seconds, unless some other means has been provided to ensure alarm receipt in that time frame. For example, a system could be devised where alarm signals would be transmitted by the equipment at the protected premises at a time other than during the normal interrogation and response sequence. However, most systems meet this requirement by completing the interrogation and response sequence within the 90-second time frame.

(b)* The maximum allowable time lapse from the occurrence of an adverse condition in any transmission channel until recording of the adverse condition is started shall not exceed 90 seconds for Type 1 and Type 2 systems, and 200 seconds for Type 3 systems. (*See 4-2.3.1.3.*)

A-4-2.3.1.2.3(b) Derived channel systems comprise Type 1 and Type 2 systems only.

The reporting of an adverse condition on a Type 3 system within 200 seconds allows for a system described in 4-2.3.1.2.3, where alarm signals would be transmitted by the equipment at the protected premises at a time other than during the normal interrogation and response sequence. This requirement ensures the interrogation and response sequence will occur at least every 200 seconds.

(c) In addition to the maximum operating time allowed for fire alarm signals, the requirements of one of the following paragraphs shall be met:

1. A system unit having more than 500 initiating device circuits shall be able to record not less than 50 simultaneous status changes in 90 seconds.

2. A system unit having fewer than 500 initiating device circuits shall be able to record not less than 10 percent of that total number of simultaneous status changes within 90 seconds.

These requirements ensure that the portion of the multiplex system that processes and records status changes can do so with sufficient speed to handle a reasonable volume of signal traffic.

4-2.3.1.3 System Classification. Active multiplex systems are divided into three categories based upon their ability to perform under adverse conditions of their transmission channels. System classifications are as follows:

(a) A Type 1 system shall have dual control as described in 4-2.2.4. An adverse condition on a trunk or leg facility shall not prevent the transmission of signals from any other trunk or leg facility, except those normally dependent on the por-

tion of the transmission channel in which the adverse condition has occurred. An adverse condition limited to a leg facility shall not interrupt normal service on any trunk or other leg facility. The requirements of 4-2.2.1, 4-2.2.2, and 4-2.2.3 shall be met by Type 1 systems.

To meet the requirements for Type 1, an active multiplex system must somehow isolate each leg and trunk from other legs and trunks. This is normally accomplished by using a device called a "closed window bridge." Such a device is used wherever two or more trunks or two or more legs converge. Coupling circuitry within the closed window bridge allows signals to pass, but keeps interference from a fault on one leg or trunk from adversely affecting the receipt of signals from another leg or trunk. The bridge may be supplied by and located in the wire center of the public telephone utility. The bridge could also be located at a protected premises. For example, fire alarm systems for the individual stores in a shopping mall and for the mall common areas could be multiplexed to a supervising station through a bridge located at an equipment room in the mall.

Type 1 systems also employ dual control, as described in 4-2.2.4. This gives the system an alternate transmission path should the primary trunk fail.

(b) A Type 2 system shall have the same requirements as a Type 1 system, except that dual control of the primary trunk facility shall not be required.

Type 2 systems also employ closed window bridges to provide isolation between trunks and legs. Dual control need not be provided for Type 2 systems. Thus, a Type 2 system has no alternate transmission path should the primary trunk fail.

(c) A Type 3 system shall automatically indicate and record at the supervising station the occurrence of an adverse condition on the transmission channel between a protected premises and the supervising station. The requirements of 4-2.2, except for 4-2.2.4, shall be met.

Type 3 systems have no requirement for isolation between legs and trunks. They commonly employ what is called an "open window bridge." While this device allows the coupling of signals wherever two or more legs or two or more trunks converge, an adverse condition on one leg or trunk may affect the operation of other legs or trunks.

Table 4-2.3.1.4

	System Type		
	Type 1	Type 2	Type 3
A. Trunks			
Maximum number of fire alarm service initiating device circuits per primary trunk facility	5120	1280	56
Maximum number of leg facilities for fire alarm service per primary trunk facility	512	128	64
Maximum number of leg facilities for all types of fire alarm service per secondary trunk facility*	128	128	128
Maximum number of all types of initiating device circuits per primary trunk facility in any combination*	10,240	2560	512
Maximum number of leg facilities for all types of fire alarm service per primary trunk facility in any combination*	1024	256	128
B. System Units at the Supervising Station			
Maximum number of all types of initiating device circuits per system unit*	10,240**	10,240**	10,240**
Maximum number of fire protecting buildings and premises per system unit	512**	512**	512**
Maximum number of fire fire alarm service initiating device circuits per system unit	5120**	5120**	5120**
C. Systems Emitting from Subsidiary Station	Same as B	Same as B	Same as B

*Includes every initiating device circuit, i.e., waterflow, fire alarm, supervisory, guard, burglary, hold-up, etc.
**Paragraph 4-2.3.1.5 applies.

4-2.3.1.4 **System Loading Capacities.** The capacities of active multiplex systems are based on the overall reliability of the signal receiving, processing, display, and recording equipment at the supervising and subsidiary stations, and the capability to transmit signals during adverse conditions of the signal transmission facilities. Table 4-2.3.1.4 establishes the allowable capacities.

The loading of trunks depends on the capability of the type of system. Since a Type 1 system has a redundant primary trunk (dual control) and isolation between legs and trunks, it has the greatest permitted trunk capacity. A Type 2 system does not have dual control, but does have isolation between legs and trunks. Its trunk capacity is less than that of a Type 1, but more than that of a Type 3. A Type 3 system has neither dual control nor isolation between legs

and trunks, so its trunk capacity is the least of the three types.

4-2.3.1.5 **Exceptions to Loading Capacities Listed in Table 4-2.3.1.4.** Where the signal receiving, processing, display, and recording equipment is duplicated at the supervising station and a switch-over can be accomplished in not more than 30 seconds with no loss of signals during this period, the capacity of a system unit shall be unlimited.

Part B of Table 4-2.3.1.4 is modified by this requirement. However, to meet this requirement an active multiplex system would have to employ complete redundancy of all critical components and be able to complete a switch-over in 30 seconds with no loss of signals. Those systems that meet this requirement generally process all incoming signals in tandem. That is, the standby unit is fully functioning at all times and simply continues to function normally if the main unit fails. The only actual "change-over" would occur for those incidental peripheral devices for which redundancy was not required.

4-2.3.2 Digital Alarm Communicator Systems.

4-2.3.2.1 Digital Alarm Communicator Transmitter (DACT).

4-2.3.2.1.1 A DACT shall be connected to the public switched telephone network upstream of any private telephone system at the protected premises. In addition, special attention is required to ensure that this connection shall be made only to a loop start telephone circuit and not to a ground start telephone circuit.

Exception: If public cellular telephone service is utilized as a secondary means of transmission, the requirements of this paragraph shall not apply.

The DACT is connected to the public switched telephone network so that it may seize the line to which it is connected, effectively disconnecting any private telephone equipment beyond its point of connection. This arrangement gives the DACT control over the line at all times.

A loop-start telephone line is one where voltage is continuously supplied from the first telephone company wire center. The vast majority of residential telephone connections use loop-start lines. In contrast, almost all business telephone connections, particularly those employing private branch exchange (PBX) connections, use ground-start lines. In order to obtain dial tone and operating power on a ground-start line, one side of the line must be momentarily connected to earth ground. The equipment that uses ground-start lines is designed so that it makes this connection whenever a user requests dial tone. Since no voltage is present on a ground-start line when it is not in use, the DACT cannot monitor the integrity of the line as it can with a loop-start line.

Functionally, a DACT can signal over a ground-start line and frequently does so when used as part of a burglar alarm system. However, the DACT can only monitor a loop-start line for integrity.

The exception is necessary since public cellular telephone systems do not use telephone lines. Thus, when the pub-lic cellular telephone system is used as a secondary means of signal transmission, the requirements of this subsection do not apply to the cellular portion of the system.

4-2.3.2.1.2 All information exchanged between the DACT at the protected premises and the digital alarm communicator receiver (DACR) at the supervising or subsidiary station shall be by digital code or equivalent. Signal repetition, digital parity check, or some equivalent means of signal verification shall be used.

The functional requirements of this subsection effectively rule out the use of an analog or digital voice tape dialer to transmit fire alarm signals. This device dials a predetermined telephone number and then plays a voice message, such as, "There is a fire at 402 Spruce Street." Over the years, there have been reports of voice tape dialers endlessly repeating their message, tying up vital emergency telephone lines in public fire service communications centers. Their use is strictly forbidden by the code.

4-2.3.2.1.3* A DACT shall be capable of seizing the telephone line (going off-hook) at the protected premises, disconnecting an outgoing or incoming telephone call, and preventing its use for outgoing telephone calls until signal transmission has been completed. A DACT shall not be connected to a party line telephone facility.

In order to ensure reliability for transmission of fire alarm, supervisory, and trouble signals, this requirement gives the DACT exclusive control over the telephone lines to which it is connected.

A-4-2.3.2.1.3 In order to give the DACT the ability to disconnect an incoming call to the protected premises, telephone service must be of the type that provides for timed-release disconnect. In some telephone systems (step-by-step offices), timed-release disconnect may not be provided.

4-2.3.2.1.4 A DACT shall have the means to satisfactorily obtain an available dial tone, dial the number(s) of the DACR, obtain verification that the DACR is ready to receive signals, transmit the signal, and receive acknowledgment that the DACR has accepted that signal. In no event shall the time from going off-hook to on-hook exceed 90 seconds per attempt.

This subsection describes the normal sequence of operation for a DACT. Upon initiation of an alarm, a supervisory, or a trouble signal, the DACT seizes the line, obtains dial tone, dials the number of the DACR, receives a "hand-shake" signal from the DACR, transmits its data, receives an acknowledgment signal — sometimes called the "kiss off" signal — from the DACR, and hangs up. This calling and verification sequence must take no longer than 90 seconds to complete.

4-2.3.2.1.5* A DACT shall have suitable means to reset and retry if the first attempt to complete a signal transmission sequence is unsuccessful. A failure to complete connection shall not prevent subsequent attempts to transmit an alarm if such alarm is generated from any other initiating device circuit or signaling line circuit, or both. Additional attempts shall be made until the signal transmission sequence has been completed to a minimum of five and a maximum of ten attempts.

If the maximum number of attempts to complete the sequence is reached, an indication of the failure shall be made at the premises.

If the DACT attempts to complete a transmission sequence as described in 4-2.3.2.1.4 and is unsuccessful, it must make at least five attempts to complete the sequence. However, it must not make more than ten attempts. This helps prevent a malfunctioning DACT from tying up the DACR.

A-4-2.3.2.1.5 A DACT may be programmed to originate calls to the DACR telephone lines (numbers) in any alternating sequence. The sequence can consist of single or multiple calls to one DACR telephone line (number), followed by single or multiple calls to a second DACR telephone line (number), or any combination thereof that is consistent with the minimum/maximum attempt requirements in 4-2.3.2.1.5.

4-2.3.2.1.6 A DACT shall be connected to two separate means of transmission at the protected premises. The DACT shall be capable of selecting the operable means of transmission in the event of failure of the other. The primary means of transmission shall be a telephone line (number) connected to the public switched network.

If the DACT detects that one of the two transmission means has failed (loss of voltage on a wire line, loss of one-way radio alarm service, or loss of cellular telephone service), it must switch to the other operable means.

Formal Interpretation 87-1
Reference: 4-2.3.2.1.6

Question: NFPA 72, 4-2.3.2.1.6, states: "A DACT shall be connected to two separate means of transmission at the protected premises." To comply with this section, would it be required that the utility company install two (2) separate incoming phone lines, each one entering the building at a separate location?
Answer: No.

Issue Edition: NFPA 71-1987
Reference: 5-2.6
Issue Date: April 1988 ■

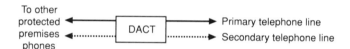

Figure 4.4   Connections to a DACT.

4-2.3.2.1.6.1 The secondary means of transmission shall be permitted to be one of the following:

(a) A one-way radio system utilized in accordance with 4-2.3.2.3.

(b) Public cellular telephone service. A verification signal shall be transmitted at least once a month.

(c) A telephone line (number).

Neither a one-way radio alarm system nor public cellular telephone service may be used for the primary transmission means of a digital alarm communicator system.

4-2.3.2.1.6.2 The first transmission attempt shall utilize the primary means of transmission.

Transmission reliability studies have shown that the public switched telephone line has a somewhat higher reliability than public cellular telephone service or one-way radio alarm service. The requirement in 4-2.3.2.1.6.2 ensures that the first call attempt will use the public switched telephone line.

4-2.3.2.1.7* Failure of either of the telephone lines (numbers) at the protected premises shall be annunciated at the protected premises, and a trouble signal shall be transmitted to the supervising or subsidiary station over the other line (number). Transmission shall be initiated within 4 minutes of detection of the fault. If public cellular telephone service is used as the secondary means of transmission, loss of cellular service shall be considered a failure.

As important as it is to monitor the integrity of the transmission means, it is equally important to avoid nuisance trouble signals. The permissible 4-minute delay in transmitting a trouble signal allows for momentary, or even somewhat longer, interruptions in the transmission path such as might occur during a storm.

A-4-2.3.2.1.7 Most failures of a telephone line may be detected by supervising the presence of the telephone line voltage. A loss of voltage indicates failure. Where the telephone line is also used for telephone communication, the voltage will drop when the telephone is in use. The presence of current will also indicate a normal line condition during this period.

4-2.3.2.1.8 Each DACT shall be programmed to call a second DACR line (number) should the signal transmission sequence to the first called line (number) be unsuccessful.

To help avoid a possible disarrangement of the transmission path on the receiving end of the digital alarm communicator system, this requirement specifies that the DACT must call a second number if calling the first number does not result in completion of the transmission.

4-2.3.2.1.9 If long distance telephone service (including WATS) is used, the second telephone number shall be provided by a different long distance service provider, where available.

This requirement is designed to help overcome a total long distance network failure by a particular long distance network provider. Unfortunately, there have been several significant network failures in the last 5 years. At least one of these was caused by a fire in a telephone company wire center that also housed an important switching machine for a long distance network provider. Other long distance network failures of up to several hours' duration have occurred due to various switching machine anomalies.

4-2.3.2.1.10 Each DACT shall automatically initiate and complete a test signal transmission sequence to its associated DACR

at least once every 24 hours. A successful signal transmission sequence of any other type within the same 24-hour period shall be considered sufficient to fulfill the requirement to verify the integrity of the reporting system, if signal processing is automated so that 24-hour delinquencies shall be individually acknowledged by supervising station personnel.

At least once every 24 hours each DACT must initiate a signal to verify the end-to-end integrity of the digital alarm communicator system. Where the receiving or processing equipment at the supervising station has sufficient intelligence to automatically keep track of signal traffic, any incoming signal from a particular DACT may serve to satisfy this requirement as long as one signal is received during every 24-hour period.

4-2.3.2.1.11* If DACTs are programmed to call a telephone line (number) that is call forwarded to the line (number) of the DACR, a means shall be implemented to verify the integrity of the call forwarding feature every 4 hours.

A-4-2.3.2.1.11 Since call forwarding requires equipment at a telephone company central office that might occasionally interrupt the call forwarding feature, a signal should be initiated whereby the integrity of the forwarded telephone line (number) that is being called by DACTs is verified every 4 hours. This may be accomplished by a single DACT either in service or used solely for verification that automatically initiates and completes a transmission sequence to its associated DACR every 4 hours. A successful signal transmission sequence of any other type within the same 4-hour period should be considered sufficient to fulfill this requirement.

Call forwarding should not be confused with WATS or 800 service. The latter, differentiated by dialing the 800 prefix, is a dedicated service used mainly for its toll-free feature; all calls are predetermined to terminate at a fixed telephone line (number) or to a dedicated line.

Occasionally a supervising station will maintain one or more telephone numbers in a local calling area that are call forwarded to an actual number connected to the DACR. When this practice is employed, the integrity of the call forward instruction must be verified every 4 hours. This is accomplished most often by programming the automatic test signal of six of the DACTs in the service area that use the call forwarded number so that their test signals will be initiated 4 hours apart over the course of a 24-hour period.

4-2.3.2.2 Digital Alarm Communicator Receiver (DACR).

4-2.3.2.2.1 Equipment.

4-2.3.2.2.1.1 Spare DACRs shall be provided in the supervising or subsidiary station and shall be able to be switched in place of a failed unit within 30 seconds after detection of failure.

NOTE: One spare DACR shall be permitted to serve as a backup for up to five DACRs in use.

The fact that a spare DACR is present does not by itself satisfy the requirements of this section. In order to meet the switching requirement in 4-2.3.2.2.1.1, personnel on duty in the supervising station must be trained to accomplish the switch-over. Adequate instructions must be provided. Preferably, the connections to the unit should terminate in a fashion that permits rapid, error-free reconnection to the spare unit.

4-2.3.2.2.1.2 The number of incoming telephone lines to a DACR shall be limited to eight lines.

Exception: Where the signal receiving, processing, display, and recording equipment at the supervising or subsidiary station is duplicated and a switchover can be accomplished in less than 30 seconds with no loss of signal during this period, the number of incoming lines to the unit is unlimited.

The limit of eight incoming lines to a single DACR is intended to eliminate the possibility of overloading a DACR's ability to receive and process signals promptly. In some fully automated supervising stations, facilities may exist to take advantage of the exception.

4-2.3.2.2.2 Transmission Channel.

4-2.3.2.2.2.1* The DACR equipment at the supervising or subsidiary station shall be connected to a minimum of two separate incoming telephone lines (numbers). If the lines (numbers) are in a single hunt group, they shall be individually accessible; otherwise, separate hunt groups are required. These lines (numbers) are to be used for no other purpose than receiving signals from DACTs. These lines (numbers) shall be unlisted.

Hunt groups provided by some older telephone company central office equipment may have the potential for lock-ing on to a defective line, thus rendering all lines in the hunt group inoperative. The set of requirements in 4-2.3.2.2.2.1 helps ensure that the design of the digital alarm communicator system receiving network affords a high degree of reliability.

A-4-2.3.2.2.2.1 The timed-release disconnect considerations as outlined in A-4-2.3.2.1.3 apply to the telephone lines (numbers) connected to a DACR at the supervising station.

It may be necessary to consult with appropriate telephone service personnel to ensure that numbers assigned to the DACR can be individually accessed even though they may be connected in rotary (a hunt group).

4-2.3.2.2.2.2 Failure of any telephone line (number) connected to a DACR due to loss of line voltage shall be annunciated visually and audibly in the supervising station.

The telephone lines connected to the DACR must be loop-start lines so that voltage is normally present. This voltage is monitored as a means of ensuring that the line is operable up to the first public telephone company wire center.

4-2.3.2.2.2.3* The loading capacity for a hunt group shall be in accordance with Table 4-2.3.2.2.2.3 or be capable of demonstrating a 90 percent probability of immediately answering the incoming call.

(a) Each supervised burglar alarm (open/close) or each suppressed guard tour transmitter shall reduce the allowable DACTs as follows:

    1. up to a 4-line hunt group, by 10
    2. up to a 5-line hunt group, by 7
    3. up to a 6-line hunt group, by 6
    4. up to a 7-line hunt group, by 5
    5. up to an 8-line hunt group, by 4.

(b) Each guard tour transmitter shall reduce the allowable DACTs as follows:

    1. up to a 4-line hunt group, by 30
    2. up to a 5-line hunt group, by 21
    3. up to a 6-line hunt group, by 18
    4. up to a 7-line hunt group, by 15
    5. up to an 8-line hunt group, by 12.

A-4-2.3.2.2.2.3 In determining system loading, Table 4-2.3.2.2.2.3 may be used, or it may be demonstrated that there

Table 4-2.3.2.2.2.3

	Number of Lines in Hunt Group				
	1	2	3	4	5 to 8
System Loading at the Supervising Station					
With DACR lines processed in parallel					
Number of initiating circuits	N/A	5000	10,000	20,000	20,000
Number of DACTs*	N/A	500	1500	3000	3000
With DACR lines processed serially (put on hold, then answered one at a time)					
Number of initiating circuits	N/A	3000	5000	6000	6000
Number of DACTs*	N/A	300	800	1000	1000

*Table 4-2.3.2.2.2.3 is based on an average distribution of calls and an average connected time of 30 seconds for a message. The loading figures in the table presume that the lines are in a hunting group (i.e., DACT can access any available line). Note that a single-line DACR is NOT ACCEPTABLE (N/A) for any of the listed configurations.

is a 90 percent probability of incoming line availability. Table 4-2.3.2.2.2.3 is based on an average distribution of calls and an average connected time of 30 seconds for a message. Therefore, when it is proposed to use Table 4-2.3.2.2.2.3 to determine system loading, if any factors are disclosed that will extend DACR connect time so as to increase the average connect time, this will dictate that the alternate method of determining system loading be used. Higher (or possibly lower) loadings may be appropriate in some applications. Some factors that may increase (or decrease) the capacity of a hunt group are listed below.

(a) Shorter (or longer) average message transmission time.

(b) The use of audio monitoring (listen-in) slow scan video or other similar equipment may significantly increase the connected time for a signal and reduce effective hunt group capacity.

(c) The clustering of active burglar alarm signals may generate high peak loads at certain hours.

(d) Inappropriate scheduling of 24-hour test signals may generate excessive peak loads.

Demonstration of a 90-percent probability of incoming line availability can be accomplished by the following in-service monitoring of line activity:

1. Incoming lines are assigned to telephone hunt groups. When a DACT calls the main number of a hunt group, it can connect to any currently available line in that hunt group.

2. The receiver continuously monitors the "available" status of each line. A line is available if it is waiting for an incoming call. A line is unavailable for any of the following reasons:

(i) Currently processing a call

(ii) Line in trouble

(iii) Audio monitoring (listen-in) in progress

(iv) Any other condition that makes the line input unable to accept calls.

3. The receiver monitors the "available" status of the hunt group. A hunt group is available if any line in it is available.

4. A message is emitted by the receiver if a hunt group is unavailable for more than 1 minute in 10. This message references the hunt group and the degree of overload.

The loading of a DACR is an important feature in the overall reliability of a digital alarm communicator system. Two options for determining loading capacity are available to system designers: using Table 4-2.3.2.2.2.3 or ensuring 90 percent availability. The latter option is usually used by a larger-capacity supervising station that employs a computer-based automation system to oversee the handling of signals. Such a system can monitor traffic to ensure that the necessary reliability is maintained as loading increases due to the addition of new customers.

4-2.3.2.2.2.4* A signal shall be received on each individual incoming DACR line at least once every 24 hours.

A-4-2.3.2.2.2.4 The verification of the 24-hour DACR line test should be done early enough in the day to allow repairs to be made by the telephone company.

Depending on the number of lines involved and the design and complexity of the particular hunt group arrangements, these tests may be performed automatically by the supervising station automation system, or the supervising station operators may initiate the test signals manually while sequentially creating a busy signal on each line in a hunt group.

4-2.3.2.2.2.5 The failure to receive a test signal from the protected premises shall be treated as a trouble signal. (*See 4-3.6.1.4.*)

The daily test signal serves to verify the end-to-end functioning of the system. It monitors the integrity of the system and guards against the catastrophic loss of both telephone lines connected to the DACT, as well as the malfunctioning of an entire hunt group at the DACR. In larger supervising stations, the automation system oversees the test signals. In smaller supervising stations, a manual logging system may be used to keep track of the test signals.

4-2.3.2.3 Digital Alarm Radio System (DARS).

Paragraph 4-2.3.2.1.6.1(a) permits a one-way radio system that complies with the requirements of 4-2.3.2.3 to be used as a secondary means of transmitting signals from a DACT.

4-2.3.2.3.1 In the event that any DACT signal transmission is unsuccessful, the information shall be transmitted by means of the digital alarm radio transmitter (DART). The DACT shall continue its normal transmission sequence as required by 4-2.3.2.1.5.

Exception: Simultaneous status change reporting by both the DACT and DART shall be permitted.

When a DARS is provided as the secondary transmission path for a DACT, the DACT must still continue to attempt to complete the call to the DACR.

4-2.3.2.3.2 Failure of the telephone line at the protected premises shall result in a trouble signal being transmitted to the supervising station by means of the DART within 4 minutes of detection of the fault.

4-2.3.2.3.3 The DARS shall be capable of demonstrating a minimum of 90 percent probability of successfully completing each transmission sequence.

To fulfill this requirement, radio propagation studies that would satisfy the specified 90 percent reliability factor must be completed. Such studies would be similar to those required for a one-way radio alarm system. See 4-2.3.5.2 and A-4-2.3.5.2.

4-2.3.2.3.4 Transmission sequences shall be repeated a minimum of five times. The DART transmission shall be permitted to be terminated in less than five sequences if the DACT successfully communicates to the DACR.

A sufficient number of attempts to complete the signal transmission must be made in order to ensure overall system reliability.

4-2.3.2.3.5 Each DART shall automatically initiate and complete a test signal transmission sequence to its associated digital alarm radio receiver (DARR) at least once every 24 hours. A successful DART signal transmission sequence of any other type within the same 24-hour period shall be considered sufficient to fulfill the requirement to test the integrity of the reporting system, if signal processing is automated so that 24-hour delinquencies must be individually acknowledged by supervising station personnel.

When a DACT is connected to a single telephone line as the primary transmission path and to a DARS as the secondary transmission path, it must conduct a test at least once every 24 hours for each transmission path.

4-2.3.2.4 Digital Alarm Radio Transmitter (DART). A DART shall transmit a digital code or equivalent by use of radio transmission to its associated digital alarm radio receiver (DARR). Signal repetition, digital parity check, or some equivalent means of signal verification shall be used. The DART shall comply with applicable FCC rules consistent with its operating frequency.

This requirement ensures that digital information is used to communicate the status of the fire alarm system at the protected premises. It precludes the use of some type of voice information transmission.

4-2.3.2.5 Digital Alarm Radio Receiver (DARR).

4-2.3.2.5.1 Equipment.

4-2.3.2.5.1.1 A spare DARR shall be provided in the supervising station and shall be able to be switched in place of a failed unit within 30 seconds after detection of failure.

4-2.3.2.5.1.2 Facilities shall be provided at the supervising station for the following supervisory and control functions of subsidiary and repeater station radio receiving equipment. This shall be accomplished via a supervised circuit where the radio equipment is remotely located from the supervising or subsidiary station. The following conditions shall be supervised at the supervising station:

(a) Failure of ac power supplying the radio equipment
(b) Receiver malfunction
(c) Antenna and interconnecting cable malfunction
(d) Indication of automatic switchover of the DARR
(e) Data transmission line between the DARR and the supervising or subsidiary station.

Monitoring the integrity of these functions helps ensure overall system reliability. Where a large supervising station is equipped with a computer-based automation system, that system will perform most or all of these functions.

4-2.3.2.6 Derived Local Channel.

4-2.3.2.6.1 When a DACT is connected to a telephone line (number) that is also supervised for adverse conditions by derived local channel, a second telephone line (number) shall not be required.

Some public telephone companies offer so-called "cut line" detection to supervising station alarm system providers. This service uses derived local channel equipment to detect adverse conditions on a telephone line. When such a service is used, this subsection permits a DACS to operate with a single telephone line connected to each DACT.

4-2.3.2.6.2 Failure of the telephone line (number) at the protected premises shall be automatically indicated and recorded at the supervising station in accordance with 4-2.2.3.

This requirement ensures that if the derived local channel system detects an adverse condition, the signal will be passed on to the supervising station, rather than signal only at the public telephone company's wire center.

4-2.3.3 McCulloh Systems.

McCulloh systems are the oldest form of transmission between a protected premises and a supervising station. Coded transmitters at a protected premises are connected in series with transmitters at other protected premises and with receiving equipment at the supervising station. Continuous metallic circuit continuity must be maintained in order for the dc current to flow from the power supply at the supervising station, out over the series circuit, and through the coded contacts of the transmitters at the protected premises. Where the public telephone company does not wish to maintain circuits that will offer continuous metallic circuit continuity, an alternative exists (see 4-2.3.2.6) that converts the McCulloh-type system into a multiplex system between public telephone utility company wire centers.

Initiation of an alarm, supervisory, or trouble signal at the protected premises actuates the associated transmitter. As the code wheel of the actuated transmitter turns, it alternately breaks the circuit and connects the circuit to earth ground. Under normal circumstances, the breaking of the circuit operates receiving equipment that records the coded pulses at the supervising station. These pulses are converted, either manually or by a computer-based automation system at the supervising station, to information that gives the location of the protected premises. If the circuit between the protected premises and the supervising station is impaired by a single open fault or single ground fault, then the signal produced by the turning of the code wheel transmits through earth ground.

4-2.3.3.1 Transmitters.

4-2.3.3.1.1 A coded alarm signal from a transmitter shall consist of not less than three complete rounds of the number or code transmitted.

4-2.3.3.1.2* A coded fire alarm box shall produce not less than three signal impulses for each revolution of the coded signal wheel or equivalent device.

A-4-2.3.3.1.2 The following suggested coded signal assignments for a building having four floors and basements are provided as a guide:

Location	Coded Signal
4th Floor	2-4
3rd Floor	2-3
2nd Floor	2-2
1st Floor	2-1
Basement	3-1
Sub-Basement	3-2

4-2.3.3.1.3 Circuit-adjusting means for emergency operating shall be permitted to either be automatic or be provided through manual operation upon receipt of a trouble signal.

Original McCulloh supervising stations required operator action to condition a circuit impaired by either an open fault or a ground fault to receive subsequent signals. As these stations have become equipped with computer-based automation systems, the conditioning of the circuits is now often done automatically by the interface equipment.

4-2.3.3.1.4 Equipment shall be provided at the supervising or subsidiary station on all circuits extending from the supervising or subsidiary station utilized for McCulloh systems for making the following tests:

(a) Current on each circuit under normal conditions
(b) Current on each side of the circuit with the receiving equipment conditioned for an open circuit.

NOTE: The current readings in test (a) above should be compared with the normal readings to determine if a change in the circuit condition has occurred. A zero current reading in test (b) above indicates that the circuit is clear of a foreign ground.

By taking a current reading on each McCulloh circuit, it is sometimes possible to detect a strap across the circuit (short circuit), as long as the strap is not at the protected premises. If there is sufficient resistance between the location of the strap and the protected premises, the loss of that resistance when the strap is applied will result in an increase in current that should be apparent to the operator taking the reading. In most cases a strap alone is not enough to disable the transmission of signals, since the McCulloh transmitter will also connect the circuit to ground with each pulse of the code wheel. If the person placing the strap across the circuit also disconnects the circuit beyond the strap, then the McCulloh transmitters located beyond the strap and the open circuit fault will not be able to signal to the supervising station.

A foreign ground — an unintentional connection of the circuit to ground — can adversely affect the circuit's ability to transmit a signal. These current readings help detect the presence of foreign grounds so that they may be located and cleared. Typical foreign grounds include such things as a tree branch rubbing against an aerial portion of the circuit or water filling a conduit containing a portion of the circuit where the insulation has degraded.

4-2.3.3.2 **Transmission Channels.**

4-2.3.3.2.1 Circuits between the protected premises and the supervising or subsidiary station that are essential to the actuation or operation of devices initiating a signal indicative of fire shall be so arranged that the occurrence of a single break or single ground fault will not prevent transmission of an alarm.

Exception No. 1: Circuits wholly within the supervising or subsidiary station.

Exception No. 2: Carrier system portion of circuits.

An important feature of a McCulloh system is its ability to continue to function even with a single open fault or a single ground fault on the circuit. Upon receipt at the supervising station of a trouble signal that indicates a single open fault or single ground fault, the receiving equipment is conditioned automatically or manually to receive any signals transmitted on that circuit through earth ground. Exceptions No. 1 and No. 2 exclude this requirement from circuits that are completely within the supervising station and from the portion of the circuit described in 4-2.3.3.2.6 that does not have continuous metallic continuity.

4-2.3.3.2.2 The occurrence of a single break or a single ground fault on any circuit shall not of itself cause a false signal that may be interpreted as an alarm of fire. Where such single fault prevents the normal functioning of any circuit, its occurrence shall be indicated automatically at the supervising station by a trouble signal compelling attention and readily distinguishable from signals other than those indicative of an abnormal condition of supervised parts of a fire suppression system.

This classic requirement dictates that the signals produced by a single open fault or a single ground fault on any circuit associated with the McCulloh system must not produce a false fire alarm signal. It also requires that such faults produce a trouble signal. While 4-2.3.3.2.2 seems to permit a trouble signal to indicate both a fault and a fire extinguishing system supervisory off-normal condition, paragraph 1-5.4.7(b) requires a distinct supervisory off-normal signal. Thus, the permission to combine the trouble and supervisory off-normal signals is limited to McCulloh systems.

4-2.3.3.2.3 The circuits and devices shall be arranged to receive and record a signal readily identifiable as to location of origin, and provisions shall be made for equally identifiable transmission to the public fire service communication center.

4-2.3.3.2.4 Multipoint transmission channels between the protected premises and the supervising or subsidiary station and within the protected premises, consisting of one or more coded transmitters and associated system unit(s), shall meet the requirements of either 4-2.3.3.2.5 or 4-2.3.3.2.6.

4-2.3.3.2.5 When end-to-end metallic continuity is present, proper signals shall be received from other points under any one of the following transmission channel fault conditions at one point on the line:

(a) Open
(b) Ground
(c)* Wire-to-wire short

A-4-2.3.3.2.5(c) Though rare, it is understood that the occurrence of a wire-to-wire short on the primary trunk facility near the supervising station could disable the transmission system without immediate detection.

(d) Open and ground.

The traditional McCulloh system does have end-to-end metallic continuity. Most often the circuit between the protected premises and the supervising station is leased from the public telephone company, which does not supply any power for the circuit, but simply provides a pair of wires. The telephone company usually refers to such a circuit as a "PL circuit" (private line circuit). The fact that end-to-end metallic continuity must be maintained somewhat limits the electrical distance between the protected premises and the supervising station, because the supervising station is supplying the power to operate the system.

4-2.3.3.2.6 When end-to-end metallic continuity is not present, the nonmetallic portion of transmission channels shall meet all of the following requirements:

(a) Two nonmetallic channels or one channel plus a means for immediate transfer to a standby channel shall be provided for each transmission channel, a maximum of eight transmission channels being associated with each standby channel, or over one channel, provided service is limited to one plant.

(b) The two nonmetallic channels (or one channel with standby arrangement) for each transmission channel shall be provided in one of the following ways, in descending order of preference:

1. Over separate facilities and separate routes
2. Over separate facilities in the same route
3. Over the same facilities in the same route.

(c) Failure of a nonmetallic channel or any portion thereof shall be indicated immediately and automatically in the supervising station.

(d) Proper signals shall be received from other points under any one of the following fault conditions at one point on the metallic portion of the transmission channel:

1. Open
2. Ground
3.* Wire-to-wire short.

As public telephone companies have moved away from communications technology that uses end-to-end metallic continuity, the availability of PL circuits has significantly diminished. When such circuits were scheduled to be eliminated between certain telephone company wire centers, the utility would sometimes provide an alternative with features described in this subsection. Also, some alarm service providers use this method to transport signals from a somewhat remote area to the supervising station.

A-4-2.3.3.2.6(d)(3) Though rare, it is understood that the occurrence of a wire-to-wire short on the primary trunk facility near the supervising station could disable the transmission system without immediate detection.

4-2.3.3.3 Loading Capacity of McCulloh Circuits.

The loading capacities discussed in this subsection have been part of the NFPA signaling standards for well over 60 years. The capacities are based on a concept of limiting the number of signals that might be lost under various adverse conditions, including a clash of simultaneous signals coming from two or more protected premises on the same McCulloh circuit. Virtually every loading requirement for McCulloh systems contained in the *National Fire Alarm Code* has its root in these numbers.

4-2.3.3.3.1 The number of transmitters connected to any transmission channel shall be limited to avoid interference. The total number of code wheels or equivalent connected to a single transmission channel shall not exceed 250. Alarm signal transmission channels shall be reserved exclusively for fire alarm signal transmitting service, except as provided in 4-2.3.3.3.4.

4-2.3.3.3.2 The number of waterflow switches permitted to be connected to actuate a single transmitter shall not exceed five switches.

4-2.3.3.3.3 The number of supervisory switches permitted to be connected to actuate a single transmitter shall not exceed 20.

4-2.3.3.3.4 Combined alarm and supervisory transmission channels shall comply with the following:

(a) Where both sprinkler supervisory signals and fire or waterflow alarm signals are transmitted over the same transmission channel, provision shall be made to obtain either alarm signal precedence or sufficient repetition of the alarm signal to prevent the loss of any alarm signal.

(b) Other signal transmitters (burglar, industrial processes, etc.) on an alarm transmission channel shall not exceed five.

4-2.3.3.3.5* Where signals from manual fire alarm boxes and waterflow alarm transmitters within a building are transmitted over the same transmission channel and are operating at the same time, there shall be no interference with the fire box signals. Provision of the shunt noninterfering method of operation is acceptable for this performance.

With a McCulloh coded-type manual fire alarm box connected electrically first on the McCulloh circuit, operation of the box places a short circuit or shunt across the McCulloh circuit while disconnecting the McCulloh transmitters that are electrically downstream from the manual fire alarm box. This effectively, if somewhat crudely, prevents another transmitter from interfering with the signal produced by the manual box.

A-4-2.3.3.3.5 Verify by test at time of system acceptance.

4-2.3.3.3.6 One alarm transmission channel shall serve not more than 25 plants. A plant may consist of one or more buildings under the same ownership, and the circuit arrangement shall be such that an alarm signal will not be received from more than one transmitter within a plant at a time. If such noninterfering is not provided, each building shall be considered a plant.

The routing of the McCulloh circuit throughout a large, multiple-building facility could have a significant effect on the loading of the circuit. If a noninterfering shunt arrangement was employed and the circuit was routed throughout the facility, preferably from the most important building to the least important building, that circuit could serve more buildings than if such a level of care was not excercised in the routing of the circuit.

4-2.3.3.3.7 One sprinkler supervisory transmission channel circuit shall serve not more than 25 plants. A plant may consist of one or more buildings under the same ownership.

4-2.3.3.3.8 Connections to a guard supervisory transmission channel or to a combination manual fire alarm and guard transmission channel shall be limited so that not more than 60 scheduled guard report signals will be transmitted in any 1-hour period. Patrol scheduling shall be such as to avoid interference between guard report signals.

When this requirement was incorporated into the NFPA signaling standards, the recording of guard supervisory signals was largely done manually. It was determined that the maximum signal traffic an operator reasonably could be expected to handle was one signal per minute, or sixty signals per hour.

4-2.3.4 Two-Way RF Multiplex Systems.

A two-way RF multiplex system is a traditional multiplex fire alarm system that uses a licensed two-way radio system to transmit signals from the protected premises to the supervising station. Essentially, the radio portion of the fire alarm system is transparent to the operation of the supervising station fire alarm system. In this regard, the operational requirements for two-way RF multiplex systems are virtually identical to those for active multiplex systems.

4-2.3.4.1 The maximum end-to-end operating time parameters allowed for a two-way RF multiplex system are as follows:

(a) The maximum allowable time lapse from the initiation of a single fire alarm signal until it is recorded at the supervising station shall not exceed 90 seconds. When any number of subsequent fire alarm signals occur at any rate, they shall be recorded at a rate no slower than one every additional 10 seconds.

Paragraph 4-2.3.4.1(a) effectively ensures that an interrogation and response sequence will be completed at least every 90 seconds, unless some other means has been provided to ensure alarm receipt in that time frame. For example, a system could be devised where alarm signals would be transmitted by the equipment at the protected premises at a time other than during the normal interrogation and response sequence. However, all currently listed systems meet this requirement by completing the interrogation and response sequence within the 90-second time frame.

(b) The maximum allowable time lapse from the occurrence of an adverse condition in any transmission channel until recording of the adverse condition is started shall not exceed 90 seconds for Type 4 and Type 5 systems. (*See 4-2.3.4.4.*)

As in 4-2.3.4.1(a), 4-2.3.4.1(b) also effectively ensures that an interrogation and response sequence will be completed at least every 90 seconds.

(c) In addition to the maximum operating time allowed for fire alarm signals, the requirements of one of the following paragraphs shall be met:

1. System units having more than 500 initiating device circuits shall be able to record not less than 50 simultaneous status changes in 90 seconds.
2. System units having fewer than 500 initiating device circuits shall be able to record not less than 10 percent of that total number of simultaneous status changes within 90 seconds.

These requirements ensure that the portion of the multiplex system that processes and records status changes can do so with sufficient speed to handle a reasonable volume of signal traffic.

4-2.3.4.2 Facilities shall be provided at the supervising station for the following supervisory and control functions of the supervising or subsidiary station, and repeater station radio transmitting and receiving equipment. This shall be accomplished via a supervised circuit where the radio equipment is remotely located from the system unit.

(a) The following conditions shall be supervised at the supervising station:

1. RF transmitter in use (radiating)
2. Failure of ac power supplying the radio equipment
3. RF receiver malfunction
4. Indication of automatic switchover.

(b) Independent deactivation of either RF transmitter shall be controlled from the supervising station.

These supervisory functions help ensure the continuity of signal transmission between the protected premises and the supervising station. In addition, one must remember that an interrogation and response sequence is also being transmitted and received between the protected premises and the supervising station every 90 seconds.

4-2.3.4.3 Transmission Channel.

4-2.3.4.3.1 The RF multiplex transmission channel shall terminate in a RF transmitter/receiver at the protected premises and in a system unit at the supervising or subsidiary station.

With this system, each protected premises will have its own RF transmitter/receiver unit. The supervising station also has an RF transmitter/receiver unit. The interrogation and response sequence takes place between these units. This system is similar to an active multiplex system where each protected premises has a transponder, and the supervising station has an active multiplex system control unit.

4-2.3.4.3.2 Operation of the transmission channel shall conform to the requirements of this code whether channels are private facilities, such as microwave, or leased facilities furnished by a communication utility company. When private signal transmission facilities are utilized, the equipment necessary to transmit signals shall also comply with requirements for duplicate equipment or replacement of critical components, as described in 4-2.4.2.

This requirement ensures that the system will comply with the requirements of the code, even if the facilities are leased from a communication utility company. It further ensures continuity of operations by requiring that critical assemblies either be redundant or be able to be replaced with on-premises spares, and that service be restored within 30 minutes.

4-2.3.4.4* Two-way RF multiplex systems are divided into two categories based upon their ability to perform under adverse conditions. System classifications are of two types.

(a) A Type 4 system shall have two or more control sites configured as follows:

1. Each site shall have a RF receiver interconnected to the supervising or subsidiary station by a separate channel.

2. The RF transmitter/receiver located at the protected premises shall be within transmission range of at least two RF receiving sites.

3. The system shall contain two RF transmitters, either:

(i) Located at one site with the capability of interrogating all of the RF transmitters/receivers on the premises, or

(ii) Dispersed with all of the RF transmitters/receivers on the premises having the capability to be interrogated by two different RF transmitters.

4. Each RF transmitter shall maintain a status that permits immediate use at all times. Facilities shall be provided in the supervising or subsidiary station to operate any off-line RF transmitter at least once every 8 hours.

5. Any failure of one of the RF receivers shall in no way interfere with the operation of the system from the other RF receiver. Failure of any receiver shall be annunciated at the supervising station.

6. A physically separate channel is required between each RF transmitter or RF receiver site, or both, and the system unit.

These requirements essentially create a two-way RF multiplex system that has redundancy of critical components. Such a system would normally be used where a rather high

volume of traffic or unusual transient RF propagation problems were anticipated.

(b) A Type 5 system shall have a single control site configured as follows:

1. A minimum of one RF receiving site.
2. A minimum of one RF transmitting site.

NOTE: The sites above can be co-located.

A-4-2.3.4.4 The intent of the plurality of control sites is to safeguard against damage caused by lightning and to minimize the effect of interference on the receipt of signals.

This appendix item refers to the plurality of control sites (each of which contains a transmitter/receiver unit) of a Type 4 two-way RF multiplex system.

4-2.3.4.5 Loading Capacities.

4-2.3.4.5.1 The loading capacities of two-way RF multiplex systems are based on the overall reliability of the signal receiving, processing, display, and recording equipment at the supervising or subsidiary station and the capability to transmit signals during adverse conditions of the transmission channels. Table 4-2.3.4.5.1 establishes the allowable loading capacities.

The loading of a two-way RF multiplex system depends on the capability of the type of system. Since a Type 4 system has a redundant transmitter/receiver at different locations exerting control over the interrogation and response sequence between the protected premises and the supervising station, it has the greatest permitted system loading. A Type 5 system does not have dual transmitters/receivers in control of the system, so its trunk capacity is more limited than that of a Type 4 system.

4-2.3.4.5.2 Exceptions to Loading Capacities Listed in Table 4-2.3.4.5.1. Where the signal receiving, processing, display, and recording equipment is duplicated at the supervising station and a switch-over can be accomplished in not more than 30 seconds with no loss of signals during this period, the capacity of a system unit shall be unlimited.

Part B of Table 4-2.3.4.5.1 is modified by this requirement. However, to meet this requirement, a two-way RF

Table 4-2.3.4.5.1

	System Type	
	Type 4	Type 5
A. Trunks		
Maximum number of fire alarm service initiating device circuits per primary trunk facility	5120	1280
Maximum number of leg facilities for fire alarm service per primary trunk facility	512	128
Maximum number of leg facilities for all types of fire alarm service per secondary trunk facility*	128	128
Maximum number of all types of initiating device circuits per primary trunk facility in any combination	10,240	2560
Maximum number of leg facilities for types of fire alarm service per primary trunk facility in any combination*	1024	256
B. System Units at the Supervising Station		
Maximum number of all types of initiating device circuits per system unit*	10,240**	10,240**
Maximum number of fire protected buildings and premises per system unit	512**	512**
Maximum number of fire alarm service initiating device circuits per system	5120**	5120**
C. Systems Emitting from Subsidiary Station	Same as B	Same as B

*Includes every initiating device circuit, i.e., waterflow, fire alarm supervisory, guard, burglary, hold up, etc.

**Paragraph 4-2.3.4.5.2 applies.

multiplex system would have to employ complete redundancy of all critical components and be able to complete a switch-over in 30 seconds with no loss of signals. Those systems that meet this requirement generally process all incoming signals in tandem. That is, the standby unit is fully functioning at all times and simply continues to function normally if the main unit fails. The only actual "change-over" would occur for those incidental peripheral devices for which redundancy was not required.

4-2.3.5 One-Way Private Radio Alarm Systems.

In creating an RF transmission system without an interrogation and response sequence to monitor the integrity of the transmission of signals between the protected premises and the supervising station, the requirements for digital alarm communicator systems were adapted heavily.

4-2.3.5.1 The requirements of this section for a radio alarm repeater station receiver (RARSR) shall be satisfied if signals from each radio alarm transmitter (RAT) are received and supervised, in accordance with this chapter, by at least two independently powered, independently operating, and separately located RARSR.

The one-way radio alarm system consists of an RF transmitter at the protected premises that is connected to the protected premises control unit and that can transmit alarm, supervisory, and trouble signals to at least two receivers. The receivers relay the received signal to the supervising station by RF or wired transmission means. This system allows the use of either a private system operated by a single alarm service provider or a multi-user system operated by a one-way radio network provider. Most systems in use are of the latter type.

4-2.3.5.2* The end-to-end operating time parameters allowed for a one-way radio alarm system shall be as follows:

(a) There shall be a 90 percent probability that the time between the initiation of a single fire alarm signal until it is recorded at the supervising station shall not exceed 90 seconds.

(b) There shall be a 99 percent probability that the time between the initiation of a single fire alarm signal until it is recorded at the supervising station shall not exceed 180 seconds.

(c) There shall be a 99.999 percent probability that the time between the initiation of a single fire alarm signal until it is recorded at the supervising station shall not exceed 7.5 minutes (450 seconds), at which time the RAT shall cease transmitting.

When any number of subsequent fire alarm signals occur at any rate, they shall be recorded at an average rate no slower than one every additional 10 seconds.

(d) In addition to the maximum operating time allowed for fire signals, the system shall be able to record not less than

12 simultaneous status changes within 90 seconds at the supervising station.

A-4-2.3.5.2 It is intended that each RAT communicate with two or more independently located RARSRs. The location of such RARSRs should be such that they do not share common facilities.

NOTE: All probability calculations required for the purposes of Chapter 4 should be made in accordance with established communications procedures, should assume the maximum channel loading parameters specified, and should further assume that 25 RATs are actively in alarm and are being received by each RARSR.

Because this system does not have an interrogation and response sequence to verify the operating capability of the communication channel and all equipment associated with it, it must rely on other means to achieve an acceptable level of operational integrity. The probabilities specified in 4-2.3.5.2 help ensure that level of integrity. To achieve the probabilities, the system functions similarly to a digital alarm communicator transmitter that makes a given number of attempts to connect to the digital alarm communicator receiver, after which it stops attempting to complete the call so as not to tie up the receiver.

4-2.3.5.3 Supervision.

4-2.3.5.3.1 Equipment shall be provided at the supervising station for the supervisory and control functions of the supervising or subsidiary station, and repeater station radio transmitting and receiving equipment. This shall be accomplished via a supervised circuit where the radio equipment is remotely located from the system unit. The following conditions shall be supervised at the supervising station:

(a) Failure of ac power supplying the radio equipment

(b) RF receiver malfunction

(c) Indication of automatic switchover (if applicable).

The specified supervisory functions help ensure the continuity of signal transmission between the protected premises and the supervising station.

4-2.3.5.3.2 Protected Premises.

4-2.3.5.3.2.1 Interconnections between elements of transmitting equipment, including any antennas, shall be supervised to either cause an indication of failure at the protected premises or transmit a trouble signal to the supervising station.

4-2.3.5.3.2.2 Where these elements are physically separated, the wiring or cabling between them shall be protected by conduit.

These two subsections address a serious point of potential failure of the one-way radio alarm system: loss of antenna or the loss of connection between the transmitter and the antenna. In some systems, the transmitter is connected directly to the antenna. In others, the antenna is located at a point in the building more advantageous for successful transmission of a signal. These requirements ensure that loss of antenna or its connection will at least be annunciated locally, and that the conductors to a remote antenna will be mechanically protected by conduit.

4-2.3.5.4 Transmission Channel.

4-2.3.5.4.1 The one-way RF transmission channel shall originate with a one-way RF transmitting device at the protected premises and shall terminate at the RF receiving system of a RARSR capable of receiving transmissions from such transmitting devices.

A receiving network transmission channel shall terminate at a RARSR at one end, and either with another RARSR or a radio alarm supervising station receiver (RASSR) at the other end.

This subsection permits the architecture necessary to develop a network to handle a large number of radio alarm transmitters. The network interconnections can use multiple RARSRs that in turn repeat the received signals to other RARSRs until the signals ultimately reach a RASSR. Along each segment of the transmission path, signals must always be received by at least two RARSRs.

4-2.3.5.4.2 Operation of receiving network transmission channels shall conform to the requirements of this code whether channels are private facilities, such as microwave, or leased facilities furnished by a communication utility company. Where private signal transmission facilities are utilized, the equipment necessary to transmit signals shall also comply with requirements for duplicate equipment or replacement of critical components as described in 4-2.4.2.

The system shall provide information indicating the quality of the received signal for each RARSR supervising each RAT in accordance with 4-2.3.5 and shall provide information at the supervising station if such signal quality falls below the minimum signal quality levels set forth in 4-2.3.5.

Each RAT shall be installed in such a manner so as to provide a signal quality over at least two independent one-way RF transmission channels, of the minimum quality level specified that satisfies the performance requirements in 4-2.2.2 and 4-2.4

This requirement ensures that the system will comply with the code, even if the facilities are leased from a communications utility company or other one-way radio network service provider. It further ensures continuity of operations by requiring either that critical assemblies be redundant, or that they can be replaced with on-premises spares and service can be restored within 30 minutes.

The system must also monitor the quality of the transmitted signal, including the various operating time parameters specified in 4-2.3.5.2. One way of doing this is to provide each RAT with a clock and include the time of first transmission and the current time in the signal along with the alarm, supervisory, or trouble data being transmitted.

4-2.3.5.5 Nonpublic one-way radio alarm systems shall be divided into two categories based upon the following number of RASSRs present in the system:

(a) A Type 6 system shall have one RASSR and at least two RARSRs.

(b) A Type 7 system shall have more than one RASSR and at least two RARSRs.

In a Type 7 system, if more than one RARSR is out of service and as a result any RATs are no longer being supervised, then the affected supervising station shall be notified.

In a Type 6 system, if any RARSR is out of service, a trouble signal shall be annunciated at the supervising station.

A pattern set previously for Type 1, 2, and 3 active multiplex systems and for Type 4 and 5 two-way RF multiplex systems is no longer valid, namely, that the most apparently capable type of system has the lower number in its type designation. Ironically, this makes little difference since the loading is the same for both types of one-way radio alarm systems.

4-2.3.5.6 The loading capacities of one-way radio alarm systems are based on the overall reliability of the signal receiving, processing, display, and recording equipment at the supervis-

ing or subsidiary station and the capability to transmit signals during adverse conditions of the transmission channels. Table 4-2.3.5.6 establishes the allowable loading capacities.

Table 4-2.3.5.6

	System Type	
	Type 6	Type 7
A. Radio Alarm Repeater Station Receiver (RARSR)		
Maximum number of fire alarm service initiating device circuits per RARSR	5120	5120
Maximum number of RATs for fire	512	512
Maximum number of all types of initiating device circuits per RARSR in any combination*	10,240	10,240
Maximum number of RATs for all types of fire alarm service per RARSR in any combination*†	1024	1024
B. System Units at the Supervising Station		
Maximum number of all types of initiating device circuits per system unit*	10,240**	10,240**
Maximum number of fire protected buildings and premises per system unit	512**	512**
Maximum number of fire alarm service initiating device circuits per system unit	5120**	5120**

*Includes every initiating device circuit, i.e., waterflow, fire alarm, supervisory, guard, burglary, hold-up, etc.

**Paragraph 4-2.3.5.7 applies.

†Each supervised BA (open/close) or each suppressed guard tour transmitter shall reduce the allowable RATs by 5.

Each guard tour transmitter shall reduce the allowable RATs by 15.

Each two-way protected premises radio transmitter shall reduce the allowable RATs by 2.

4-2.3.5.7 Exceptions to Loading Capacities Listed in Table 4-2.3.5.6. Where the signal receiving, processing, display, and recording equipment is duplicated at the supervising station and a switch-over can be accomplished in not more than 30 seconds with no loss of signals during this period, the capacity of a system unit is unlimited.

Part B of Table 4-2.3.5.6 is modified by this requirement. However, to meet this requirement, a one-way radio alarm system would have to employ complete redundancy of all critical components at the supervising station and be able to complete a switch-over in 30 seconds with no loss of signals. Those systems that meet this requirement generally process all incoming signals in tandem. That is, the standby unit is fully functioning at all times and simply continues to function normally if the main unit fails. The only actual "change-over" would occur for those incidental peripheral devices for which redundancy was not required.

4-2.3.6 Directly-Connected Noncoded Systems.

4-2.3.6.1 Circuits for transmission of alarm signals between the fire alarm control unit or the transmitter in the protected premises and the supervising station shall be arranged so as to comply with either of the following provisions:

(a) These circuits shall be arranged so that the occurrence of a single break or single ground fault will not prevent the transmission of an alarm signal. Circuits complying with this paragraph shall be automatically self-adjusting in the event of either a single break or a single ground fault and shall be automatically self-restoring in the event that the break or fault is corrected.

(b) These circuits shall be arranged so as to normally be isolated from ground (except for reference ground detection) and so that a single ground fault will not prevent the transmission of an alarm signal. Circuits complying with this paragraph shall be provided with a ground reference circuit so as to detect and indicate automatically the existence of a single ground fault, unless a multiple ground-fault condition that would prevent alarm operation will be indicated by an alarm or by a trouble signal.

The vast majority of the circuits employed for transmitting signals from a protected premises to a supervising station over directly connected noncoded systems meet the requirements of 4-2.3.6.1(b). Only one manufacturer offers a system that powers the circuit by means of a float-charged set of batteries with a center tap connection to earth ground. Two sets of alarm relays are connected across the circuit at the supervising station and reference to earth ground. Loss of one side of the circuit would not prevent the transmission of the signal over the remaining side of the circuit and earth ground.

In most cases, the circuits for directly-connected noncoded systems are leased from the public telephone company utility. [See commentary following A-4-2.3.3.2.5(c).] The fact that end-to-end metallic continuity must be maintained somewhat limits the electrical distance between the protected premises and the supervising station, because either the supervising station or the protected premises is supplying the power to operate the system.

4-2.3.6.2 Circuits for transmission of supervisory signals shall be separate from alarm circuits. These circuits within the protected premises and between the protected premises and the supervising station shall be arranged as described in 4-2.3.6.1(a) or 4-2.3.6.1(b).

Exception: Where the reception of alarm signals and supervisory signals at the same supervising station is permitted by the authority having jurisdiction, the supervisory signals do not interfere with the alarm signals, and alarm signals have priority, the same circuit between the protected premises and the supervising station shall be permitted to be used for alarm and supervisory signals.

4-2.3.6.3 The occurrence of a single break or a single ground fault on any circuit shall not of itself cause a false signal that may be interpreted as an alarm of fire.

4-2.3.6.4 The requirements of 4-2.3.6.1 and 4-2.3.6.2 shall not apply to the following circuits:

(a) Circuits wholly within the supervising station,
(b) Circuits wholly within the protected premises extending from one or more automatic fire detectors or other noncoded initiating devices other than water flow devices to a transmitter or control unit, or
(c) Power supply leads wholly within the building or buildings protected.

These requirements clarify that the named circuits need not have the operational capability of the circuits extending between the protected premises and the supervising station.

4-2.3.6.5 Loading Capacity of Circuits.

4-2.3.6.5.1 The number of initiating devices connected to any signaling circuit and the number of plants that shall be permitted to be served by a signal circuit shall be determined by the authority having jurisdiction and shall not exceed the limitations specified in 4-2.3.6.5.

NOTE: A plant may consist of one or more buildings under the same ownership.

4-2.3.6.5.2 A single circuit shall not serve more than one plant.

NOTE: Where a single plant involves more than one gate entrance or involves a number of buildings, separate circuits may be required so that the alarm to the supervising station will indicate the area to which the fire department should be dispatched.

One of the unique features of a directly-connected non-coded system is that it may serve only one plant.

4-2.3.7 **Private Microwave Radio Systems.**

4-2.3.7.1 Where a private microwave radio is used as the transmission channel, appropriate supervised transmitting and receiving equipment shall be provided at supervising, subsidiary, and repeater stations.

The requirements in 4-2.3.7.1 were originally developed in conjunction with AT&T and were based on standardized microwave relay link requirements to ensure normal network reliability. In most cases, a private microwave radio system would be used to transport signals from a subsidiary station to a supervising station, where a rather high volume of signal traffic is expected.

4-2.3.7.2 Where more than 5 protected buildings or premises or 50 initiating devices or initiating device circuits are being serviced by a private radio carrier, the supervising, subsidiary, and repeater station radio facilities shall meet all of the following:

(a) Dual supervised transmitters, arranged for automatic switching from one to the other in case of trouble, shall be installed. Where the transmitters are located where someone is always on duty, switchboard facilities shall be permitted to be manually operated if the switching can be carried out within 30 seconds. Where the transmitters are located where no one is normally on duty, the circuit extending between the supervising station and the transmitters shall be a supervised circuit.

(b)* Transmitters shall be operated on a two-to-one time ratio basis within each 24 hours.

A-4-2.3.7.2(b) Transmitters should be operated alternately, 16 hours on, 16 hours off.

(c) Dual receivers shall be installed with a means for selecting a usable output from one of the two receivers. The failure of one shall in no way interfere with the operation of the other. Failure of either receiver shall be annunciated.

4-2.3.7.3 Means shall be provided at the supervising station for the supervision and control of supervising, subsidiary, and repeater station radio transmitting and receiving equipment. This shall be accomplished via a supervised circuit when the radio equipment is remote from the supervising station.

(a) The following conditions shall be supervised at the supervising station:

1. Transmitter in use (radiating)
2. Failure of ac power supplying the radio equipment
3. Receiver malfunction
4. Indication of automatic switchover.

(b) It shall be possible to independently deactivate either transmitter from the supervising station.

4-2.4 **Display and Recording.**

The requirements in this subsection specify the content and nature of the display and recording of signals received from a protected premises at a supervising station. The requirements take into account a reasonable quantity of signal traffic, as well as certain ergonomic necessities for interfacing electronically reproduced signals with one or more human operators.

4-2.4.1* Any status changes that occur in an initiating device or in any interconnecting circuits or equipment from the location of the initiating device(s) to the supervising station shall be presented in a form to expedite prompt operator interpretation. Status change signals shall provide the following information:

(a) *Type of Signal.* Identification of the type of signal to show whether it is an alarm, supervisory, delinquency, or trouble signal.

(b) *Condition.* Identification of the signal to differentiate between an initiation of an alarm, supervisory, delinquency, or trouble signal, and a restoration to normal from one or more of these conditions.

(c) *Location.* Identification of the point of origin of each status change signal.

A-4-2.4.1   The signal information may be provided in coded form. Records may be used to interpret these codes.

4-2.4.2*   If duplicate equipment for signal receiving, processing, display, and recording is not provided, the installed equipment shall be so designed that any critical assembly can be replaced from on-premises spares and the system restored to service within 30 minutes. A critical assembly is one in which a malfunction will prevent the receipt and interpretation of signals by the supervising station operator.

Exception: Proprietary and remote station systems.

A-4-2.4.2   In order to expedite repairs, it is recommended that spare modules, such as printed circuit boards, CRT displays, printers, etc., be stocked at the supervising station.

This requirement ensures that a malfunction in a critical assembly, which is defined in 4-2.4.2, will be properly repaired, most often by being replaced by an on-premises spare. An assembly too complex to be so repaired would need to have a duplicate.

4-2.4.3*   Any method of recording and display or indication of change of status signals shall be permitted, providing all of the following conditions are met:

(a) Each change of status signal requiring action to be taken by the operator shall result in an audible signal and not less than two independent methods of identifying the type, condition, and location of the status change.

(b) Each change of status signal shall be automatically recorded. The record shall provide the type of signal, condition, and location as required by 4-2.4.1 in addition to the time and date the signal was received.

(c) Failure of an operator to acknowledge or act upon a change of status signal shall not prevent subsequent alarm signals from being received, indicated or displayed, and recorded.

(d) Change of status signals requiring action to be taken by the operator shall be displayed or indicated in a manner that clearly differentiates them from those that have been acted upon and acknowledged.

(e) Each incoming signal to a DACR or DARR shall cause an audible signal that persists until manually acknowledged.

Exception: Test signals (see 4-2.3.2.1.10) received at a DACR or DARR shall be permitted to be excepted from this requirement.

A-4-2.4.3   For all forms of transmission, the maximum time to process an alarm signal should be 90 seconds. The maximum time to process a supervisory signal should be 4 minutes. The time to process an alarm or supervisory signal is defined as that time from which a signal is received to the time that retransmission or subscriber contact is initiated.

When the level of traffic in a supervising station system reaches a magnitude such that delayed response is possible, even though the loading tables or loading formulas of this code are not exceeded, it is envisioned that it will be necessary to employ an enhanced method of processing.

For example, in a system where a single DACR instrument provided fire and burglar alarm service is connected to multiple telephone lines, it is conceivable that during certain periods of the day, fire alarm signals may be delayed by the security signaling traffic such as opening and closing signals. Such an enhanced system would be one that, upon receipt of signal would:

(a) Automatically process the signals, differentiating between those that require immediate response by supervising station personnel and those that need only be logged

(b) Automatically provide relevant subscriber information to assist supervising station personnel in their response

(c) Maintain a timed, unalterable log of the signals received and the response of supervising station personnel to such signals.

4-2.5   **Testing and Maintenance.** Testing and maintenance of communication methods shall be in accordance with the requirements of Chapter 7.

Section 4-2 covers the requirements for various transmission technologies. Starting with Section 4-3, Chapter 4 turns to stating the requirements for various supervising station fire alarm systems.

4-3 Fire Alarm Systems for Central Station Service.

NOTE: The requirement of Chapters 1, 3, 5, 6, 7, and Section 4-2 shall apply to central station fire alarm systems unless they conflict with the requirements of this section.

4-3.1 **Scope.** This section describes the general requirements and use of fire alarm systems to provide central station service.

4-3.2 **General.**

4-3.2.1 These systems include the central station physical plant, exterior communications channels, subsidiary stations, and signaling equipment located at the protected premises.

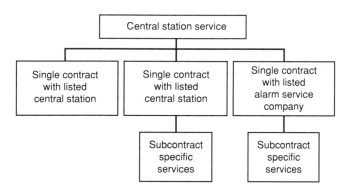

Figure 4.5 Subscriber contracts for central station services.

4-3.2.2* This section applies to central station service, which consists of the following elements: installation of fire alarm transmitters; alarm, guard, supervisory and trouble signal monitoring; retransmission; associated record keeping and reporting; testing and maintenance; and runner service. These services shall be provided under contract to a subscriber by one of the following:

(a) A listed central station that provides all of the elements of central station service with its own facilities and personnel.

(b) A listed central station that provides as a minimum the signal monitoring, retransmission, and associated record keeping and reporting with its own facilities and personnel and that may subcontract all or any part of the installation, testing and maintenance, and runner service.

(c) A listed fire alarm service — local company that provides the installation, and testing and maintenance with its own facilities and personnel and that subcontracts the monitoring, retransmission, and associated record keeping and reporting to a listed central station. The required runner service shall be provided by the listed fire alarm service — local company with its own personnel or the listed central station with its own personnel.

A-4-3.2.2 There are related types of contract service that often are provided from or controlled by a central station, but that are neither anticipated by nor consistent with the provisions of 4-3.2.2. Although 4-3.2.2 does not preclude such arrangements, a central station company is expected to recognize, provide for, and preserve the reliability, adequacy, and integrity of those supervisory and alarm services intended to be in accordance with the provisions of 4-3.2.2.

Central station service consists of eight distinct elements. Four of these exist at the protected premises (installation, testing, maintenance, and runner service) and four exist at the supervising station (management or operation of the system, monitoring of signals from the protected premises, retransmission of signals, and record keeping). This subsection provides for three ways of accomplishing central station service: a central station may provide all eight elements; a central station may provide the four elements at the supervising station and subcontract one or more of the four elements at the protected premises; a listed local company fire alarm service may provide the four elements at the protected premises and subcontract the four elements at the supervising station to a central station. In this latter arrangement, if the local company fire alarm service does not provide the runner service, then runner service must be provided by the central station.

Most central station service fits the second category, where the central station provides the four elements at the supervising station and subcontracts one or more of the elements at the protected premises. Most commonly, the central station subcontracts part of the installation — notably the installation of waterflow alarm and sprinkler supervisory devices, which are subcontracted to a sprinkler system contractor — while providing all other elements at the protected premises.

4-3.2.3 It shall be conspicuously indicated by the prime contractor that the fire alarm system providing service at a protected premises complies with all the requirements of this code by providing a means of third party verification, as specified in 4-3.2.3.1 or 4-3.2.3.2.

The code allows certification and placarding as the two ways that providers of central station service can verify that the service complies with all aspects of Section 4-3. Unless an authority having jurisdiction specifies one of these two methods, the prime contractor — either the central station or the listed local company fire alarm service — may choose the method of verification.

4-3.2.3.1 The installation shall be certificated.

4-3.2.3.1.1 Fire alarm systems providing service that complies with all requirements of this code shall be certified by the organization that has listed the central station, and a document attesting to this certification shall be located on or near the fire alarm system control unit or, if no control unit exists, on or near a fire alarm system component.

4-3.2.3.1.2 A central repository of issued certification documents, accessible to the authority having jurisdiction, shall be maintained by the organization that has listed the central station.

The certification system is operated by the organization that has listed the central station. The organization produces a document, or certificate, for each protected premises, which indicates compliance with the requirements of the code. Underwriters Laboratories Inc. (UL) has a central station fire alarm certificate program that meets the intent of this requirement. In this program, a central station applies to UL for issuance of a certificate for the protected premises. UL issues the certificate and annually inspects a statistically-significant sampling of certificated installations at each listed central station. UL maintains a computer-operated database of certificated installations. This database is available for inspection by authorities having jurisdiction by means of UL's certificate verification service (ULCVS).

4-3.2.3.2 The installation shall be placarded.

> **Formal Interpretation 89-1**
> Reference: 4-3.2.3.2
>
> *Question:* Is placarding (which is not defined as is certification) intended to be an independent method of verification by the central station with no third party agency being involved as assurance?
> *Answer:* No.
>
> *Issue Edition:* NFPA 71-1989
> *Reference:* 1-2.3.1
> *Issue Date:* June 22, 1992 ■

4-3.2.3.2.1 Fire alarm systems providing service that complies with all requirements of this code shall be conspicuously marked by the central station to indicate compliance. The marking shall be by one or more securely affixed placards that meet the requirements of the organization that has listed the central station and requires the placard.

4-3.2.3.2.2 The placard(s) shall be 20 sq in. (130 cm²) or larger, shall be located on or near the fire alarm system control unit or, if no control unit exists, on or near a fire alarm system component, and shall identify the central station by name and telephone number.

When placarding is chosen as the method to verify compliance, a placard is prepared stating compliance and is placed on or near the fire alarm system control unit or, if no control unit exists, on or near a significant component of the fire alarm system.

4-3.2.4* Fire alarm system service not complying with all requirements of Section 4-3 shall not be designated as central station service.

A-4-3.2.4 It is the responsibility of the prime contractor to remove all compliance markings (certification markings or placards) when a service contract goes into effect that conflicts in any way with the requirements of 4-3.2.4.

Many fire alarm systems installed at protected premises are connected to a remote location that monitors signals from those systems. Relatively few such arrangements meet the requirements of Section 4-3 and should not be called "central station service." Only service that incorporates all eight elements of central station service and is provided by listed alarm service providers who will design, specify, install, test, maintain, and use the system in accordance with the requirements of Section 4-3 should be called "central station service."

If a change is made that invalidates the designation "central station service," the prime contractor is responsible for removing the certificate or placard or other designation that the system is providing central station service.

4-3.2.5* For the purpose of Section 4-3, the subscriber shall notify the prime contractor in writing of the identity of the authority(ies) having jurisdiction.

A-4-3.2.5 The prime contractor should be aware of statutes, public agency regulations, or certifications regarding fire alarm systems that may be binding on the subscriber. The prime contractor should identify for the subscriber which agencies could be an authority having jurisdiction and, where possible, advise the subscriber of any requirements or approvals being mandated by these agencies.

The subscriber has the responsibility for notifying the prime contractor of those private organizations that are being designated as an authority having jurisdiction. The subscriber also has the responsibility to notify the prime contractor of changes in the authority having jurisdiction, such as where there is a change in insurance companies. Although the responsibility is primarily the subscriber's, the prime contractor should also take responsibility to seek out these "private" authorities having jurisdiction through the subscriber. The prime contractor has the responsibility for maintaining current records on the authority(ies) having jurisdiction for each protected premises.

The most prevalent public agency involved as an authority having jurisdiction with regard to fire alarm systems is the local fire department or fire prevention bureau. These are normally city or county agencies with statutory authority and may be required to approve fire alarm system installations. At the state level, the fire marshal's office would be most likely to serve as the public regulatory agency.

The most prevalent private organizations involved as authorities having jurisdiction are insurance companies. Others include insurance rating bureaus, insurance brokers and agents, and private consultants. It is important to note that these organizations have no statutory authority and become authorities having jurisdiction only when designated by the subscriber.

With both public and private concerns to satisfy, it is not uncommon to find multiple authorities having jurisdiction involved with a particular protected premises. It is necessary to identify all authorities having jurisdiction in order to obtain all the necessary approvals for a central station fire alarm system's installation.

Because the authority having jurisdiction plays such an important role in the provision of central station service, it is essential that all of the authorities having jurisdiction involved at a particular protected premises be identified. This responsibility rests with the subscriber. However, the subscriber would normally only be aware of any private authorities having jurisdiction involved at a particular property, while the prime contractor would be aware of any additional public authorities having jurisdiction over a particular property. Thus, the resolution of this important requirement must be a joint effort.

4-3.3 **Supervising Station Facilities.**

4-3.3.1 The central station building or that portion of a building occupied by a central station shall conform to the construction, fire protection, restricted access, emergency lighting, and power facilities requirements of the latest edition of ANSI/UL 827, *Central Stations for Watchman, Fire Alarm and Supervisory Service.*

4-3.3.2 Subsidiary station buildings or those portions of buildings occupied by subsidiary stations shall conform to the construction, fire protection, restricted access, emergency lighting, and power facilities requirements of the latest edition of ANSI/UL 827, *Central Stations for Watchman, Fire Alarm and Supervisory Service.*

This requirement and those that follow in 4-3.3.2.1 through 4-3.3.7.2 reflect the fact that a subsidiary station is an unstaffed location. Usually, a subsidiary station serves a particular geographic area, effectively concentrating signals from many protected premises and transmitting those concentrated signals on to a supervising station. A malfunction at a subsidiary station can substantially impair the successful transmission of signals from the properties it serves. Thus, these requirements help ensure the overall operational reliability of the subsidiary station and the transmission path between the subsidiary station and the supervising station.

4-3.3.2.1 All intrusion, fire, power, and environmental control systems for subsidiary station buildings shall be monitored by the central station in accordance with 4-3.3.

4-3.3.2.2 The subsidiary facility shall be inspected at least monthly by central station personnel for the purpose of verifying the operation of all supervised equipment, all telephones, battery conditions, and all fluid levels of batteries and generators.

4-3.3.2.3 In the event of the failure of equipment at the subsidiary station or the communication channel to the central station, a backup shall be operational within 90 seconds. Restoration of a failed unit shall be accomplished within 5 days.

Paragraph 4-3.3.2.3 requires redundant equipment that can be remotely placed into operation in the event of a failure.

4-3.3.2.4 There shall be continuous supervision of each communication channel between the subsidiary station and the central station.

4-3.3.2.5 When the communication channel between the subsidiary station and the supervising station fails, the communication shall be switched to an alternate path. Public switched telephone network facilities shall be used only as the alternate path.

Public telephone utility "dial up, make good" service initiated at the supervising station could be used to re-establish the communication channel between the subsidiary station and the supervising station.

4-3.3.2.6 In the subsidiary station, there shall be either a cellular telephone or an equivalent communication path that is independent of the telephone cable between the subsidiary station and the serving wire center.

This requirement ensures that service personnel will be able to establish communication with the supervising station upon arrival at a totally impaired subsidiary station.

4-3.3.2.7 A plan of action to provide for restoration of services specified by this code shall exist for each subsidiary station.

This requirement ensures that a written plan will be formulated for restoration. It is reasonable to expect the organization that lists a central station utilizing a subsidiary station to ask for review of such a plan.

4-3.3.2.7.1 This plan shall provide for restoration of services within 4 hours of any impairment causing loss of signals from the subsidiary station to the central station.

4-3.3.2.7.2 There shall be an exercise to demonstrate the adequacy of the plan at least once a year.

As with all emergency plans, it is important to test this plan's accuracy and validity. By performing this annual exercise, the personnel who will have to implement the plan during an emergency have the opportunity to become thoroughly familiar with exactly what is expected of them. Such an exercise also helps keep the plan up-to-date when changes occur at either the subsidiary station or the supervising station.

4-3.4 Equipment.

4-3.4.1 The central station and all subsidiary stations shall be so equipped to receive and record all signals in accordance with 4-2.4. Circuit-adjusting means for emergency operation shall be permitted to either be automatic or be provided through manual operation upon receipt of a trouble signal. Computer aided alarm and supervisory signal processing hardware and software shall be listed for the specific application.

Because the supervising station employs specially trained personnel (*see 4-3.5*), this subsection permits the circuit-adjusting means to be operated automatically or manually. It also requires computer aided alarm and supervisory signal processing hardware and software to be listed specifically for the processing of signals for central station service.

4-3.4.2 Power supplies shall comply with the requirements of Chapter 1.

4-3.4.3 Transmission means shall comply with the requirements of Section 4-2.

4-3.4.4* Two independent means shall be provided to retransmit a fire alarm signal to the appropriate public fire service communication center.

NOTE: The use of a universal emergency number 911 (public safety answering point) does not meet the intent of this code for the principal means of retransmission.

A-4-3.4.4 Two telephone lines at the central station connected to the public switched telephone network, each having its own telephone instrument connected, and two telephone numbers available at the public fire service communication center to which a central station operator may retransmit an alarm meets the intent of this requirement.

Retransmitting fire alarm signals to the public fire service communication center is a very important element of central station service. The central station must have a reasonably secure means to effect this communication. Because the central station may not be located in the same community as the protected premises, because most 911 and enhanced 911 emergency telephone systems are answered at a public service answering point (PSAP) that is staffed with personnel who are not a part of the public fire department, and because the vast majority of 911 calls are for nonfire emergencies, it seems prudent to avoid the bottleneck that sometimes occurs at the PSAP. This can be done by requiring the central station to dial the 7-digit fire-reporting number for the public fire service communication center serving the protected premises.

4-3.4.4.1 Where the principal means of retransmission is not equipped to permit the center to acknowledge receipt of each fire alarm report, both means shall be used to retransmit.

4-3.4.4.2* Where required by the authority having jurisdiction, one of the means shall be supervised so that interruption of retransmission circuit (channel) communication integrity will result in a trouble signal at the central station.

A-4-3.4.4.2 The following methods have been used successfully for supervising retransmission circuits (channels):

(a) An electrically supervised circuit (channel) provided with suitable code sending and automatic recording equipment.

(b) A supervised circuit (channel) providing suitable voice transmitting, receiving, and automatic recording equipment. The circuit may be a telephone circuit that:

1. Cannot be used for any other purpose;

2. Is provided with a two-way ring down feature for supervision between the fire department communications center and the central station;

3. Is provided with terminal equipment located on the premises at each end; and

4. Is provided with 24-hour standby power provided.

Exception: Local on-premises circuits need not be supervised.

(c) Radio facilities using transmissions over a supervised channel with supervised transmitting and receiving equipment. Circuit continuity ensured at intervals not exceeding 8 hours by any means is satisfactory.

When the public switched telephone network has proven to be unreliable or for some other reason, such as an extreme hazard to property or life existing at one or more protected premises, the authority having jurisdiction may require the central station to have a communication channel with each public fire service communication center to which it must retransmit fire alarm signals.

4-3.4.4.3 The retransmission means shall be tested in accordance with Chapter 7.

4-3.4.4.4 The retransmission signal and the time and date of retransmission shall be recorded at the central station.

While this requirement does not mandate that the time and date be recorded automatically, in most cases where computer-based automation systems are managing the receipt and retransmissions of signals, recording of the time and date is performed automatically.

In addition, based on a somewhat broad interpretation of 4-3.4.1, most central stations record the telephone call to the public fire service communication center. This tape recording, along with records of signals received at the central station, is often quite helpful to investigators when they attempt to reconstruct the sequence of events that occurred during a major fire.

4-3.5 Personnel.

4-3.5.1 The central station shall have sufficient personnel (a minimum of two persons) on duty at the central station at all times to ensure attention to signals received.

This paragraph, modified for this version of the code, requires that two operators be on duty at all times at the central station. The code previously permitted one of the operators to function as a runner, as long as he or she was in constant contact, most likely by radio, with the central station. By mandating that two operators be present, the code maximizes the likelihood that at least one of the operators will be fully alert to incoming signals.

4-3.5.1.1 Operation and supervision shall be the primary functions of the operators, and no other interest or activity shall take precedence over the protective service.

The code expects the operators to have no other duties that would distract from the prompt, effective handling of signals.

4-3.6 Operations.

4-3.6.1 Disposition of Signals.

4-3.6.1.1 Alarm signals initiated by manual fire alarm boxes, automatic fire detectors, waterflow from the automatic sprinkler system, or actuation of other fire suppression systems or equipment shall be treated as fire alarms.

The central station shall:

(a)* Immediately retransmit the alarm to the public fire service communication center

A-4-3.6.1.1(a) Use of the term "immediately" in this context is intended to mean "without unreasonable delay." Routine handling should take a maximum of 90 seconds from receipt of an alarm signal by the central station until the initiation of retransmission to the public fire service communication center.

(b) Dispatch a runner or technician to the protected premises to arrive within 1 hour after receipt of signal when equipment needs to be manually reset by the prime contractor

(c) Notify the subscriber by the quickest available method

(d) Provide notice to the subscriber and/or authority having jurisdiction, if required.

Exception: When the alarm signal results from a prearranged test, it is not necessary to take the actions required by (a) and (c).

The importance of promptly and accurately handling fire alarm signals cannot be overemphasized. Potentially, these signals may already be delayed up to 90 seconds by the particular transmission technology chosen for the system.

The runner or technician need respond only when the equipment at the protected premises must be reset manually by the prime contractor. If provisions can be made for the equipment to reset automatically, or the subscriber or some other trained individual is allowed to reset the equipment, then the runner or technician does not need to respond.

Telephone notification to the subscriber is generally the quickest available method. The "notice" mentioned in 4-3.6.1.1(d) is usually a written notice.

4-3.6.1.2 Upon failure to receive a guard's regular signal within a 15-minute maximum grace period, the central station shall:

(a) Communicate without unreasonable delay with personnel at the protected premises

(b) If communications cannot be established, dispatch a runner to the protected premises to arrive within 30 minutes of the delinquency

(c) Report all delinquencies to the subscriber and/or authority having jurisdiction, if required.

If the personnel at the protected premises cannot be contacted promptly, then the runner should be dispatched to investigate why the guard missed a signal. Potentially, the runner could arrive on site up to 45 minutes from the time when the signal would normally be received. There are cases in which a responding runner has found the guard injured or ill and, by summoning medical assistance, has saved the guard's life.

4-3.6.1.2.1 Failure of the guard to follow a prescribed route in transmitting signals shall be handled as a delinquency.

Guard's tour supervision by a central station mandates a compulsory tour arrangement, where the guard must follow a prescribed route. This can be done either by monitoring every reporting station along the route at the central station, or by an arrangement whereby the guard must proceed from station to station in a fixed sequence in order to eventually operate periodic master stations that transmit a signal to the central station.

4-3.6.1.3* Upon receipt of a supervisory signal from a sprinkler system, other fire suppression system, or other equipment, the central station shall:

(a)* Communicate immediately with person(s) designated by the subscriber

A-4-3.6.1.3(a) Use of the term "immediately" in this context is intended to mean "without unreasonable delay." Routine handling should take a maximum of 4 minutes from receipt of a supervisory signal by the central station until initiation of communication with person(s) designated by the subscriber.

(b) Dispatch a runner or maintenance person (arrival time not to exceed 1 hour) to investigate, unless abnormal condition is restored to normal in accordance with a scheduled procedure determined by (a) above

(c) Notify the fire department and/or law enforcement agency, if required

(d) Notify the authority having jurisdiction when sprinkler systems or other fire suppression systems or equipment have been wholly or partially out of service for 8 hours

(e) When service has been restored, provide notice, if required, to the subscriber and/or the authority having jurisdiction as to the nature of the signal, time of occurrence, and restoration of service when equipment has been out of service for 8 hours or more.

Exception: When the supervisory signal results from a prearranged test, it is not necessary to take the actions required by (a), (c), and (e).

A-4-3.6.1.3 It is anticipated that the central station will first attempt to notify designated personnel at the protected premises. When such notification cannot be made, it may be appropriate to notify law enforcement and/or the fire depart-

ment. For example, if a valve supervisory signal is received where protected premises are not occupied, it may be appropriate to notify police.

It is important to handle supervisory signals promptly and accurately. These signals may indicate that a vital protective system is out of service.

Only when the central station operator cannot resolve the supervisory signal by contacting designated personnel does the runner or technician need to respond.

The quickest available method to notify the subscriber is generally by telephone. The "notice" mentioned in 4-3.6.1.3(e) is generally a written notice.

4-3.6.1.4 Upon receipt of trouble signals or other signals pertaining solely to matters of equipment maintenance of the fire alarm systems, the central station shall:

(a)* Communicate immediately with persons designated by the subscriber

A-4-3.6.1.4(a) Use of the term "immediately" in this context is intended to mean "without unreasonable delay." Routine handling should take a maximum of 4 minutes from receipt of a trouble signal by the central station until initiation of the investigation by telephone.

(b) If necessary, dispatch personnel to arrive within 4 hours to initiate maintenance

(c) Provide notice, if required, to the subscriber and/or the authority having jurisdiction as to the nature of the interruption, time of occurrence, and restoration of service, when the interruption is more than 8 hours.

It is important to handle trouble signals promptly and accurately. These signals indicate that the fire alarm system itself is wholly or partly out of service.

The personnel dispatched to arrive within 4 hours must be capable of initiating repairs. This generally means that a technician, rather than a runner, must respond.

Telephone notification to the subscriber is generally the quickest available method. In 4-3.6.1.4(c), the "notice" is generally a written notice.

4-3.6.1.5 All test signals received shall be recorded to indicate date, time, and type.

(a) Test signals initiated by the subscriber, including those for the benefit of an authority having jurisdiction, shall be acknowledged by central station personnel whenever the subscriber or authority inquires.

(b)* Any test signal not received by the central station shall be investigated immediately and appropriate action taken to reestablish system integrity.

A-4-3.6.1.5(b) Use of the term "immediately" in this context is intended to mean "without unreasonable delay." Routine handling should take a maximum of 4 minutes from receipt of a trouble signal by the central station until initiation of the investigation by telephone.

(c) The central station shall dispatch personnel to arrive within 1 hour when protected premises equipment must be manually reset after testing.

It is crucial that test signals be handled promptly and accurately. These signals help ensure that the fire alarm system is functioning as it was intended to function. Likewise, it is important to cooperate with any authority having jurisdiction who inquires regarding test signals.

The runner or technician need respond only when personnel from the central station must reset equipment. If provisions can be made for the equipment to reset automatically, or the subscriber or some other trained individual is allowed to reset the equipment, then the runner or technician does not need to respond.

4-3.6.2 **Record Keeping and Reporting.**

4-3.6.2.1 Complete records of all signals received shall be retained for at least 1 year.

4-3.6.2.2 The central station shall make arrangements to furnish reports of signals received to the authority having jurisdiction in a form acceptable to it.

When an authority having jurisdiction requests reports from a central station, that authority has the right to ask for the report in a useful and useable form.

4-3.7 **Testing and Maintenance.** Testing and maintenance for central station service shall be performed in accordance with Chapter 7.

4-4 Proprietary Supervising Station Systems.

NOTE: The requirements of Chapters 1, 3, 5, 6, 7, and Section 4-2 shall apply to proprietary fire alarm systems, except where they conflict with the requirements of this section.

4-4.1 Scope. Section 4-4 describes the operational procedures for the supervising facilities of proprietary fire alarm systems. It provides the minimum requirements for the facilities, equipment, personnel, operation, and testing and maintenance of the proprietary supervising station.

4-4.2 General.

4-4.2.1 Proprietary supervising stations shall be located at the protected property and operated by trained, competent personnel in constant attendance who are responsible to the owner of the protected property. (See 4-4.5.3.)

The key to understanding a proprietary fire alarm system is to recognize it as a management tool that helps the owner of a protected property oversee the built-in fire protection systems in order to mitigate hazards at that facility.

4-4.2.2 The protected property shall be either a single property or noncontiguous properties under one ownership.

From a single proprietary supervising station, an owner may oversee the protection features at one or more of his or her properties. These properties need not be contiguously located.

4-4.2.3* Section 4-4 recognizes the interconnection of other systems to make the premises safer in the event of fire or other emergencies indicative of hazards to life or property.

In addition to the information contained in Section 4-4, refer to 3-8.15, Elevator Recall for Fire Fighters' Service, to Section 3-9, Fire Safety Control Functions, and to Section 3-10, Suppression System Actuation.

A-4-4.2.3 The following functions are in Appendix A to provide guidelines for utilizing building systems and equipment in addition to proprietary fire alarm equipment to provide life safety and property protection.

Building functions that may be initiated or controlled during a fire alarm condition include, but are not limited to, the following:

(a) Elevator operation consistent with ANSI A17.1, *Safety Code for Elevators, Dumbwaiters, Escalators, and Moving Walks.*

(b) Unlocking stairwell and exit doors. Refer to NFPA 80, *Standard for Fire Doors and Fire Windows*, and NFPA *101, Life Safety Code.*

(c) Release of fire and smoke dampers. Refer to NFPA 90A, *Standard for the Installation of Air Conditioning and Ventilating Systems*, and NFPA 90B, *Standard for the Installation of Warm Air Heating and Air Conditioning Systems.*

(d) Monitoring and initiating of self-contained automatic fire extinguishing systems and equipment. Refer to NFPA 11, *Standard for Low Expansion Foam and Combined Agent Systems*; NFPA 11A, *Standard for Medium- and High-Expansion Foam Systems*; NFPA 12, *Standard on Carbon Dioxide Extinguishing Systems*; NFPA 12A, *Standard on Halon 1301 Fire Extinguishing Systems*; NFPA 12B, *Standard on Halon 1211 Fire Extinguishing Systems*; NFPA 13, *Standard for the Installation of Sprinkler Systems*; NFPA 14, *Standard for the Installation of Standpipe and Hose Systems*; NFPA 15, *Standard for Water Spray Fixed Systems for Fire Protection*; and NFPA 17, *Standard for Dry Chemical Extinguishing Systems.*

(e) Lighting control necessary to provide essential illumination during fire alarm conditions. Refer to NFPA 70, *National Electrical Code*, and NFPA *101, Life Safety Code.*

(f) Emergency shutoff of hazardous gas.

(g) Control of building environmental heating, ventilating, and air conditioning equipment to provide smoke control. Refer to NFPA 90A, *Standard for the Installation of Air Conditioning and Ventilating Systems.*

(h) Control of process, data processing, and similar equipment as necessary during fire alarm conditions.

4-4.3 Supervising Station Facilities.

4-4.3.1 The proprietary supervising station shall be located in a fire-resistive, detached building or in a suitable cut-off room and shall not be near or exposed to the hazardous parts of the premises protected.

The requirements of 4-4.3.1 are intended to maintain a high degree of physical integrity for the supervising station.

4-4.3.2 Access to the proprietary supervising station shall be restricted to those persons directly concerned with the implementation and direction of emergency action and procedure.

It is important that the supervising station not become a congregating place for guards, fire fighters, or other personnel, whose presence may interfere with the operators giving proper attention to signal traffic. If the supervising station is part of a guard house where vehicles and personnel are admitted to the premises, it is important to provide some segregation to ensure that operators will be able to effectively and efficiently handle the signal traffic.

4-4.3.3 The proprietary supervising station, as well as remotely located power rooms for batteries or engine-driven generators, shall be provided with portable fire extinguishers that comply with the requirements of NFPA 10, *Standard for Portable Fire Extinguishers.*

It is important for personnel to have the means to handle a small fire in the supervising station or in the power rooms for batteries or for engine-driven generators. It is important to reference the provisions of NFPA 600, *Standard on Industrial Fire Brigades*, to ensure that personnel are properly organized and trained to safely use the fire extinguishers provided.

4-4.3.4 The proprietary supervising station shall be provided with an automatic emergency lighting system. The emergency source shall be independent of the primary lighting source.

4-4.3.5 Where 25 or more protected buildings or premises are connected to a subsidiary station, both of the following shall be provided at the subsidiary station:

(a) Automatic means for receiving and recording signals under emergency-staffing conditions
(b) A telephone.

When a proprietary supervising station receives signals from a very large premises or from several noncontiguous premises, it may be necessary to have a subsidiary station

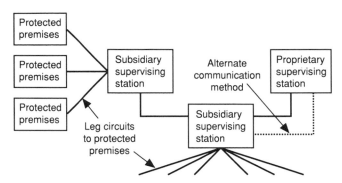

Figure 4.6 Proprietary supervising station facilities — subsidiary stations.

receive some signals and then pass the signals on to the supervising station. The requirements of 4-4.3.5 ensure that, if necessary, the subsidiary station can be staffed and operated independently of the supervising station.

4-4.4 Equipment.

4-4.4.1 This section shall apply to signal-receiving equipment in a proprietary supervising station.

4-4.4.2 Provision shall be made to designate the building in which a signal originates. The floor, section, or other subdivision of the building shall be designated at the proprietary supervising station or at the building protected, except that the authority having jurisdiction shall be permitted to waive this detailed designation where the area, height, or special conditions of occupancy make it unessential. This detailed designation shall utilize indicating appliances acceptable to the authority having jurisdiction.

In order to effectively manage the built-in fire protection features of the protected premises, the information received by the proprietary fire alarm system must have sufficient detail. This subsection requires that the information contain such detail, subject to possible waiver by the authority having jurisdiction.

4-4.4.3 The proprietary supervising station shall have, in addition to a recording device, two different means for alerting the operator when each signal is received indicating a change of state of any connected initiating device circuit. One of these shall be an audible signal and shall persist until manually acknowledged.

This shall include the receipt of alarm signals, supervisory signals, and trouble signals including signals indicating restoration to normal.

The requirements of 4-4.4.3 recognize that the operators in the supervising station may have other duties. In order to alert the operators to incoming signals, two means of notification are required.

4-4.4.4 Where suitable means is provided in the proprietary supervising station to readily identify the type of signal received, a common audible indicating appliance shall be permitted to be used for alarm, supervisory, and trouble indication.

Subsection 1-5.4.7 requires distinctive signals. The requirements of 4-4.4.4 modify this requirement by permitting a common audible indicating appliance in the supervising station as long as the type of signal can be otherwise readily identifiable.

4-4.4.5 At a proprietary supervising station, an audible trouble signal shall be permitted to be silenced provided the act of silencing it shall not prevent it from operating immediately upon receipt of a subsequent trouble signal.

4-4.4.6 All signals received by the proprietary supervising station that show a change in status shall be automatically and permanently recorded, including time and date of occurrence. This record shall be in a form that will expedite operator interpretation in accordance with any one of the following:

(a) In the event that a visual display is used that automatically provides change of status information for each individual signal, including type and location of occurrence, any form of automatic permanent visual record shall be acceptable. The recorded information shall include the content described above. The visual display shall show status information content at all times and shall be distinctly different after the operator has manually acknowledged each signal. Acknowledgment shall cause recorded information indicating time and date of acknowledgment.

This type of visual display is essentially an annunciator that continually shows the status of every point in the system that generates a signal. At a glance, the operator can see the status of every point. This permits any type of permanent visual record to be used, most often a logging-type printer.

(b) In the event that a visual display is not provided, signal content information shall be automatically recorded on duplicate permanent visual recording instruments.

One recording instrument shall be used for recording all incoming signals, while the other shall be used for fire, supervisory, and trouble signals only. Failure to acknowledge a signal shall not prevent subsequent signals from recording. Restoration of the signaling device to its prior or normal condition shall be recorded.

Two printers may be used to satisfy the requirements of 4-4.4.6(b). The printers should be arranged as described above — one printer to record all signals, the other to record only fire alarm, supervisory, and trouble signals. Since the operator will need to read the output of these printers in order to take action, the output should be formatted for ease of use.

(c) In the event that a system combines the use of a sequential visual display and recorded permanent visual presentation, the signal content information shall be displayed and recorded. The visual information component shall be either retained on the display until manual acknowledgment or periodically repeated at intervals not greater than 5 seconds, for durations of 2 seconds each, until manually acknowledged. Each new displayed status change shall be accompanied by an audible indication that shall persist until manual acknowledgment of the signal is performed.

There shall be a means provided for the operator to redisplay status of initiating device circuits that have been acknowledged but not yet restored to a normal condition. If the system retains the signal on the visual display until manually acknowledged, subsequent recorded presentations shall not be inhibited upon failure to acknowledge. Fire alarm signals shall be segregated on a separate visual display in this configuration unless given priority status on the common visual display.

The visual display unit described in 4-4.4.6(c) presents one or more lines of information, but does not display the status of all points in the system at the same time. The operator must scroll through the display once all signals have been acknowledged. To help the operator give proper precedence to fire alarm signals, they must either appear on a separate display or be given priority status on a common display. A permanent visual record is still required, but the type of printer is not specified.

4-4.4.7 The maximum elapsed time from sensing a fire alarm at an initiating device or initiating device circuit until it is recorded or displayed at the proprietary supervising station shall not exceed 90 seconds.

Paragraph 4-4.4.7 adds a higher level of performance than does 4-2.3.2.1.5, 4-2.3.2.3.4, or 4-2.3.5.2(a), (b), and (c). The subsection further reinforces the requirements in 4-2.3.1.2.3(a) and 4-2.3.4.1(a). Whatever transmission technology is chosen, it must guarantee delivery of the fire alarm signal within 90 seconds of actuation of the initiating device.

4-4.4.8 To facilitate the prompt receipt of fire alarm signals from systems handling other types of signals that may produce multiple simultaneous status changes, the requirements of either of the following shall be met:

(a) In addition to the maximum processing time for a single alarm, the system shall record simultaneous status changes at a rate not slower than either a quantity of 50, or 10 percent of the total number of initiating device circuits connected, within 90 seconds, whichever number is smaller, without loss of any signal.

(b) In addition to the maximum processing time, the system shall display or record fire alarm signals at a rate not slower than one every 10 seconds, regardless of the rate or number of status changes occurring, without loss of any signals.

Exception: Where fire alarm, waterflow alarm, sprinkler supervisory signals, and their associated trouble signals are the only signals processed by the system, the rate of recording shall not be slower than one round of code every 30 seconds.

This subsection reinforces the requirement for active multiplex transmission systems in 4-2.3.1.2.3(c) and imposes that requirement on other transmission technologies. The exception mitigates the severity of the requirement for other technologies when the proprietary fire alarm system handles only fire alarm signals, waterflow alarm signals, sprinkler supervisory signals, and their associated trouble signals.

4-4.4.9 Trouble signals required in 1-5.8 and their restoration to normal shall be automatically indicated and recorded at the proprietary supervising station within 200 seconds.

This subsection reinforces, in part, the requirements in 4-2.3.1.2.3(b) for active multiplex transmission systems and imposes those requirements on other transmission technologies.

4-4.4.10 The recorded information for the occurrence of any trouble condition of signaling line circuit, leg facility, or trunk facility that prevents receipt of alarm signals at the proprietary supervising station shall be such that the operator is able to determine the presence of the trouble condition. Trouble conditions in a leg facility shall not affect or delay receipt of signals at the proprietary supervising station from other leg facilities on the same trunk facility.

When active multiplex transmission technology is specified, the requirements of 4-4.4.10 effectively mandate that either Type 1 or Type 2 active multiplex transmission systems be used. The specific implementation of other transmission technologies would need to be individually analyzed to determine if it could meet the performance requirements of this subsection.

4-4.5 Personnel.

4-4.5.1 At least two operators, one of whom shall be permitted to be a runner, shall be on duty at all times.

Exception: Where the means for transmitting alarms to the fire department is automatic, at least one operator shall be on duty at all times.

4-4.5.2 When the runner is not in attendance at the proprietary supervising station, the runner shall establish two-way communications with the station at intervals not exceeding 15 minutes.

4-4.5.3 The primary duties of the operator(s) shall be to monitor signals, operate the system, and take such action as shall be required by the authority having jurisdiction. The operator(s) shall not be assigned any additional duties that would take precedence over the primary duties.

The code expects the operators to have no other duties that would distract from the prompt, effective handling of signals.

4-4.6 Operations.

4-4.6.1 All communication and transmission channels between the proprietary supervising station and the protected premises master control unit (panel) shall be operated manually or automatically once every 24 hours to verify operation.

4-4.6.1.1 When a communication or transmission channel fails to operate, the operator shall immediately notify the person(s) identified by the owner or authority having jurisdiction.

4-4.6.2 All operator controls at the proprietary supervising station(s) designated by the authority having jurisdiction shall be operated at each change of shift.

4-4.6.3 If operator controls fail, the operator shall immediately notify the person(s) identified by the owner or authority having jurisdiction.

The requirements of 4-4.6.1, 4-4.6.1.1, 4-4.6.2, and 4-4.6.3 ensure the operational continuity of the supervising station. By exercising communication channels and operating controls, potential failure modes will be identified more quickly. Also, by promptly contacting designated persons when a failure occurs, the operators will help ensure that repairs will be started as soon as possible.

4-4.6.4 Indication of a fire shall be promptly retransmitted to the public fire service communications center or other locations acceptable to the authority having jurisdiction, indicating the building or group of buildings from which the alarm has been received.

4-4.6.5* The means of retransmission shall be acceptable to the authority having jurisdiction and shall be in accordance with Sections 4-3, 4-5, 4-6, or 4-7.

Exception: Secondary power supply capacity shall be as required in Chapter 1.

A-4-4.6.5 It is the intent of this code that the operator within the proprietary supervising station should have a secure means of immediately retransmitting any signal indicative of a fire to the public fire department communication center. Automatic retransmission using an approved method installed in accordance with Sections 4-3, 4-4, 4-5, 4-6, and 4-7 is no doubt the best method for proper retransmission. However, a manual means may be used, consisting of either a manual connection following the requirements of Sections 4-3, 4-5, and 4-7, or, for proprietary supervising stations serving only contiguous properties, in the form of a municipal fire alarm box installed within 50 ft (15 m) of the proprietary supervising station in accordance with Section 4-6.

This subsection requires that the proprietary supervising station retransmit signals to the public fire service communication center by means of a central station fire alarm system, a remote station fire alarm system, or an auxiliary fire alarm system. It also permits a proprietary supervising station covering a contiguous property to retransmit signals to the public fire service communication center by means of a municipal fire alarm box.

4-4.6.6* Retransmission by coded signals shall be confirmed by two-way voice communication indicating the nature of the alarm.

A-4-4.6.6 No matter what type of retransmission facility is used, telephone communication between the proprietary supervising station and the fire department should be available at all times and should not depend on a switchboard operator.

Even though 4-4.6.5 requires retransmission to the public fire service communication center by means of a central station fire alarm system, a remote station fire alarm system, or an auxiliary fire alarm system, it is possible for the authority having jurisdiction to permit the use of an ordinary telephone, similar to the use permitted in 4-3.4.4.

4-4.6.7 Dispositions of Signals.

4-4.6.7.1 Alarms. Upon receipt of a fire alarm signal, the proprietary supervising station operator shall initiate action to:

(a) Immediately notify the fire department, the plant fire brigade, and such other parties as the authority having jurisdiction may require

(b) Promptly dispatch a runner to the alarm location (Travel time shall not exceed 1 hour.)

(c) Restore the system to its normal operating condition as soon as possible after disposition of the cause of the alarm signal.

4-4.6.7.2 Guard's Tour Delinquency. If a regular signal is not received from a guard within a 15-minute maximum grace period,

or if a guard fails to follow a prescribed route in transmitting the signals (if a prescribed route has been established), it shall be treated as a delinquency signal. When a guard's tour delinquency occurs, the proprietary supervising station operator shall initiate action to:

(a) Communicate at once with the protected areas or premises by telephone, radio, calling back over the system circuit, or other means acceptable to the authority having jurisdiction

(b) Dispatch a runner to investigate the delinquency, if communications with the guard cannot be promptly established. (Travel time shall not exceed one-half hour.)

4-4.6.7.3 Supervisory Signals. Upon receipt of sprinkler system and other supervisory signals, the proprietary supervising station operator shall initiate action to:

(a) Where required, communicate immediately with the designated person(s) to ascertain the reason for the signal

(b) Where required, dispatch a runner or maintenance person (travel time not to exceed 1 hour) to investigate, unless supervisory conditions are promptly restored to normal

(c) Where required, notify the fire department

(d) Where required, notify the authority having jurisdiction when sprinkler systems are wholly or partially out of service for 8 hours or more

(e) Where required, provide written notice to the authority having jurisdiction as to the nature of the signal, time of occurrence, and restoration of service, when equipment has been out of service for 8 hours or more.

4-4.6.7.4 Trouble Signals. Upon receipt of trouble signals or other signals pertaining solely to matters of equipment maintenance of the fire alarm system, the proprietary supervising station operator shall initiate action to:

(a) Where required, communicate immediately with the designated person(s) to ascertain reason for the signal

(b) Where required, dispatch a runner or maintenance person (travel time not to exceed 1 hour) to investigate

(c) Where required, notify the fire department

(d) Where required, notify the authority having jurisdiction when interruption of normal service will exist for 4 hours or more

(e) Where required, provide written notice to the authority having jurisdiction as to the nature of the signal, time of occurrence, and restoration of service, when equipment has been out of service for 8 hours or more.

The requirements in 4-4.6.7 are similar to those in 4-3.6.1 with the following notable differences: upon receipt of an alarm, a runner must always be dispatched; it may be required to notify the fire department upon receipt of a supervisory signal or a trouble signal so they will know protection is impaired; the runner response time on a trouble signal is 1 hour; and the authority having jurisdiction must be notified if an interruption to service that produces a trouble signal persists for 4 hours.

4-4.6.8 Record Keeping and Reporting.

4-4.6.8.1 Complete records of all signals received shall be retained for at least 1 year.

4-4.6.8.2 The proprietary supervising station shall make arrangements to furnish reports of signals received to the authority having jurisdiction, in a form acceptable to it.

When an authority having jurisdiction requests reports from a proprietary supervising station, that authority has the right to ask for the report in a form that is useful and useable.

4-4.7 Testing and Maintenance. Testing and maintenance of proprietary fire alarm systems shall be performed in accordance with Chapter 7.

4-5 Remote Supervising Station Fire Alarm Systems.

NOTE: The requirements of Chapters 1, 3, 5, 6, 7, and Section 4-2 shall apply to remote supervising station fire alarm systems, except where they conflict with the requirements of this section.

4-5.1 Scope. This section describes the installation, maintenance, testing, and use of a remote supervising station fire alarm system that serves properties under various ownership from a remote supervising station where trained competent personnel are in constant attendance. It covers the minimum requirements for the remote supervising station physical facilities, equipment, operating personnel, response, retransmission, signals, reports, and testing.

Remote station service was originally conceived as a means of transmitting fire alarm, supervisory, and trouble signals to the public fire service communication center when the municipality did not have a public fire reporting system, and when central station service was not available. Its use has been somewhat broadened as more and more public fire service communication centers refuse to receive private fire alarm signals that are reported by means other than the public voice telephone network.

4-5.2 General.

4-5.2.1 Remote supervising station fire alarm systems provide an automatic audible and visible indication of alarm and, when required, of supervisory and trouble conditions at a location remote from the protected premises and a manual or automatic permanent record of these conditions.

Notice that paragraph 4-5.2.1 does not require automatic permanent visual record of signals received. However, a manually written log must be kept.

4-5.2.2 This section does not require the use of audible signal notification appliances other than those required at the remote supervising station. If it is desired to provide fire alarm evacuation signals in the protected premises, the alarm signals, circuits, and controls shall comply with the provisions of Chapter 3 and Chapter 6 in addition to the provisions of this section.

This subsection is a reminder that evacuation signals are found in Chapters 3 and 6 of the code.

4-5.2.3 The loading capacities of the remote supervising station equipment for any approved method of transmission shall be as designated in Section 4-2.

Remote station service has the full range of technologies available, as long as they are provided in compliance with the requirements of Section 4-2.

4-5.3* Supervising Station Facilities.

A-4-5.3 As a minimum, the room or rooms containing the remote supervising station equipment should have a 1-hour fire rating, and the entire structure should be protected by an alarm system complying with Chapter 3.

The recommended fire rating of the remote supervising station is not a requirement.

4-5.3.1 Where a remote supervising station connection is used to transmit an alarm signal, the signal shall be received at the public fire service communications center, at a fire station, or at the similar governmental agency that has a public responsibility for taking prescribed action to ensure response upon receipt of a fire alarm signal.

Exception: Where such an agency is unwilling to receive alarm signals or will permit the acceptance of another location by the authority having jurisdiction, such alternate location shall have personnel on duty at all times trained to receive the alarm signal and immediately retransmit it to the fire department.

The requirements of 4-5.3.1 and its exception allow any location acceptable to the authority having jurisdiction to act as the remote supervising station. For example, an authority having jurisdiction could permit a listed central supervising station to receive these signals. This would constitute remote station service, but not central station service.

4-5.3.2 Supervisory and trouble signals shall be handled at a constantly attended location having personnel on duty trained to recognize the type of signal received and to take prescribed action. This shall be permitted to be a location different from that at which alarm signals are received.

In some installations, fire alarm signals are transmitted to the public fire service communication center, and supervisory signals and trouble signals are transmitted to another location acceptable to the authority having jurisdiction.

4-5.3.3 Where locations other than the public fire service communication center are used for the receipt of signals, access to receiving equipment shall be restricted in accordance with requirements of the authority having jurisdiction.

This requirement helps ensure the security and operational integrity of the remote station receiving equipment.

4-5.4 Equipment.

4-5.4.1 Signal-receiving equipment shall indicate receipt of each signal both audibly and visibly.

4-5.4.1.1 Audible signals shall meet the requirements of Chapter 6 for the private operating mode.

See 6-3.2.

4-5.4.1.2 Means for silencing alarm, supervisory, and trouble signals shall be provided and shall be so arranged that subsequent signals shall re-sound.

Silencing one signal must not prevent a subsequent signal from causing the audible notification appliance to resound.

4-5.4.1.3 A trouble signal shall be received when the system or any portion of the system at the protected premises is placed in a bypass or test mode.

The requirement in 4-5.4.1.3 prevents any type of so-called silent disconnect switch at the protected premises. While a disconnect switch is permitted, operation of that switch must produce a trouble signal at the remote supervising station.

4-5.4.1.4 An audible and visible indication shall be provided upon restoration from any off-normal condition.

It is not sufficient to indicate restoration to normal by merely extinguishing a lamp; the audible notification appliance at the remote supervising station must also resound.

4-5.4.1.5 Where suitable visible means are provided in the remote supervising station to readily identify the type of signal received, a common audible notification appliance shall be permitted to be used.

4-5.4.2 Power supplies shall comply with the requirements of Chapter 1.

Exception: In a remote supervising station fire alarm system where the alarm and supervisory signals are transmitted over a listed supervised one-way radio system, 24 hours of secondary (standby) power shall be permitted in lieu of 60 hours, as required in 1-5.2.5, at the radio alarm repeater station receivers (RARSR), provided that personnel are dispatched to arrive within 4 hours after detection of failure to initiate maintenance.

Many of the one-way radio alarm system sites that house an RARSR have a very minimal footprint in equipment rooms where space is leased at a very high cost, such as at the top of a high-rise building. This small footprint does not give adequate room for 60 hours of standby batteries. The exception permits the use of 24 hours of standby power as long as a technician arrives within 4 hours of the receipt of a trouble signal from the RARSR.

4-5.4.3 Transmission means shall comply with the requirements of Section 4-2.

The full range of transmission technology is permitted, as long as it can meet the requirements of Section 4-5.

4-5.4.4 Retransmission of an alarm signal, where required, shall be by one of the following methods, listed in descending order of preference:

(a) A dedicated circuit that is independent of any switched telephone network. This circuit shall be permitted to be used for voice or data communication.

(b) A one-way (outgoing only) telephone at the remote supervising station that utilizes the public switched telephone network. This telephone shall be used primarily for voice transmission of alarms to a telephone at the public fire service communications center, which cannot be used for outgoing calls.

(c) A private radio system using the fire department frequency where permitted by the fire department.

(d) Other methods acceptable to the authority having jurisdiction.

The vast majority of remote supervising stations will use a retransmission method that complies with the requirements of 4-5.4.4(b).

4-5.5 Personnel. Sufficient personnel shall be available at all times to receive alarm signals at the remote supervising station and to take immediate appropriate action. Duties pertaining to other than operation of the remote supervising station receiving

and retransmitting equipment shall be permitted subject to the approval of the authority having jurisdiction.

While it is recognized that operators at the remote supervising station may have other duties, by requiring approval of the authority having jurisdiction, the requirement attempts to limit the extent to which these other duties might interfere with the proper handling of signals.

4-5.6 Operations.

4-5.6.1 Where the remote supervising station is at a location other than the public fire service communication center, alarm signals shall be immediately retransmitted to the public fire service communications center.

4-5.6.2 Upon receipt of an alarm, supervisory, or trouble signal by the remote supervising station other than the public fire service communications center, it shall be the responsibility of the operator on duty to immediately notify the owner or the owner's designated representative.

By promptly contacting designated persons when a failure occurs, the operators will help ensure that repairs will be started as soon as possible.

4-5.6.3 A permanent record of the time, date, and location of all signals and restorations received; the action taken thereon; and the results of all tests shall be maintained for at least 1 year and made available to the authority having jurisdiction. These records shall be permitted to be made by manual means.

Most often these required records will be in the form of a manually maintained log book.

4-5.6.4 All operator controls at the remote supervising station shall be operated at the beginning of each shift or change in personnel, and the status of all off-normal conditions noted and recorded.

The requirements of 4-5.6.4 help ensure the operational continuity of the supervising station. By exercising operating controls, potential failure modes will be more quickly identified.

4-5.7 **Testing and Maintenance.** Testing and maintenance for remote supervising stations shall be performed in accordance with Chapter 7.

4-6 Public Fire Alarm Reporting Systems.

4-6.1 **Scope.** This section covers the general requirements and use of public fire alarm reporting systems. These systems include the equipment necessary to effect the transmission and reception of fire alarms or other emergency calls from the public.

Section 4-6 is somewhat different than the other sections of this chapter in that it does not deal solely with connection from a protected premises to a supervising station. Rather, it deals with means provided by a municipality or other governmental entity for the public to initiate an alarm signal. Such means are intended principally to transmit signals indicative of a fire emergency (*see 4-6.2.2*), or those relating to trouble with the public fire alarm reporting system. The transmission of supervisory signals is not contemplated by this section.

In order to use the four types of systems described in this section for the transmission of fire alarm systems from a protected premises to the public fire service communication center, it would be necessary to provide an auxiliary fire alarm system as described in Section 4-7. Such a system provides the interface between a fire alarm system installed in accordance with the provisions of Chapter 3 of the code and the public fire alarm reporting system described in Section 4-6.

4-6.2 **General Fundamentals.**

4-6.2.1 Where implemented at the option of the authority having jurisdiction, a public fire alarm reporting system shall be designed, installed, operated, and maintained to provide the maximum practicable reliability for transmission and receipt of fire alarms.

4-6.2.2 It shall be permissible for a public fire alarm reporting system, as described herein, to be used for the transmission of other signals or calls of a public emergency nature, provided such transmission does not interfere with the transmission and receipt of fire alarms.

Such a system is principally concerned with transmitting an alarm of fire to the public fire service communication center. However, with certain types of public fire reporting systems, it would be possible to give a voice or electronic request for emergency medical or police response. These calls are permitted as long as they do not interfere with the transmission of a fire alarm signal.

4-6.2.3 Alarm systems shall be Type A or Type B. A Type A system shall be provided when the number of all alarms required to be transmitted over the dispatch circuits exceeds 2500 per year.

NOTE: Where a Type A system is required, automatic transmission of alarms from boxes by use of electronic equipment is permissible, only if the following requirements are satisfied:

(a) Reliable facilities are provided for the automatic receipt, storage, retrieval, and transmission of alarms in the order received; and
(b) Override capability is provided to the operator(s) so that manual transmission and dispatch facilities are instantly available.

In a Type A system, operators at the public fire service communication center receive signals from the public fire reporting system. They then retransmit these signals to the fire stations designated to respond. In a Type B system, signals initiated by a fire alarm box on the public fire reporting system are automatically retransmitted to all fire stations and other locations connected to the system. Subsection 4-6.2.3 requires systems that transmit over 2500 alarms on the dispatch circuits to be Type A systems.

4-6.2.4 Any portion(s) of a public fire alarm reporting system used to effect the auxiliarized protection of a structure or multiple of structures shall be listed as compliant with Chapter 3 and Section 4-7.

4-6.3 **Management and Maintenance.** (*See Chapter 7.*)

4-6.4 **Equipment and Installation.**

4-6.4.1 Means for actuation of alarms by the public shall be conspicuous and readily accessible for easy operation.

4-6.4.2 Public fire alarm reporting systems as defined in this chapter, shall, in their entirety, be subject to a complete operational acceptance test upon completion of system installation.

Said test(s) shall be made in accordance with the requirements of the authority having jurisdiction. However, in no case shall the operational functions tested be less than those stipulated in Chapter 7. Like tests shall be performed on any alarm reporting devices as identified in this chapter that are added subsequent to the installation of the initial system.

4-6.4.3 Publicly accessible boxes shall be recognizable as such. Boxes shall have operating instructions plainly marked on the exterior surface.

4-6.4.4 The actuating device shall be readily available and of such design and so located as to make the method of its use apparent.

4-6.4.5 Publicly accessible boxes shall be as conspicuous as possible. Their color shall be distinctive.

4-6.4.6 All publicly accessible boxes mounted on support poles shall be identified by a wide band of distinctive colors or adequate signs placed 8 ft (2.44 m) above the ground and visible from all directions whenever possible.

4-6.4.7* Indicating lights of a distinctive color, visible for at least 1500 ft (460 m), shall be installed over publicly accessible boxes in mercantile and manufacturing areas. Equipping the street light nearest the box with a distinctively colored light shall be acceptable.

A-4-6.4.7 Indicating Lights.

(a) Current supply for designating lamps at street boxes should preferably be secured at lamp locations from the local electric utility company.

(b) Alternating current power may be superimposed on metallic fire alarm circuits for supplying designating lamps or for control or actuation of equipment devices for fire alarm or other emergency signals, provided:

1. Voltages between any wire and ground or between one wire and any other wire of the system shall not exceed 150 volts; the total resultant current in any line circuit shall not exceed ¼ amp.

2. Coupling capacitors, transformers, chokes, coils, etc., shall be rated for 600-volt working voltage and have a break-

down voltage of at least twice the working voltage plus 1000 volts.

3. There is not interference with fire alarm service under any conditions.

While superimposing box light power on the fire alarm circuit was popular in the 1930s and 1940s, it is almost unheard of today. One somewhat popular building fire alarm scheme does impose alternating current for audible fire alarm notification appliances on the direct current manual fire alarm initiating device circuit of a shunt-type master fire alarm box. The system is marketed by the Gamewell Company under the name of "dual alarm."

4-6.4.8 Boxes shall be securely mounted on poles, pedestals, or structural surfaces as directed by the authority having jurisdiction.

4-6.4.9 Concurrent operation of at least four boxes shall not result in the loss of an alarm.

This requirement means that the boxes must be designed so that they can sense that a common box circuit is busy, retain their signal until the circuit is clear, and then transmit the signal. In meeting this requirement, manufacturers must create boxes that are noninterfering, successive, or the equivalent.

4-6.5 Design of Boxes. (*See Chapter 5.*)

Box design information can be found in 5-9.2.

4-6.6* Location of Boxes. Location of publicly accessible boxes shall be designated by the authority having jurisdiction. Schools, hospitals, nursing homes, and places of public assembly shall have a box located at or near the main entrance, as directed by the authority having jurisdiction.

In most cases the authority having jurisdiction will be municipal fire officials, though they may be somewhat influenced by the Municipal Grading Schedule of the Insurance Services Office.

A-4-6.6 If the intent is for complete coverage, then it will not be necessary to travel in excess of one block or 500 ft (150 m) to reach a box. In residential areas, it will not be necessary to travel in excess of 2 blocks or 800 ft (240 m) to reach a box.

4-6.7 **Power Supply.**

4-6.7.1 **General.**

4-6.7.1.1 Batteries, motor-generators, or rectifiers shall be sufficient to supply all connected circuits without exceeding the capacity of any battery or overloading any generator or rectifier, so that circuits developing grounds or crosses with other circuits each may be supplied by an independent source to the extent required by 4-6.7.1.8(b).

4-6.7.1.2 Provision shall be made in the operating room for supplying any circuit from any battery, generator, or rectifier. Enclosed fuses shall be provided at points where supplies for individual circuits are taken from common leads. Necessary switches, testing, and signal transmitting and receiving devices shall be provided to permit the isolation, control, and test of each circuit, to at least 10 percent of the total number of box and dispatch circuits, but never less than 2 circuits.

These requirements help ensure maximum reliability for the public fire alarm reporting system. The intention of the last phrase is that there will be enough equipment, such as transmitting and receiving equipment, so that no one device or appliance serves more than 10 percent of the circuits, and so that there will always be at least two of every device or appliance.

4-6.7.1.3 If common-current source systems are grounded, the ground shall not exceed 10 percent of resistance of any connected circuit and be located at one side of the battery. Visual and audible indicating devices shall be provided for each box and dispatch circuit to give immediate warning of ground leakage endangering operability.

Because a large percentage of the cable plant for the public fire reporting system is installed throughout the municipality as either aerial or underground construction, there is great concern that foreign grounds will render a portion of a circuit inoperable. For this reason, much attention is given to the prompt discovery of excess current leakage to ground.

4-6.7.1.4 Local circuits at communication centers shall be supplied either in common with box circuits or coded radio-receiving system circuits or by a separate power source. The source of power for local circuits required to operate the essential features of the system shall be supervised.

Power for circuits and equipment within the public fire service communication center must be supervised so that loss of this power results in a trouble signal.

4-6.7.1.5 Visual and audible means to indicate a 15 percent or greater reduction of normal power supply (rated voltage) shall be provided.

When power for the public fire reporting system or for local circuits drops 15 percent or more below the normal rated voltage, such reduction must result in a trouble signal.

4-6.7.1.6 The forms and arrangements of power supply shall be classified as described in the paragraphs below.

NOTE: If the electrical service/capacity of the equipment required under NFPA 1221, *Standard for the Installation, Maintenance, and Use of Public Fire Service Communication Systems*, 2-1.6, is adequate to satisfy the needs of equipment in Section 4.6, said equipment need not be duplicated.

The wide range of alternatives for power supplies offered by the forms described in 4-6.7.1.7, 4-6.7.1.8, and 4-6.7.1.9 gives the designer of a public fire alarm reporting system a great deal of flexibility in executing his or her design.

4-6.7.1.7 **Form 2.** These forms shall be permissible for Type A systems only. Box circuits shall be served in multiple by:

(a)* *Form 2A.* A rectifier or motor-generator powered from a single source of alternating current, with a floating storage battery having a 24-hour standby capacity.

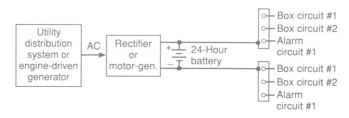

Figure A-4-6.7.1.7(a)   Form 2A.

(b)* *Form 2B.* A rectifier or motor-generator powered from two sources of alternating current, with a floating storage battery having a 4-hour standby capacity.

NOTE: For Forms 2A, 2B, and 2C, these arrangements are permissible but are not recommended where circuits are wholly or partly open-wire because of the possibility of trouble from multiple grounds.

4-6.7.1.8 **Form 3.** Each box circuit or coded radio receiving system shall be served by:

(a)* *Form 3A.* A rectifier or motor-generator powered from a single source of alternating current with a floating storage battery having a 60-hour standby capacity.

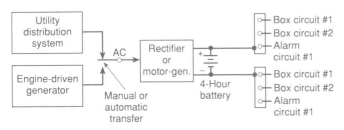

Figure A-4-6.7.1.7(b)(1) Form 2B-1.

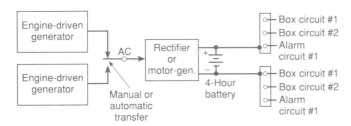

Figure A-4-6.7.1.7(b)(2) Form 2B-2.

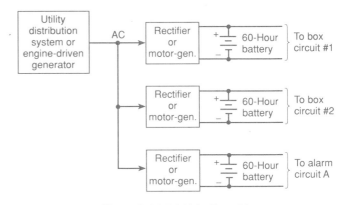

Figure A-4-6.7.1.8(a) Form 3A.

(c)* *Form 2C.* A duplicate rectifier or motor-generator powered from two sources of alternating current with transfer facilities to apply power from the secondary source to the system within 30 seconds (*see NFPA 1221, Standard for the Installation, Maintenance, and Use of Public Fire Service Communication Systems*). Each rectifier or motor-generator shall be capable of powering the entire system.

(b)* *Form 3B.* A rectifier or motor-generator powered from two sources of alternating current with a floating storage battery having a 24-hour standby capacity.

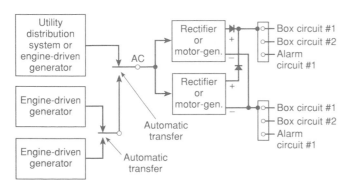

Figure A-4-6.7.1.7(c) Form 2C.

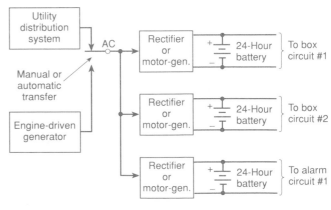

Figure A-4-6.7.1.8(b)(1) Form 3B-1.

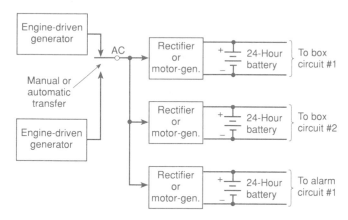

Figure A-4-6.7.1.8(b)(2) Form 3B-2.

4-6.7.1.9 Form 4. Each box circuit or coded radio receiving system shall be served by:

(a)* *Form 4A.* An inverter powered from a common rectifier receiving power by a single source of alternating current with a floating storage battery having a 24-hour standby capacity.

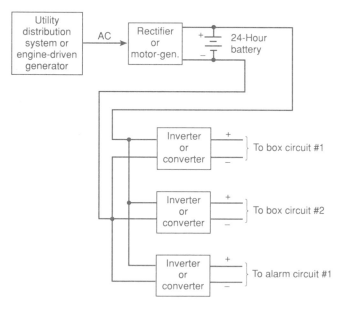

Figure A-4-6.7.1.9(a) Form 4A.

(b)* *Form 4B.* An inverter powered from a common rectifier receiving power from two sources of alternating current with a floating storage battery having a 4-hour standby capacity.

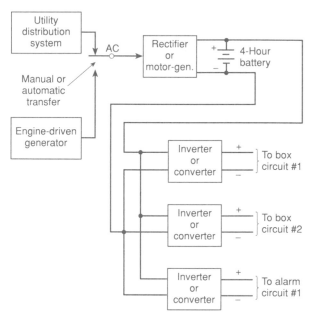

Figure A-4-6.7.1.9(b)(1) Form 4B-1.

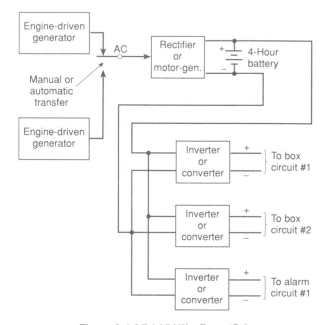

Figure A-4-6.7.1.9(b)(2) Form 4B-2.

NOTE: For Form 4A and Form 4B, it is permissible to distribute the system load between two or more common rectifiers and batteries.

(c)* *Form 4C.* A rectifier, converter, or motor-generator receiving power from two sources of alternating current with transfer facilities to apply power from the secondary source to the system within 30 seconds (*see NFPA 1221, Standard for the Installation, Maintenance, and Use of Public Fire Service Communication Systems*).

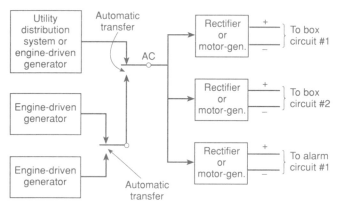

Figure A-4-6.7.1.9(c)   Form 4C.

4-6.7.2 Rectifiers, Converters, Inverters, and Motor-Generators.

4-6.7.2.1 Rectifiers shall be supplied through an isolating transformer taking energy from a circuit not to exceed 250 volts.

4-6.7.2.2 Complete, ready-to-use spare units or spare parts shall be available in reserve.

4-6.7.2.3 One spare rectifier shall be provided for each ten required for operation, but in no case less than one.

4-6.7.2.4 Leads from rectifiers or motor-generators, with storage battery floating, shall have fuses rated at not less than 1 amp and not more than 200 percent of maximum connected load. Where not provided with battery floating, the fuse shall be not less than 3 amps.

These fuse sizing requirements provide circuits with sufficient protection without being so sensitive that the circuit will operate unnecessarily, affecting the overall reliability of the public fire alarm reporting system.

4-6.7.3 Engine-Driven Generator Sets.

The requirements in this subsection are designed to help ensure the continuity of supplied power and thus the reliability of the public fire alarm reporting system.

4-6.7.3.1 The provisions of 4-6.7.3 shall apply to generators driven by internal combustion engines.

4-6.7.3.2 The installation of such units shall conform to the provisions of NFPA 37, *Standard for the Installation and Use of Stationary Combustion Engines and Gas Turbines*, and NFPA 110, *Standard for Emergency and Standby Power Systems*, except as restricted by the provisions of 4-6.7.3.

4-6.7.3.3 The engine-driven generator shall be located in an adequately ventilated cutoff area of the building housing the communication center equipment. The area housing the unit shall be used for no other purpose except storage of spare parts or equipment. Exhaust fumes shall be discharged directly outside the building.

4-6.7.3.4 Liquid fuel shall be stored in outside underground tanks and gravity feed shall not be used. Sufficient fuel shall be available for 24 hours of operation at full load if a reliable source of fuel supply is available, at any time, on 2 hours notice. If a source of supply is not reliable or readily available, or if special arrangements must be made for refueling as necessary, a supply sufficient for 48 hours of operation at full load shall be maintained.

4-6.7.3.5 Liquefied petroleum gas and natural gas installations shall meet the requirements of NFPA 54, *National Fuel Gas Code*, and NFPA 58, *Standard for the Storage and Handling of Liquefied Petroleum Gases*.

4-6.7.3.6 The unit, as a minimum, shall be of sufficient capacity to supply power to operate all fire alarm facilities and emergency lighting of the operating rooms or communication building.

4-6.7.3.7 A separate storage battery on automatic float charger shall be provided for starting the engine-driven generator.

4-6.7.3.8 Where more than one engine-driven generator is provided, each shall be provided with a separate fuel line and transfer pump.

4-6.7.4 Float-Charged Batteries.

The requirements in 4-6.7.4 are intended to help ensure the continuity of the supplied power and thus the reliability of the public fire alarm reporting system.

4-6.7.4.1 Batteries shall be of the storage type; primary batteries (dry cells) shall not be used. All cells shall be of the sealed type. Lead-acid batteries shall be in jars of glass or other suitable transparent materials; other types of batteries shall be in containers suitable for the purpose.

4-6.7.4.2 Batteries shall be located in the same building as the operating equipment, preferably on the same floor, readily accessible for maintenance and inspection. The battery room shall be aboveground, except as permitted by NFPA 1221, *Standard for the Installation, Maintenance, and Use of Public Fire Service Communication Systems*, 2-1.1.2, and shall be ventilated to prevent accumulation of explosive gas mixtures; special ventilation is required only for unsealed cells.

Unless the facility has been constructed specifically for use below grade, an aboveground location must be selected.

4-6.7.4.3 Batteries shall be mounted to provide effective insulation from the ground and from other batteries. The mounting shall be suitably protected against deterioration, and consideration shall be given to stability, especially in geographic areas subject to seismic disturbance.

As stated previously, the requirements for power supplies in 4-6.7.1.6 through 4-6.7.4.3 all help ensure the overall reliability of the power supply portion of the public fire alarm reporting system.

4-6.8 Requirements for Metallic Systems and Metallic Interconnections.

4-6.8.1 Circuit Conductors.

4-6.8.1.1 Wires shall be terminated so as to provide good electrical conductivity and to prevent breaking from vibration or stress.

4-6.8.1.2 Circuit conductors on terminal racks shall be identified and isolated from conductors of other systems whenever possible and shall be suitably protected from mechanical injury.

4-6.8.1.3 Except as otherwise provided herein, exterior cable and wire shall conform to International Municipal Signal Association specifications or their equivalent.

The International Municipal Signal Association (IMSA) publishes wire and cable specifications for use in the installation of public fire reporting systems. IMSA can be reached at P.O. Box 539, Newark, NY 14513, (315)331-2182.

4-6.8.1.4 If a municipal box is installed inside a building, it shall be placed as close as practical to the point of entrance of the circuit, and the exterior wire shall be installed in conduit or electrical metallic tubing, in accordance with Chapter 3 of NFPA 70, *National Electrical Code*.

Exception: This requirement shall not apply to coded radio box systems.

The intent of 4-6.8.1.4 is to limit the exposure of the public fire alarm reporting system circuit to a hostile fire within a building. If that circuit was installed so that it ran extensively throughout the building, there is a greater chance that a fire could burn through a portion of the circuit before an alarm was transmitted, thus rendering the protection useless.

4-6.8.2 Cables.

4-6.8.2.1 General.

4-6.8.2.1.1 Cables that meet the requirements of NFPA 70, *National Electrical Code*, Article 310, for installation in wet locations shall be satisfactory for overhead or underground installation, except that direct-burial cable shall be specifically approved for the purpose.

4-6.8.2.1.2 Paper or pressed pulp insulation shall not be considered satisfactory for an emergency service such as a fire alarm system, except that cables containing conductors with such insulation shall be acceptable if pressurized with dry air or nitrogen. Loss of pressure in cables shall be indicated by a visual or audible warning system located where someone who can interpret the pressure readings and who has authority to have the indicated abnormal condition corrected is in constant attendance.

Certain cable constructions installed in the 1940s, particularly for telephone and other communications purposes, use copper wire insulated with paper or pressed pulp materials. These cables must remain dry or the insulation materials will degrade. To ensure dryness, it is common practice to pressurize the cable with dry nitrogen and monitor the cable for leakage. Such cable would not normally be used in new installations.

4-6.8.2.1.3 Natural rubber-sheathed cable shall not be used where it may be exposed to oil, grease, or other substances or conditions that may tend to deteriorate the cable sheath. Braided-sheathed cable shall be used only inside of buildings where run in conduit or metal raceways.

4-6.8.2.1.4 Other municipally controlled signal wires shall be permitted to be installed in the same cable with fire alarm wires. Cables controlled by or containing wires of private signaling organizations shall be permitted to be used for fire alarm purposes only by permission of the authority having jurisdiction.

Occasionally, municipalities that maintain their own governmental service telephone system will run the wiring for that service in the same cable as the public fire alarm reporting system. Conversely, parallel telephone-type systems and, on rare occasions, other types of public fire reporting systems may consist of interconnecting wiring leased from the public telephone utility or even from a private agency, such as Western Union. The requirements of this subsection apply to such cases.

4-6.8.2.1.5 Signaling wires that, because of the source of current supply, might introduce a hazard shall be protected and supplied as required for lighting circuits.

See NFPA 70, *National Electrical Code*, Articles 760 and 800.

4-6.8.2.1.6 All cables with all taps and splices made shall be tested for insulation resistance when installed, but before connection to terminals. Such tests shall indicate an insulation resis-

tance of at least 200 megohms per mile between any one conductor and all others, the sheath, and ground.

This unique code entry requires cables and splices to be tested with a megohm meter (megger) to ensure the strength of the insulation. This test must be made before any devices or appliances are connected to the cable plant.

4-6.8.2.2 Underground Cables.

4-6.8.2.2.1 Underground cables in duct or direct burial shall be brought aboveground only at points where liability of mechanical injury or of disablement from heat incident to fires in adjacent buildings is minimized.

4-6.8.2.2.2 Cables shall be in duct systems and manholes containing low-tension fire alarm system conductors only, except low-tension secondary power cables shall be permitted. If in duct systems or manholes containing power circuit conductors in excess of 250 volts to ground, fire alarm cables shall be located as far as possible from such power cables and shall be separated from them by a noncombustible barrier or by such other means as may be practicable to protect the fire alarm cables from injury.

4-6.8.2.2.3 All cables installed in manholes shall be properly racked and marked for identification.

4-6.8.2.2.4 All conduits or ducts entering buildings from underground duct systems shall be effectively sealed against moisture or gases entering the building.

4-6.8.2.2.5 Cable joints shall be located only in manholes, fire stations, and other locations where proper accessibility is provided and where there is little liability of injury to the cable due to either falling walls or operations in the buildings. Cable joints shall be made to provide and maintain conductivity, insulation, and protection at least equal to that afforded by the cables that are joined. Cable ends shall be sealed against moisture.

4-6.8.2.2.6 Direct-burial cable, without enclosure in ducts, shall be laid in grass plots, under sidewalks, or in other places where the ground is not apt to be opened for other underground construction. If splices are made, such splices shall, where practicable, be accessible for inspection and tests. Such cables shall be buried at least 18 in. (0.5 m) deep and, where crossing streets or other areas likely to be opened for other underground construction, shall be in duct or conduit or be covered by creosoted planking of at least 2 in. by 4 in. (50 mm by 100 mm) with half-round grooves, spiked or banded together after the cable is installed.

To maintain the overall operational integrity of the public fire alarm reporting system, the requirements of 4-6.8.2.2.1 through 4-6.8.2.2.6 help ensure that underground cables are not exposed to undue potential for mechanical injury.

4-6.8.2.3 Aerial Construction.

4-6.8.2.3.1 Fire alarm wires shall be run under all other wires except communication wires. Suitable precautions shall be provided where passing through trees, under bridges, over railroads, and at other places where injury or deterioration is possible. Wires and cables shall not be attached to a crossarm carrying electric light and power wires, except circuits carrying up to 220 volts for municipal communication use. Such 220-volt circuits shall be tagged or otherwise identified.

4-6.8.2.3.2 Aerial cable shall be supported by messenger wire of adequate tensile strength, except as permitted in 4-6.8.2.3.3.

4-6.8.2.3.3 Two-conductor cable shall be messenger-supported unless it has conductors of No. 20 AWG or larger size and has mechanical strength equivalent to No. 10 AWG hard-drawn copper.

4-6.8.2.3.4 Single wire shall meet International Municipal Signal Association specifications and shall not be smaller than No. 10 Roebling gauge if of galvanized iron or steel, No. 10 AWG if of hard-drawn copper, No. 12 AWG if of approved copper-covered steel, or No. 6 AWG if of aluminum. Span lengths shall not exceed manufacturers' recommendations.

4-6.8.2.3.5 Wires to buildings shall contact only intended supports and shall enter through an approved weatherhead or suitable sleeves slanting upward and inward. Drip loops shall be formed on wires outside of buildings.

4-6.8.2.4 Leads Down Poles.

4-6.8.2.4.1 Leads down poles shall be protected against mechanical injury. Any metallic covering shall form a continuous conducting path to ground. Installation, in all cases, shall prevent water from entering the conduit or box.

4-6.8.2.4.2 Leads to boxes shall have 600-volt insulation approved for wet locations, as defined in NFPA 70, *National Electrical Code.*

To maintain the overall operational integrity of the public fire alarm reporting system, the requirements of 4-6.8.2.3, 4-6.8.2.4, and their related subsections help ensure that aerial cables and leads down poles are not exposed to undue potential for mechanical injury or electrical failure.

4-6.8.2.5 Wiring Inside Buildings.

4-6.8.2.5.1 At the communication center, conductors shall extend as directly as possible to the operating room in conduits, ducts, shafts, raceways, or overhead racks and troughs of a type of construction affording protection against fire and mechanical injury.

4-6.8.2.5.2 All conductors inside buildings shall be in conduit, electrical tubing, metal molding, or raceways. Installation shall be in accordance with NFPA 70, *National Electrical Code.*

4-6.8.2.5.3 Conductors shall have an approved insulation; the insulation or other outer covering shall be flame-retardant and moisture-resistant.

4-6.8.2.5.4 Conductors shall be installed as far as possible without joints. Splices shall be permitted only in junction or terminal boxes. Wire terminals, splices, and joints shall conform to NFPA 70, *National Electrical Code.*

4-6.8.2.5.5 Conductors bunched together in a vertical run connecting two or more floors shall have a flame-retardant covering sufficient to prevent the carrying of fire from floor to floor. This requirement shall not apply if the conductors are encased in a metallic conduit or located in a fire-resistive shaft having fire stops at each floor.

4-6.8.2.5.6   Where cables or wiring are exposed to unusual fire hazards, they shall be properly protected.

4-6.8.2.5.7   Cable terminals and cross-connecting facilities shall be located in or adjacent to the operations room.

4-6.8.2.5.8   Where signal conductors and electric light and power wires are run in the same shaft, they shall be separated by at least 20 in. (50 mm), or either system shall be encased in a non-combustible enclosure.

To maintain the overall operational integrity of the public fire alarm reporting system, the requirements of 4-6.8.2.5 and its related subsections help ensure that wiring inside a building is not exposed to undue potential for mechanical injury or electrical failure, and that it will not contribute to the spread of fire in a building.

4-6.9   Facilities for Signal Transmission.

4-6.9.1   Circuits.

4-6.9.1.1   General.

4-6.9.1.1.1   ANSI/IEEE C2, *The National Electrical Safety Code*, shall be used as a guide for the installation of outdoor circuitry.

This document, developed by a committee from the Institute of Electrical and Electronic Engineers, is used by public and private electric company utilities, public and private telephone company utilities, and public and private community antenna television company utilities. ANSI/IEEE C2 describes the placement and spacing of outdoor aerial cable installations to ensure safe operation of the associated systems.

4-6.9.1.1.2   In all installations, first consideration shall be given to continuity of service. Particular attention shall be given to liability of mechanical injury; disablement from heat incident to a fire; injury by falling walls; and damage by floods, corrosive vapors, or other causes.

4-6.9.1.1.3   Open local circuits within single buildings are permitted in accordance with Chapter 3.

Circuits of the protected premises fire alarm system that are not a part of a public fire alarm reporting system must be installed in accordance with Chapter 3 of the code.

4-6.9.1.1.4   All circuits shall be so routed as to permit ready tracing of circuits for trouble.

4-6.9.1.1.5   Circuits shall not pass over, under, through, or be attached to buildings or property not owned by or under the control of the authority having jurisdiction or the agency responsible for maintaining the system, except where the circuit is terminated in a box on the premises.

The circuits for many of the original public fire alarm reporting systems in various East Coast municipalities were strung throughout a city from building to building. Engineers soon learned that during fires in those buildings, the circuits would be damaged, placing the operational integrity of the public fire alarm reporting system in jeopardy.

4-6.9.1.2   Box Circuits.

4-6.9.1.2.1   If a box is installed inside a building, it shall be placed as close as is practical to the point of entrance of the circuit, and the exterior wire shall be installed in conduit or electrical metallic tubing in accordance with Chapter 3 of NFPA 70, *National Electrical Code*.

The intent of 4-6.9.1.2.1 is to limit the exposure of the public fire alarm reporting system circuit to a hostile fire within a building. If that circuit was installed so that it ran extensively throughout the building, there is a greater chance that a fire could burn through a portion of the circuit before an alarm was transmitted, thus rendering the protection useless.

4-6.9.1.2.2   Accessible and reliable means, available only to the authority having jurisdiction or the agency responsible for maintaining the public fire alarm reporting system, shall be provided for disconnecting the auxiliary loop to the box inside the building, and definite notification shall be given to occupants of the building when the interior box is not in service.

If a connection is made to the public fire alarm reporting system in accordance with the requirements of Section 4-7, a means must be provided to disconnect the protected premises connection. This means must only be available to the authority having jurisdiction over the public fire alarm reporting system.

4-6.9.1.3 Tie Circuits.

Tie circuits connect the public fire service communication center with a subsidiary communication center. For example, in a large municipality, a subsidiary communication center might be used to concentrate signals from a particular neighborhood before transmission to the public fire service communication center. Also, in some cities where several boroughs have their own public fire service communication center (New York City, for example), tie circuits may be used to interconnect the centers so signals can be handled even if one of the centers is impaired.

4-6.9.1.3.1 A separate tie circuit shall be provided from the communication center to each subsidiary communication center.

4-6.9.1.3.2 The tie circuit between the communication center and the subsidiary communication center shall not be used for any other purpose.

4-6.9.1.3.3 In a Type B wire system, where all boxes in the system are of succession type, it shall be permissible to use the tie circuit as a dispatch circuit, to the extent permitted by NFPA 1221, *Standard for the Installation, Maintenance, and Use of Public Fire Service Communication Systems*.

4-6.9.1.4* Circuit Protection.

A-4-6.9.1.4 All requirements for circuit protection do not apply to coded radio reporting systems. These systems do not use metallic circuits.

Circuit protection is intended to limit equipment damage caused when transient currents are applied to the circuits of the public fire alarm reporting system. One common source of such transients is lightning.

4-6.9.1.4.1 General.

4-6.9.1.4.1.1 The protective devices shall be located close to or be combined with the cable terminals.

4-6.9.1.4.1.2 Lightning arresters suitable for the purpose shall be provided. Lightning arresters shall be marked with the name of the manufacturer and model designation.

4-6.9.1.4.1.3 All lightning arresters shall be connected to a suitable ground in accordance with NFPA 70, *National Electrical Code*.

4-6.9.1.4.1.4 All fuses shall be plainly marked with their rated ampere capacity. All fuses rated over 2 amps shall be of the enclosed type.

4-6.9.1.4.1.5 Circuit protection required at the communication center shall be provided in every building housing communication center equipment.

4-6.9.1.4.1.6 Each conductor entering a fire station from partially or entirely aerial lines shall be protected by a lightning arrester.

4-6.9.1.4.2 Communication Center.

4-6.9.1.4.2.1 All conductors entering the communication center shall be protected by the following devices, in the order named, starting from the exterior circuit:

(a) A fuse rated at 3 amps minimum to 7 amps maximum, and not less than 2000 volts
(b) A lightning arrester
(c) A fuse or circuit breaker, rated at $\frac{1}{2}$ amp.

4-6.9.1.4.2.2 The $\frac{1}{2}$-amp protection on the tie circuits shall be omitted at subsidiary communication centers.

4-6.9.1.4.3 Protection on Aerial Construction.

4-6.9.1.4.3.1 At junction points of open aerial conductors and cable, each conductor shall be protected by a lightning arrester of weatherproof type. There shall also be a connection between the lightning arrester ground, any metallic sheath, and messenger wire.

4-6.9.1.4.3.2 Aerial open-wire and non-messenger-supported 2-conductor cable circuits shall be protected by a lightning arrester at intervals of approximately 2000 ft (610 m).

4-6.9.1.4.3.3 Lightning arresters, other than air-gap or self-restoring type, shall not be installed in fire alarm circuits.

4-6.9.1.4.3.4 All protective devices shall be accessible for maintenance and inspection.

4-6.10 Power.

4-6.10.1 Requirements for Constant-Current Systems.

The coded wired public fire alarm reporting system normally operates at a constant current of nominally 100 milliamperes. The requirements of 4-6.10.1.1 through 4-6.10.1.4 help ensure that such a system will maintain a high level of operational integrity.

4-6.10.1.1 Means shall be provided for manually regulating current in box circuits so that operating current is maintained within 10 percent of normal throughout changes in external circuit resistance from 20 percent above to 50 percent below normal.

4-6.10.1.2 The voltage supplied to maintain normal line current on box circuits shall not exceed 150 volts, measured under no-load conditions, and shall be such that the line current will not be reduced below safe operating value by the simultaneous operation of four boxes.

4-6.10.1.3 Visual and audible means to indicate a 20 percent or greater reduction in the normal current in any alarm circuit shall be provided. All devices connected in series with any alarm circuit shall function properly when the alarm circuit current is reduced to 70 percent of normal.

4-6.10.1.4 Sufficient meters shall be provided to indicate the current in any box circuit and the voltage of any power source. Meters used in common for several circuits shall be provided with

cut-in devices designed to reduce the probability of cross-connecting circuits.

4-6.11 Receiving Equipment — Facilities for Receipt of Box Alarms.

The requirements of the following set of subsections are based on the premise that the signals transmitted over the public fire alarm reporting system will be received at the public fire service communication center and automatically recorded in a manner that provides a permanent visual record of the signals. At the same time, the operators will be alerted to incoming signals by an audible notification appliance. The signal will indicate the exact location from which it is transmitted. This is often done by assigning a unique number to each public fire alarm box. A chart or an interface to a computer-aided dispatching system — the requirements of which are covered in NFPA 1221 — will translate the box number to an exact location.

4-6.11.1 General.

4-6.11.1.1 Alarms from boxes shall be automatically received and recorded at the communication center.

4-6.11.1.2 A permanent visual record and an audible signal shall be required to indicate the receipt of an alarm. The permanent record shall indicate the exact location from which the alarm is being transmitted.

NOTE: The audible signal device may be common to several box circuits and arranged so that the fire alarm operator can manually silence the signal temporarily by a self-restoring switch.

4-6.11.1.3 Facilities shall be provided that will automatically record the date and time of receipt of each alarm, except that only the time need be automatically recorded in voice recordings.

4-6.11.2 Visual Recording Devices.

4-6.11.2.1 A device for producing a permanent graphic recording of all alarm, supervisory, trouble, and test signals received and/or retransmitted shall be provided at each communication center for each alarm circuit and tie circuit. If each circuit is served by a dedicated recording device, then the number of reserve recording devices required on site shall be equal to at least 5 percent of the circuits in service and in no case less than 1 percent. If two or more circuits are served by a common recording device, then a reserve recording device shall be available on site for each circuit connected to a common recorder.

4-6.11.2.2 In a Type B wire system, one such device shall be installed in each fire station and at least one in the communication center.

4-6.12 Supervision.

The requirements of 4-6.12.1 through 4-6.12.5 help ensure that any impairment to the circuits or power supplies of the public fire alarm reporting system will result in a trouble signal at the public fire service communication center.

4-6.12.1 To ensure reliability, wired circuits upon which transmission and receipt of alarms depend shall be under constant electrical supervision to give prompt warning of conditions adversely affecting reliability.

4-6.12.2 The power supplied to all required circuits and devices of the system shall be supervised.

4-6.12.3 Trouble signals shall actuate a sounding device located where there is always a responsible person on duty.

4-6.12.4 Trouble signals shall be distinct from alarm signals and shall be indicated by both a visual light and an audible signal.

NOTE 1: The audible signal may be common to several supervised circuits.

NOTE 2: A switch for silencing the audible trouble signal is permitted if the visual signal remains operated until the silencing switch is restored to its normal position.

4-6.12.5 The audible signal shall be responsive to faults on any other circuits that may occur prior to restoration of the silencing switch to normal.

4-6.13 Coded Wired Reporting Systems.

The following subsections (4-6.13, 4-6.14, 4-6.15, and 4-6.16) give the requirements for each of the four types of public fire alarm reporting systems: coded wired, coded radio, telephone (series), and telephone (parallel).

4-6.13.1 For a Type B system, the effectiveness of noninterference and succession functions between box circuits shall be no less than between boxes in any one circuit. The disablement of any metallic box circuit shall cause a warning signal in all other circuits, and, thereafter, the circuit or circuits not otherwise broken shall be automatically restored to operative condition.

In a Type B coded wired system, signals on one box circuit or alarm circuit are "repeated" on to the other box and alarm circuits. By repeating these signals, other boxes that might be actuated while the first signals are being transmitted will sense that the circuit is busy and wait until it is clear before transmitting. This ensures the proper functioning of the noninterfering and successive features.

4-6.13.2 Box circuits shall be sufficient in number and so laid out that the areas that would be left without box protection in case of disruption of a circuit would not exceed that covered by 20 properly spaced boxes where all or any part of the circuit is of aerial open-wire, or 30 properly spaced boxes where the circuit is entirely in underground or messenger-supported cable.

4-6.13.3 Where all boxes on any individual circuit and associated equipment are designed and installed to provide for receipt of alarms through the ground in event of a break in the circuit, it is permissible for the circuit to serve twice the above figures for aerial open-wire and cable circuits, respectively.

Most coded wired systems are designed so that if a fire alarm box senses that the circuit has an open fault, it will idle for one round and then connect the box to earth ground. The receiving equipment at the public fire service communication center will also connect itself to earth ground once it senses an open circuit. This "conditioning" of the circuit will allow the box to transmit its signal through earth ground. Naturally, if two open faults exist on the circuit, the boxes isolated between the faults cannot transmit a signal.

4-6.13.4 The installation of additional boxes in an area served by the number of properly spaced boxes indicated above shall not constitute geographical overloading of a circuit.

Once the boxes have been properly spaced throughout an area, the addition of more boxes (e.g., when master fire alarm boxes are added to allow the connection of protected premises fire alarm systems to the public fire service communication center) will not be considered to be overloading the circuit.

4-6.13.5 Sounding devices for signals shall be provided for box circuits.

NOTE 1:   In a Type A system, it is satisfactory to use a common sounding device for more than one circuit, and it should be installed at the communication center.

NOTE 2:   In a Type B system, a sounding device is to be installed in each fire station at the same location as the recording device for that circuit, except that at the communication center, a common sounding device is permitted.

4-6.14 Coded Radio Reporting Systems.

4-6.14.1 Radio Box Channel (Frequency).

4-6.14.1.1 The number of boxes permitted on a single frequency shall be governed by the following:

(a) For systems utilizing one-way transmission in which the individual box automatically initiates the required message (*see 4-6.14.1.4*) using circuitry integral to the boxes, not more than 500 boxes shall be permitted on a single frequency.

(b) For systems utilizing a two-way concept in which interrogation signals (*see 4-6.14.1.4*) are transmitted to the individual boxes from the communication center on the same frequency used for receipt of alarms, not more than 250 boxes shall be permitted on a single frequency. If interrogation signals are transmitted on a frequency different from that used for receipt of alarms, not more than 500 boxes shall be permitted on a single frequency.

(c) A specific frequency shall be designated for both fire and other fire-related or public safety alarm signals, and supervisory signals (test and tamper). All acknowledgment and other signals shall utilize a separate frequency.

This requirement prevents the public fire service communication center from using the radio frequency assigned to the boxes for normal two-way or one-way radio communication. Such a use might inadvertently interfere with receipt of signals from the boxes.

4-6.14.1.2 Where box message signals to the communication center or acknowledgment of message receipt signals from the communication center to the box are repeated, associated repeating facilities shall conform to the requirements indicated in NFPA 1221, *Standard for the Installation, Maintenance, and Use of Public Fire Service Communication Systems*, 3-4.1.2.

4-6.14.1.3 All coded radio box systems shall provide constant monitoring of the frequency in use. Both an audible and visual indication of any sustained carrier signal (when in excess of 15 seconds' duration) shall be provided for each receiving system at the communication center.

A sustained carrier that would interfere with the transmission of signals from the boxes would be analogous to an open fault or ground fault on a coded wired public fire alarm reporting system.

4-6.14.1.4 Each coded radio box shall automatically transmit a message at least once during each 24-hour period.

The 24-hour test signal safeguards against the catastrophic failure of a single fire alarm box.

4-6.14.2 Metallic Interconnections. Accessible and reliable means, available only to the agency responsible for maintaining the public fire alarm reporting system, shall be provided for disconnecting the auxiliary loop to the box inside the building, and definite notification shall be given to occupants of the building when the interior box is not in service.

If a connection is made to the public fire alarm reporting system in accordance with the requirements of Section 4-7, a means must be provided to disconnect the protected premises connection. This means must only be available to the authority having jurisdiction over the public fire alarm reporting system.

4-6.14.3 Receiving Equipment — Facilities for Receipt of Box Alarms.

4-6.14.3.1 Type A System.

4-6.14.3.1.1* For each frequency used, two separate receiving networks, each including an antenna, audible alerting device, receiver, power supply, signal processing equipment, a means of providing a permanent graphic recording of the incoming message that is both timed and dated, and other associated equipment shall be provided and shall be installed at the communication center. Facilities shall be so arranged that a failure of either receiving network will not affect the receipt of messages from boxes.

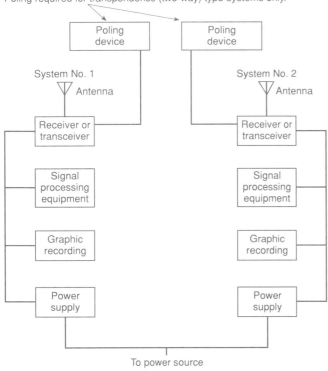

Poling required for transpondence (two-way) type systems only.

Figure A-4-6.14.3.1.1

These requirements for redundant receiving equipment help ensure the overall reliability of the coded radio public fire alarm reporting sytem.

4-6.14.3.1.2 Where the system configuration is such that a polling device is incorporated into the receiving network to allow remote/selective initiation of box tests (*see Chapter 7*), a separate such device shall be included in each of the two required receiving networks. Further, the polling devices shall be configured for automatic cycle initiation in their primary operating mode, capable of continuous self-monitoring, and integrated into the network(s) to provide automatic switchover and operational continuity in the event of failure of either device.

Some coded radio systems provide for an interrogation and response sequence initiated from the public fire service communication center to ensure the operational integrity of the system. When such an arrangement is provided, the requirements of 4-6.14.3.1.2 indicate the need for redundancy to further ensure the integrity of the system.

4-6.14.3.1.3 Test signals from boxes shall not be required to include the date as part of their permanent recording, providing that the date is automatically printed on the recording tape at the beginning of each calendar day.

4-6.14.3.2 Type B System.

4-6.14.3.2.1 For each frequency used, a single complete receiving network shall be permitted in each fire station, providing the communication center conforms to 4-6.14.3.1.1. If the jurisdiction maintains in operation two or more alarm reception points, one receiving network shall be permitted to be at each alarm reception point.

4-6.14.3.2.2 If alarm signals are transmitted to a fire station from the communication center using the coded radio-type receiving equipment in the fire station to receive and record the alarm message, a second receiving network conforming to 4-6.14.3.2.1 shall be provided at each fire station, and that receiving network shall employ a frequency other than that used for the receipt of box messages.

4-6.14.4 Power. Power shall be provided in accordance with 4-6.7.

4-6.14.5 Testing. (*See Chapter 7.*)

4-6.14.6 Supervision. Radio repeaters upon which receipt of alarms depend shall be provided with dual receivers and transmitters. Failure of the primary transmitter or receiver shall cause an automatic switchover to the secondary receiver and transmitter.

Exception: If the repeater controls are located where someone is always on duty, manual switchover shall be permitted if it can be completed within 30 seconds.

4-6.15 Telephone (Series) Reporting Systems.

Such systems are sometimes added to all or a portion of an existing coded wired reporting system to give that system the capability of transmitting and receiving voice alarms. In this case, the telephone (series) system uses the same cable plant as the coded wired system.

4-6.15.1 A permanent visual recording device installed in the communication center shall be provided to record all incoming box signals. A spare recording device shall be provided for five or more box circuits.

This permanent visual recording device records the date, time, and box number, but not the content of the voice message. See 4-6.15.4.

4-6.15.2 A second visual means of identifying the calling box shall be provided.

4-6.15.3 Audible signals shall indicate all incoming calls from box circuits.

4-6.15.4 All voice transmissions from boxes for emergencies shall be recorded with the capability of instant playback.

Special audio recording equipment has been designed that not only provides an audio log of signal content from the boxes, but also can be instantly recycled to the beginning of each message. This allows operators at the public fire service communication center to review messages when the content is not clear.

4-6.15.5 A voice recording facility shall be provided for each operator handling incoming alarms to eliminate the possibility of interference.

4-6.15.6 Box circuits shall be sufficient in number and so laid out that the areas that would be left without box protection in case of disruption of a circuit would not exceed that covered by 20 properly spaced boxes where all or any part of the circuit is of aerial open-wire, or 30 properly spaced boxes where the circuit is entirely in underground or messenger-supported cable.

4-6.15.7 Where all boxes on any individual circuit and associated equipment are designed and installed to provide for receipt of alarms through the ground in event of a break in the circuit, it shall be permissible for the circuit to serve twice the above figures for aerial open-wire and cable circuits, respectively.

Some telephone (series) systems are designed so that if a fire alarm box senses that the circuit has an open fault, it will connect itself to earth ground. The receiving equipment at the public fire service communication center will also connect itself to earth ground once it senses an open circuit. This "conditioning" of the circuit will allow the box to transmit its signal through earth ground. Naturally, if two open faults exist on the circuit, the boxes isolated between the faults cannot transmit a signal.

4-6.15.8 The installation of additional boxes in an area served by the number of properly spaced boxes indicated above shall not constitute geographical overloading of a circuit.

Once the boxes have been properly spaced throughout an area, the addition of more boxes (e.g., when master fire alarm boxes are added to allow the connection of protected premises fire alarm systems to the public fire service communication center) will not be considered to be overloading the circuit.

4-6.16 Telephone (Parallel) Reporting Systems.

Telephone (parallel) public fire alarm reporting systems are normally leased from the public telephone company utility. In most systems, each box has its own circuit that leads back to the public fire service communication center. However, in some cases, each box is connected to a concentrator-identifier, which is in turn connected to the public fire service communication center.

4-6.16.1 Box Circuits.

4-6.16.1.1 If a municipal box is installed inside a building, it shall be placed as close as practical to the point of entrance of the circuit, and the exterior wire shall be installed in conduit or electrical metallic tubing, in accordance with Chapter 3 of NFPA 70, *National Electrical Code.*

4-6.16.1.2 Accessible and reliable means, available only to the authority having jurisdiction or the agency responsible for maintaining the public fire alarm reporting system, shall be provided for disconnecting the box inside the building, and definite notification shall be given to occupants of the building when the interior box is not in service.

If a connection is made to the public fire alarm reporting system in accordance with the requirements of Section 4-7 of this code, a means must be provided to disconnect the protected premises connection. This means must only be available to the authority having jurisdiction over the public fire alarm reporting system.

4-6.16.1.3 A separate circuit shall be provided for each box.

4-6.16.1.4 Where a concentrator-identifier or similar device is employed, at least two tie circuits for the first 40 boxes connected shall be provided to the communication center. A tie circuit shall be provided for each 40 additional boxes, or fraction thereof, connected to the concentrator-identifier.

NOTE: These tie circuits are not to be used for any other purpose or function.

A concentrator-identifier is used to connect a group of boxes in a particular geographic area to the public fire service communication center. This device is used where the length of circuit from each box to the communication center would be excessively long. The concentrator-identifier is connected to the communication center by means of at least two tie circuits. The concentrator-identifier precisely indicates at the communication center exactly which box has been operated.

4-6.16.1.5 Power shall be provided in accordance with Section 4-6.7

4-6.16.2 Receiving Equipment — Facilities for Receipt of Box Alarms.

4-6.16.2.1 The box circuits shall be terminated:

(a) Directly on a console or switchboard located in the communication center, or

(b) In concentrator-identifier equipment located in a subsidiary communication center.

NOTE: The audible signal device may be common to several box circuits and arranged so that the operator can manually silence the signal temporarily with a self-restoring switch.

4-6.16.2.2 All voice transmissions from boxes for emergencies shall be recorded with the capability of instant playback.

Special audio recording equipment has been designed that not only provides an audio log of signal content from the boxes, but also can be instantly rewound to the beginning of each message. This allows operators at the public fire service communication center to review messages when the content is not clear.

4-6.16.2.3 A means of voice recording shall be provided for each operator handling incoming alarms to eliminate the possibility of interference.

4-6.16.2.4 Either a continuous line test or periodic (up to 6 minutes) automatic line tests shall detect an open, short, ground, or leakage condition. If one of these conditions occurs, a visual and audible trouble signal shall be actuated where there is an operator on duty.

The equipment supplied by the public telephone company utility either provides continuous supervision of each circuit for the faults stated in 4-6.16.2.4 or it provides a test for these faults no less frequently than every 6 minutes.

4-7   Auxiliary Fire Alarm Systems.

NOTE: The requirements of Chapters 1, 3, 5, 6, 7, and Section 4-2 shall apply to auxiliary fire alarm systems, except where they conflict with the requirements of this section.

4-7.1 Scope. This section describes the equipment and circuits necessary to connect a protected premises (*see Chapter 3*) to a public fire alarm reporting system (*see Section 4-6*).

4-7.2 General.

4-7.2.1 An auxiliary fire alarm system shall be used only in connection with a public fire alarm reporting system that is suitable for the service. A system satisfactory to the authority having jurisdiction shall be considered as meeting this requirement.

If there is no public fire alarm reporting system in a community, there can be no auxiliary fire alarm system. An auxiliary system is dependent upon the public fire alarm reporting system to transport signals from the protected premises to the public fire service communication center.

4-7.2.2 Permission for the connection of an auxiliary fire alarm system to a public fire alarm reporting system and acceptance of the type of auxiliary transmitter, its actuating mechanism, circuits, and components connected thereto, shall be obtained from the authority having jurisdiction.

4-7.2.3 An auxiliary fire alarm system shall be maintained and supervised by a responsible person or corporation.

Proper maintenance of an auxiliary system requires careful coordination with those who operate and maintain the public fire alarm reporting system.

4-7.2.4 Section 4-7 does not require the use of audible alarm signals other than those necessary to operate the auxiliary fire alarm system. If it is desired to provide fire alarm evacuation signals in the protected property, the alarms, circuits, and controls shall comply with the provisions of Chapter 3, in addition to the provisions of Section 4-7.

An auxiliary system is not in itself designed to notify occupants of a fire alarm. If such notification is required, then a protected premises fire alarm system with notification appliances must be installed in accordance with Chapter 3.

4-7.3 Communication Center Facilities. The communication center facilities shall be in accordance with the requirements of Section 4-6.

4-7.4 Equipment.

4-7.4.1 Types of Systems. There are three types of auxiliary fire alarm systems in use, and these are described in (a), (b), and (c) below.

The detailed requirements of 4-7.4.1(a) through (c) describe the three types of auxiliary systems: local energy, shunt, and parallel telephone. In each case, the requirements help ensure the operational integrity of the particular type of system. The shunt system is generally considered the least desirable.

(a)* The local energy type [Figure A-4-7.4.1(a)(1)] is electrically isolated from the public fire alarm reporting system and has its own power supply. The tripping of the transmitting device does not depend on the current in the system. In a wired circuit, whether or not the alarm will be received by the communication center if the circuit is accidently opened depends on the design of the transmitting device and the associated communication center equipment, i.e., whether or not the system is designed to receive alarms through manual or automatic ground operational facilities. In a radio box type system, whether or not the alarm will be received by the communication center depends on the proper operation of the radio transmitting and receiving equipment.

1. Local energy systems shall be permitted to be of coded or noncoded type.

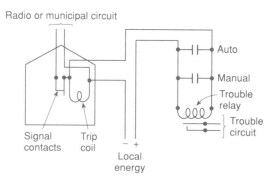

Figure A-4-7.4.1(a)(1)

2. Power supply sources for local energy systems shall conform to Chapter 1.

(b)* The shunt type [Figure A-4-7.4.1(b)(1)] is electrically connected to and is an integral part of the public fire alarm report-

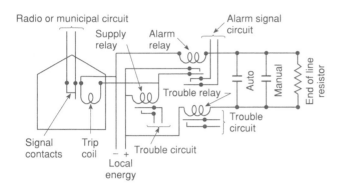

Figure A-4-7.4.1(a)(2)

ing system. A ground fault on the auxiliary circuit is a fault on the public fire alarm reporting system circuit, and an accidental opening of the auxiliary circuit will send a needless (or false) alarm to the communication center. An open circuit in the transmitting device trip coil will not be indicated either at the protected property or at the communication center; also, if an initiating device is operated, an alarm will not be transmitted but an open circuit indication will be given at the communications center. If a public fire alarm reporting system circuit is open when a connected shunt type system is operated, the transmitting device will not trip until the public fire alarm reporting system circuit returns to normal, at which time the alarm will be transmitted unless the auxiliary circuit is first returned to a normal condition.

A local system made into an auxiliary system by the addition of a relay whose coil is energized by a local power supply and whose normally closed contacts trip a shunt type master box shall not be permitted [Figure A-4-7.4.1(b)(2)].

1. Shunt systems shall be noncoded with respect to any remote electrical tripping or actuating devices.

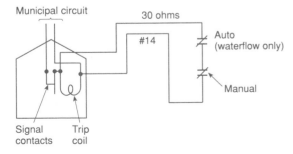

Figure A-4-7.4.1(b)(1)

2. All conductors of the shunt circuit shall be installed in accordance with Article 346, for rigid conduit, or Article 348, for electrical metallic tubing, of NFPA 70, *National Electrical Code.*

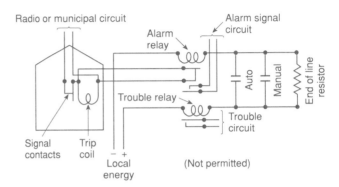

Figure A-4-7.4.1(b)(2)

3. Both sides of the shunt circuit shall be in the same conduit.

4. Where an auxiliary transmitter is located within a private premise, it shall be installed in accordance with 4-6.9.1.

5. Where a shunt loop is used, it shall not exceed a length of 750 ft (230 m) and shall be in conduit.

6. Conductors of the shunt circuits shall not be smaller than No. 14 AWG and shall be insulated as prescribed in NFPA 70, *National Electrical Code*, Article 310.

7. The power for shunt-type systems shall be provided by the public fire alarm reporting system.

8. Additional design restrictions for shunt systems shall be found in laws or ordinances.

(c)* A parallel telephone type system [Figure A-4-7.4.1(c)] is a system in which alarms are transmitted over a circuit directly connected to the annunciating switchboard at the public fire service communication center and terminated at the protected property by an end-of-line device.

Such auxiliary systems are for connection to public fire alarm reporting systems of the type in which each alarm box annunciates at the communication center by individual circuit.

NOTE: The essential difference between the local energy or parallel telephone types and the shunt-type system is that accidental opening of the alarm initiating circuits will cause an alarm on the shunt type system only.

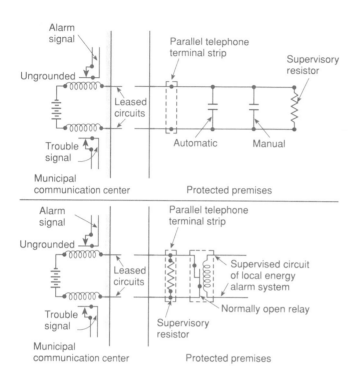

Figure A-4-7.4.1(c)

1. Parallel telephone systems shall be noncoded with respect to any remote electrical tripping or actuating devices.

2. Two methods of parallel telephone systems shall be permitted to be used:

(i) The circuits are extended beyond the entrance termination point to actuating devices with the supervisory device beyond the last actuating device in the circuit; or

(ii) The supervisory device for the circuit is located at the entrance termination point. The tripping relay shall be located immediately adjacent to the supervisory device and shall be connected thereto with conductors not smaller than No. 14 AWG in conduit.

3. Nonvoice circuits connected to a parallel telephone system shall be indicated with distinctive and different color from voice circuits and shall be grouped in a reserved separate section of the receiving equipment with adequate written warning that no voice is to be expected on these alarms and that the fire department must be dispatched on alarm light indications.

4-7.4.2 The interface of the three types of auxiliary fire alarm systems with the four types of public fire alarm reporting systems shall be in accordance with Table 4-7.4.2.

Table 4-7.4.2

	Local Energy-Type	Shunt-Type	Parallel-Type
Coded Wired	Yes	Yes	No
Coded Radio	Yes	No	No
Telephone Series	Yes	No	No
Telephone Parallel	No	No	Yes

4-7.4.3 The application of the three types of auxiliary fire alarm systems shall be limited to the initiating devices specified in Table 4-7.4.3.

Table 4-7.4.3

	Local Energy-Type	Shunt-Type	Parallel-Type
Manual Fire Alarm	Yes	Yes	Yes
Waterflow or Actuation of Extinguishing System	Yes	Yes	Yes
Automatic Detection Devices	Yes	No	Yes

4-7.4.4 Location of Transmitting Devices.

4-7.4.4.1 Auxiliary systems shall be arranged so that one auxiliary transmitter does not serve more than 100,000 sq ft (9290 m²) total area, unless otherwise permitted by the authority having jurisdiction.

4-7.4.4.2 A separate auxiliary transmitter shall be provided for each building or where permitted by the authority having jurisdiction for each group of buildings of single ownership or occupancy.

4-7.4.4.3 The same box shall be permitted to be used as a public fire alarm reporting system box and as a transmitting device for an auxiliary system where permitted by the authority having jurisdiction, provided that the box is located at the outside of the entrance to the protected property.

NOTE: The fire department may require the box to be equipped with a signal light to differentiate between automatic and manual operation, unless local outside alarms at the protected property would serve the same purpose.

4-7.4.4.4 The transmitting device shall be located as required by the authority having jurisdiction.

4-7.4.4.5 The system shall be so designed and arranged that a single fault on the auxiliary system shall not jeopardize operation of the public fire alarm reporting system and shall not, in case of a single fault on either the auxiliary or public fire alarm reporting system, transmit a false alarm on either system.

Exception: Shunt systems. [See 4-7.4.1(b).]

4-7.5 **Personnel.** Personnel necessary to receive and act on signals from auxiliary fire alarm systems shall be in accordance with the requirements of Section 4-6 and NFPA 1221, *Standard for the Installation, Maintenance, and Use of Public Fire Service Communication Systems.*

4-7.6 **Operations.** Operations for auxiliary fire alarm systems shall be in accordance with the requirements of Section 4-6 and NFPA 1221, *Standard for the Installation, Maintenance, and Use of Public Fire Service Communication Systems.*

4-7.7 **Testing and Maintenance.** Testing and maintenance of auxiliary fire alarm systems shall be in accordance with the requirements of Chapter 7.

5

Initiating Devices

Contents, Chapter 5

5-1 General
  5-1.1 Scope
  5-1.2 Application
    5-1.2.3 Flame Detector
    5-1.2.4 Spark/Ember Detector
  5-1.3 Installation and Required Location of Detection
    Devices
  5-1.4 Connection to the Fire Alarm System
5-2 Heat-Sensing Fire Detectors
  5-2.3 Operating Principles
    5-2.3.1 Fixed Temperature Detector
      5-2.3.1.2 Thermal Lag
    5-2.3.2 Rate Compensation Detector
    5-2.3.3 Rate-of-Rise Detector
  5-2.4 Classification and Sensitivity
  5-2.5 Location
  5-2.6 Temperature
  5-2.7 Spacing
    5-2.7.1 Smooth Ceiling Spacing
      5-2.7.1.1 Irregular Areas
      5-2.7.1.2 High Ceilings
    5-2.7.2 Solid Joist Construction
    5-2.7.3 Beam Construction
    5-2.7.4 Sloped Ceilings
      5-2.7.4.1 Peaked
      5-2.7.4.2 Shed
5-3 Smoke-Sensing Fire Detectors
  5-3.1 General
  5-3.3 Principles of Detection
    5-3.3.1 Ionization Smoke Detection
    5-3.3.2 Photoelectric Light-Scattering Smoke Detection
    5-3.3.3 Photoelectric Light Obscuration Smoke Detection
    5-3.3.4 Cloud Chamber Smoke Detection
  5-3.4 Sensitivity
  5-3.5 Location and Spacing
    5-3.5.1 General
      5-3.5.1.2 Stratification
    5-3.5.2 Spot-Type Smoke Detectors
    5-3.5.3 Projected Beam-Type Smoke Detectors
    5-3.5.4 Sampling-Type Smoke Detectors
    5-3.5.5 Smooth Ceiling Spacing
      5-3.5.5.1 Spot-Type Detectors

5-3.5.5.2 Projected Beam-Type Detectors

5-3.5.6 Solid Joist Construction

5-3.5.7 Beam Construction

5-3.5.8 Sloped Ceilings

    5-3.5.8.1 Peaked

    5-3.5.8.2 Shed

5-3.5.9 Raised Floors and Suspended Ceilings

5-3.5.10 Partitions

5-3.6 Heating, Ventilating, and Air Conditioning (HVAC)

5-3.7 Special Considerations

    5-3.7.2 Spot-Type Detectors

    5-3.7.3 Projected Beam-Type Detectors

    5-3.7.4 Air Sampling-Type Detectors

    5-3.7.5 High Rack Storage

    5-3.7.6 High Air Movement Areas

        5-3.7.6.1 General

        5-3.7.6.2 Location

        5-3.7.6.3 Spacing

5-4 Radiant Energy-Sensing Fire Detectors

5-4.1 General

    5-4.1.2 Radiant Energy

5-4.2 Definitions and Operating Principles

    5-4.2.1 Definitions

    5-4.2.2 Operating Principles of Flame Detectors

    5-4.2.3 Operating Principles of Spark/Ember Detectors

5-4.3 Fire Characteristics and Detector Selection

5-4.4 Spacing Considerations

    5-4.4.1 General Rules

    5-4.4.2 Spacing Considerations for Flame Detectors

5-4.5 Spacing Considerations for Spark/Ember Detectors

5-4.6 Other Considerations

5-5 Gas-Sensing Fire Detectors

5-5.5 Operating Principles

    5-5.5.1 Semiconductor

    5-5.5.2 Catalytic Element

5-5.6 Location and Spacing

    5-5.6.1 General

        5-5.6.1.2 Stratification

    5-5.6.4 Smooth Ceiling Spacing

        5-5.6.4.1 Spot-Type Detectors

    5-5.6.5 Solid Joist Construction

    5-5.6.6 Beam Construction

    5-5.6.7 Sloped Ceilings

        5-5.6.7.1 Peaked

        5-5.6.7.2 Shed

5-5.6.8 Suspended Ceilings

5-5.6.9 Partitions

5-5.7 Heating, Ventilating, and Air Conditioning (HVAC)

5-5.8 Special Considerations

5-6 Other Fire Detectors

5-6.5 Location and Spacing

5-6.6 Special Considerations

5-7 Sprinkler Waterflow Alarm-Initiating Devices

5-8 Detection of the Operation of Other Automatic Extinguishing Systems

5-9 Manually Actuated Alarm-Initiating Devices

    5-9.1.1 Mounting

    5-9.1.2 Distribution

    5-9.2 Publicly Accessible Fire Service Boxes (Street Boxes)

        5-9.2.12 Coded Radio Street Boxes

        5-9.2.13 Power Source

        5-9.2.14 Design of Telephone Street Boxes (Series or Parallel)

5-10 Supervisory Signal-Initiating Devices

5-10.1 Control Valve Supervisory Signal-Initiating Device

5-10.2 Pressure Supervisory Signal-Initiating Device

5-10.3 Water Level Supervisory Signal-Initiating Device

5-10.4 Water Temperature Supervisory Signal-Initiating Device

5-10.5 Room Temperature Supervisory Signal-Initiating Device

5-11 Smoke Detectors for Control of Smoke Spread

5-11.4 Purposes

5-11.5 Application

    5-11.5.1 Area Detectors Within Smoke Compartments

    5-11.5.2 Smoke Detection for the Air Duct System

        5-11.5.2.1 Supply Air System

        5-11.5.2.2 Return Air System

5-11.6 Location and Installation of Detectors in Air Duct Systems

5-11.7 Smoke Detectors for Door Release Service

    5-11.7.4 Number of Detectors Required

    5-11.7.5 Location

There are several concepts that provide the basis for many of the requirements in this chapter of the *National Fire Alarm Code*. This chapter covers manual fire alarm boxes, waterflow alarm devices, and system supervisory devices.

Nothing can be done to fight a fire until someone or something knows the fire exists. The sooner the fire is detected, the better since small fires are easier to extinguish than large fires.

The only fire detector with a brain is a human being. Automatic fire detectors are not very smart and have a finite likelihood of rendering an unwanted alarm signal. The more sensitive the detector, the higher the probability of an unwanted alarm. Every requirement in Chapter 5 of the *National Fire Alarm Code* stems from the need for speed and surety of response to a fire with minimal probability that an alarm signal will result from some nonfire circumstance.

Fire detection devices do not respond to fire but to some change in the ambient conditions created by a fire in the immediate vicinity of the detector. A heat detector responds to an increase in the ambient temperature. A smoke detector responds to the introduction of smoke into the air within the detector. A flame detector responds to the influx of radiant energy that travels from the fire to the detector. In each case, either heat, smoke, or light must travel from the fire to the detector.

Naturally, in order to achieve the objective of quick and reliable response, we must take into account the time required for the travel of fire products. Many of the placement and spacing criteria established by this code for automatic fire detectors are intended to provide detectors in sufficient quantity so that there is a detector in the path of heat, smoke, or radiant energy regardless of where the fire is located. The criteria established for the installation of each type of detection device are based upon years of experience and the best research data currently available.

5-1 General.

5-1.1 Scope. This chapter covers minimum requirements for performance, selection, use, and location of automatic fire detection devices, sprinkler waterflow detectors, manually activated fire alarm stations, and supervisory signal initiating devices, including guard tour reporting used to ensure timely warning for the purposes of life safety and the protection of a building, space, structure, area, or object.

NOTE: For detector requirements in a household system, refer to Chapter 2.

It is crucial to understand the scope limitations of Chapter 5. When this code is referenced in laws or ordinances, the requirements of the code assume the "effect of law." The scope statement determines whether the provisions of this chapter are applicable to a device in question.

The term "initiating device" has been broadened to cover not only traditional fire detection devices, but also other devices that monitor some condition related to fire safety. This includes sprinkler system flow switches, pressure switches, valve tamper switches, manual alarm stations, municipal fire alarm boxes, and any signaling switches used to monitor special extinguishing systems. The requirements in Section 5-1 are applied to all monitoring devices covered in 5-1.1 that provide information, either in the form of a switch transition or electronic code, to a fire alarm control panel.

5-1.2 Application.

5-1.2.1 The material in this chapter is intended for use by persons knowledgeable in the application of fire detection and fire alarm systems/services.

5-1.2.2 Automatic and manual initiating devices contribute to life safety, fire protection, and property conservation only when used in conjunction with other equipment. The interconnection of these devices with control equipment configurations, and power supplies or with output systems responding to external actuation is detailed elsewhere in this code and others.

Chapter 5 of the *National Fire Alarm Code* only covers the requirements relevant to the installation of fire alarm and supervisory initiating devices when they are required by some other code or standard, such as NFPA *101*, *Life Safety Code*. Chapter 5 establishes the selection and placement criteria that determine the necessary number and type of detectors, but it does not address which types of facilities need initiating devices. The requirement for detection or some form of initiating device is established in the codes and standards that cover a specific class of occupancy or, in some cases, a class of fire protection system. Once this requirement has been established by some other code or standard, the reader then refers to this code for the specifics of selection and placement.

For example, NFPA 664, *Standard for the Prevention of Fires and Explosions in Wood Processing and Woodworking Facilities*, requires the use of spark/ember detectors in certain specific instances. The reader of NFPA 664 must then refer to this chapter of the *National Fire Alarm Code* for the relevant installation requirements for spark/ember detectors. In a second example, Section 28 of NFPA *101*, which covers industrial facilities, requires a fire alarm system when there are more than 100 people on site or more than 25 people on a floor other than the ground floor. The fire protection designer of such an industrial site must then refer to NFPA 72 for the relevant requirements for that fire alarm system. The designer must refer to Section 5-1 for the determination of the type, quantity, and placement of the fire detection devices. The designer must also refer to Section 5-1 for the installation requirements for flow switches, pressure switches, and other initiating devices that may be required by NFPA 13, *Standard for the Installation of Sprinkler Systems*, or NFPA 12, *Standard on Carbon Dioxide Extinguishing Systems*, or other such standards.

5-1.3 Installation and Required Location of Detection Devices.

This portion of Chapter 5 begins with some basic requirements that apply to all initiating devices, regardless of type. These requirements have come from years of experience relating to the installation of heat, smoke, or radiant energy-sensing detectors. However, with the broadening of the scope of this chapter to cover supervisory switches, manual fire alarm boxes, and other types of initiating devices, these general requirements are applicable to those types of initiating devices also.

5-1.3.1 Where subject to mechanical damage, detectors shall be protected.

While this paragraph refers to detectors, the prudent designer and installer would apply this requirement to every component of the fire alarm system. Nevertheless, the cause of many unwanted alarms or system failures has been found to be the result of damage to a detector or other initiating

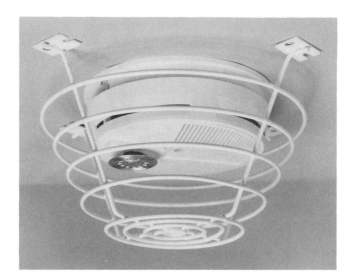

Figure 5.1(a)   Combination smoke and heat detector with protective cover.

Figure 5.1(b)   Rate-of-rise heat detector with protective cover.

device. See Figures 5.1(a) and 5.1(b) for examples of protected detectors.

Mechanical damage is not necessarily limited to catastrophic destruction. Mechanical damage can occur over an extended period of time from vibration, extremes in temperature, corrosive atmospheres, or excessive humidity. The designer and installer must be sure that the initiating device will be appropriate for the environment in which it is to be installed.

5-1.3.2 In all cases, detectors shall be supported independently of their attachment to the circuit conductors.

This requirement is intended to be applied to all types of initiating devices. The copper used in the installation of conductors is not formulated to serve as a mechanical support. Copper, by its very nature, fatigues over time if placed under a mechanical stress. This fatigue results in increasing brittleness and increasing electrical resistance. Ultimately, the fatigued conductor either breaks or its resistance becomes too high to allow the initiating device circuit to function properly. In either case, the operation of the circuit is impaired, and a tragic loss of life could conceivably result because of fire alarm system failure.

Initiating devices should always be mounted in the manner shown in the manufacturer's instructions. The requirements for listing include a method for mounting that adequately supports the initiating device so that no mechanical stresses are applied to the circuit conductors. When the instructions show use of a backbox, then it is a requirement of the listing, and the specific type of backbox shown must be used. If not shown, the use of a backbox is determined by field conditions and the requirements of NFPA 70, *National Electrical Code.*

5-1.3.3 Detectors shall not be recessed in any way into the mounting surface, unless they have been tested and listed for such recessed mounting.

Recessing fire detectors has an adverse effect on the ability of the detector to perform as intended. Consider first the case of a heat detector that has been recessed, contrary to this code. A heat detector must absorb heat from the fire before it can respond. Approximately 95–98 percent of the heat a detector receives is carried to the detector in air currents created by the fire. This is called convection. However, when a detector is recessed, it is removed from the flow of air, and, consequently, the quantity of heat it receives per unit of time is reduced. This slows down response, allowing the fire to grow larger before it is detected. A heat detector also receives a small percentage of radiated heat. If the detector is recessed, this radiated heat energy cannot strike the detector. The result is very slow response and a fire that grows very large before it is detected.

Other modes of fire detection are also less sensitive if they are recessed. Smoke detectors depend upon air movement to convey smoke from the fire to the detector. Smoke detectors are typically mounted on the ceiling to take advantage of the air movement created by the fire. The best research available today still supports the theory that there is a thin layer of air immediately below the ceiling surface that is not as involved in the airflow created by the fire plume and, consequently, contains less smoke. If a smoke detector is recessed, this relatively clean air immediately below the ceiling surface could blanket the detector, slowing down the entry of smoke into the detector. The result is slow response to the fire.

5-1.3.4 Detectors shall be installed in all areas where required by the appropriate NFPA standard or the authority having jurisdiction. Each installed detector shall be accessible for periodic maintenance and testing. Where total coverage is required, this shall include all rooms, halls, storage areas, basements, attics, lofts, spaces above suspended ceilings, and other subdivisions and accessible spaces, and inside all closets, elevator shafts, enclosed stairways, dumbwaiter shafts, and chutes. Inaccessible areas shall not be required to be protected by detectors unless they contain combustible material, in which case they shall be made accessible and be protected by detector(s).

Exception No. 1: Detectors may be omitted from combustible blind spaces where any of the following conditions prevail:

(a) Where the ceiling is attached directly to the underside of the supporting beams of a combustible roof or floor deck.

(b) Where the concealed space is entirely filled with a noncombustible insulation. In solid joist construction, the insulation need fill only the space from the ceiling to the bottom edge of the joist of the roof or floor deck.

(c) Where there are small concealed spaces over rooms, provided any space in question does not exceed 50 sq ft (4.6 m²) in area.

(d) In spaces formed by sets of facing studs or solid joists in walls, floors, or ceilings where the distance between the facing studs or solid joists is less than 6 in. (150 mm).

Exception No. 2: Detectors may be omitted from below open grid ceilings where all of the following conditions prevail:

(a) The openings of the grid are ¼ in. (6.4 mm) or larger in the least dimension.

<table>
<tr><td>

Formal Interpretation 78-2
Reference: 5-1.3.4

In 5-1.3.4 it states "Where total coverage is required, this shall include all rooms, . . . spaces above suspended ceilings, . . . and chutes." I am interested in the "spaces above suspended ceilings."
There are many buildings which have suspended ceilings — acoustic tile exposed T-bar ceilings.

Question 1: Does the above mean that detectors are required on the underside of the suspended ceiling for area protection of the room and also detectors required above the ceiling to protect space between ceiling and roof or space between ceiling and floor above?
Answer: Yes.

Question 2: Is there any criteria to consider that would not require detectors in the spaces above suspended ceilings?
Answer: Yes. If the space contained no combustible material as defined by NFPA 220 and the ceiling tiles were secured to their T-bar by clips or other methods of fixing such as in an approved fire resistant ceiling-roof assembly or, if the authority having jurisdiction does not require total coverage.

Issue Edition: NFPA 72E-1978
Reference: 2-6.5
Issue Date: September 1980 ■

</td><td>

Formal Interpretation 78-3
Reference: 5-1.3.4

Question 1: Does Exception No. 1 apply only to the last sentence of 5-1.3.4, referring to "inaccessible areas which contain combustible material"?
Answer: Yes.

Question 2: If the answer to Question 1 is "yes," does this indicate that inaccessible areas which do not contain combustible material need not have detectors?
Answer: Yes.

Question 3: If the answer to Question 2 is "yes," does the installation of small access doors for servicing of smoke or fire dampers make the space above an otherwise nonaccessible ceiling accessible within the intent of the code?
Answer: Yes.

Issue Edition: NFPA 72E-1978
Reference: 2-6.5
Issue Date: September 1980 ■

</td></tr>
</table>

(b) The thickness of the material does not exceed the least dimension.

(c) The openings constitute at least 70 percent of the area of the ceiling material.

Exception No. 3: Concealed, accessible spaces above suspended ceilings, used as a return air plenum meeting the requirements of NFPA 90A, Standard for the Installation of Air Conditioning and Ventilating Systems, where equipped with smoke detection at each connection from the plenum to the central air handling system.

As is true elsewhere in this chapter, the term "detector" in 5-1.3.4 should be interpreted by the reader to include other types of initiating devices as well. The first requirement of 5-1.3.4 provides correlation between the *National Fire Alarm Code* and other codes and standards. Initiating devices must be used wherever they are required by another code or standard. The question of how many and how they should be installed is established here. Note that the authority having jurisdiction may require initiating devices in areas where they are not necessarily required by other codes or standards. If the authority having jurisdiction makes such a requirement, those initiating devices must also be installed in a manner consistent with this code.

Subsection 5-1.3.4 requires that all detectors be accessible for maintenance and testing. The prudent designer or installer should interpret this to apply to all initiating devices. The term "accessible" is often subject to debate. For example, if detectors are mounted on the ceiling of an auditorium, one individual can assert that they are accessible with a scaffold while another can disagree, asserting that erecting a scaffold precludes the use of the facility for its intended purpose and is, therefore, not a viable alternative. The accessibility of a detector or other initiating device will ultimately be defined by the ability to perform maintenance at the required frequency, as outlined in Chapter 7.

The remainder of 5-1.3.4 deals with the concept of total coverage. The terms "total coverage" and "partial coverage" were introduced by NFPA *101*, *Life Safety Code*, and by other codes and standards with the phrases "complete smoke detection system" and "partial smoke detection system." Total coverage, a complete smoke detection system, is required by the *Life Safety Code* for a number of types of occupancies. However, it is this subsection of the *National Fire Alarm Code* that defines exactly what total coverage entails. It requires that detectors be placed within all volumes, combustible or not. Inaccessible, noncombustible spaces do not need detectors. Inaccessible, combustible spaces must be made accessible and be provided with detection. However, engineering judgment is necessary.

There are three exceptions to the new requirement that provide guidance to help identify areas that do not require detectors. The first exception includes a number of blind, boxed-in spaces that are common in stud-wall, curtain-wall, and frame construction, which, if not excepted from this requirement, would result in detectors being placed within walls, etc. The basis for this exception is that these spaces contain limited combustibles, and the probability of an ignition originating in these spaces is remote.

The second exception pertains to open-grid ceilings where all of the listed criteria are met. In most facilities where there are suspended ceilings, the above-ceiling space contains combustibles, does not comply with Exception No. 1, and must therefore be equipped with detection. The ceiling does not represent a significant barrier to the movement of smoke and fire gases. If the open-grid ceiling does not meet all of the criteria in Exception No. 2 and complete coverage is required, detectors must be placed in the above-ceiling space.

The third exception addresses above-ceiling spaces that are used as a return air plenum and are in compliance with the requirements of NFPA 90A *and* are equipped with smoke detection at the exhaust duct connection, again per NFPA 90A. This exception is not just for when detectors have been installed to comply with NFPA 90A, but is an alternative to regularly spaced area detectors. The detectors in the exhaust duct will not be effective under "no airflow"

conditions, and area detection is still recommended. The relevant sections of NFPA 90A define the types and quantities of combustible materials that may be included within the return air plenum.

5-1.3.5* Detectors shall be required underneath open loading docks or platforms and their covers, and for accessible underfloor spaces of buildings without basements.

Exception: By permission of the authority having jurisdiction, detectors may be omitted when all of the following conditions prevail:

(a) The space is not accessible for storage purposes or entrance of unauthorized persons and is protected against accumulation of windborne debris.

(b) The space contains no equipment such as steam pipes, electric wiring, shafting, or conveyors.

(c) The floor over the space is tight.

(d) No flammable liquids are processed, handled, or stored on the floor above.

This subsection pertains to heat, smoke, radiant energy, or fire-gas detectors. There have been fire losses that were far worse than they should have been simply because detectors were not placed in the types of spaces referenced here. The exception allows the authority having jurisdiction to waive this requirement only when all of the referenced conditions in (a) through (d) exist.

A-5-1.3.5 Detectors may be required under large benches, shelves, or tables, and inside cupboards or other enclosures.

5-1.3.6 Where codes, standards, laws, or authorities having jurisdiction require the protection of selected areas only, the specified areas shall be protected in accordance with this code.

This paragraph effectively establishes that whenever and wherever an initiating device is installed, regardless of its ultimate purpose or function, it must be installed in accordance with this code.

5-1.4* Connection to the Fire Alarm System.

A-5-1.4 Refer to Figures A-5-1.4(a) and (b) for proper connections of automatic fire detectors to fire alarm systems initiating device circuits and power supply circuits.

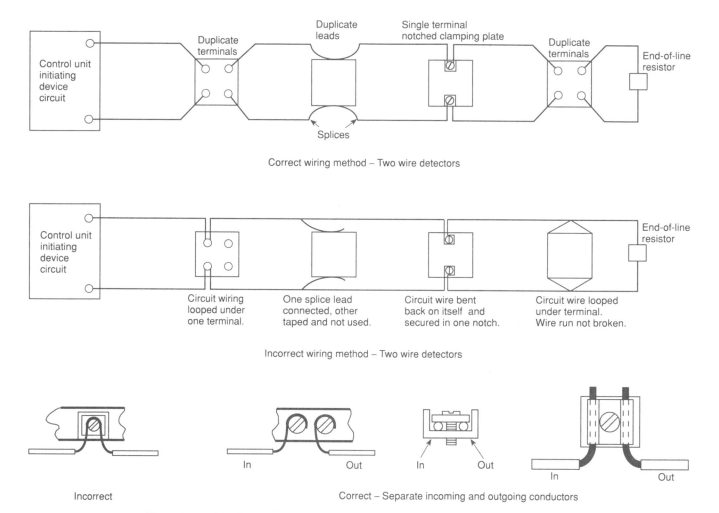

Correct wiring method – Two wire detectors

Incorrect wiring method – Two wire detectors

Figure A-5-1.4(a)   *Correct wiring methods — four-wire detectors with separate power supply.*

It is helpful to review the equivalent circuit inside the detector when considering Figures A-5-1.4(a) and (b). The "generic" 4-wire detector has power supply terminals, terminals for a normally closed trouble contact, and terminals for a normally open alarm contact. Figure 5.2 shows the equivalent schematic of the detector (initiating device). The numbering of the terminals shown here is strictly illustrative and will probably not be consistent with the numbering of commercially available detectors.

Using the designations in Figure 5.2, the operating potential (voltage) for the detector is supplied to terminals 7 and 8. Within the detector there is a connection from terminal 7 to terminal 4, and from terminal 8 to terminal 3. Terminals 4 and 3 are wired to terminals 7 and 8, respectively, providing operating potential (voltage) to the subsequent detectors. The application of operating potential (voltage) in the proper polarity closes a normally closed trouble contact (n.c.) between terminals 5 and 6. Within the detector there is a jumper between terminals 1 and 2. Thus, under normal operational conditions, terminals 1, 2, 5, and 6 provide a circuit path for the supervisory current. Between terminals 1 and 6 there is a normally open alarm contact (n.o.) that closes when the detector senses the by-products of a fire.

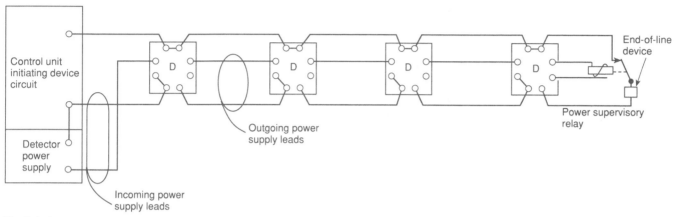

D = Detector

Illustrates 4-wire smoke detector employing a 3-wire connecting arrangement. One side of power supply is connected to one side of initiating device circuit. Wire run broken at each connection to smoke detector to provide supervision.

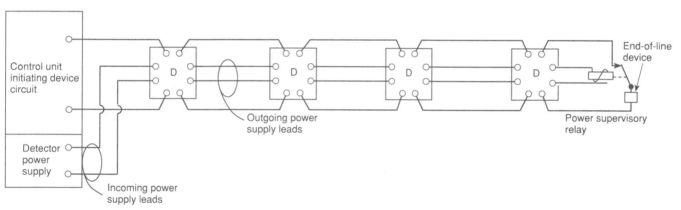

D = Detector

Illustrates 4-wire smoke detector employing a 4-wire connecting arrangement. Incoming and outgoing leads or terminals for both initiating device and power supply connections. Wire run broken at each connection to provide supervision.

Figure A-5-1.4(b)

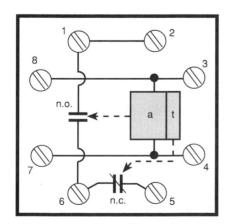

Figure 5.2 *The equivalent circuit of a generic 4-wire detector. The circuitry (shaded area) operates the trouble contact (t) and the alarm contact (a). See commentary following A-5-1.4 for explanation.*

The normally closed contacts allow a supervisory current to flow from the control panel into terminal 6, out terminal 5, on through each detector, through the end-of-line device, and back through terminals 1 and 2 of each detector to the control panel. If an initiating device (detector) loses its source of operating potential (voltage), its trouble contact between terminals 5 and 6 opens, interrupting the current flow. If an initiating device senses a fire, the alarm contact between terminals 1 and 6 closes, bypassing the end-of-line device, which increases the current flowing through the initiating device circuit. The control panel interprets the larger flow of current as a fire alarm.

The ability to use the three-wire format or the four-wire format is determined by the initiating device input circuit of the fire alarm control panel, not the detector. Some control panels use one side of the power supply as part of the initiating device circuit, others do not. The system must be wired according to the instructions provided by the manufacturer of the fire alarm system control panel. In addition, the only detectors that may be connected to a fire alarm control panel are those that have been listed as being compatible with the specific make and model control panel.

5-1.4.1 Duplicate terminals or leads, or equivalent, shall be provided on each initiating device for the express purpose of connecting into the fire alarm system to provide supervision of the connections. Such terminals or leads are necessary to ensure that the wire run is broken and that the individual connections are made to the incoming and outgoing leads or other terminals for signaling and power.

Exception: Initiating devices that provide equivalent supervision.

Traditionally, fire alarm system control panels have used a small supervisory current to recognize a break in a conductor or the removal of a detector from the circuit. Under normal conditions, the supervisory current flows through the circuit. When a detector is removed or a conductor is broken, the current path is interrupted and the flow of current stops. The control panel translates this into a trouble signal.

It has been common practice in the electrical trade when installing initiating devices to remove a short section of insulation from the conductor and to loop the wire beneath the screw terminal without ever cutting the conductor. This is an unacceptable method of installation. If this method is employed, the connection to the initiating device (detector) could loosen over time, and the control panel would not be able to recognize this as a break in the circuit. Subsection 5-1.4.1 was incorporated into the code to preclude this practice.

Recently, systems using "smart" detectors and even "smarter" control panels have been introduced. A micro-

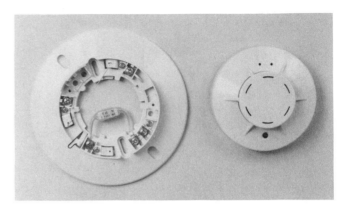

Figure 5.3(a) Smoke detector with base showing incoming and outgoing terminals.

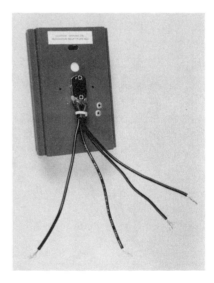

Figure 5.3(b) Manual fire alarm box showing incoming and outgoing leads.

computer in the control panel maintains a list of the "names" of all of the initiating devices in the system. It sequentially addresses each device by name and verifies the response from that device. In this way, the control panel recognizes when an initiating device fails to respond, indicating either a device failure or a break in the wiring. This method does not depend upon the continuous flow of current. Consequently, these systems are exempt from the duplicate terminal requirement.

5-2 Heat-Sensing Fire Detectors.

5-2.1 Fire detectors that sense heat produced by burning substances are usually referred to as heat detectors. Heat is both the added energy that causes substances to rise in temperature as well as the energy produced by a burning substance.

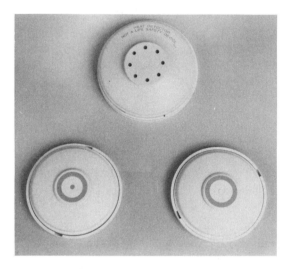

Figure 5.4 Typical "low-profile" rate-of-rise and fixed-temperature heat detectors.

We think we know what heat is when we feel it, but to define it in precise engineering terms is not easy. However, we must understand the relationship between heat and temperature if we are to apply heat-sensing detectors properly. Heat is energy and can be quantified in terms of an amount, whereas temperature is a measure of intensity and is quantified in terms of extent. Think of heat as sugar and temperature as sweetness. If we add sugar to coffee, we make it sweeter; if we add heat, we make an object hotter — we raise its temperature. Heat detectors are devices that change in some way when the temperature at the detector achieves a particular level, or set-point. That increase in temperature is due to the absorption of heat from a fire, either through convection or radiation.

5-2.2 Heat detectors shall be installed in all areas where required either by the appropriate NFPA standard or the authority having jurisdiction.

5-2.3 Operating Principles.

Heat detectors are available in two general types: spot-type, which are devices that occupy a specific spot or point, and line-type, which are linear devices that extend over a distance, sensing temperature along their entire length.

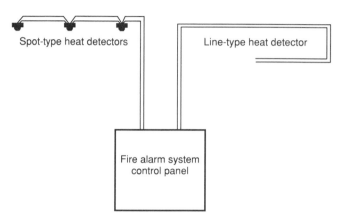

Figure 5.5 The two types of heat detector, spot-type and line-type.

Heat detectors operate on one or more of three different principles. These operating principles are categorized as fixed temperature, rate compensation, and rate-of-rise. Each principle has its performance advantages and can be employed in either a spot-type device or a line-type device. See Figures 5.4 through 5.8.

Figure 5.6 Typical spot-type fixed-temperature heat detector.

Figure 5.7(a) Line-type heat detectors.

Figure 5.7(b) Typical line-type heat detector installed in cable tray applications.

Figure 5.8 Rate compensation heat detector — horizontal mounting.

Finally, there are a number of different technologies that can be used to detect the heat from a fire. Some of these technologies lend themselves to a specific type of heat detector or operating principle. Other technologies can be used with a number of different principles and types. These technologies include:

1. Expanding bimetallic components
2. Eutectic solders
3. Eutectic salts
4. Melting insulators
5. Thermistors
6. Temperature-sensitive semiconductors
7. Expanding air volume
8. Expanding liquid volume
9. Temperature-sensitive resistors
10. Thermopiles

The code has been written to allow the development and use of new technologies. The reader must make special note not to confuse the terms "type" and "principle" with "technology," the method employed to achieve heat detection.

5-2.3.1 Fixed Temperature Detector.

5-2.3.1.1 A fixed temperature detector is a device that will respond when its operating element becomes heated to a predetermined level.

5-2.3.1.2 **Thermal Lag.** Where a fixed temperature device operates, the temperature of the surrounding air will always be higher than the operating temperature of the device itself. This difference between the operating temperature of the device and the actual air temperature is commonly referred to as thermal lag and is proportional to the rate at which the temperature is rising.

The concept of thermal lag has a long history. It was first applied to sprinkler heads. It is also relevant to heat detectors.

The First Law of Thermodynamics states that heat always flows from objects of higher temperature to objects of lower temperature. However, the flow of heat takes time. There is a thermal resistance between the heat source and its surroundings that determines the rate at which heat can flow. The "R" rating of an insulated wall is a measure of thermal resistance. The amount of heat that can flow between two objects per unit of time is determined by the difference in the objects' temperature, the mass of each object, the specific heat of each object, the area being exposed, and the thermal resistance between the objects.

In the context of a fire where the air is being heated very rapidly, the temperature of the air in the vicinity of the heat detector will rise much more rapidly than the actual temperature of the detector because of the time that it takes for enough heat to flow into the detector to raise its temperature. The detector is always "catching up." This results in thermal lag. Thermal lag is either a time lag or the difference in temperature between the air and the heat detector, depending upon the context.

5-2.3.1.3 Typical examples of fixed temperature-sensing elements are:

(a) *Bimetallic.* A sensing element comprised of two metals having different coefficients of thermal expansion arranged so that the effect will be deflection in one direction when heated and in the opposite direction when cooled.

(b) *Electrical Conductivity.* A line-type or spot-type sensing element whose resistance varies as a function of temperature.

(c) *Fusible Alloy.* A sensing element of a special composition (eutectic) metal, which melts rapidly at the rated temperature.

(d) *Heat-Sensitive Cable.* A line-type device whose sensing element comprises, in one type, two current-carrying wires separated by heat-sensitive insulation that softens at the rated temperature, thus allowing the wires to make electrical contact. In

another type, a single wire is centered in a metallic tube, and the intervening space filled with a substance that, at a critical temperature, becomes conductive, thus establishing electrical contact between the tube and the wire.

(e) *Liquid Expansion.* A sensing element comprising a liquid capable of marked expansion in volume in response to temperature increase.

5-2.3.2 **Rate Compensation Detector.**

5-2.3.2.1 A rate compensation detector is a device that will respond when the temperature of the air surrounding the device reaches a predetermined level, regardless of the rate of temperature rise.

5-2.3.2.2 A typical example is a spot-type detector with a tubular casing of a metal that tends to expand lengthwise as it is heated and an associated contact mechanism that will close at a certain point in the elongation. A second metallic element inside the tube exerts an opposing force on the contacts, tending to hold them open. The forces are balanced in such a way that on a slow rate of temperature rise, there is more time for heat to penetrate to the inner element, which inhibits contact closure until the total device has been heated to its rated temperature level. However, on a fast rate of temperature rise, there is not as much time for heat to penetrate to the inner element, which exerts less of an inhibiting effect so that contact closure is obtained when the total device has been heated to a lower level. This, in effect, compensates for thermal lag.

See Figure 5.9.

5-2.3.3 **Rate-of-Rise Detector.**

5-2.3.3.1 A rate-of-rise detector is a device that will respond when the temperature rises at a rate exceeding a predetermined amount.

5-2.3.3.2 Typical examples are:

(a) *Pneumatic Rate-of-Rise Tubing.* A line-type detector comprising small diameter tubing, usually copper, that is installed on the ceiling or high on the walls throughout the protected area. The tubing is terminated in a detector unit containing diaphragms and associated contacts set to actuate at a predetermined pressure. The system is sealed except for calibrated vents that compensate for normal changes in temperature.

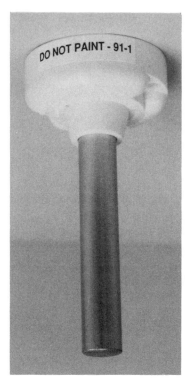

Figure 5.9 *Typical rate compensation heat detector — vertical mounting.*

(b) *Spot-Type Pneumatic Rate-of-Rise Detector.* A device consisting of an air chamber, diaphragm, contacts, and compensating vent in a single enclosure. The principle of operation is the same as that described in 5-2.3.3.2(a). (See Fig. 5.10)

(c) *Thermoelectric Effect Detector.* A device whose sensing element comprises a thermocouple or thermopile unit that produces an increase in electric potential in response to an increase in temperature. This potential is monitored by associated control equipment, and an alarm is initiated when the potential increases at an abnormal rate.

(d) *Electrical Conductivity Rate-of-Change Detector.* A line-type or spot-type sensing element whose resistance changes due to a change in temperature. The rate of change of resistance is monitored by associated control equipment, and an alarm is initiated when the rate of temperature increase exceeds a preset value.

5-2.4 Classification and Sensitivity.

5-2.4.1 Heat detectors of the fixed-temperature or rate-compensated spot-pattern type shall be classified as to the tem-

Figure 5.10 *Typical spot-type combination rate-of-rise fixed-temperature heat detector.*

perature of operation and marked with the appropriate color code. (*See Table 5-2.4.1.*)

Spot-type heat detectors are the most popular type of heat detector in general use. In Table 5-2.4.1 there are specific criteria for the nominal temperature rating versus the maximum expected normal temperature for the location of the detector. Note that it is necessary to provide for at least a 20°F difference between the temperature setting of the detector and the maximum expected normal temperature. This requirement technically applies only to low temperature detectors, not to detectors with other temperature classifications. However, it is still good practice to follow the 20°F difference recommendation. Also note that it is crucial not to select a temperature setting any higher than necessary. The higher the setting, the longer it will take to achieve an alarm.

5-2.4.1.1 Where the overall color of a detector is the same as the color code marking required for that detector, either one of the following arrangements, applied in a contrasting color and visible after installation, shall be employed:

(a) A ring on the surface of the detector

See Figure 5.11(a).

(b) The temperature rating in numerals at least ⅜ in. (9.5 mm) high.

See Figure 5.11(b).

Table 5-2.4.1

Temperature Classification	Temp. Rating Range °F	Temp. Rating Range °C	Max. Ceiling Temp. °F	Max. Ceiling Temp. °C	Color Code
Low*	100 to 134	39 to 57	20 below**	11	Uncolored
Ordinary	135 to 174	58 to 79	100	38	Uncolored
Intermediate	175 to 249	80 to 121	150	66	White
High	250 to 324	122 to 162	225	107	Blue
Extra High	325 to 399	163 to 204	300	149	Red
Very Extra High	400 to 499	205 to 259	375	191	Green
Ultra High	500 to 575	260 to 302	475	246	Orange

For SI Units: $C = \frac{5}{9} (F - 32)$.

*Intended only for installation in controlled ambient areas. Units shall be marked to indicate maximum ambient installation temperature.

**Maximum ceiling temperature has to be 20°F (11°C) or more below detector rated temperature.

NOTE: The difference between the rated temperature and the maximum ambient should be as small as possible to minimize the response time.

Figure 5.11(a) Heat detectors with color coded rings.

Figure 5.11(b) Heat detector with temperature marked with numerals.

The unified color coding of heat detectors facilitates inspections, making it possible to identify the temperature rating of a ceiling-mounted heat detector while standing on the floor. The color code for heat detectors is very similar to that used for sprinkler heads, as described in NFPA 13, 2-2.3.1.

5-2.4.2* A heat detector integrally mounted on a smoke detector shall be listed or approved for not less than 50-ft (15-m) spacing.

See Figure 5.12.

A-5-2.4.2 The linear space rating is the maximum allowable distance between heat detectors. The linear space rating is also a measure of the heat detector response time to a standard test fire where tested at the same distance. The higher the rating, the faster the response time. This code recognizes only those heat detectors with ratings of 50 ft (15 m) or more.

This paragraph originated in NFPA 74, *Standard for the Installation, Maintenance, and Use of Household Fire Warning Equipment*. There are many common smoke detectors designed for household applications that are equipped with an integral heat sensor. In order for the heat sensor portion of the detector to comply with the code, it must have a 50-foot spacing factor.

5-2.5 Location.

5-2.5.1* Spot-type heat detectors shall be located on the ceiling not less than 4 in. (100 mm) from the side wall or on the side walls between 4 in. (100 mm) and 12 in. (300 mm) from the ceiling. (*See Figure A-5-2.5.1.*)

Figure 5.12   Smoke detector with 50-foot listed heat detection.

Exception No. 1: In the case of solid open joist construction, detectors shall be mounted at the bottom of the joists.

Exception No. 2: In the case of beam construction where beams are less than 12 in. (300 mm) in depth and less than 8 ft (2.4 m) on center, detectors may be installed on the bottom of beams.

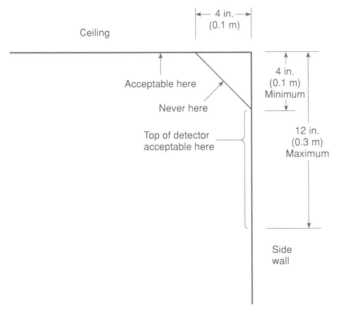

NOTE: Measurements shown are to the closest edge of the detector.
Figure A-5-2.5.1   Example of proper mounting for detectors.

Section 1-4 defines a ceiling as the upper surface of a space, regardless of height. Subsection 5-1.3.4 defines which spaces must be considered for detector placement. This sub-

section requires heat detectors to be located on the ceiling of those spaces or on the side walls within 12 inches of the ceiling. These locations derive maximum benefit from the upward flow of hot air from a fire. However, the best currently available research data supports the theory that there is a dead air space where the walls meet the ceiling in a typical room. Figure A-5-2.5.1 shows this dead air space extending 4 inches in from the wall and 4 inches down from the ceiling. Consequently, the code excludes detectors from that area.

Exception No. 1 to 5-2.5.1 addresses ceilings with solid open joist construction. A joist is defined in Section 1-4 (under the definition of "solid joist construction") as being a solid structural or nonstructural member extending down from the ceiling more than 4 inches and with less than 3 feet between adjacent joists. The narrow spacing between joists (usually 16 inches) creates air pockets between them. These air pockets not only slow the rate of smoke and heat spread but also cause air to flow below the bottoms of the joists. This necessitates the placement of the detectors on the bottom of the joists.

5-2.5.2   Line-type heat detectors shall be located on the ceiling or on the side walls not more than 20 in. (500 mm) from the ceiling.

Exception No. 1: In the case of solid open joist construction, detectors shall be mounted at the bottom of the joists.

Exception No. 2: In the case of beam construction where beams are less than 12 in. (300 mm) in depth and less than 8 ft (2.4 m) on center, detectors may be installed on the bottom of beams.

Subsection 5-2.5.2 does not prohibit the installation of a line-type detector inside the dead air space. However, the prudent designer, even when using a line-type heat detector, should always follow the requirements of 5-2.7 and avoid that region.

5-2.6*   **Temperature.** Detectors having fixed-temperature or rate-compensated elements shall be selected in accordance with Table 5-2.4.1 for the maximum ceiling temperature that can be expected.

A-5-2.6   A heat detector with a temperature rating somewhat in excess of the highest normally expected ambient temperature is specified in order to avoid the possibility of premature operation of the heat detector to nonfire conditions.

5-2.7* Spacing.

A-5-2.7 In addition to the special requirements for heat detectors installed on ceilings with exposed joists, reduced spacing may also be required due to other structural characteristics of the protected area, possible drafts, or other conditions that may affect detector operation.

The spacing criteria established by 5-2.7 determine how many detectors of a given type are necessary to properly protect a given area. The designer has an alternative method at his or her disposal, namely using the analytical method described in Appendix B of this code. Either method is acceptable. Because Appendix B is optional, it is recommended that the designer obtain the approval or acceptance of the authority having jurisdiction.

The number of detectors required is a function of the spacing factor, S, of the detector to be used. The spacing is established through a series of fire tests and is indicative of the relative sensitivity of the detector. It is critical to note that the spacings derived from the fire tests relate heat detectors to the response of an ordinary sprinkler head. Detector performance is defined relative to the distance at which it could detect the same fire that fused a standard sprinkler head in 2 minutes ($\pm$ 10 seconds) located 10 feet from a fire. For example, when a heat detector mounted 50 feet away from a fire activates before an ordinary sprinkler head mounted 10 feet from the same fire, the heat detector receives a 50-foot listed spacing rating.

The number of detectors necessary for a given application also depends on the ceiling height, the type of ceiling (whether it has exposed joists or beams), and other features that may effect the flow of air or the accumulation of heat from a fire. All of these factors enter into the spacing design rules that follow.

5-2.7.1* Smooth Ceiling Spacing.
One of the following rules shall apply:

(a) The distance between detectors shall not exceed their listed spacing, and there shall be detectors within a distance of one-half the listed spacing, measured at a right angle, from all walls or partitions extending to within 18 in. (460 mm) of the ceiling; or

(b) All points on the ceiling shall have a detector within a distance equal to 0.7 times the listed spacing (0.7S). This will be useful in calculating locations in corridors or irregular areas.

A-5-2.7.1 Maximum linear spacings on smooth ceilings for spot-type heat detectors are determined by full-scale fire tests. These tests assume that the detectors are to be installed in a pattern of one or more squares, each side of which equals the maximum spacing as determined in the test. This is illustrated in Figure A-5-2.7.1(c). The detectors to be tested are placed at one corner of the square, which is the furthest distance it can be from the fire while still within the square. Thus the distance from the detector ("D") to the fire ("F") is always the test spacing multiplied by 0.7 and can be set up in the following tables:

Test Spacing	Maximum Test Distance from Fire to Detector (0.7 × D)
50 × 50 ft	35 ft
40 × 40 ft	28 ft
30 × 30 ft	21 ft
25 × 25 ft	17.5 ft
20 × 20 ft	14 ft
15 × 15 ft	10.5 ft

For SI Units: 1 ft = 0.305 m.

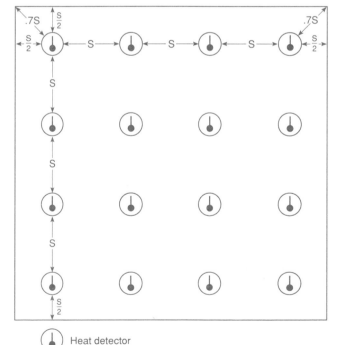

Heat detector

S   Spacing between detectors

Figure A-5-2.7.1(a) *Spot-type detectors.*

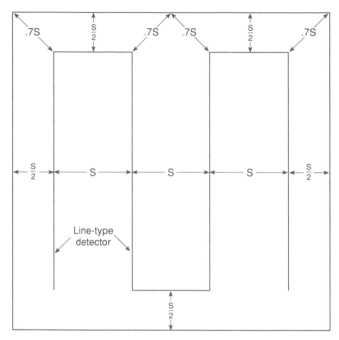

Figure A-5-2.7.1(b) *Line-type detectors — spacing layouts, smooth ceiling.*

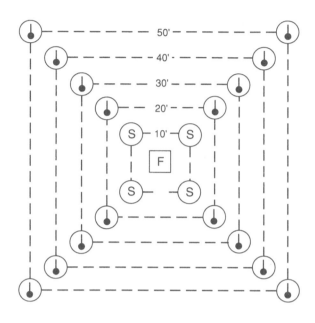

Legend

F — Test fire, denatured alcohol, 190-proof. Pan located approximately 3 ft (0.9 m) above floor.
S — Indicates normal sprinkler spacings on 10-ft (3-m) schedules.
⊥ — Indicates normal heat detector spacing on various spacing schedules.

For SI Units: 1 ft = 0.305 m.

Figure A-5-2.7.1(c) *Fire test layout.*

Once the correct maximum test distance has been determined, then it is valid to interchange the positions of the fire ("F") and the detector ("D"). The detector is now in the middle of the square, and what the listing actually says is that the detector is adequate to detect a fire that occurs anywhere within that square — even out to the farthest corner.

In laying out detector installations, designers speak in terms of rectangles, as building areas are generally rectangular in shape. The pattern of heat spread from a fire source, however, is not rectangular in shape. On a smooth ceiling, heat will spread out in all directions, in an ever-expanding circle. Thus, the coverage of a detector is not in fact a square, but rather a circle whose radius is the linear spacing multiplied by 0.7.

This is graphically illustrated in Figure A-5-2.7.1(d). With the detector as the center, by rotating the square, an infinite number of squares can be laid out, the corners of which will plot a circle whose radius is 0.7 times the listed spacing. The detector will cover any of these squares and, consequently, any point within the confines of the circle.

So far this explanation has considered squares and circles. In practical applications, very few areas turn out to be exactly square, and circular areas are rare indeed. Designers deal generally with rectangles of odd dimensions and corners of rooms

Formal Interpretation
Reference: 5-2.7.1

Statement of Problem: Paragraph 5-2.7.1 indicates acceptable effective spacing for irregular areas but does not specify that it does (or does not) refer to smooth ceiling only. The explanation in A-5-2.7.1 refers to smooth ceiling, but only in the context as to how the "U.L. listed" spacing is reached. Insurance authorities are not providing a consistent interpretation, apparently, as there is some indication that all will not allow the "irregular area" increase in anything except smooth ceiling.

Question: Does 5-2.7.1 apply to all types of ceiling construction with suitable reductions for type?
Answer: Yes.

Issue Edition: NFPA 72E-1974
Reference: 3-5.1.1
Issue Date: February 1976 ■

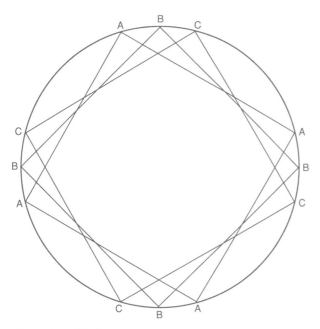

Figure A-5-2.7.1(d) *A detector will cover any square laid out in the confines of a circle whose radius is 0.7 times the listed spacing.*

or areas formed by wall intercepts, where spacing to one wall is less than one-half the listed spacing. To simplify the rest of this explanation, consider the use of a detector with a listed spacing of 30 ft by 30 ft (9.1 m by 9.1 m). The principles derived will be equally applicable to other types.

Figure A-5-2.7.1(e) illustrates the derivation of this concept. A detector is placed in the center of a circle with a radius of 21 ft (0.7 × 30 ft) [6.4 m (0.7 × 9.1 m)]. A series of rectangles with one dimension less than the permissible maximum of 30 ft.(9.1 m) is constructed within the circle. The following conclusions can be drawn:

(a) As the smaller dimension decreases, the longer dimension can be increased beyond the linear maximum spacing of the detector with no loss in detection efficiency.

(b) A single detector will cover any area that will fit within the circle. For a rectangle, a single properly located detector will suffice if the diagonal of the rectangle does not exceed the diameter of the circle.

(c) Relative detector efficiency will actually be increased, because the area coverage in sq ft is always less than the 900 sq ft (83.6 m²) permissible if the full 30 ft by 30 ft (9.1 m by 9.1 m) square were to be utilized. The principle illustrated here allows equal linear spacing between the detector and the fire,

with no recognition for the effect of reflection from walls or partitions, which in narrow rooms or corridors will be of additional benefit. For detectors that are not centered, the longer dimension should always be used in laying out the radius of coverage.

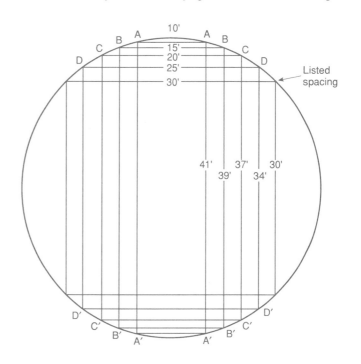

Rectangle	A = 10' x 41' = 410 sq ft
	B = 15' x 39' = 585 sq ft
	C = 20' x 37' = 740 sq ft
	D = 25' x 34' = 850 sq ft
Listed spacing	= 30' x 30' = 900 sq ft

For SI Units: 1 ft = 0.305 m.

Figure A-5-2.7.1(e) *Detector spacing, rectangular areas.*

Areas so large that they exceed the rectangular dimensions given in Figure A-5-2.7.1(e) require additional detectors. Often proper placement of detectors can be facilitated by breaking down the area into multiple rectangles of the dimensions that fit most appropriately. [*See Figure A-5-2.7.1(f).*] For example, see Figure A-5-2.7.1(e). A corridor 10 ft (3 m) wide and up to 82 ft (25 m) long can be covered with two 30-ft (9.1-m) detectors. An area 40 ft (12.2 m) wide and up to 74 ft (22.6 m) long can be covered with four detectors. Irregular areas will take more careful planning to make sure that no spot on the ceiling is more than 21 ft (6.4 m) away from a detector. These points can be determined by striking arcs from the remote corner. Where any part of the area lies beyond the circle with a radius of 0.7 times the listed spacings, additional detectors are required.

5-2.7.1.1* **Irregular Areas.** For irregularly shaped areas, the spacing between detectors may be greater than the listed spacing, provided the maximum spacing from a detector to the furthest point of a side wall or corner within its zone of protection is not greater than 0.7 times the listed spacing. (*See Figure A-5-2.7.1.1.*)

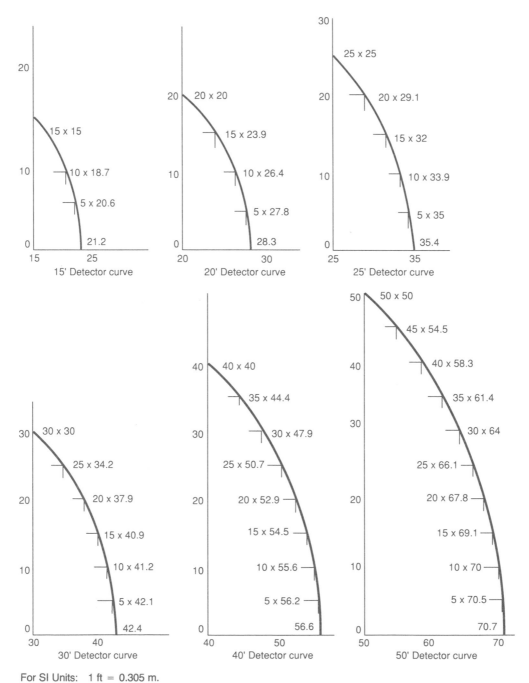

For SI Units: 1 ft = 0.305 m.

Figure A-5-2.7.1(f) *Typical rectangles for detector curves of 15 - 50 ft.*

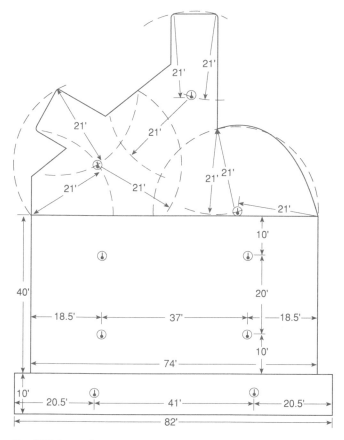

For SI Units: 1 ft = 0.305 m.

Figure A-5-2.7.1.1 *Detector spacing layout, irregular areas.*

5-2.7.1.2* High Ceilings.

On ceilings 10 ft (3 m) to 30 ft (9.1 m) high, heat detector linear spacing shall be reduced in accordance with Table 5-2.7.1.2.

Exception: Table 5-2.7.1.2 does not apply to the following detectors, which rely on the integration effect:

(a) Line-type electrical conductivity detectors. [See 5-2.3.1.3(b).]

(b) Pneumatic rate-of-rise tubing. [See 5-2.3.3.2(a).]

(c) Series connected thermoelectric effect detectors. [See 5-2.3.3.2(c).]

In these cases, the manufacturer's recommendations shall be followed for appropriate alarm point and spacing.

NOTE: Table 5-2.7.1.2 provides for spacing modifications to take into account different ceiling heights for generalized fire conditions. An alternative design method that allows a designer to take into account ceiling height, fire size, and ambient temperature is provided in Appendix B.

Table 5-2.7.1.2

Ceiling Height Above (ft)	Up to	Percent of Listed Spacing
0	10	100
10	12	91
12	14	84
14	16	77
16	18	71
18	20	64
20	22	58
22	24	52
24	26	46
26	28	40
28	30	34

For SI Units: 1 ft = 0.305 m.

A-5-2.7.1.2 Both 5-2.7.1.2 and Table 5-2.7.1.2 are constructed to provide essentially the equivalent detector performance on higher ceilings [to 30 ft (9.1 m) high] as that which would exist with detectors on a 10-ft (3-m) ceiling.

The Fire Detection Institute Fire Test Report (*see references in Appendix C*), used as a basis for Table 5-2.7.1.2, does not include data on integration-type detectors. Pending development of such data, the manufacturer's recommendations provide guidance.

The reduction of spacing with increased ceiling height places detectors closer to the location of the fire, thus requiring the hot air and radiated heat to travel a shorter distance before encountering a detector. As hot air rises from the fire, it gives off energy and actually cools. The spacing factor for a given detector is a rough measure of how far the air can travel from the standard test fire before it has cooled too much to be reliably detected. Increasing the ceiling height has a very significant effect on that factor.

The inverse square law predicts that when the distance between the fire and the detector is doubled, the amount of radiated heat reaching the detector will be reduced by a factor of 4. This also contributes to the need to reduce the spacing as the ceiling height is increased.

It is important to note that Table 5-2.7.1.2 covers ceiling heights up to 30 feet. This is the highest ceiling for which the Technical Committee had test data. (*See references in Appendix B.*) Where ceilings higher than 30 feet are encountered, the designer must act with the knowledge that those conditions are beyond the limits of the testing that provided the basis for the requirements of the code. There is the temptation to extrapolate for higher ceiling heights. However, a theoretical basis for doing so has yet to be reviewed by the Technical Committee. This is currently an area of considerable research that may yield important new insights in the near future.

The code neither prohibits nor permits the use of heat detectors on ceilings higher than 30 feet. However, it should be understood that a much larger fire will be necessary to activate the detectors on higher ceilings, and this delayed detection may not meet the protection goals of the owner or the intent of the code. The final decision as to whether or not this would be acceptable rests with the authority having jurisdiction.

5-2.7.2* **Solid Joist Construction.** The spacing of heat detectors, where measured at right angles to the solid joists, shall not exceed 50 percent of the smooth ceiling spacing allowable under 5-2.7.1 and 5-2.7.1.1. (*See Figure A-5-2.7.2.*)

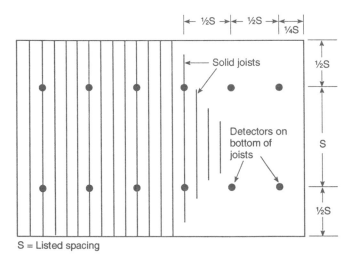

S = Listed spacing

Figure A-5-2.7.2 Detector spacing layout, solid joist construction.

Subsection 5-2.5 establishes the requirement to locate heat detectors on the bottom of joists. In subsection 5-2.7.2, the effect of joists on detector spacing is established. The hot gases and smoke from a fire rise vertically in a plume until it impinges upon the ceiling. There the hot air changes direction and begins moving horizontally across the ceiling. When the joists are running parallel to the direction of travel of the hot gases, they have little effect on the speed with which the hot gases move across the ceiling. However, when the

joists are perpendicular to the direction of gas travel from the fire to the detector, they slow down the plume velocity. This arrangement necessitates a closer spacing.

Remember that joists are solid members extending more than 4 inches down from the ceiling and are installed on centers of less than 3 feet. If the solid members extending down from the ceiling are on 3-foot centers or larger, they are beams. Also remember that bar-joists have no effect on spacing unless the top cord is greater than 4 inches.

5-2.7.3* **Beam Construction.** A ceiling shall be treated as a smooth ceiling if the beams project no more than 4 in. (100 mm) below the ceiling. If the beams project more than 4 in. (100 mm) below the ceiling, the spacing of spot-type heat detectors at right angles to the direction of beam travel shall be not more than two-thirds the smooth ceiling spacing allowable under 5-2.7.1 and 5-2.7.1.1. If the beams project more than 18 in. (460 mm) below the ceiling and are more than 8 ft (2.4 m) on center, each bay formed by the beams shall be treated as a separate area.

Open-web beams and trusses have little effect on the passage of air currents caused by fire. Generally, they are not considered in determining the proper spacing of detectors unless the solid part of the top cord extends more than 4 inches down from the ceiling.

A-5-2.7.3 Location and spacing of heat detectors should consider beam depth, ceiling height, beam spacing, and fire size.

(a) If the ratio of beam depth (D) to ceiling height (H) (D/H) is greater than 0.10 and the ratio of beam spacing (W) to ceiling height (H) (W/H) is greater than 0.40, heat detectors should be located in each beam pocket.

(b) If either the ratio of beam depth to ceiling height (D/H) is less than 0.10 or the ratio of beam spacing to ceiling height (W/H) is less than 0.40, heat detectors should be installed on the bottom of the beams.

5-2.7.4 **Sloped Ceilings.**

5-2.7.4.1* **Peaked.** A row of detectors shall first be spaced and located at or within 3 ft (0.9 m) of the peak of the ceiling, measured horizontally. The number and spacing of additional detectors, if any, shall be based on the horizontal projection of the ceiling in accordance with the type of ceiling construction. (*See Figure A-5-2.7.4.1.*)

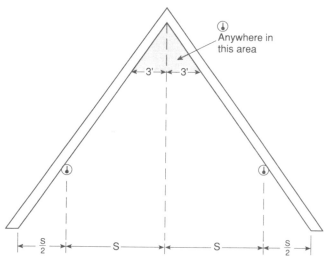

S – Detector spacing
Ⓣ – Heat detector

For SI Units: 1 ft = 0.305 m.
Figure A-5-2.7.4.1 *Detector spacing layout, sloped ceilings (peaked type).*

5-2.7.4.2* **Shed.** Sloped ceilings having a rise greater than 1 ft in 8 ft (1 m in 8 m) horizontally shall have a row of detectors located on the ceiling within 3 ft (0.9 m) of the high side of the ceiling measured horizontally, spaced in accordance with the type of construction. Remaining detectors, if any, shall then be located in the remaining area on the basis of the horizontal projection of the ceiling. (*See Figure A-5-2.7.4.2.*)

5-2.7.4.3 For a roof slope of less than 30 degrees, all detectors shall be spaced utilizing the height at the peak. For a roof slope of greater than 30 degrees, the average slope height shall be used for all detectors other than those located in the peak.

Subsections 5-2.5 and 5-2.7 address the location and spacing of heat detectors. The spacing adjustments are cumulative in nature. For example, if detectors are to be installed on a ceiling that is both higher than 10 feet and has joists or beams, both spacing reductions must be applied. First, the reduction to the design spacing due to ceiling height is applied, and then additional reductions are made pursuant to the requirements to accommodate joists or beams.

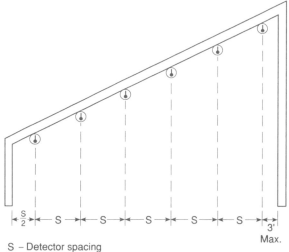

S – Detector spacing
Ⓣ – Heat detector

For SI Units: 1 ft = 0.305 m.
Figure A-5-2.7.4.2 *Detector spacing layout, sloped ceilings (shed type).*

5-3 Smoke-Sensing Fire Detectors.

5-3.1 General.

5-3.1.1* The purpose of Section 5-3 is to provide information to assist in design and installation of reliable early warning smoke detection systems for protection of life and property.

A-5-3.1.1 The addition of a heat detector to a smoke detector does not enhance its performance as an early warning device.

The location and spacing criteria in Section 5-3 are based on smoke detectors that have been listed by a testing laboratory as having passed specific performance tests. To pass the tests, the detectors must respond to nominal smoke obscuration of 1–4 percent, depending on the type of smoke and type of detector. In most fires, these devices respond sooner than either sprinkler systems or heat detectors. In flaming fire tests, smoke detectors activate long before typical heat detectors. The difference in the speed of response is even more dramatic with low-energy smoldering fires. This difference in the speed of response is the basis for concluding that the addition of a heat detector to a smoke detector adds little to overall fire detection performance.

5-3.1.2 Section 5-3 covers general area application of smoke detectors in ordinary indoor locations.

This paragraph limits the applicability of the requirements and recommendations of Section 5-3 to "general area application . . . in ordinary indoor locations." Whether or not a hazard area falls into this category is up to the authority having jurisdiction.

There are some authorities having jurisdiction that will establish additional requirements for specific types of occupancies above and beyond the requirements of Section 5-3. The designer may also choose closer spacings in areas where there are extremely valuable assets, such as in a data center.

Finally, the common interpretation of 5-3.1.2 usually does not include special compartments such as switch gear enclosures or aircraft lavatories. While detectors may be used in these and similar locations, the designer should use engineering judgment.

5-3.1.3 For information on use of smoke detectors for control of smoke spread, refer to Section 5-11.

Early in the second half of this century there were several fires in high-rise buildings that demonstrated the futility of trying to evacuate all of the occupants. A new strategy of protecting in place or using refuge zones was developed. Concurrently, it became well known that smoke inhalation was the principal cause of death associated with fires. If occupants were to be protected in place, it was critical to be able to automatically control the flow of smoke with the heating, ventilating, and air conditioning (HVAC) system. Smoke detectors are employed for that purpose, and their use for such applications is covered in Section 5-11.

5-3.1.4 For additional guidance in the application of smoke detectors for flaming fires of various sizes and growth rates in areas of various ceiling heights, refer to Appendix B.

Traditionally, the application of smoke detectors has been for early warning. Appendix B allows the use of smoke

detectors for flaming fire detection, in which case the smoke detector acts like an expensive heat detector. In some applications, the use of smoke detectors for flaming fire detection allows extension of the normal 30-foot spacing of smoke detectors. This spacing allowance should not be confused with the spacing normally used in life safety or early warning applications.

Testing performed under the auspices of the Fire Detection Institute was the basis for a new method for predicting detector response. This testing gave rise to the calculated procedure outlined in Appendix B for smoke detector applications in flaming fire detection scenarios. It provides a more analytical and precise method of determining detector spacing.

5-3.2* Smoke detectors shall be installed in all areas where required either by the appropriate NFPA standard or by the authority having jurisdiction.

A-5-3.2 The person designing an installation should keep in mind that in order for a smoke detector to respond, the smoke must travel from the point of origin to the detector. In evaluating any particular building or location, likely fire locations should first be determined. From each of these points of origin, paths of smoke travel should be determined. Wherever practical, actual field tests should be conducted. The most desired location for smoke detectors would be the common points of intersection of smoke travel from fire locations throughout the building.

NOTE: This is one of the reasons that specific spacing is not assigned to smoke detectors by the testing laboratories.

5-3.3 Principles of Detection.

5-3.3.1 Ionization Smoke Detection. Ionization smoke detection is based on the principle of using a small amount of radioactive material to ionize the air between two differentially charged electrodes. This gives the sensing chamber an effective measurable electrical conductance. Where smoke particles enter the ionization volume, they decrease the conductance of the air by reducing ion mobility. The conductance signal is processed and used to convey an alarm condition where the signal meets preset criteria.

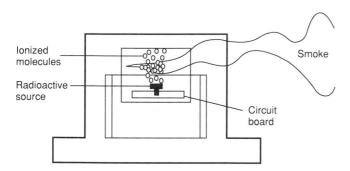

Figure 5.13 *Schematic operation of an ionization smoke detector.*

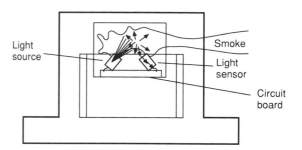

Figure 5.15 *Schematic operation of a photoelectric smoke detector.*

5-3.3.1.1 Ionization detection is more responsive to invisible (less than one micron in size) particles produced by most flaming fires. It is somewhat less responsive to the larger particles typical of most smoldering fires.

5-3.3.1.2 Smoke detectors utilizing the ionization principle are usually of the spot type.

Figure 5.14 *Typical ionization smoke detector.*

5-3.3.2* **Photoelectric Light-Scattering Smoke Detection.** Photoelectric light-scattering smoke detection is based on the principle of a light source and a photosensitive sensor arranged so that the principal portion of rays from the light source do not normally fall on the photosensitive sensor. Where smoke particles enter the light path, some of the light is scattered by reflection and refraction onto the sensor. The scattered light signal is processed and used to convey an alarm condition where the signal meets preset criteria.

A-5-3.3.2 Most light-scattering detectors use a high intensity pulsed light source with silicon photodiode or phototransistor light

sensors, resulting in excellent response to most smoldering fires and good response to most flaming fires.

5-3.3.2.1 Photoelectric light-scattering detection is more responsive to visible (more than one micron in size) particles produced by most smoldering fires. It is somewhat less responsive to the smaller particles typical of most flaming fires. It is also less responsive to black smoke than to lighter colored smoke.

5-3.3.2.2 Smoke detectors utilizing the light-scattering principle are usually of the spot type.

Figure 5.16 *Typical photoelectric smoke detector.*

5-3.3.3* **Photoelectric Light Obscuration Smoke Detection.** Photoelectric light obscuration smoke detection is based on the principle of reduction of light transmission between a light source and a photosensitive sensor onto which the principal portion of the source emissions are focused. Where smoke particles enter the light path, some of the light is scattered and some absorbed, thereby reducing the light reaching the receiving sensor. The receiving sensor signal is processed and used to convey an alarm condition where the signal meets preset criteria.

A-5-3.3.3 Projected beam detectors respond to the sum of the smoke obscuration in the beam path along its entire length between the transmitting unit and the receiving unit. A reduction in the received light initiates an alarm signal. A total and sudden loss of received light initiates a trouble signal indicating beam blockage or the need for service. Some projected beam detectors have signal-processing circuits to compensate for transient conditions and the effect of dust on sensitivity.

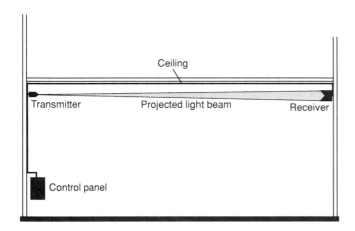

Figure 5.17(a)   A projected-beam light obscuration detector uses line-type smoke detection technology.

5-3.3.3.1 The response of photoelectric light obscuration smoke detectors is usually not affected by the color of smoke.

5-3.3.3.2 Smoke detectors utilizing the light obscuration principle are usually of the line type. These detectors are commonly called projected beam smoke detectors.

5-3.3.4 **Cloud Chamber Smoke Detection.** Cloud chamber smoke detection is usually of the sampling type. An air sample is drawn from the protected areas into a high humidity chamber within the detector. After the humidity of the sample has been raised, the pressure is lowered slightly. If smoke particles are present, the moisture in the air condenses on them, forming a cloud in the chamber. The density of this cloud is then measured by a photoelectric principle. The density signal is processed and

Figure 5.17(b)   Projected-beam smoke detector.

used to convey an alarm condition where the signal meets preset criteria.

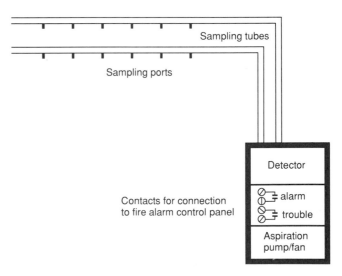

Figure 5.18   Air sampling-type smoke detection technology.

5-3.4 **Sensitivity.**

5-3.4.1 Smoke detectors shall be marked with their normal production sensitivity (percent per foot obscuration), measured as required by the listing. The production tolerance around the normal sensitivity shall also be indicated.

Because the mission of most smoke detection systems is the protection of human life, the response of a smoke detector is usually defined in human terms. The percent per foot obscuration method of measuring sensitivity relates to a person's ability to see well enough to escape from a fire. Smoke is composed of both visible and invisible particulate matter. However, the portion of the smoke that is invisible has little immediate impact on an individual's ability to escape before untenable conditions develop.

5-3.4.2 Smoke detectors that have provision for field adjustment of sensitivity shall have an adjustment range of not less than 0.6 percent per ft obscuration. If the means of adjustment is on the detector, a method shall be available to restore the detector to its factory calibration. Detectors that have provision for program controlled adjustment of sensitivity shall be permitted to only be marked with their programmable sensitivity range.

The adjustment of detector sensitivity over a range of less than 0.6 percent per foot has little, if any, practical benefit. Even when smoke detectors are used for property conservation (as in data centers), the difference in response represented by an adjustment range of less than 0.6 percent per foot is minor.

There are some smoke detectors that have a feature allowing the adjustment of detector sensitivity to accommodate the immediate ambient conditions in the area of the detector. Other smoke detectors send a voltage or current back to the control panel that is proportional to the smoke-sensing signal in the detector. The trip point of the detector is a voltage or current level stored in the control panel memory. In either case, there may be occasion to adjust the detector sensitivity, either at the detector or control panel. The means to restore the detector to its factory sensitivity must be provided, and the detector must be labeled, showing the sensitivity range. In some cases, the adjustment feature may be used between cleaning intervals to maintain stability. Naturally, after the unit has been cleaned, it is desirable to restore it to its original design sensitivity. The maintenance of smoke detectors is covered in Chapter 7.

5-3.5 Location and Spacing.

5-3.5.1* **General.** The location and spacing of smoke detectors shall result from an evaluation based on the guidelines detailed in this code and on engineering judgment. Ceiling shape and surfaces, ceiling height, configuration of contents, burning characteristics of combustible material present, ventilation, and the ambient environment are some of the conditions that shall be considered.

A-5-3.5.1 For operation, all types of smoke detectors depend on smoke entering the sensing chamber or light beam. Where sufficient concentration is present, operation is obtained. Since the detectors are usually mounted on the ceiling, response time depends on the nature of the fire. A hot fire will rapidly drive the smoke up to the ceiling. A smoldering fire, such as in a sofa, produces little heat; therefore, the time for smoke to reach the detector will be increased.

These general criteria are far less specific than those established for heat detectors. Heat detectors are used to detect fires that liberate significant quantities of energy, creating their own air currents. The energy from the fire propels the hot air/smoke mixture across the ceiling. Under these circumstances it is possible to generate models through fluid flow physics and thermodynamics. The Fire Detection Institute has been an important guiding force in the development of these models. These models predict that smoke detectors provide response well before heat detectors. This prediction has been verified experimentally.

However, under smoldering, low-energy-output fire conditions the smoke is conveyed on the existing air currents, with little, if any, propulsion from the fire. This makes the prediction of flow far more difficult and, consequently, the location and spacing of detectors more subject to the judgment of the designer.

5-3.5.1.1 Where the intent is to protect against a specific hazard, the detector(s) shall be permitted to be installed closer to the hazard in a position where the detector will readily intercept the smoke.

The designer is allowed to add detectors where he/she expects the pre-existing, normal air currents will convey the smoke from an early-stage fire. Usually the design process begins by locating detectors so that they will provide general area protection. Then additional detectors are added or positions adjusted to take into account known or anticipated ignition sources.

5-3.5.1.2* **Stratification.** The possible effect of smoke stratification at levels below the ceiling shall be considered.

Paragraph 5-3.5.1.2 addresses the potential for stratification. The design of a smoke detection system must address both the high-energy-output fire and the low-energy-output fire. In areas of high ceilings, this often necessitates layers of detectors or combining detectors to address all of the possible fire scenarios.

A-5-3.5.1.2 **Stratification.** Stratification of air in a room may hinder air containing smoke particles or gaseous combustion products from reaching ceiling-mounted smoke or fire-gas detectors.

Stratification occurs when air containing smoke particles or gaseous combustion products is heated by smoldering or burning material and, becoming less dense than surrounding cooler air, rises until it reaches a level at which there is no longer a difference in temperature between it and the surrounding air.

Stratification may also occur when evaporative coolers are used, because moisture introduced by these devices may condense on smoke, causing it to fall toward the floor. Therefore, to ensure rapid response, smoke detectors may need to be installed on sidewalls or at locations below the ceiling.

In installations where detection of smoldering or small fires is desired and where the possibility of stratification exists, consideration should be given to mounting a portion of the detectors below the ceiling. In high ceiling areas, projected beam-type or air sampling-type detectors at different levels should also be considered.

The objective of detecting the fire before it has achieved a high-energy output requires additional insight into the placement of detectors. The high-energy-output flaming fire produces a fire plume that propels smoke and hot air upward. The larger the fire, the higher the plume will extend and the greater the air velocity within the plume. In the low-energy-output smoldering fire [the type of ignition often encountered in residential (homes, hotels, apartments), institutional (hospitals, nursing homes, schools), and commercial (offices, stores) occupancies], significant quantities of smoke may be produced before the development of an energetic fire plume. This smoke may lack the energy to rise up to ceiling-mounted smoke detectors. This situation must be addressed in any fire alarm system designed for residential, institutional, or commercial occupancies. The addition of

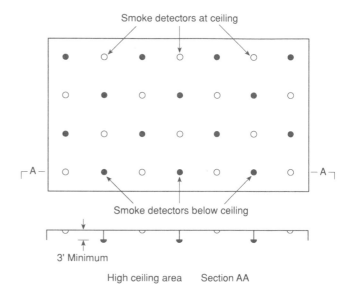

For SI Units: 1 ft = 0.305 m.

Figure A-5-3.5.1.2 *Smoke detector layout accounting for stratification.*

smoke detectors at some distance below the ceiling does not eliminate the requirement for ceiling-mounted detectors.

5-3.5.2* **Spot-Type Smoke Detectors.** Spot-type smoke detectors shall be located on the ceiling not less than 4 in. (100 mm) from a sidewall to the near edge or, if on a sidewall, between 4 in. (100 mm) and 12 in. (300 mm) down from the ceiling to the top of the detector. (*See Figure A-5-2.1.*)

Exception No. 1: See 5-3.5.1.2.

Exception No. 2: See 5-3.5.6.

Exception No. 3: See 5-3.5.7.

This location requirement is based upon the concept that in the event of a high-energy-output fire — one that is immediately life threatening — there will be a fire plume, and that plume will convey smoke to ceiling-mounted detectors due to the thermal lift from the fire. This logic is identical to that used in locating heat detectors (*see 5-2.5*). Subsection 5-3.5.2 instructs us to locate smoke detectors on the ceiling or on the side walls within 12 inches of the ceiling. These locations derive maximum benefit from the upward flow of smoke from a fire. However, the best currently available research data supports the theory that there is a dead air space where the walls meet the ceiling in a typical room.

Figure A-5-2.5.1 shows this dead air space extending 4 inches in from the wall and 4 inches down from the ceiling. Consequently, the code excludes detectors from that area.

A-5-3.5.2 In high ceiling areas, such as atriums, where spot-type smoke detectors are not accessible for periodic maintenance and testing, projected beam-type or air sampling-type detectors should be considered where access can be provided.

Paragraph A-5-3.5.2 clarifies the perceived conflict between 5-3.5.2, which instructs us to locate detectors on the ceiling or within 12 inches of it, and 5-1.3.4, which requires that all initiating devices, including smoke detectors, be installed in such a manner that they can be main-

tained. Atriums and other areas with exceptionally high ceilings (auditoriums, gymnasiums, exhibit halls, storage facilities, and some manufacturing facilities) represent very difficult situations for the use of spot-type smoke detection because of the problems that arise due to stratification, maintenance concerns, and smoke dissipation. This paragraph advises the designer to consider either air-sampling or projected beam-type photoelectric light obscuration smoke detection as alternatives.

5-3.5.2.1* To minimize dust contamination of smoke detectors where installed under raised room floors and similar spaces, they shall only be mounted in an orientation for which they have been listed. (*See Figure A-5-3.5.2.1.*)

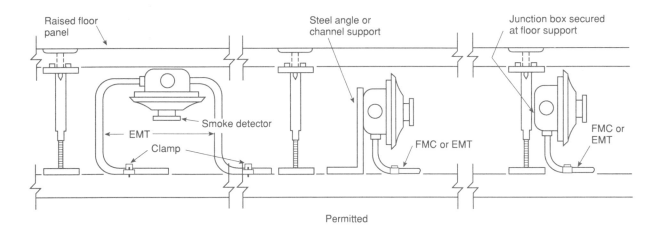

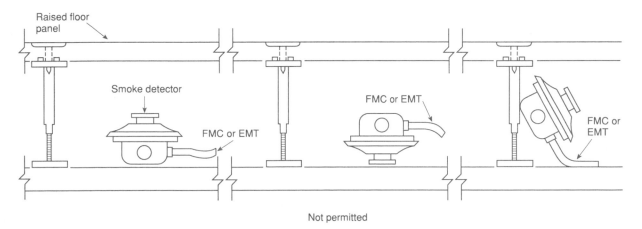

Figure A-5-3.5.2.1 *Mounting installations — permitted (top) and not permitted (bottom).*

The fast-moving air in a data center underfloor has sufficient energy to suspend dust. As that air enters the detector, it slows down and the suspended dust settles in the detector. The accumulation of dust within a smoke detector has a similar effect to that of smoke. In an ionization detector, the dust impedes the flow of current within the chamber. In a photoelectric detector, the dust increases the reflectance within the chamber. Thus, dust causes each type of detector to become more sensitive, increasing the likelihood of false alarms. The orientations shown in Figure A-5-3.5.2.1 minimize the possibility of dust falling into the detector from the floor and also minimize the effect of air-conveyed dust on the detector.

There are other concerns that reinforce the benefits of positioning detectors as shown in Figure A-5-3.5.2.1. Since the cables used to connect computer components lie on the underfloor slab, air movement will be concentrated in the upper half of the underfloor volume. Placing the detector in the upper half of the underfloor improves the system's ability to respond to an early-stage fire.

Another reason for positioning detectors as shown in Figure A-5-3.5.2.1 is that detectors mounted in the upper half of the underfloor volume are far less likely to be damaged as new cables are installed or old cables are rerouted through the underfloor space. If water-cooled computers are in use, the detectors are less likely to become wet if the computer cooling system leaks. Also, when there is no airflow, the detectors will be in the best orientation for detection. Finally, Figure A-5-3.5.2.1 shows the detectors in the orientation for which they have been tested and listed.

5-3.5.3 Projected Beam-Type Smoke Detectors. Projected beam-type smoke detectors (*see 5-3.3.3.1*) shall normally be located with their projected beams parallel to the ceiling and in accordance with the manufacturer's documented instructions.

Exception No. 1: See 5-3.5.1.2.

Exception No. 2: Beams may be installed vertically or at any angle needed to afford protection of the hazard involved. (Example: Vertical beams through the open shaft area of a stairwell where there is a clear vertical space inside the handrails.)

5-3.5.3.1 The beam length shall not exceed the maximum permitted by the equipment listing.

Projected-beam smoke detection systems have limitations on both the minimum and maximum beam length necessary to achieve detection. The projected-beam smoke detector must be able to identify a low concentration of smoke distributed along a substantial portion of the beam and a high concentration of smoke localized in a short segment of the beam. Each manufacturer obtains a listing from a testing laboratory that sets the upper and lower limits on the beam length. Failure to observe these limits could result in an unstable detector or the failure to detect a fire.

5-3.5.3.1.1 Where mirrors are used with projected beams, they shall be installed in accordance with the manufacturer's documented instructions.

5-3.5.4 Sampling-Type Smoke Detector. Each sampling port of a sampling-type smoke detector shall be treated as a spot-type detector for the purpose of location and spacing. Maximum air sample transport time from the farthest sampling point shall not exceed 120 seconds.

An air sampling-type smoke detector is defined in Section 1-4. The air transport time criterion places an effective limit on the design of the fan and the maximum distance from the detector to the farthest sampling port, as well as the size and layout of the sampling tubes. The manufacturer's listing and instructions will define any limitations. Some air sampling-type smoke detectors have a means to detect changes in airflow, providing some measure of supervision of the tubing or piping network.

5-3.5.5 Smooth Ceiling Spacing.

5-3.5.5.1 Spot-Type Detectors. On smooth ceilings, spacing of 30 ft (9.1 m) shall be permitted to be used as a guide. In all cases, the manufacturer's documented instructions shall be followed. Other spacing shall be permitted to be used depending on ceiling height, different conditions, or response requirements. (*See Appendix B for detection of flaming fires.*)

Because of the difficulty in modeling the flow of smoke in low energy output fire scenarios, the code cannot provide definitive spacings for other modes of detection. Additional research is being funded and managed under the auspices of the National Fire Protection Research Foundation and the Fire Detection Institute. It is hoped that this research will provide the basis for more definitive spacing requirements in the near future.

An analytical method based on temperature rise data from the first phase of the Fire Detection Institute research is included in Appendix B. This method is very useful in aiding in detector spacing and placement for flaming fire scenarios and has become a very important tool for the fire protection systems designer.

5-3.5.5.1.1* For smooth ceilings, all points on the ceiling shall have a detector within a distance equal to 0.7 times the selected spacing.

A-5-3.5.5.1.1 This will be useful in calculating locations in corridors or irregular areas. (*See A-5-2.7.1 and Figure A-5-2.7.1.1.*) For irregularly shaped areas, the spacing between detectrs may be greater than the selected spacing, provided the maximum spacing from a detector to the farthest point of a sidewall or corner within its zone of protection is not greater than 0.7 times the selected spacing (0.7S). (*See Figure A-5-2.7.1.1.*)

The concepts behind the spacing of smoke detectors follow directly from the concepts developed for heat detectors. Subsection A-5-2.7.1 develops the concepts that enable us to determine the area that will be covered by a detector. That area can vary in shape as long as the distance from the detector to the farthest point to be covered by the detector does not exceed 0.7 times the spacing factor. See A-5-2.7.1 and its associated figures.

5-3.5.5.2* **Projected Beam-Type Detectors.** For location and spacing of projected beam-type detectors, the manufacturer's documented installation instructions shall be followed. (*See Figure A-5-3.5.5.2.*)

A-5-3.5.5.2 On smooth ceilings, a spacing of not more than 60 ft (18.3 m) between projected beams and not more than one-half that spacing between a projected beam and a sidewall (wall parallel to the beam travel) may be used as a guide. Other spacing may be determined depending on ceiling height, airflow characteristics, and response requirements.

In some cases, the light beam projector will be mounted on one end wall, with the light beam receiver mounted on the opposite wall. However, it is also permissible to suspend the projec-

tor and receiver from the ceiling at a distance from the end walls not exceeding one-quarter the selected spacing. For an illustration of this, see Figure A-5-3.5.5.2.

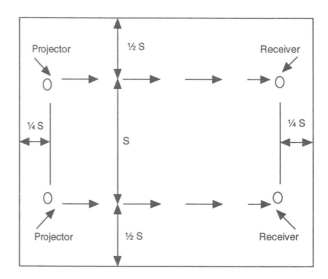

Figure A-5-3.5.5.2 *Maximum distance that ceiling-suspended light projector and receiver may be positioned from end wall is ¹/₄ selected spacing (S).*

Notice the similarity between the installation and spacing concept developed for line-type heat detectors and line-type (projected beam) smoke detectors. The logic behind the design rules is consistent. Just as a line-type heat detector can be thought of as a row of spot-type heat detectors, it is often helpful to think of a projected beam detector as the equivalent to a row of spot-type detectors when developing a spacing strategy. The distance between the projected beams is analogous to the distance between rows of spot-type detectors. Obviously, in high ceiling areas where stratification is probable and a serious concern, projected beams can be positioned at several levels.

5-3.5.6* **Solid Joist Construction.**

A-5-3.5.6 Detectors are placed at reduced spacings at right angles to joists or beams in an attempt to ensure that detection time is equivalent to that which would be experienced on a flat ceiling. It takes longer for the combustion products (smoke or heat) to travel at right angles to beams or joists, because of the phenomenon wherein a plume from a relatively hot fire with significant thermal lift tends to fill the pocket between each beam or joist before moving to the next one.

Though it is true that this phenomenon may not be significant in a small smoldering fire where there is only enough thermal lift to cause stratification at the bottom of the joists, reduced spacing is still recommended to ensure that detection time is equivalent to that which would exist on a flat ceiling, even in the hotter type of fire.

The logic behind these recommendations for the spacing of smoke detectors on ceilings with joists or beams is the same as is used in the requirements for heat detectors. Remember that joists are effectively defined as solid projections of more than 4 inches in depth, placed on centers of less than 3 feet.

5-3.5.6.1 Ceiling construction where joists are 8 in. (200 mm) or less in depth shall be considered equivalent to a smooth ceiling. Spot-type detectors shall be mounted on the bottom of the joists. (*See also 5-3.5.1.2.*)

5-3.5.6.2 If joists exceed 8 in. (200 mm) in depth, the spacing of spot-type detectors in the direction perpendicular to the joists shall be reduced by one third. If the projected light beams of line-type detectors run perpendicular to the joists, no spacing reduction shall be necessary; however, if the projected light beams are parallel to the joists, the spacing between light beams shall be reduced. Spot-type detectors shall be mounted on the bottom of the joists. (*See also 5-3.5.1.2.*)

While the concept is the same, the dimension criteria regarding the depth of the joists is different for smoke detectors and heat detectors. When heat detectors are used, wherever the joists are more than 4 inches deep, a spacing reduction of 50 percent is required. When smoke detectors are used, a spacing reduction is not required until the joist depth exceeds 8 inches, when a spacing reduction of $\frac{1}{3}$ is required.

5-3.5.7 Beam Construction.

The similarity in concept between the use of heat detectors and the use of smoke detectors extends to the design guideline for ceilings with beams. The downward project-

ing beams create an impediment to the horizontal flow of smoke beneath the ceiling, slowing down the velocity of the smoke. Where this occurs the spacing must be reduced to achieve consistently acceptable response times.

5-3.5.7.1 Ceiling construction where beams are 8 in. (200 mm) or less in depth shall be considered equivalent to a smooth ceiling. (*See also 5-3.5.1.2.*)

These dimension criteria are different from those for heat detectors.

5-3.5.7.2 If beams are over 8 in. (200 mm) in depth, the spacing of spot-type detectors in the direction perpendicular to the beams shall be reduced. The spacing of projected light beam detectors run perpendicular to the ceiling beams need not be reduced; however, if the projected light beams are run parallel to the ceiling beams, the spacing shall be reduced. (*See also 5-3.5.1.2.*)

There is no definitive reduction percentage as there is regarding heat detectors (*see 5-2.7.3*). The designer must use engineering judgment. Research is currently being conducted with the objective of developing definitive reduction factors for smoke detectors in beamed ceilings.

5-3.5.7.3 If beams are less than 12 in. (305 mm) in depth and less than 8 ft (2.4 m) on center, spot-type detectors shall be permitted to be installed on the bottom of beams.

This paragraph is consistent with the requirements of paragraph 5-2.5.1 pertaining to heat detectors. The air currents that convey smoke and heat are usually strong enough to ensure that the concentration of smoke at a detector mounted on the bottom of a beam less than 12 inches deep will be sufficient to achieve an alarm. As the smoke plume from a fire impinges upon the ceiling and begins expanding horizontally, a beam becomes a dam of sorts. Time and the expenditure of smoke energy forces the smoke downward far enough to spill over the dam (actually flow under the dam), encountering the smoke detector in the process. Energetic fires that produce large thermal outputs quickly force smoke across beams due to the large amounts of thermal energy available. Low-energy-output fires provide less energy, and the movement across beams is slower.

When beams are less than 12 inches deep, detectors are permitted (not required) to be mounted on the beam bottoms. If the beam is 12 inches deep or greater, detectors must be mounted on the ceiling between the beams.

5-3.5.7.4* If the beams exceed 18 in. (460 mm) in depth and are more than 8 ft (2.4 m) on center, each bay shall be treated as a separate area requiring at least one spot-type or projected beam-type detector.

Deep beams on wide spacings form bays. Before the smoke can travel further across the ceiling, it must fill the bay. When the beams are more than 18 inches deep, the volume is large enough that it demands its own detector.

A-5-3.5.7.4 To detect flaming fires (strong plumes), detectors should be installed as follows:

(a) If the ratio of the beam depth (D) to ceiling height (H) (D/H) is greater than 0.10 and the ratio of beam spacing (W) to ceiling height (H) (W/H) is greater than 0.40, detectors should be located in each beam pocket.

(b) If either the ratio of beam depth to ceiling height (D/H) is less than 0.10 or the ratio of beam spacing to ceiling height (W/H) is less than 0.40, detectors should be installed on the bottom of the beams.

To detect smoldering fires (weak or no plumes), detectors should be installed as follows:

(a) If air mixing into beam pockets is good (e.g., air-flow parallel to long beams) and condition (a) exists as above, detector should be located in each beam pocket.

(b) If air mixing into beam pockets is limited or condition (b) exists as above, detectors should be located on the bottom of the beams.

Research on plumes and ceiling jets indicates that the radius of a plume where it impinges on the ceiling is approximately 20 percent of the ceiling height above the fire source (p. 2H) and the minimum depth of the ceiling jet (at its turning point) is approximately 10 percent of the ceiling height above the fire source (y. 0.10H). For ceilings with beams deeper than the jet depth and spaced wider than the plume width, detectors will respond faster in the beam pocket because they will be in either the plume or ceiling jet. For ceilings with beams of less depth than ceiling jet or spaced closer than the plume width, detector response will

not be enhanced by placing detectors in each beam pocket, and the detectors may perform better on (for spot-type detectors) or below (for beam detectors) the bottom of the beams.

Where plumes are weak, ventilation and mixing into the beam pockets will determine detector response. Where beams are closely spaced and airflow is perpendicular to the beam, mixing into the beam pocket is limited and detectors will perform better on or below the bottom of the beams.

5-3.5.8 Sloped Ceilings.

The concepts for smoke detector placement and spacing in applications with sloped ceilings is derived from the location and spacing criteria developed for heat detectors.

5-3.5.8.1 **Peaked.** Detectors shall first be spaced and located within 3 ft (0.9 m) of the peak, measured horizontally. The number and spacing of additional detectors, if any, shall be based on the horizontal projection of the ceiling. (*See Figure A-5-2.7.4.1.*)

5-3.5.8.2 **Shed.** Detectors shall first be spaced and located within 30 ft (0.9 m) of the high side of the ceiling, measured horizontally. The number and spacing of additional detectors, if any, shall be based on the horizontal projection of the ceiling. (*See Figure A-5-2.7.4.2.*)

5-3.5.9 **Raised Floors and Suspended Ceilings.** In under-floor spaces and above-ceiling spaces that are not HVAC plenums, detector spacing shall be in accordance with Section 5-3.5.

Subsection 5-1.3.4 requires detection (when total coverage is required by the authority having jurisdiction or other codes) in all accessible spaces (combustible or noncombustible and in inaccessible combustible spaces). The spaces beneath raised floors and above suspended ceilings usually fall into that category and, hence, require detection using the same location and spacing concepts as the occupied portion of a building.

5-3.5.10 **Partitions.** Where partitions extend upward to within 18 in. (460 mm) of the ceiling, they will not influence the spacing. Where the partition extends to within less than 18 in. (460 mm) of the ceiling, the effect of smoke travel shall be considered in reduction of spacing.

From the information presented in A-5-3.5.7.4 on ceiling jet thickness, the reader should see that a partition extending to within 18 inches of the ceiling will very likely affect the ceiling jet, restricting the horizontal flow of smoke across the ceiling. Research that quantifies the effect has not yet been conducted. Consequently, the designer is instructed to consider it on a qualitative basis. It should also be noted that this subsection has different guidance than that for heat detectors.

5-3.6 Heating, Ventilating, and Air Conditioning (HVAC).

5-3.6.1* In spaces served by air-handling systems, detectors shall not be located where air from supply diffusers could dilute smoke before it reaches the detectors. Detectors shall be located to intercept the air flow toward the return air opening(s). This may require additional detectors, since placing detectors only near return air openings may leave the balance of the area with inadequate protection when the air-handling system is shut down.

A-5-3.6.1 Detectors should not be located in a direct air-flow nor closer than 3 ft (900 mm) from an air supply diffuser.

This paragraph recommends a separation of at least 3 feet between an air supply diffuser and the detector. There may be situations where 3 feet is not adequate, depending upon the air velocity and diffuser size.

5-3.6.2 In under-floor spaces and above-ceiling spaces that are used as HVAC plenums, detectors shall be listed for the anticipated environment. (*See 5-3.7.1.1.*) Detector spacings and locations shall be selected based upon anticipated airflow patterns and fire type.

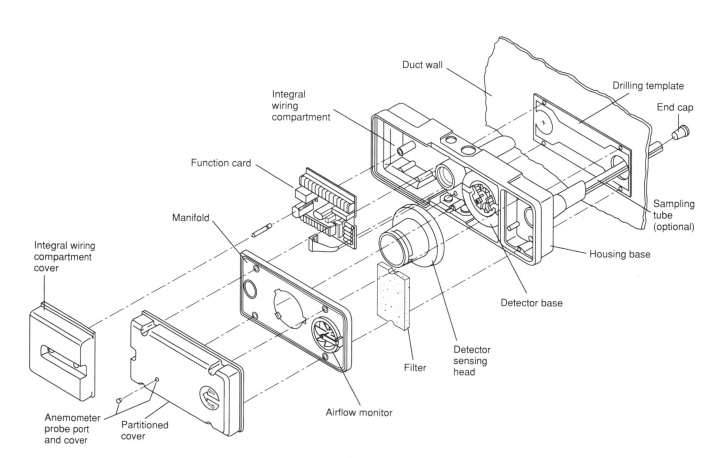

Figure 5.19(a) *Exploded view of a duct smoke detector.*

Figure 5.19(b)   Duct-type smoke detector.

In order to cool a room to 70°F, it may be necessary to introduce extremely frigid air. Heating a room sometimes requires superheated air. Consequently, HVAC plenums usually have ambient conditions that are far more extreme than the spaces they support. Smoke detectors are electronic sensors whose operation is affected by the ambient temperature, the relative humidity, and, especially in the case of spot-type ionization detectors, the velocity of the air around the detector. Not all smoke detectors are listed for the range of conditions found in HVAC plenums.

5-3.6.2.1 Detectors placed in environmental air ducts or plenums shall not be used as a substitute for open area detectors. (*See Section 5-11, Table A-5-3.7.1.1, A-5-11.1, and A-5-11.2.*) Where open area protection is required, 5-3.5 shall apply.

Smoke may not be drawn into the duct or plenums when the ventilating system is shut down. Further, when the ventilating system is operating, the detector(s) may be less responsive to a fire condition in the room of fire origin due to dilution by clean air.

5-3.7 Special Considerations.

5-3.7.1 The selection and placement of smoke detectors shall take into consideration both the performance characteristics of the detector and the areas into which the detectors will be installed to prevent nuisance alarm or nonoperation after installation. Some of the considerations are as follows.

5-3.7.1.1* The installation of smoke detectors shall take into consideration the range of environmental conditions present. Smoke detectors shall be intended for installation in areas where the normal ambient conditions are not likely to exceed the following:

    (a) A temperature of 100°F (38°C), or fall below 32°F (0°C); or

    (b) A relative humidity of 93 percent; or

    (c) An air velocity of 300 fpm (1.5 mps).

Exception: Detectors specifically designed for use in ambients exceeding the limits of (a) through (c) and listed for the temperature, humidity, and air velocity conditions expected.

A-5-3.7.1.1 Product-listing standards include tests for temporary excursions beyond normal limits. In addition to temperature, humidity, and velocity variations, smoke detectors should operate reliably under such common environmental conditions as mechanical vibration, electrical interference, and other environmental influences. Tests for these conditions are also conducted by the testing laboratories in their listing program. In those cases in which environmental conditions approach the limits shown in Table A-5-3.7.1.1, consult the detector manufacturer for additional information and recommendations.

Table A-5-3.7.1.1   Environmental Conditions that Influence Detector Response

Detection Protection	Air Velocity >300'/min	Atm. Pressure Above Sea Level	>3000' Humidity >93%	Temp. <32°F >100°F	Color of Smoke
Ion	X	X	X	X	O
Photo	O	O	X	X	X
Beam	O	O	X	X	O
Air Sampling	O	O	X	X	O

X = May affect detector response.   O = Generally does not affect detector response.

Table A-5-3.7.1.2(a)   Common Sources of Aerosols and Particulate Matter Moisture

Moisture

Live steam	Excessive tobacco smoke
Steam tables	Heat treating
Showers	Corrosive atmospheres
Humidifiers	Dust or lint
Slop sink	Linen/bedding handling
Humid outside air	Sawing, drilling, and grinding
Water spray	Pneumatic transport
	Textile and agricultural processing

Combustion Products and Fumes

Cooking equipment	**Engine Exhaust**
Ovens	
Dryers	Gasoline forklift trucks
Fireplaces	Diesel trucks and locomotives
Exhaust hoods	Engines not vented to the outside
Cutting, welding, and brazing	
	Heating element with abnormal conditions
Machining	
Paint spray	Dust accumulations
Curing	Improper exhaust
Chemical fumes	Incomplete combustion
Cleaning fluids	

smoke, moisture, dust or fumes, and electrical or mechanical influences.

A-5-3.7.1.2   Smoke detectors may be affected by electrical and mechanical influences and by aerosols and particulate matter found in protected spaces. Location of detectors should be such that the influences of aerosols and particulate matter from sources such as those in Table A-5-3.7.1.2(a) are minimized. Similarly, the influences of electrical and mechanical factors shown in Table A-5-3.7.1.2(b) should be minimized. While it may not be possible to totally isolate environmental factors, an awareness of these factors during system layout and design will favorably affect detector performance.

Table A-5-3.7.1.2(b)   Sources of Electrical and Mechanical Influences on Smoke Detectors

Electrical Noise and Transients	Airflow
Vibration or shock	Gusts
Radiation	Excessive velocity
Radio frequency	Power supply
Intense light	
Lightning Electrostatic discharge	

Different detection technologies are affected differently by these environmental extremes. Different makes and models within each group may be affected more or less than others. It is beyond the scope of this handbook to identify these effects beyond the generalities presented here. However, the reader must recognize that some detector designs are inherently more forgiving than others. The tests performed in the process of listing ascertain that a detector meets minimum performance criteria. There may be design features in specific devices that allow them to be used in extreme environments. The manufacturer should be consulted when such an application is contemplated.

These environmental limits may require the designer to consider alternative detection modes. While smoke detection may be preferable from the early warning standpoint, heat or radiant energy detection may be a better choice due to the range of environmental conditions.

5-3.7.1.2*   To avoid nuisance alarms, the location of smoke detectors shall take into consideration normal sources of

In applications where the factors outlined in Tables A-5-3.7.1.2(a) and A-5-3.7.1.2(b) cannot be sufficiently limited to allow reasonable stability and response times, alternate modes of fire detection should be considered.

5-3.7.1.3   Detectors shall not be installed until after the construction clean-up of all trades is complete and final.

Exception: Where required by the authority having jurisdiction for protection during construction.

Detectors that have been installed prior to final clean-up by all trades shall be cleaned or replaced per Chapter 7.

Many needless alarms have been caused by the early installation of smoke detectors. This subsection forbids that practice unless the authority having jurisdiction requires it and the detectors are cleaned after all construction trades have finished their work. In all other cases, smoke detectors are not allowed to be installed until all finish work is complete.

5-3.7.2 Spot-Type Detectors.

5-3.7.2.1 Smoke detectors having a fixed temperature element as part of the unit shall be selected in accordance with Table 5-2.4.1 for the maximum ceiling temperature that can be expected in service.

5-3.7.2.2* Holes in the back of a detector shall be covered by a gasket, sealant, or equivalent, and the detector shall be mounted so that air flow from inside or around the housing will not prevent the entry of smoke during a fire or test condition.

A-5-3.7.2.2 Airflow through holes in the rear of a smoke detector may interfere with smoke entry to the sensing chamber. Similarly, air from the conduit system may flow around the outside edges of the detector and again interfere with smoke reaching the sensing chamber. Additionally, holes in the rear of a detector provide a means for entry of dust, dirt, and insects, each of which can adversely affect the detector's performance.

The conditions listed in A-5-3.7.2.2 have been encountered frequently enough to warrant inclusion of the requirements in 5-3.7.2.2 into the code. However, this list cannot be assumed to be exhaustive. It is important for the reader to recognize that there may be special considerations unique to a specific application. The designer should be aware of any factor in the protected area that could contribute to unwanted alarms or could prevent the successful conveyance of smoke to the detector.

5-3.7.3 Projected Beam-Type Detectors.

5-3.7.3.1 Projected beam-type detectors and mirrors shall be firmly mounted on stable surfaces so as to prevent false or erratic operation due to movement. The beam shall be so designed that small angular movements of the light source or receiver do not prevent operation due to smoke and do not cause nuisance alarms.

Contrary to popular belief, buildings move. Portions of buildings vibrate due to passing traffic on nearby streets. They sway due to wind or uneven thermal expansion; even tides can cause oceanfront buildings to flex. Buildings are designed to flex, however, and the design

of fire alarm systems utilizing projected beam smoke detection must accommodate this fact. The manufacturers of these detection devices provide installation instructions that address the potential for this type of difficulty. Most manufacturers do not allow the use of mirrors due to the physical instability of mounting surfaces and building movement.

5-3.7.3.2 Since the projected beam-type unit will not operate for alarm [but will give a trouble signal (*see A-5-3.3.3*)] where the light path to the receiver is abruptly interrupted or obscured, the light path shall be kept clear of opaque obstacles at all times.

5-3.7.4 Air Sampling-Type Detectors.

In addition to the cloud chamber type of smoke detector, two varieties of aspirating-type air-sampling smoke detectors exist. These detectors are essentially photoelectric smoke detectors with aspirating fans and a control unit. The apparatus as a whole constitutes a smoke detector. These detectors are used in a variety of applications where the designer is concerned with the effects of high airflow on smoke detection. Because of their sensitivity ranges, air-sampling detectors are also used in areas housing very valuable equipment. See Figures 5.20 and 5.21.

5-3.7.4.1* To ensure proper performance, a sampling pipe network shall be designed to include details of the sampling network based on and supported by sound fluid dynamic principles and calculations showing flow characteristics of the pipe network and for each sampling point.

A-5-3.7.4.1 Air Sampling-Type Detectors. A single pipe network has a shorter transport time than a multiple pipe network of similar length pipe; however, a multiple pipe system provides a faster smoke transport time than a single pipe system of the same total length. As the number of sampling holes in a pipe increases, the smoke transport time increases. Where practical, pipe run lengths in a multiple pipe system should be nearly equal or the system should be otherwise pneumatically balanced.

The manufacturers of this class of smoke detection unit provide engineering guidelines in their installation manuals

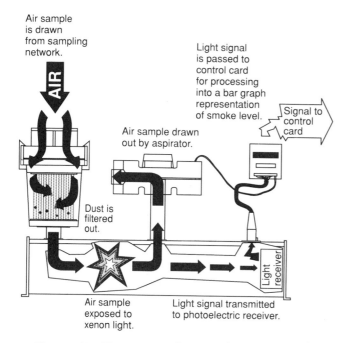

Air sample is drawn from sampling network.

AIR

Light signal is passed to control card for processing into a bar graph representation of smoke level.

Signal to control card

Air sample drawn out by aspirator.

Dust is filtered out.

Light receiver

Air sample exposed to xenon light.

Light signal transmitted to photoelectric receiver.

Figure 5.20 How an optical air-sampling system works.

that ensure that the products meet the criteria of 5-3.7.4.1. These guidelines are evaluated by testing laboratories as part of the listing procedure.

The generalities in A-5-3.7.4.1 give the designer some basis for deciding the type of piping network that best serves the application under consideration.

5-3.7.4.2* Air sampling detectors shall give a trouble signal where the air flow is outside the manufacturer's specified range. The sampling ports and inline filter (if used) shall be kept clear in accordance with manufacturer's documented instructions.

A-5-3.7.4.2 The air sampling-type detector system should be able to withstand dusty environments by either air filtering or electronic discrimination of particle size. The detector should be capable of providing optimal time delays of alarm outputs to eliminate nuisance alarms due to transient smoke conditions. The detector should also provide facilities for the connection of monitoring equipment for the recording of background smoke level information necessary in setting alert and alarm levels and delays.

5-3.7.4.3 Air sampling network piping and fittings shall be airtight and permanently fixed. Sampling piping shall be conspic-

uously identified as "SMOKE DETECTOR SAMPLING PIPE," with a warning not to disturb or alter.

5-3.7.5* High Rack Storage. *[See Figures A-5-3.7.5(a) and (b).]* Detection systems are often installed in addition to suppression systems. Where smoke detectors are installed for early warning in high rack storage areas, it shall be necessary to consider installing detectors at several levels in the racks to ensure quicker response to smoke. Where detectors are installed to actuate a suppression system, see NFPA 231C, *Standard for Rack Storage of Materials.*

Fire protection for high rack storage warehouses is a particularly difficult problem. The reasons for this are that the fuel load per unit of floor area is extremely high, the accessibility of the fuel is relatively low, and the combustibility of the materials in any given rack can vary from nominally noncombustible to flammable. Early detection and rapid extinguishment is critical. There have been a number of catastrophic total losses in high rack storage facilities in the past decade.

A-5-3.7.5 High Rack Storage. For most effective detection of fire in high rack storage areas, detectors should be located on the ceiling above each aisle and at intermediate levels in the racks. This is necessary to detect smoke that may be trapped in the racks at an early stage of fire development, when insufficient thermal energy is released to carry the smoke to the ceiling. Earliest detection of smoke is achieved by locating the intermediate level detectors adjacent to alternate pallet sections as shown in Figures A-5-3.7.5(a) and (b). Detector manufacturer's recommendations and engineering judgment should be followed for specific installations.

A projected beam-type detector may be used in lieu of a single row of individual spot-type smoke detectors.

Sampling ports of an air sampling-type detector may be located above each aisle to provide coverage equivalent to the location of spot-type detectors. Manufacturer's recommendations and engineering judgment should be followed for specific installation.

5-3.7.6 High Air Movement Areas.

5-3.7.6.1 General. The purpose and scope of 5-3.7.6 are to provide location and spacing guidance for smoke detectors in high air movement areas for early warning of fire.

Exception: Detectors provided for the control of smoke spread are covered by the requirements of Section 5-11.

(a)

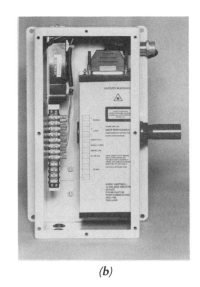

(b)

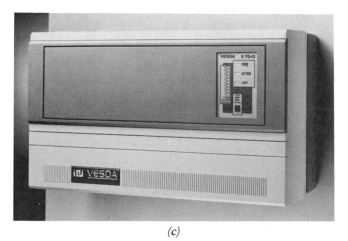

(c)

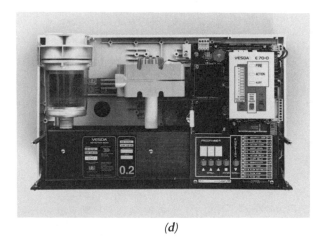

(d)

Figure 5.21 *Typical air sampling-type smoke detectors with cover in place [(a) and (c)] and with cover removed [(b) and (d)].*

The most regularly encountered example of a high-air-movement area is the data center (computer room), specifically its underfloor and above-ceiling plenums. This is by no means the only area that falls into this category. In general, areas where the air velocity across the detector exceeds 300 feet per minute (1.5 meters per second) are considered high-air-movement ambients. Table 5-3.7.6.3 and Figure 5-3.7.6.3 do not apply to underfloor or ceiling plenums, but the principles do apply. Any time high airflow is encountered, consideration should be given to reducing the spacing of spot-type detectors or utilizing detectors that are not as affected by high airflow.

5-3.7.6.2 Location. Smoke detectors shall not be located directly in the air stream of supply registers.

5-3.7.6.3 Spacing. Smoke detector spacing depends upon the movement of air within the room (including both supplied and recirculated air), which can be designated as minutes per air change or air changes per hour. Spacing shall be in accordance with Table 5-3.7.6.3 and Figure 5-3.7.6.3.

Exception: Air sampling or projected beam smoke detectors installed in accordance with the manufacturer's documented instructions.

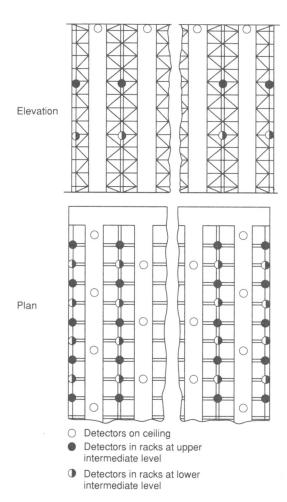

Elevation

Plan

○ Detectors on ceiling
● Detectors in racks at upper
  intermediate level
◑ Detectors in racks at lower
  intermediate level

Figure A-5-3.7.5(a)   *For solid storage (closed rack) in which transverse and longitudinal flue spaces are irregular or nonexistent, as for slatted or solid shelved storage.*

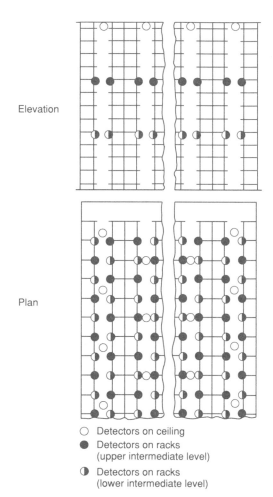

Elevation

Plan

○ Detectors on ceiling
● Detectors on racks
  (upper intermediate level)
◑ Detectors on racks
  (lower intermediate level)

Figure A-5-3.7.5(b)   *For palletized storage (open rack) or no shelved storage in which regular transverse and longitudinal flue spaces are maintained.*

Because of the very high value of a data center, when a smoke detection system is being designed for such a site, it is common for spacing to be reduced. This spacing may be derived from Table 5-3.7.6.3 and Figure 5-3.7.6.3.

The reduced spacing for spot-type detectors is based upon the concerns over the dilution of smoke. Data has not yet been presented to the Technical Committee that would enable the committee to determine the effect of high air movement on projected beam or air sampling detectors. The designer must depend upon the data provided by the manufacturer.

Table 5-3.7.6.3

Minutes/Air Change	Air Changes/Hour	Sq Ft/Detector
1	60	125
2	30	250
3	20	375
4	15	500
5	12	625
6	10	750
7	8.6	875
8	7.5	900
9	6.7	900
10	6	900

For SI Units: 1 sq ft = 0.0929 m².

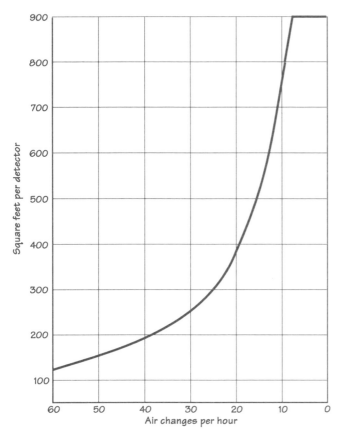

Figure 5-3.7.6.3 *High air movement areas (not to be used for under-floor or above-ceiling spaces).*

5-4 Radiant Energy-Sensing Fire Detectors.

This section was entitled "Flame Detectors" until the 1990 edition of NFPA 72E, when very substantial revisions were made. The title was changed to "Radiant Energy-Sensing Fire Detectors" to encompass both flame detectors and spark/ember detectors.

5-4.1 General.

5-4.1.1 The purpose and scope of Section 5-4 are to provide standards for the selection, location, and spacing of fire detectors that sense the radiant energy produced by burning substances. These detectors are categorized as flame detectors and spark/ember detectors.

5-4.1.1.1 **Flame Detector.** *(See 5-4.2.1, definition of Flame Detector.)*

5-4.1.1.2 **Spark/Ember Detector.** *(See 5-4.2.1, definition of Spark/Ember Detectors.)*

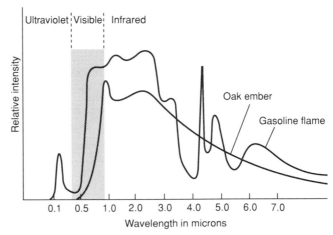

Figure 5.22 *Emission spectra of Class A and Class D combustibles.*

5-4.1.2 **Radiant Energy.** For the purpose of this code, radiant energy includes the electromagnetic radiation emitted as a by-product of the combustion reaction, which obeys the laws of optics. This includes radiation in the ultraviolet, visible, and infrared spectrum emitted by flames or glowing embers. These portions of the spectrum are distinguished by wavelengths as follows:

Ultraviolet	0.1 to 0.35 microns
Visible	0.36 to 0.75 microns
Infrared	0.76 to 220 microns
(1.0 micron = 1000 nanometers = 10,000 Angstroms)	

The committee selected this definition of radiant energy to distinguish between heat detectors and optical detectors that sense sparks, embers, and flames.

5-4.2 **Definitions and Operating Principles.**

5-4.2.1 **Definitions.**

Ember.* A particle of solid material that emits radiant energy due either to its temperature or the process of combustion on its surface. *(See definition of Spark.)*

A-5-4.2.1 **Ember.** Class A and Class D combustibles will burn as embers under conditions where the typical flame associated with fire does not necessarily exist. This glowing combustion yields radiant emissions in radically different parts of the radiant energy spectrum than flaming combustion. Specialized detectors, specifically designed to detect those emission, should be used in applications where this type of combustion is expected. In general, flame detectors are not intended for the detection of embers.

Field of View. The solid cone extending out from the detector within which the effective sensitivity of the detector is at least 50 percent of its on-axis, listed, or approved sensitivity.

Flame. A body or stream of gaseous material involved in the combustion process and emitting radiant energy at specific wavelength bands determined by the combustion chemistry of the fuel. In most cases, some portion of the emitted radiant energy is visible to the human eye.

Flame Detector Sensitivity. The distance along the optical axis of the detector at which the detector will detect a fire of specified size and fuel within a given time frame.

Flame Detectors. Radiant energy fire detectors that are intended to detect flames and are designed to operate in environments where sunlight or other ambient lighting is assumed.

Spark.* A moving ember.

A-5-4.2.1 **Spark.** The overwhelming majority of applications involving the detection of Class A and Class D combustibles with radiant energy-sensing detectors involves the transport of particulate solid materials through pneumatic conveyor ducts or mechanical conveyors. It is common in the industries that include such hazards to call a moving piece of burning material a "spark" and systems for the detection of such fires "spark detection systems."

Editions of NFPA 72E prior to 1990 did not address sparks at all. This definition and the associated Appendix A text is necessary for the reader to recognize the importance of the distinction between flame detectors and spark/ember detectors.

Spark/Ember Detector Sensitivity. The number of watts (or fractions of watts) of radiant power from a point source radiator applied as a unit step signal at the wavelength of maximum detector sensitivity, necessary to produce an alarm signal from the detector within the specified response time.

Spark/Ember Detectors. Radiant energy fire detectors that are designed to detect sparks or embers, or both. These devices are normally intended to operate in dark environments and in the infrared part of the spectrum.

Wavelength.* The distance between the peaks of a sinusoidal wave. All radiant energy can be described as a wave having a wavelength. Wavelength serves as the unit of measure for distinguishing between different parts of the spectrum. Wavelengths are measured in microns (uM), nanometers (nM), or angstroms (Å).

A-5-4.2.1 **Wavelength.** The concept of wavelength is extremely important in selecting the proper detector for a particular application. There is a precise interrelation between the wavelength of light being emitted from a flame and the combustion chemistry producing the flame. Specific sub-atomic, atomic, and molecular events yield radiant energy of specific wavelengths. For example, ultraviolet photons are emitted as the result of the complete loss of electrons or very large changes in electron energy levels. During combustion, molecules are violently torn apart by the chemical reactivity of oxygen, and electrons are released in the process, recombining at drastically lower energy levels, thus giving rise to ultraviolet radiation. Visible radiation is generally the result of smaller changes in electron energy levels within the molecules of fuel, flame intermediates, and products of combustion. Infrared radiation comes from the vibration of molecules or parts of molecules when they are in the superheated state associated with combustion. Each chemical compound exhibits a group of wavelengths at which it is resonant. These wavelengths constitute the chemical's infrared spectrum, which is usually unique to that chemical.

This interrelationship between wavelength and combustion chemistry affects the relative performance of various types of detectors to various fires.

This appendix material concerning the definition of wavelength brings to light an extremely important concept: The detector must be matched to the radiant emissions expected from the fire.

5-4.2.2 **Operating Principles of Flame Detectors.**

5-4.2.2.1 Ultraviolet flame detectors typically use a vacuum photodiode Geiger-Muller tube to detect the ultraviolet radiation that is produced by a flame. The photodiode allows a burst of current to flow for each ultraviolet photon that hits the active area of the tube. When the number of current bursts per unit time reaches a predetermined level, the detector initiates an alarm.

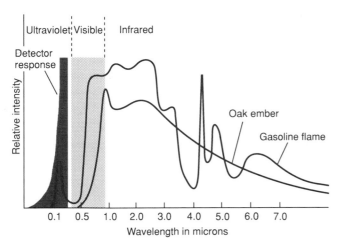

Figure 5.23 *The spectral response of a UV flame detector superimposed on the spectrum of "typical" radiators.*

5-4.2.2.2 A single wavelength infrared flame detector uses one of several different photocell types to detect the infrared emissions in a single wavelength band that are produced by a flame. These detectors generally include provisions to minimize alarms from commonly occurring infrared sources such as incandescent lighting or sunlight.

5-4.2.2.3 An ultraviolet/infrared (UV/IR) flame detector senses ultraviolet radiation with a vacuum photodiode tube and a selected

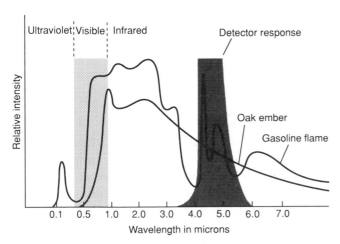

Figure 5.24 *The spectral response of a single wavelength infrared flame detector superimposed on the spectrum of "typical" radiators.*

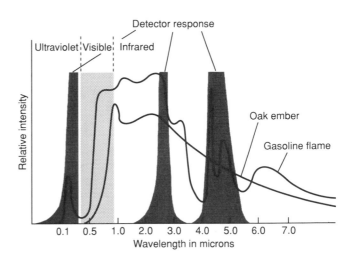

NOTE: Some UV/IR flame detectors require radiant emissions at .1 microns (UV) and 2.5 microns (IR). Other UV/IR flame detectors require radiant emissions at .1 microns (UV) and nominal 4.7 microns (IR).

Figure 5.25 *The spectral response of an ultraviolet/infrared (UV/IR) flame detector superimposed on the spectrum of "typical" radiators.*

wavelength of infrared radiation with a photocell and uses the combined signal to indicate a fire. These detectors require both types of radiation to be present before an alarm signal is initiated.

5-4.2.2.4 A multiple wavelength infrared (IR/IR) flame detector senses radiation at two or more narrow bands of wavelengths in the infrared spectrum. These detectors electronically compare the emissions between the band and initiate a signal where the relationship between the two bands indicates a fire.

See Figure 5.26.

5-4.2.3 **Operating Principles of Spark/Ember Detectors.** A spark/ember-sensing detector usually uses a solid state photodiode or phototransistor to sense the radiant energy emitted by embers, typically between 0.5 and 2.0 microns in normally dark environments. These detectors can be made extremely sensitive (microwatts), and their response times can be made very short (microseconds).

See Figure 5.27.

5-4.3 **Fire Characteristics and Detector Selection.**

5-4.3.1* The type and quantity of radiant energy-sensing fire detectors shall be determined based upon an analysis of the

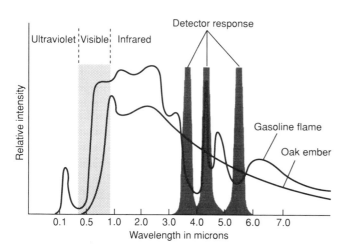

NOTE: Some IR/IR flame detectors compare radiant emissions at 4.3 microns (IR) to a reference at nominal 3.8 microns (IR). Other IR/IR flame detectors use a nominal 5.6 microns (IR) reference.

Figure 5.26 *The spectral response of a multiple wavelength infrared (IR/IR) flame detector superimposed on the spectrum of "typical" radiators.*

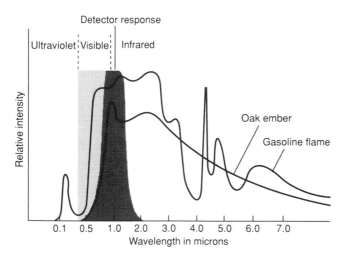

Figure 5.27 *The spectral response of an infrared spark/ember detector superimposed on the spectrum of "typical" radiators.*

hazard, including the burning characteristics of the fuel, the fire growth rate, the environment, the ambient conditions, and the capabilities of the extinguishing media and equipment.

Both flame detectors and spark/ember detectors are routinely installed outdoors where they are exposed to the weather and fluctuations in temperature. Special attention must be given to the temperature range limits provided by the manufacturer to ensure that the detector has been qualified for the anticipated extremes.

A-5-4.3.1 The radiant energy from a flame or spark/ember is comprised of emissions in various bands of the ultraviolet, visible, and infrared portions of the spectrum. The relative quantities of radiation emitted in each part of the spectrum are determined by the fuel chemistry, the temperature, and the rate of combustion. The detector should be matched to the characteristics of the fire.

Almost all materials that participate in flaming combustion will emit ultraviolet radiation to some degree during flaming combustion, whereas only carbon-containing fuels will emit significant radiation at the 4.35 micron (carbon dioxide) band used by many detector types to detect a flame.

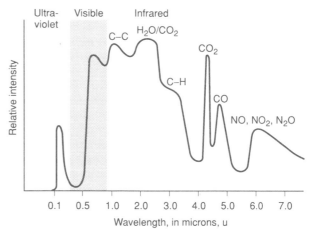

Figure A-5-4.3.1 *Spectrum of a "typical" flame (free burning gasoline).*

The radiant energy emitted from an ember is determined primarily by the fuel temperature (Plank's Law Emissions) and the emissivity of the fuel. Radiant energy from an ember is primarily infrared and, to a lesser degree, visible in wavelength. In general, embers do not emit ultraviolet energy in significant quantities (0.1 percent of total emissions) until the ember achieves temperatures of 2000°K (1727°C or 3240°F). In most cases, the emissions will be included in the band of 0.8 to 2.0 microns, corresponding to temperatures of approximately 750°F (398°C) to 1830°F (1000°C).

Most radiant energy detectors have some form of qualification circuitry within them that uses time to help distinguish between spurious, transient signals and legitimate fire alarms. These circuits become very important when one considers the anticipated fire scenario and the ability of the detector to respond to that anticipated fire. For example, a detector that utilizes an integration circuit or a timing circuit to respond to the flickering light from a fire may not respond well to a deflagration resulting from the ignition of accumulated combustible vapors and gases, or where the fire is a spark that is traveling up to 100 meters per second past the detector. Under these circumstances, a detector that has a high speed response capability would be most appropriate. On the other hand, in applications where the development of the fire will be slower, a detector that utilizes time for the confirmation of repetitive signals would be appropriate. Consequently, the fire growth rate should be considered in selecting the detector. The detector performance should be selected to respond to the anticipated fire.

The radiant emissions are not the only criteria to be considered. The medium between the anticipated fire and the detector is also very important. Different wavelengths of radiant energy are absorbed with varying degrees of efficiency by materials suspended in the air or that may accumulate on the optical surfaces of the detector. Generally, aerosols and surface deposits reduce the sensitivity of the detector. The detection technology utilized should take into account those normally occurring aerosols and surface deposits to minimize the reduction of system response between maintenance intervals. Note that the smoke evolved from the combustion of middle and heavy fraction petroleum distillates is highly absorptive in the ultraviolet end of the spectrum. Where using this type of detection, the system should be designed to minimize the interference of smoke on the response of the detection system.

The environment and ambient conditions anticipated in the area to be protected will impact on the choice of detector. All detectors have limitations on the range of ambient temperatures over which they will respond, consistent with their tested or approved sensitivities. The designer should make certain that the detector is compatible with the range of ambient temperatures anticipated in the area in which it is installed. In addition, rain, snow, and ice will attenuate both ultraviolet and infrared radiation to varying degrees. Where anticipated, provisions should be made to protect the detector from accumulations of these materials on the optical surfaces.

5-4.3.2 The selection of the radiant energy-sensing detectors shall be based upon:

(a) The matching of the spectral response of the detector to the spectral emissions of the fire or fires to be detected; and

(b) Minimizing the possibility of spurious nuisance alarms from nonfire sources inherent to the hazard area. (*See A-5-4.3.1.*)

Subsection 5-4.3 gives general guidance that pertains to all radiant energy-sensing fire detectors. It reveals that embers do not provide emissions in a form that flame detectors are able to detect. The hazard analysis must, therefore, begin with the determination as to whether the combustible will burn in the solid phase as an ember or in the gaseous phase as a flame. That determination then points the designer toward the spark/ember detector (for solid phase combustion) or the flame detector (for gas phase combustion).

5-4.4 Spacing Considerations.

5-4.4.1 General Rules.

5-4.4.1.1* Radiant energy-sensing fire detectors shall be employed consistent with the listing or approval and the inverse square law, which defines the fire size vs. distance curve for the detector.

The inverse square law relates the size of the fire, the detector sensitivity, and the distance between the fire and the detector. It enables the engineer to compute with considerable precision how large the fire must get before there is enough radiant energy hitting the detector to cause an alarm.

A-5-4.4.1.1 All optical detectors respond according to the following theoretical equation:

$$S = \frac{Kpe^{\zeta d}}{d^2}$$

Where:

k = proportionality constant for the detector
p = radiant power emitted by the fire
e = Naperian logarithm base (2.7183)
ζ = the extinction coefficient of air
d = the distance between the fire and the detector
S = radiant power reaching the detector.

The sensitivity (S) would typically be measured in nanowatts. This equation yields a family of curves similar to the one shown in Figure A-5-4.4.1.1.

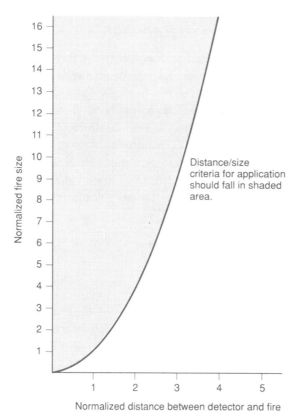

Figure A-5-4.4.1.1 *Generalized fire size vs. distance.*

The curve defines the maximum distance at which the detector consistently detects a fire of defined size and fuel. Detectors should only be employed in the shaded area beneath the curve.

Under the best of conditions, with no atmospheric absorption, the radiant power reaching the detector is reduced by a factor of four if the distance between the detector and the fire is doubled. For the consumption of the atmospheric extinction, the exponential term, Zeta (ζ) is added to the equation. Zeta is a measure of the clarity of the air at the wavelength under consideration. Zeta will be affected by humidity, dust, and any other contaminants in the air that are absorbent at the wavelength in question. Zeta generally has values between $-.001$ and $-.1$ for normal ambient air.

5-4.4.1.2 Detectors shall be used in sufficient quantity and positioned so that no point requiring detection in the hazard area is obstructed or outside the field of view of at least one detector.

A flame detector or spark/ember detector cannot detect what it cannot see.

The requirements in 5-4.4.1.2 pertain to both flame and spark/ember detectors. There are aspects unique to either flame or spark/ember detectors that are covered in 5-4.2.2 and 5-4.2.3, respectively.

5-4.4.2 Spacing Considerations for Flame Detectors.

5-4.4.2.1* The location and spacing of detectors shall be the result of an engineering evaluation, taking into consideration:

(a) The size of the fire that is to be detected
(b) The fuel involved
(c) The sensitivity of the detector
(d) The field of view of the detector
(e) The distance between the fire and the detector
(f) The radiant energy absorption of the atmosphere
(g) The presence of extraneous sources of radiant emissions
(h) The purpose of the detection system
(i) The response time required.

The spacing of flame detectors as a group must address each one of these criteria. In addition, the positioning and aiming of each individual detector must also be reviewed in light of each of these criteria.

A-5-4.4.2.1 The types of application for which flame detectors are suitable are:

(a) High ceiling, open spaced buildings such as warehouses and aircraft hangers

(b) Outdoor or semi-outdoor areas where winds or draughts may prevent smoke reaching a heat or smoke detector

(c) Risks where rapidly developing flaming fires may occur, such as aircraft hangers, petrochemical production, storage and transfer areas, natural gas installations, paint shops, solvent areas, etc.

(d) Spot protection of high fire risk machinery or installations, often coupled with an automatic gas extinguishing system

(e) Environments that are unsuitable for other types of detectors.

Some extraneous sources of radiant emissions that have been identified as interfering with the stability of flame detectors include:

(a) Sunlight

(b) Lightning

(c) X-rays

(d) Gamma rays

(e) Cosmic rays

(f) Ultraviolet radiation from arc welding

(g) Electromagnetic interference (EMI, RFI)

(h) Hot objects

(i) Artificial lighting.

5-4.4.2.2 The system design shall specify the size of the flaming fire of given fuel that is to be detected.

5-4.4.2.3* In applications where the fire to be detected could occur in an area not on the optical axis of the detector, the distance shall be reduced or detectors added to compensate for the angular displacement of the fire in accordance with the manufacturer's documented instructions.

A-5-4.4.2.3 The greater the angular displacement of the fire from the optical axis of the detector, the larger the fire must become before it is detected. This phenomenon establishes the field of view of the detector. Figure A-5-4.4.2.3 shows an example of the effective sensitivity versus angular displacement of a flame detector.

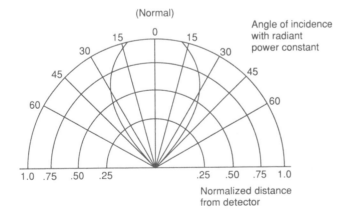

Figure A-5-4.4.2.3   Normalized sensitivity vs. angular displacement.

5-4.4.2.4* In applications in which the fire to be detected is of a fuel different than the test fuel used in the process of listing

or approval, the distance between the detector and the fire shall be adjusted consistent with the fuel specificity of the detector as established by the manufacturer.

A-5-4.4.2.4 Virtually all radiant energy-sensing detectors exhibit some kind of fuel specificity. Different fuels when burned at uniform rates (joules/second or watts) will emit different levels of radiant power in the ultraviolet, visible, and infrared portions of the spectrum. Under free-burn conditions, a fire of given surface area but of different fuels will burn at different rates (joules/second or watts) and emit varying levels of radiation in each of the major portions of the spectrum. Most radiant energy detectors designed to detect flame are qualified based upon a defined fire under specific conditions. Where employing these detectors for fuels other than the defined fire, the designer should make certain that the appropriate adjustments to the maximum distance between the detector and the fire are made consistent with the fuel specificity of the detector.

In an effort to make flame detectors more sensitive yet more immune to unwanted alarms, designers began designing detectors that concentrated on very specific features of the flame spectrum, the emissions of the flame across the range of wavelengths from ultraviolet to infrared. In concept, such flame detectors infer that a flame exists if an emission of a specific wavelength is detected. However, one fuel emits a different radiant intensity at a given wavelength than another fuel. This gives rise to detectors that are fuel specific. There are cases where a flame detector may be several times more sensitive to one fuel than another. The designer is advised to obtain spectra of potential fuels and response curves from the detector manufacturer to make certain the detector will respond to the fuel(s) involved.

5-4.4.2.5 Since flame detectors are essentially line of sight devices, special care shall be taken to ensure that their ability to respond to the required area of fire in the zone that is to be protected will not be compromised by the presence of intervening structural members or other opaque objects or materials.

5-4.4.2.6* Provisions shall be made to sustain detector window clarity in applications where airborne particulates and aerosols coat the detector window between maintenance intervals and affect sensitivity.

A-5-4.4.2.6 The means by which this requirement has been satisfied include:

(a) Lens clarity monitoring and cleaning where a contaminated lens signal is rendered

(b) Lens air purge.

5-4.5 Spacing Considerations for Spark/Ember Detectors.

5-4.5.1* The location and spacing of detectors shall be the result of an engineering evaluation, taking into consideration:

(a) The size of the spark or ember that is to be detected

(b) The fuel involved

(c) The sensitivity of the detector

(d) The field of view of the detector

(e) The distance between the fire and the detector

(f) The radiant energy absorption of the atmosphere

(g) The presence of extraneous sources of radiant emissions

(h) The purpose of the detection systems

(i) The response time required.

A-5-4.5.1 Spark/ember detectors are installed primarily to detect sparks and embers that may, if allowed to continue to burn, precipitate a much larger fire or explosion. Spark/ember detectors are typically mounted on some form of duct or conveyor, monitoring the fuel as it passes by. Usually, it is necessary to enclose the portion of the conveyor where the detectors are located as these devices generally require a dark environment. Extraneous sources of radiant emissions that have been identified as interfering with the stability of spark/ember detectors include:

(a) Ambient light

(b) Electromagnetic interference (EMI, RFI)

(c) Electrostatic discharge in the fuel stream.

Figure 5.28 shows typical applications where spark/ember detectors are used. Note that the detectors are located at a point along the duct or conveyor, monitoring the cross-section of the duct or conveyor at that one point.

5-4.5.2* The system design shall specify the size of the spark or ember of given fuel that the detection system is to detect.

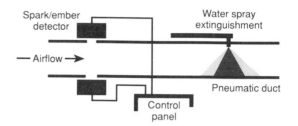

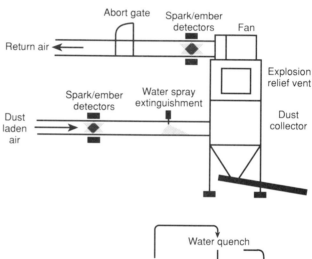

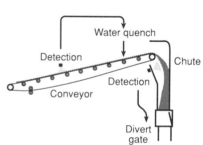

Figure 5.28 *Spark/ember detectors are usually used on conveyance ducts and conveyors to detect embers in particulate solids as they are transported. The top drawing shows the general concept of spark/ember detectors. The middle drawing illustrates the application of spark/ember detectors to protect a dust collector. The bottom drawing illustrates the protection of a conveyor.*

A-5-4.5.2 There is a minimum ignition power (watts) for all combustible dusts. If the spark or ember is incapable of delivering that quantity of power to the adjacent combustible material (dust), an expanding dust fire will not occur. The minimum ignition power is determined by the fuel chemistry, fuel particle size, fuel concentration in air, and ambient conditions such as temperature and humidity.

5-4.5.3 Spark detectors shall be positioned so that all points within the cross section of the conveyance duct, conveyor, or chute where the detectors are located are within the field of view of at least one detector as defined in 5-4.2.1.

5-4.5.4 The location and spacing of the detectors shall be adjusted using the inverse square law, modified for the atmospheric absorption and the absorption of nonburning fuel suspended in the air in accordance with the manufacturer's documented instructions. (*See A-5-4.4.1.1.*)

5-4.5.5* In applications where the sparks to be detected could occur in an area not on the optical axis of the detector, the distance shall be reduced or detectors added to compensate for the angular displacement of the fire in accordance with the manufacturer's documented instructions.

A-5-4.5.5 The greater the displacement of the fire from the optical axis of the detector, the larger the fire must become before it is detected. This phenomenon establishes the field of view of the detector. Figure A-5-4.4.2.3 shows an example of the effective sensitivity versus angular displacement of a flame detector.

5-4.5.6* Provisions shall be made to sustain the detector window clarity in applications where airborne particulates and aerosols coat the detector window and affect sensitivity.

A-5-4.5.6 The means by which this requirement has been satisfied include:

(a) Lens clarity monitoring and cleaning where a contaminated lens signal is rendered
(b) Lens air purge.

5-4.6 Other Considerations.

5-4.6.1 Radiant energy-sensing detectors shall be protected either by way of design or installation to ensure that optical performance is not compromised.

5-4.6.2 Where necessary, radiant energy-sensing detectors shall be shielded or otherwise arranged to prevent action from unwanted radiant energy.

5-4.6.3 Where used in outdoor applications, radiant energy-sensing detectors shall be shielded or otherwise arranged in a fashion to prevent diminishing sensitivity by rain, snow, etc., and yet allow a clear field of vision of the hazard area.

5-5 Gas-Sensing Fire Detectors.

5-5.1* The purpose of Section 5-5 is to provide information to assist in application and installation of fire detectors that sense gases produced by burning substances. These detectors are hereafter referred to as fire-gas detectors. This section covers general area application of fire-gas detectors in ordinary indoor locations.

A-5-5.1 Many gases may be produced by a fire. Fire-gas detectors are instruments that are triggered into alarm by one or more fire gases. Fire-gas detectors need not be able to differentiate among the various fire gases. Depending on the material being burned and the oxygen supply available, the quantity and composition of gases given off can vary greatly.

If ordinary cellulosic material such as wood or paper is burned with an abundance of oxygen, the gases given off are primarily carbon dioxide and water vapor. If, however, the same material is burned or smolders with a limited supply of oxygen, a host of additional gases will be evolved.

5-5.2 Fire-gas detectors shall be installed in all areas where required either by the appropriate NFPA standard or by the authority having jurisdiction.

5-5.3 Fire-gas detectors shall respond to one or more of the gases produced by a fire. Gases are molecules without cohesion that are produced by a burning substance and are subject to oxidation or reduction.

This code draws a distinction between "smoke" and "fire gases." Smoke is particulate matter with particles that have dimensions larger than those normally ascribed to molecules of a gas.

5-5.4 Although some fire-gas detectors are capable of detecting combustible gases or vapors prior to ignition, such applications are not within the scope of this code.

There are gas detectors that are used largely in industrial and petrochemical applications for the detection of low-level concentrations of combustible gases (e.g., natural gas, methane, ethane, propane, ethylene, and gasoline vapor). These devices are analog sensors that provide a signal to a controller or control panel, which then displays readings in percent LFL (lower flammable limit). Section 5-5 does not cover such devices. Similarly, the detection of carbon monoxide gas from auto emissions is not covered unless the detectors have been specifically listed as fire-gas detectors. This section only covers those detectors that are specifically designed to detect gases produced by the combustion of a fuel.

5-5.5 Operating Principles.

5-5.5.1 Semiconductor. Fire-gas detectors of the semiconductor type respond to either oxidizing or reducing gases by creating electrical changes in the semiconductor. The subsequent conductivity change of the semiconductor causes actuation.

5-5.5.2 Catalytic Element. Fire-gas detectors of the catalytic element type contain a material that remains unchanged, but accelerates the oxidation of combustible gases. The resulting temperature rise of the element causes actuation.

5-5.6 Location and Spacing.

5-5.6.1* General. The location and spacing of fire-gas detectors shall result from an evaluation based on the guidelines detailed in this code and on engineering judgment. Ceiling shape and surfaces, ceiling height, configuration of contents, burning characteristics of combustible material present, ventilation, and the ambient environment are some of the conditions that shall be considered.

A-5-5.6.1 Fire-gas detectors depend on fire gases reaching the sensing element. Where sufficient concentration is present, operation is obtained. Since the detectors are usually mounted on or near the ceiling, response time depends on the nature of the fire. A hot fire will drive fire gases up to the ceiling more rapidly. A smoldering fire produces little heat, and, therefore, the detection time will be increased.

The reader will note a similarity in the requirements relevant to fire-gas detectors and those for smoke detectors. Both types of detector depend upon air currents to trans-

port the fire by-products to the detector. The overwhelming majority of the current research on fire plume dynamics is focused on the application of smoke detectors.

5-5.6.1.1 Where the intent is to provide protection from a specific hazard, the detector(s) may be installed closer to the hazard in a position where the detector will readily intercept the fire gases.

This concept is occasionally used in the protection of high-value assets and equipment.

5-5.6.1.2 Stratification. The possible effect of gas stratification at levels below the ceiling shall also be considered. (*See A-5-3.5.1.2.*)

The problem of gas stratification in low-energy-output fires is as important in the application of fire-gas detection as it is in the application of smoke detection.

5-5.6.2 Spot-type fire-gas detectors shall be located on the ceiling not less than 4 in. (100 mm) from a sidewall to the near edge or, if on a sidewall, between 4 in. (100 mm) and 12 in. (300 mm) down from the ceiling to the top of the detector. (*See Figure A-5-2.5.1.*)

Exception No. 1: See 5-5.6.1.2.

Exception No. 2: In the case of solid joist construction, detectors shall be mounted at the bottom of the joists.

Exception No. 3: In the case of beam construction where beams are less than 12 in. (300 mm) in depth and less than 8 ft (2.4 m) on center, detectors may be installed on the bottom of beams.

The spacing requirements for fire-gas detectors described here and in subsections 5-5.7 and 5-5.8 are similar to the corresponding requirements regarding the placement of smoke detectors.

5-5.6.3* Each sampling port of a sampling-type fire-gas detector shall be treated as a spot-type detector for the purpose of location and spacing.

A-5-5.6.3 Gas transport to the sensor of a fire-gas detector may occur by diffusion where migration results from concentration gradients or by sampling if pumps, fans, or aspirators are employed.

5-5.6.4 Smooth Ceiling Spacing.

5-5.6.4.1 Spot-Type Detectors. On smooth ceilings, spacing of 30 ft (9.1 m) shall be permitted to be used as a guide. In all cases, the manufacturer's recommendations shall be followed. Other spacing shall be permitted to be used depending on ceiling height, varying conditions, or response requirements.

5-5.6.5 Solid Joist Construction. (See A-5-3.5.6.)

5-5.6.5.1 Ceiling construction in which joists are 8 in. (200 mm) or less in depth shall be considered equivalent to a smooth ceiling. (See also A-5-3.5.1.2.)

5-5.6.5.2 If joists exceed 8 in. (200 mm) in depth, the spacing of spot-type detectors in the direction perpendicular to the joists shall be reduced. (See also A-5-3.5.1.2.)

5-5.6.6 Beam Construction.

5-5.6.6.1 Ceiling construction where beams are 8 in. (200 mm) or less in depth shall be considered equivalent to a smooth ceiling. (See also A-5-3.5.1.2.)

5-5.6.6.2 If beams are over 8 in. (200 mm) in depth, the spacing of spot-type detectors in the direction perpendicular to the beams shall be reduced. (See also A-5-3.5.1.2.)

5-5.6.6.3* If the beams exceed 18 in. (460 mm) in depth and are more than 8 ft (2.4 m) on center, each bay shall be treated as a separate area requiring at least one spot-type detector.

A-5-5.6.6.3 Location and spacing of fire-gas detectors should consider beam depth, ceiling height, beam spacing, and anticipated fire type and location. For ceiling configurations where mixing of air into beam pockets is inhibited by ventilation systems, detectors will perform better if installed on the bottom of beams.

To detect flaming fires (strong plumes), detectors should be installed as follows:

(a) If the ratio of the beam depth (D) to ceiling height (H) (D/H) is greater than 0.10 and the ratio of beam spacing (W) to ceiling height (H) (W/H) is greater than 0.40, detectors should be located in each beam pocket.

(b) If either ratio of beam depth to ceiling height (D/H) is less than 0.10 or the ratio of beam spacing to ceiling height (W/H) is less than 0.40, detectors should be installed on the bottom of the beams.

To detect smoldering fires (weak or no plumes), detectors should be installed as follows:

(a) If air mixing into beam pockets is good (e.g., air-flow parallel to long beams) and condition (a) exists as above, a detector should be located in each beam pocket.

(b) If air mixing into beam pockets is limited or condition (b) exists as above, detectors should be located on the bottom of the beams.

5-5.6.7 Sloped Ceilings.

5-5.6.7.1 Peaked. Detectors shall first be spaced and located within 3 ft (0.9 m) of the peak, measured horizontally. The number and spacing of additional detectors, if any, shall be based on the horizontal projection of the ceiling. (See Figure A-5-2.7.4.1.)

5-5.6.7.2 Shed. Detectors shall first be spaced and located within 3 ft (0.9 m) of the high side of the ceiling, measured horizontally. The number and spacing of additional detectors, if any, shall be based on the horizontal projection of the ceiling. (See Figure A-5-2.7.4.2.)

5-5.6.8 Suspended Ceilings. (See 5-5.6.)

5-5.6.9 Partitions. Where partitions extend upward to within 18 in. (460 mm) of the ceiling, they will not influence the spacing. Where the partition extends to within less than 18 in. (460 mm) of the ceiling, the effect on gas travel shall be considered in reduction of spacing.

These requirements are similar to the corresponding requirements regarding the placement of smoke detectors.

5-5.7 Heating, Ventilating, and Air Conditioning (HVAC).

5-5.7.1* In spaces served by air-handling systems, detectors shall not be located where air from supply diffusers could dilute fire gases before they reach the detectors. Detectors shall be located to intercept the airflow toward the return air opening(s).

A-5-5.7.1 Detectors should not be located in a direct air-flow nor closer than 3 ft (900 mm) from an air supply diffuser.

5-5.7.2 In under-floor spaces and above-ceiling spaces used as HVAC plenums, detectors shall be listed for the anticipated environment. (See 5-3.7.1.1.) Detector spacings and locations shall be selected based on anticipated air-flow patterns and fire types.

5-5.7.2.1 Detectors placed in environmental air ducts or plenums shall not be used as a substitute for open area detectors. (See Section 5-11 and associated appendix material for related information.) Where open area protection is required, 5-5.6 shall apply.

5-5.8 Special Considerations.

5-5.8.1 The selection and placement of fire-gas detectors shall take into consideration both the performance characteristics of the detector and the areas into which the detectors will be installed to prevent nuisance alarm or nonoperation after installation. Some of the considerations are as follows.

5-5.8.1.1 Fire-gas detectors may alarm in nonfire situations due to certain human activities. The use of some aerosol sprays and hydrocarbon solvents are examples. Accordingly, considerable care shall be employed when installing fire-gas detectors. They shall not be installed where, under normal conditions, concentrations of detectable gases may be present. A garage is not a place to use fire-gas detectors for fire alarm purposes because the concentration of carbon monoxide may be high enough to trigger an alarm.

5-5.8.1.2 Fire-gas detectors having a fixed temperature element as part of the unit shall be selected in accordance with Table 5-2.4.1 for the maximum ceiling temperature that can be expected in service.

5-5.8.1.3* The installation of fire-gas detectors shall take into consideration the environmental condition of the area(s). (See Table A-5-3.7.1.1.) Fire-gas detectors are intended for installation in areas where the normal ambient conditions are not likely to exceed the following:

    (a) A temperature of 100°F (38°C), or fall below 32°F (0°C); or

    (b) A relative humidity outside the range of 10 to 93 percent; or

    (c) An air velocity of 300 fpm (1.5 mps).

Exception: Detectors specifically designed for use in ambients exceeding the limits of (a) through (c) and listed for the temperature, humidity, and air velocity conditions expected.

A-5-5.8.1.3 Product-listing standards include tests for temporary excursions beyond normal limits. In addition to temperature, humidity, and velocity variations, fire-gas detectors should operate reliably under such common environmental conditions as mechanical vibration, electrical interference, and other environmental influences. These conditions are also included in tests conducted by the listing agencies.

5-6 Other Fire Detectors.

It is the intent of the code to provide for the development of new technologies and to allow the use of such technologies consistent with sound principles of fire protection engineering. The requirements in this section provide for alternate methods not explicitly described in other sections of Chapter 5.

5-6.1 Detectors in the classification of "Other Fire Detectors" are those that operate on principles different from those described in 5-2.3, 5-3.3, 5-4.3, and 5-5.5. Such detectors shall be installed in all areas where they are required either by the appropriate NFPA standard or by the authority having jurisdiction.

5-6.2 Facilities for testing or metering or instrumentation to ensure adequate initial sensitivity and adequate retention thereof, relative to the protected hazard, shall be provided. These facilities shall be employed at regular intervals.

Chapter 7 of this code outlines the required maintenance procedures and schedules for all components of a fire alarm system, including the initiating devices.

5-6.3 These detectors shall operate where subjected to the abnormal concentration of combustion effects that occur during a fire, such as water vapor, ionized molecules, or other phenomena for which they are designed. Detection is dependent upon the size and intensity of fire to provide the necessary amount of required products and related thermal lift, circulation, or diffusion for adequate operation.

5-6.4 Room sizes and contours, airflow patterns, obstructions, and other characteristics of the protected hazard shall be taken into account.

5-6.5 Location and Spacing.

5-6.5.1 The location and spacing of detectors shall be based on the principle of operation and an engineering survey of the conditions anticipated in service. The manufacturer's technical bulletin shall be consulted for recommended detector uses and locations.

5-6.5.2 Detectors shall not be spaced beyond their listed or approved maximums. Closer spacing shall be utilized where the structural or other characteristics of the protected hazard warrant.

5-6.5.3 Consideration shall be given to all factors with bearing on the location and sensitivity of the detectors, including structural features such as sizes and shapes of rooms and bays, their occupancies and uses, ceiling heights, ceiling and other obstructions, ventilation, ambient environment, stock piles, files, and fire hazard locations.

5-6.5.4 The overall situation shall be reviewed frequently to ensure that changes in structural or usage conditions that could interfere with fire detection are remedied.

5-6.6 Special Considerations. The selection and placement of detectors shall take into consideration both the performance characteristics of the detector and the areas into which the detectors will be installed to prevent nuisance alarm or nonoperation after installation.

5-7 Sprinkler Waterflow Alarm-Initiating Devices.

5-7.1 The provisions of Section 5-7 apply to devices that initiate an alarm indicating a flow of water in a sprinkler system.

5-7.2* Provisions shall be made to indicate the flow of water in a sprinkler system by an alarm signal within 90 seconds after flow of water at the alarm-initiating device equal to or greater than

that from a single sprinkler of the smallest orifice size installed in the system. Movement of water due to waste, surges, or variable pressure shall not be indicated.

A-5-7.2 The waterflow device should be field adjusted so that an alarm will be initiated in no more than 90 seconds after a sustained flow of at least 10 gpm (40 L/min).

Features that should be investigated to minimize alarm response time include elimination of trapped air in the sprinkler system piping, use of an excess pressure pump, use of pressure drop alarm-initiating devices, or a combination thereof.

Care should be taken when choosing waterflow alarm-initiating devices for hydraulically calculated looped systems and those systems using small orifice sprinklers. Such systems may incorporate a single point flow significantly less than 10 gpm (40 L/min). In such cases, additional waterflow alarm-initiating devices or use of pressure drop-type waterflow alarm-initiating devices may be necessary.

Care should be taken, where choosing waterflow alarm initiating devices for sprinklers utilizing on-off sprinklers, to ensure that an alarm will be initiated in the event of a waterflow condition. On-off sprinklers open at a predetermined temperature and close when the temperature reaches a predetermined lower temperature. With certain types of fires, waterflow may occur in a series of short bursts of 10 to 30 seconds' duration each. An alarm-initiating device with retard may not detect waterflow under these conditions. It is recommended that an excess pressure system or one that operates on pressure drop be considered to facilitate waterflow detection on sprinkler systems utilizing on-off sprinklers.

Excess pressure systems may be used with or without alarm valves. The following is a description of one type of excess pressure system with an alarm valve.

An excess pressure system with an alarm valve consists of an excess pressure pump with pressure switches to control the operation of the pump. The inlet of the pump is connected to the supply side of the alarm valve, and the outlet is connected to the sprinkler system. The pump control pressure switch is of the differential type, maintaining the sprinkler system pressure above the main pressure by a constant amount. Another switch monitors low sprinkler system pressure to initiate a supervisory signal in the event of a failure of the pump or other malfunction. An additional pressure switch may be used to stop pump operation in the event of a deficiency in water supply. Another pressure switch is connected to the alarm outlet of the alarm valve to initiate a waterflow alarm signal when waterflow exists. This type of system also inherently prevents false alarms due to water surges. The sprinkler retard chamber should be eliminated to enhance the detection capability of the system for short duration flows.

In many facilities the sprinkler system is used as both a suppression system and a detection system. The flow of water is used to initiate the alarm. In a large system with large sprinkler risers, the flow from a single head has proven hard to detect. In addition, if there is any air in the piping it will act as a "gas cushion," allowing variations in water pressure from the street to cause water to slosh back and forth in the riser. This causes false alarms. Consequently, most flow switches are equipped with a retard feature that delays the transmission of a signal until after stable waterflow has been achieved. The 90-second criterion can be a problem if on/off systems are incorporated into the sprinkler system. Clearly, the designer must be familiar with NFPA 13, *Standard for the Installation of Sprinkler Systems*.

5-7.3 Piping between the sprinkler system and a pressure actuated alarm-initiating device shall be galvanized or of nonferrous metal or other approved corrosion resistant material, not less than ³⁄₈ in. (9.5 mm) nominal pipe size.

These requirements stem from the problems associated with piping corrosion and with the mechanical strength necessary to endure the environment of the sprinkler system.

5-8 Detection of the Operation of Other Automatic Extinguishing Systems.

5-8.1* Provision shall be made to detect the operation of an automatic extinguishing system by means appropriate to the system, such as agent flow or agent pressure, by alarm-initiating devices installed in accordance with their individual listings.

A-5-8.1 Appropriate means may involve:

  (a) Foam systems: Flow of water
  (b) Pump activation
  (c) Differential pressure detectors

  (d) Halon: Pressure detector
  (e) Carbon dioxide: Pressure detector.

In any case, an alarm that activates the extinguishing system may be initiated from the detection system.

Many extinguishing systems include emergency mechanical manual release capability. This provides for the release of the extinguishing agent without the operation of the fire alarm system. There must be a means to ensure that critical functions such as warnings, cessation of fuel flow, door closure, removal of electrical power, and a host of other functions are achieved if the extinguishing system discharge is caused by the mechanical release.

Figure 5.29 Emergency manual cable release for extinguishing systems.

Discharge indication switches are the usual means of providing the extinguishing system operation signal to the fire detection control panel. Due to the critical function these switches perform, they must be listed for use with the specific make and model of extinguishing system. See Figure 5.30.

Figure 5.30 *Typical extinguishing system pressure-actuated discharge switch.*

The reader should review:

- NFPA 12, *Standard on Carbon Dioxide Extinguishing Systems*
- NFPA 12A, *Standard on Halon 1301 Fire Extinguishing Systems*
- NFPA 16, *Standard on the Installation of Deluge Foam-Water Sprinkler and Foam-Water Spray Systems*
- NFPA 17, *Standard for Dry Chemical Extinguishing Systems*
- NFPA 2001, *Standard on Clean Agent Fire Extinguishing Systems.*

5-9 Manually Actuated Alarm-Initiating Devices.

5-9.1 Manual fire alarm boxes shall be used only for fire alarm-initiating purposes. However, combination manual fire alarm boxes and guard's signaling stations shall be permitted.

This requirement stems from two concerns: credibility and reliability. If the manual fire alarm box is incorporated into some other nonfire-related assembly (with the single exception of guard's signaling stations), the probability of unwarranted operation is increased. This leads to false alarms and erodes the occupants' confidence in the system. Also, when manual fire alarms are combined with nonfire-related functions, there is an increased probability that a failure in the nonfire function will compromise the fire alarm system.

5-9.1.1 Mounting. Each manual fire alarm box shall be securely mounted. The operable part of each manual fire alarm box shall be not less than 3½ ft (1.1 m) and not more than 4½ ft (1.37 m) above floor level.

5-9.1.2 Distribution. Manual fire alarm boxes shall be distributed throughout the protected area so that they are unobstructed, readily accessible, and located in the normal path of exit from the area as follows:

(a) At least one manual fire alarm box shall be provided on each floor.

(b) Additional manual fire alarm boxes shall be provided so that travel distance to the nearest fire alarm box will not be in excess of 200 ft (61 m) measured horizontally on the same floor.

(c) For systems employing automatic fire detectors or waterflow detection devices, at least one manual fire alarm box shall be provided to initiate a fire alarm signal. This manual fire alarm box shall be located where required by the authority having jurisdiction.

See Figure 5.31 for examples of manual fire alarm boxes.

5-9.1.3* A coded manual fire alarm box shall produce at least three repetitions of the coded signal, each repetition to consist of at least three impulses.

A-5-9.1.3 Coded Signal Designations. The following suggested coded signal assignment for buildings having four floors and multiple basements is provided as a guide:

Location	Coded Signal
4th Floor	2-4
3rd Floor	2-3
2nd Floor	2-2
1st Floor	2-1
Basement	3-1
Sub-Basement	3-2

Figure 5.31   Typical manual fire alarm boxes.

Coding of the signals allows the parties responsible for the site to proceed directly to the area where the alarm was activated. There are both benefits and disadvantages to coding manual fire alarm boxes. The decision to use coded signals must be part of the overall fire prevention and protection plan for the site.

5-9.2 **Publicly Accessible Fire Service Boxes (Street Boxes).**

5-9.2.1 Street boxes, when in an abnormal condition, shall leave the circuit usable.

Street boxes are exposed to all manner of possible damage, from intentional vandalism to automobile collisions. The operability of the remainder of the system must be ensured in the event of damage to the street box.

5-9.2.2 Street boxes shall be designed so that recycling will not occur if a box actuating device is held in the actuating position and so that they will be ready to accept a new signal as soon as the actuating device is released.

Street boxes are available to the public. They must be able to operate in spite of the possible panicked actions of the person seeking to report a fire.

5-9.2.3 Street boxes, when actuated, shall give a visible or audible indication to the user that the box is operating or that the signal has been transmitted to the communication center.

NOTE: Where the operating mechanism of a box creates sufficient sound to be heard by the user, the requirements are satisfied.

5-9.2.4 The street box housing shall protect the internal components from the weather.

5-9.2.5 Doors on street boxes shall remain operable under adverse climatic conditions, including icing and salt spray.

5-9.2.6 Street boxes shall be recognizable as such. Street boxes shall have instructions for use plainly marked on their exterior surfaces.

Figure 5.32 shows a typical street box.

5-9.2.7 Street boxes shall be securely mounted on poles, pedestals, or structural surfaces as directed by the authority having jurisdiction.

5-9.2.8 Street boxes shall be as conspicuous as possible. Their color shall be distinctive, and they shall be visible from as many directions as possible. A wide band of distinctive colors visible over the tops of parked cars or adequate signs completely visible from all directions shall be applied to supporting poles.

5-9.2.9* Location-designating lights of distinctive color, visible for at least 1500 ft (460 m) in all directions, shall be installed over street boxes. The street light nearest the street box, where equipped with a distinctively colored light, shall be acceptable.

Figure 5.32 *Typical street box.*

A-5-9.2.9 Current supply for location-designating lights at street boxes should preferably be secured at lamp locations from the local electric utility company.

Alternating current power may be superimposed on metallic fire alarm circuits for supplying designating lamps, or for control or actuation of equipment devices for fire alarm or other emergency signals, provided:

(a) Voltage between any wire and ground or between one wire and any other wire of the system shall not exceed 150 volts. The total resultant current in any line circuit shall not exceed ¼ amp.

(b) Coupling capacitors, transformers, choke, coils, etc., shall be rated for 600-volt working voltage and have a breakdown voltage of at least twice the working voltage plus 1000 volts.

(c) There is no interference with fire alarm service under any conditions.

5-9.2.10 Street box cases and parts at any time accessible to users shall be of insulating materials or permanently and effectively grounded. All ground connections to street boxes shall comply with the requirements of NFPA 70, *National Electrical Code*, Article 250.

5-9.2.11 If a street box is installed inside a structure, it shall be placed as close as is practical to the point of entrance of the cir-

cuit, and the exterior wire shall be installed in conduit or electrical metallic tubing in accordance with Chapter 3 of NFPA 70, *National Electrical Code*.

5-9.2.12 **Coded Radio Street Boxes.**

With the advent of modern microprocessor-based telecommunications systems, the features available in street boxes have expanded considerably. The requirements in 5-9.2.12 ensure that the transmission of a fire alarm remains the paramount priority.

5-9.2.12.1 Coded radio street boxes shall be designed and operated in compliance with all applicable rules and regulations of the FCC, as well as with the requirements established herein.

5-9.2.12.2 Coded radio street boxes shall provide no less than three specific and individually identifiable functions to the communication center in addition to the street box number, and they shall be "test," "tamper," and "fire."

5-9.2.12.3* Coded radio street boxes shall transmit to the communication center no less than one repetition for "test," no less than one repetition for "tamper," and no less than three repetitions for "fire."

A-5-9.2.12.3 FCC Rules and Regulations, Vol. V, Part 90, March 1979: "Except for test purposes, each transmission must be limited to a maximum of 2 seconds and may be automatically repeated not more than two times at spaced intervals within the following 30 seconds; thereafter, the authorized cycle may not be reactivated for 1 minute."

5-9.2.12.4 Where multifunction coded radio street boxes are used to transmit to the communication center request(s) for emergency service or assistance in addition to those stipulated in 5-9.2.12.2, each such additional message function shall be individually identifiable.

5-9.2.12.5 Multifunction coded radio street boxes shall be so designed as to prevent the loss of supplemental or concurrently actuated messages.

5-9.2.12.6 An actuating device held or locked in the activating position shall not prevent the activation and transmission of other messages.

5-9.2.13 **Power Source.**

5-9.2.13.1 Box primary power shall be permitted to be from a utility distribution system, a photovoltaic power system, user power, or be self-powered using either an integral battery or other stored energy source, as approved by the authority having jurisdiction.

5-9.2.13.2 Self-powered boxes shall have power for uninterrupted operation for not less than a period of 6 months. Self-powered boxes shall transmit a low power warning message to the communication center for at least 15 days prior to the time the power source will fail to operate the box. This message shall be part of all subsequent transmissions.

Use of a charger to extend the life of a self-powered box shall be permitted if the charger does not interfere with box operation. The box shall be capable of operation for not less than 6 months with the charger disconnected.

5-9.2.13.3 Boxes powered by a utility distribution system shall have an integral standby, sealed, rechargeable battery capable of powering box functions for at least 60 hours in the event of primary power failure. Transfer to standby battery power shall be automatic and without interruption to box operation. Where operating from primary power, the box shall be capable of operation with a dead or disconnected battery. A local trouble indication shall activate upon primary power failure. A battery charger shall be provided in compliance with 1-5.2.11.2, except as modified herein.

Where the primary power has failed, boxes shall transmit a power failure message to the communication center as part of subsequent test messages until primary power is restored. A low battery message shall be transmitted to the communication center where the remaining battery standby time is less than 54 hours.

5-9.2.13.4 Photovoltaic power systems shall provide box operation for not less than 6 months.

Photovoltaic power systems shall be supervised. The battery shall have power to sustain operation for a minimum period of 15 days without recharging. The box shall transmit a trouble message to the communication center when the charger has failed for more than 24 hours. This message shall be part of all subsequent transmissions. Where the remaining battery standby duration is less than 10 days, a low battery message shall be transmitted to the communication center.

5-9.2.13.5 User-powered boxes shall have an automatic self-test feature.

5-9.2.14 **Design of Telephone Street Boxes (Series or Parallel).**

5-9.2.14.1 If a handset is used, the caps on the transmitter and receiver shall be secured to reduce the probability of the telephone street box being disabled due to vandalism.

5-9.2.14.2 Telephone street boxes shall be designed to permit the communication center operator to determine whether or not the telephone street box has been restored to normal condition after use.

5-10 Supervisory Signal-Initiating Devices.

5-10.1 **Control Valve Supervisory Signal-Initiating Device.** Two separate and distinct signals shall be initiated: one indicating movement of the valve from its normal position, and the other indicating restoration of the valve to its normal position. The off-normal signal shall be initiated during the first two revolutions of the hand wheel or during one-fifth of the travel distance of the valve control apparatus from its normal position. The off-normal signal shall not be restored at any valve position except normal.

Control valve supervisory signal-initiating devices have traditionally been switches specifically designed, approved, and listed for service as valve-monitoring devices. See Figure 5.33. The requirement for two distinct signals does not necessarily mean two switches. A switch that transfers when the valve begins to close and stays transferred, and then returns to normal when the valve is reopened, satisfies the requirement. The initial transfer is the first signal. The return to normal is the second signal. For example, assume the switch on the valve is a normally open contact. As the operator begins to turn the valve, the switch closes, indicating an off-normal condition. The switch stays in the closed, off-normal position as the operator continues to close the valve. When the operator opens the valve, the closed contacts transfer back to the open state as the valve is completely open. The opening of the contacts is the second distinct signal.

Figure 5.33 *Outside screw and yoke (OS & Y) valve supervisory switch.*

5-10.2 **Pressure Supervisory Signal-Initiating Device.** Two separate and distinct signals shall be initiated: one indicating that the required pressure has increased or decreased, and the other indicating restoration of the pressure to its normal value.

(a) A pressure tank supervisory signal-initiating device for a pressurized limited water supply, such as a pressure tank, shall indicate both high and low pressure conditions. A signal shall be initiated where the required pressure is increased or decreased 10 psi (70 kPa) from the normal pressure.

(b) A pressure supervisory signal-initiating device for a dry-pipe sprinkler system shall indicate both high and low pressure conditions. A signal shall be initiated where the pressure is increased or decreased 10 psi (70 kPa) from the normal pressure.

(c) A steam pressure supervisory signal-initiating device shall indicate a low pressure condition. A signal shall be initiated where the pressure reaches or exceeds 110 percent of the minimum operating pressure of the steam-operated equipment supplied.

(d) An initiating device for supervising the pressure of sources other than those specified in (a) through (c) shall be provided as required by the authority having jurisdiction.

As with supervisory initiating devices for valve operation, water pressure supervisory initiating devices can consist of a single switch.

Figure 5.34 *Pressure supervisory switch.*

5-10.3 **Water Level Supervisory Signal-Initiating Device.** Two separate and distinct signals shall be initiated: one indicating that the required water level has been lowered or raised, and the other indicating restoration to the normal level.

(a) A pressure tank signal-initiating device shall indicate both high and low level conditions. A signal shall be obtained where the water level is lowered or raised 3 in. (76 mm) from the normal level.

(b) A supervisory signal-initiating device for other than pressure tanks shall initiate a low level signal where the water level is lowered 12 in. (300 mm) below the normal level.

As with supervisory initiating devices for valve operation and water pressure, water level supervisory initiating devices can also consist of a single switch.

See Figure 5.35.

5-10.4 **Water Temperature Supervisory Signal-Initiating Device.** A temperature supervisory device for a water storage container exposed to freezing conditions shall initiate two separate and distinctive signals. One signal shall indicate that the temperature of the water has dropped to 40°F (4.4°C), and the other indicating restoration to a proper temperature.

Figure 5.35   Tank water level supervisory switch.

Water temperature supervisory initiating devices can consist of a single switch, as do supervisory initiating devices for valve operation, water pressure, and water level.

Figure 5.36   Tank water temperture supervisory switch.

5-10.5 **Room Temperature Supervisory Signal-Initiating Device.** A room temperature supervisory device shall indicate the decrease in room temperature to 40°F (4.4°C) and its restoration to above 40°F (4.4°C).

As with other supervisory initiating devices mentioned in Section 5-10, room temperature supervisory initiating devices can also consist of a single switch.

5-11   Smoke Detectors for Control of Smoke Spread.

NOTE: See NFPA *101®*, *Life Safety Code®*, for definition of smoke compartment; NFPA 90A, *Standard for the Installation of Air Con-*

Figure 5.37   Room temperature supervisory switch.

ditioning and Ventilating Systems, for definition of duct systems; and NFPA 92A, *Recommended Practice for Smoke-Control Systems*, for definition of smoke zone.

Between 1960 and late 1970 there were several fires in high-rise buildings that demonstrated the futility of trying to evacuate entire buildings in the context of fire fighting. The strategies of establishing smoke compartments and refuge zones, and managing the flow of smoke by directing it away from the occupants were developed. Experiences with many high-rise fires indicate that the proactive control of smoke with both automatic smoke detectors and HVAC systems remains a viable strategy in high-rise buildings.

5-11.1* Section 5-11 covers installation and use of all types of smoke detectors to prevent smoke spread by initiating control of fans, dampers, doors, and other equipment. Detectors for this use shall be classified as:

(a) Area detectors that are installed in the related smoke compartments

(b) Detectors that are installed in the air duct systems.

Section 5-11 does not require the installation of smoke detectors for smoke control. The purpose of the section is to describe the performance and installation requirements for smoke detectors when they are used for smoke control.

A-5-11.1 Smoke detectors located in the open area(s) are preferred to duct-type detectors because of the dilution effect in air ducts. Active smoke management systems installed in accordance with NFPA 92A, *Recommended Practice for Smoke-Control Systems,* or NFPA 92B, *Guide for Smoke Management Systems in Malls, Atria, and Large Areas,* should be controlled by total coverage open area detection.

Subsection 5-1.3.4 identifies all of the spaces that must have smoke detectors if total coverage is to be achieved.

5-11.2* Detectors that are installed in the air duct system per 5-11.1(b) shall not be used as a substitute for open area protection. Where open area protection is required, 5-3.5 shall apply.

A-5-11.2 Dilution of smoke-laden air by clean air from other parts of the building or dilution by outside air intakes may allow high densities of smoke in a single room with no appreciable smoke in the air duct at the detector location. Smoke may not be drawn from open areas where air conditioning systems or ventilating systems are shut down.

5-11.3 Smoke detectors in the related smoke compartment for open area protection are the preferred means to initiate control of smoke spread.

5-11.4 Purposes.

5-11.4.1 The purposes to which smoke detectors may be applied in order to initiate control of smoke spread are:

(a) Prevention of the recirculation of dangerous quantities of smoke within a building

(b) Selective operation of equipment to exhaust smoke from a building

(c) Selective operation of equipment to pressurize smoke compartments

(d) Operation of doors and dampers to close the openings in smoke compartments.

5-11.4.2 To prevent the recirculation of dangerous quantities of smoke, a detector approved for air duct use shall be installed on the supply side of air handling systems in accordance with NFPA 90A, *Standard for the Installation of Air Conditioning and Ventilating Systems,* and 5-11.5.2.1.

5-11.4.3 To selectively initiate the operation of equipment to control smoke spread, the requirements of 5-11.5.2.2 shall apply.

5-11.4.4 Where detectors are used to initiate the operation of smoke doors, the requirements of 5-11.7 shall apply.

5-11.4.5 Where duct detectors are used to initiate the operation of smoke dampers within ducts, the requirements of 5-11.6 shall apply.

5-11.5 Application.

5-11.5.1 Area Detectors Within Smoke Compartments. Area smoke detectors located within a smoke compartment for complete area coverage shall be permitted to be used to initiate control of smoke spread by operating doors, dampers, and other equipment where appropriate in the overall fire safety plan.

This paragraph assumes that the smoke detectors located within the smoke compartment have been installed according to the requirements of Section 5-3.

5-11.5.2 Smoke Detection for the Air Duct System.

5-11.5.2.1 Supply Air System. Where the detection of smoke in the supply air system is required by other NFPA standards, detector(s) listed for the air velocity present and located in the supply air duct downstream of both the fan and the filters shall be installed.

Exception No. 1: Where complete smoke detection is installed in the smoke compartment, installation of air duct detectors in the supply air system is not necessary if their function can be accomplished by the design of the area detection system.

Exception No. 2: Additional smoke detectors are not required to be installed in ducts where the air duct system passes through other smoke compartments not served by the duct.

The relevant NFPA standards mentioned in 5-11.5.2.1 are NFPA 90A, *Standard for the Installation of Air Conditioning and Ventilating Systems;* NFPA 92A, *Recommended Practice for Smoke Control Systems;* and NFPA 101, *Life Safety Code.* As already indicated, the preferred method of detection for smoke control is complete area smoke detection. When a detector activates in the smoke compartment, it will signal the HVAC control system to operate or control fan dampers to prevent the introduction of smoke into other smoke compartments and to vent the smoke to facilitate escape. There is the possibility that the air-handling fan and/or filters may become involved in a fire. This necessitates the use of detectors downstream from the fan and filters.

The key phrase in Exception No. 1 is "if their function can be accomplished." The purpose of supply-side smoke detection is to sense smoke that may be contaminating the area served by the duct. That smoke might come from the area via return air ducts, from outside via fresh air mixing ducts, or from a fire within the duct (such as in a filter or fan belt). If the source of smoke is not within the area served by the duct, but is from outside or from within the duct, ordinary area detection within the space would not normally be expected to serve the same function as detectors within the duct. This is because of the expected dilution of smoke-laden air as it enters the space and mixes with fresh air.

There are occasions where an air duct passes through a smoke compartment without serving the compartment. Figure A-5-11.5.2.2(c) shows this situation.

5-11.5.2.2* Return Air System. Where the detection of smoke in the return air system is required by other NFPA standards, detector(s) listed for the air velocity present shall be located at every return air opening within the smoke compartment, where the air leaves each smoke compartment, or in the duct system before the air enters the return air system common to more than one smoke compartment. [See Figures A-5-11.5.2.2(a), (b), and (c).]

Exception No. 1: Where complete smoke detection is installed in the smoke compartment, installation of air duct detectors in the return air system is not necessary if their function can be accomplished by the design of the area detection system.

Exception No. 2: Additional smoke detectors are not required to be installed in ducts where the air duct system passes through other smoke compartments not served by the duct.

A-5-11.5.2.2 Detectors listed for the air velocity present may be installed at the opening where the return air enters the common return air system. The detectors should be installed up to 12 in. (0.3 m) in front of or behind the opening and spaced according to the following opening dimensions [*see Figure A-5-11.5.2.2(a)*]:

(a) *Width.*

1. Up to 36 in. (914 mm) — One detector centered in opening

2. Up to 72 in. (1829 mm) — Two detectors located at the ¼-points of the opening

3. Over 72 in. (1829 mm) — One additional detector for each full 24 in. of opening.

(b) *Depth.* The number and spacing of the detector(s) in the depth (vertical) of the opening should be the same as those given for the width (horizontal) above.

(c) *Orientation.* Detectors should be oriented in the most favorable position for smoke entry with respect to the direction of air flow. The path of a projected beam-type detector across the return air openings should be considered equivalent in coverage to a row of individual detectors.

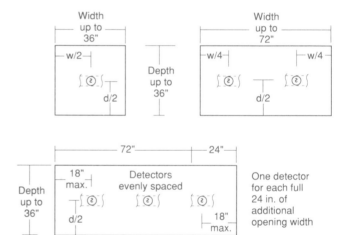

{ⓩ} Duct detector

Figure A-5-11.5.2.2(a) *Location of smoke detector(s) in return air systems for selective operation of equipment.*

In Figure A-5-11.5.2.2(c), the middle air supply duct serves the center smoke compartment. The top air supply duct serves only the left compartment and passes through the center and right compartments without serving them. Under 5-11.5.2.1, Exception No. 2, the top duct needs additional detectors and/or dampers where it passes through either the center compartment or the right compartment.

5-11.6 Location and Installation of Detectors in Air Duct Systems.

5-11.6.1 Detectors shall be listed for the purpose.

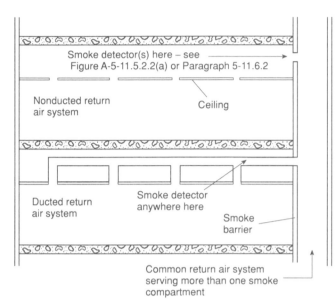

Figure A-5-11.5.2.2(b) *Location of smoke detector(s) in return air systems for selective operation of equipment.*

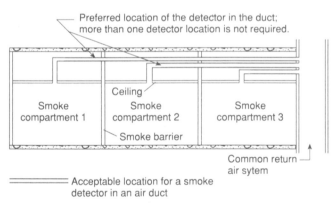

= Acceptable location for a smoke detector in an air duct

Figure A-5-11.5.2.2(c) *Detector location in a duct that passes through smoke compartments not served by the duct.*

5-11.6.2* Air duct detectors shall be securely installed in such a way as to obtain a representative sample of the air stream. This shall be permitted to be achieved by any of the following methods:

(a) Rigidly mounted within the duct

(b) Rigidly mounted to the wall of the duct with the sensing element protruding into the duct

(c) Outside the duct with rigidly mounted sampling tubes protruding into the duct

(d) Through the duct with projected light beam.

A-5-11.6.2 Where duct detectors are used to initiate the operation of smoke dampers, they should be located so that the detector is between the last inlet or outlet upstream of the damper and the first inlet or outlet downstream of the damper.

In order to obtain a representative sample, stratification and dead air space should be avoided. Such conditions may be caused by return duct openings, sharp turns or connections, as well as by long, uninterrupted straight runs. For this reason, duct smoke detectors should be located in the zone between 6 and 10 duct equivalent diameters of straight, uninterrupted run. In return air systems, the requirements of 5-11.5.2.2 take precedence over these considerations. [*See Figure A-5-11.6.2(b).*]

Note that the text above is from Appendix A and, therefore, is not a requirement of the code. While it is good practice, it is recognized that the design of the air-handling ductwork may make it impossible to achieve a distance of 6 to 10 duct widths from a bend or opening.

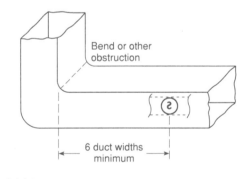

Figure A-5-11.6.2(a) *Pendant mounting air duct installation.*

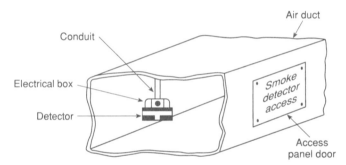

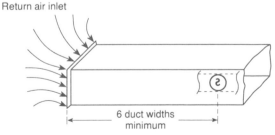

Figure A-5-11.6.2(b) *Typical duct detector placement.*

Support of the detector by the conduit or raceway containing its wires is not permitted by NFPA 70, *National Electrical Code*, unless the box is specifically listed for the purpose and installed in accordance with the listing.

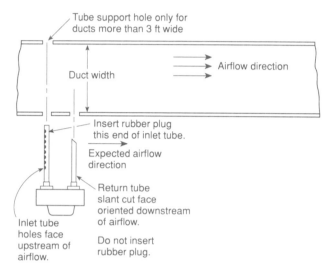

Figure A-5-11.6.2(c) *Inlet tube orientation.*

The guidance in 5-11.6.2 and A-5-11.6.2 is provided to ensure that the detectors in the air duct are suitably located to obtain an adequate sampling of air. In order to be certain that the detectors are placed where there is the highest probability that smoke will be evenly distributed throughout the duct cross-section, these location guidelines must be followed.

5-11.6.3 Detectors shall be readily accessible for cleaning and shall be mounted in accordance with the manufacturer's recommendations. If necessary, access doors or panels shall be provided.

Chapter 7 provides recommended maintenance schedules for each type of detector. It is critical that the detectors be accessible in order to facilitate cleaning, which is, in turn, critical for reliable operation of the detector.

5-11.6.4 The location of all detectors in air duct systems shall be permanently and clearly identified and recorded.

It is advisable to place permanent placards outside the first point of access, indicating that a detector is accessible

from that point. For example, the placard may be mounted on the wall beneath the ceiling tile that must be removed to access the duct. HVAC and fire alarm drawings should clearly show the actual as-built locations of the detectors. In most cases it is useful to generate one drawing that shows only the smoke detector locations.

5-11.6.5 Detectors mounted outside of a duct employing sampling tubes for transporting smoke from inside the duct to the detector shall be designed and installed to permit verification of airflow from the duct to the detector.

Sampling tubes are able to provide a flow of air through the detector enclosure due to a pressure differential that results from the flow of air across the tubes. Small errors in the orientation of the sampling tubes can reduce the pressure differential, rendering them ineffective in drawing air into the detector enclosure. In order for sampling tubes to take a representative sample of the air passing through the duct, they must be fabricated and installed in a manner consistent with their listing. Not all detectors are listed for use in a sampling tube enclosure.

5-11.6.6 Detectors shall be listed for proper operation over the complete range of air velocities, temperature, and humidity expected at the detector when the air handling system is operating.

5-11.6.7 All penetrations of a return air duct in the vicinity of detectors installed on or in an air duct shall be sealed to prevent entrance of outside air and possible dilution or redirection of smoke within the duct.

5-11.7 Smoke Detectors for Door Release Service.

5-11.7.1 Smoke door release not initiated by a fire alarm system that includes smoke detectors protecting the areas on both sides of the door affected shall be accomplished by smoke detectors applied as specified in 5-11.7.

There are two general methods of controlling doors with smoke detectors. The first is to connect area smoke detectors to a selected circuit of a fire alarm control unit and use an output circuit in the control unit to operate magnetic door release devices. When one of the area smoke detectors renders an alarm, the control unit transfers to the alarm state and energizes the output circuit that controls the door holders. The requirements for such a system are addressed in Chapter 3, Protected Premises Fire Alarm Systems.

The second method is to control the door holder mechanism directly with a dedicated smoke detector. This configuration is addressed in subsection 5-11.7.

Figure 5.38 *Typical magnetic door hold release.*

5-11.7.2 Smoke detectors listed or approved exclusively for door release service shall not be used for open area protection.

A smoke detector used concurrently for door release service and open area protection shall be acceptable if listed or approved for open area protection and installed in accordance with 5-3.5.

5-11.7.3 Smoke detectors shall be of the photoelectric, ionization, or other approved type.

5-11.7.4 Number of Detectors Required.

These placement requirements stem from the ceiling jet dynamics that have been used for guidance in placement of area smoke detection. As research continues, there may be additional insight developed for this application.

While it is not explicitly required, it is recommended that smoke detectors installed only for door release be connected to a fire alarm control unit and activate notification appliances when smoke is detected.

5-11.7.4.1 Where doors are to be closed in response to smoke flowing in either direction, the following rules shall apply.

5-11.7.4.1.1 Where the depth of wall section above the door is 24 in. (610 mm) or less, one ceiling-mounted detector shall be required on one side of the doorway only. (*See Figure 5-11.7.4.1.1, parts B and C.*)

Depth of wall section above door	Door frame mounted	Ceiling mounted	
"d"	Smoke detector listed for frame mounting or as part of closer assembly	Smoke detector ceiling mounted	
0–24" on both sides of doorway	A d = 24" Detector or detector closer mounted on either side	B d Max. of 5' min. "d" but not less than 12" One detector mounted on either side	
Over 24" on one side only	C d₁ = 20" d₂ = 30" Detector or detector closer mounted on either side	D d₁ = 20" Max. 5' min. = d₂ d₂ = 30" One detector mounted on either side	
Over 24" on both sides	E d > 24" Detector or detector closer mounted on either side	F Max. 5' min. = d Max. 5' min. = d d Two detectors required	
Over 60"	G May require additional detectors		

For SI Units: 1 in. = 25.4 mm; 1 ft = 0.305 m.
Figure 5-11.7.4.1.1

5-11.7.4.1.2 Where the depth of wall section above the door is greater than 24 in. (610 mm), two ceiling-mounted detectors shall be required, one on each side of the doorway. (*See Figure 5-11.7.4.1.1, part F.*)

This requirement is similar to the requirements regarding smoke detectors and ceilings with deep beams (*see 5-3.5.7*). The difference in the depth requirements comes from the fact that doors for smoke control are typically located in corridors that are narrower than the bays encountered in deep beam ceilings.

5-11.7.4.1.3 Where the depth of wall section above the door is 60 in. (1520 mm) or greater, additional detectors may be required as indicated by an engineering evaluation.

Since the average door height is a nominal 84 to 96 inches, the addition of 60 inches above the door results in a ceiling height of 148 inches (11.5 feet). This is above the height for which reduced spacings are required for heat detectors. The data in Appendix B indicate that when the ceiling height exceeds 10 feet, reduced spacings become advisable. Engineering judgment should be used for reducing smoke detector spacing in this application.

5-11.7.4.1.4 Where a detector is specifically listed for door frame mounting or where a listed combination or integral detector-door closer assembly is used, only one detector shall be required where installed in the manner recommended by the manufacturer.

Formal Interpretation 78-1
Reference: 5-11.7.4.1.4

Question: Does a door frame mounted combination automatic door closer incorporating a smoke detector listed by a testing laboratory for limited open area protection meet the requirements of 5-11.7.4.1.4?
Answer: Yes.

Issue Edition: NFPA 72E-1978
Reference: 9-2.2
Issue Date: February 1979 ∎

5-11.7.4.2 Where door release is intended to prevent smoke transmission from one space to another in one direction only, one detector located in the space to which smoke is to be confined shall suffice regardless of the depth of wall section above the door. Alternatively, a smoke detector conforming with 5-11.7.4.1.4 shall be used.

5-11.7.4.3 Where there are multiple doorways, additional ceiling-mounted detectors shall be required as follows.

5-11.7.4.3.1 Where the separation between doorways exceeds 24 in. (610 mm), each doorway shall be treated separately. (*See Figure 5-11.7.4.3.1.*)

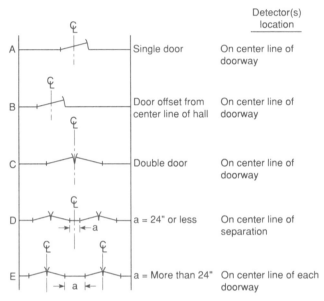

For SI Units: 1 in. = 25.4 mm.
Figure 5-11.7.4.3.1

5-11.7.4.3.2* Each group of three doorway openings shall be treated separately.

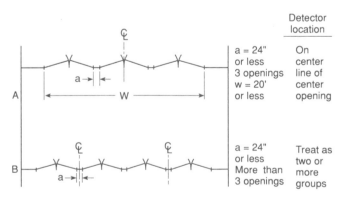

For SI Units: 1 in. = 25.4 mm; 1 ft = 0.305 m.
Figure A-5-11.7.4.3.2

5-11.7.4.3.3* Each group of doorway openings that exceeds 20 ft (6 m) in width measured at its overall extremes shall be treated separately.

For SI Units: 1 in. = 25.4 mm; 1 ft = 0.305 m.
Figure A-5-11.7.4.3.3

5-11.7.4.4   Where there are multiple doorways and listed door frame-mounted detectors or where listed combination or integral detector-door closer assemblies are used, there shall be one detector for each single or double doorway.

5-11.7.4.4.1   A double doorway is a single opening that has no intervening wall space or door trim separating the two doors. (*See Figure 5-11.7.4.3.1.*)

5-11.7.5   Location.

5-11.7.5.1   Where ceiling-mounted smoke detectors are to be installed on a smooth ceiling for a single or double doorway, they shall be located as follows. (*See Figure 5-11.7.4.3.1.*)

(a) On the centerline of the doorway, and

(b) No more than 5 ft (1.5 m) measured along the ceiling and perpendicular to the doorway (*see Figure 5-11.7.4.1.1*), and

(c) No closer than shown in Figure 5-11.7.4.1.1, parts B, D, and F.

5-11.7.5.2   Where ceiling-mounted detectors are to be installed in conditions other than those outlined in 5-11.7.5.1, engineering judgment is required.

6

Notification Appliances for Fire Alarm Systems

Contents, Chapter 6

6-1 Scope
    6-1.1 Minimum Requirements
    6-1.2 Intended Use
    6-1.5 Interconnection of Appliances
6-2 General
    6-2.1 Definitions
    6-2.2 Nameplates
    6-2.3 Physical Construction
6-3 Audible Characteristics
    6-3.1 Public Mode
    6-3.2 Private Mode
    6-3.3 Audibility
    6-3.4 Mechanical Equipment Rooms
    6-3.5 Sleeping Areas
    6-3.6 Noncoded Audible Signal Appliances
    6-3.7 Location of Audible Signal Appliances
6-4 Visible Characteristics, Public Mode
    6-4.2 Light Pulse Characteristics
    6-4.3 Appliance Photometrics
    6-4.4 Appliance Location
        6-4.4.1 Spacing Allocation for Rooms
        6-4.4.2 Spacing Allocation for Corridors
        6-4.4.3 Sleeping Areas
6-5 Visible Characteristics, Private Mode
6-6 Supplementary Visible Signaling Method
6-7 Coded Appliance Characteristics
6-8 Textual Audible Appliances
    6-8.1 Performance
    6-8.2 Loudspeaker Appliance
    6-8.3 Location of Loudspeaker Appliances
    6-8.4 Telephone Appliance
    6-8.5 Location of Telephone Appliances
6-9 Textual Visible Appliances
    6-9.2 Location

6-1 Scope.

6-1.1 Minimum Requirements. This chapter covers minimum requirements for the performance, location, and mounting required for notification appliances for fire alarm systems for the purpose of evacuation or relocation of the occupants.

It is important to note that 6-1.1 recognizes that the fire emergency plan for a particular building may require evacuation of the building or relocation of occupants from areas where they are threatened by a hostile fire to areas of refuge.

6-1.2 Intended Use. These requirements are intended to be used with other NFPA standards that deal specifically with fire alarm, extinguishment, or control systems. Notification appliances for fire alarm systems add to fire protection by providing stimuli for initiating emergency action.

This chapter relates to other chapters of the code that may require audible or visible notification appliances. For example, 3-2.3 refers to the requirements of this chapter.

6-1.3 All notification appliances or combinations thereof installed in conformity with this chapter shall be listed for the purpose for which they are used.

This requirement clearly specifies that the listing of notification appliances be use-specific. This means that the listing of an appliance must relate to the exact manner in which it will be used.

Figure 6.1 Typical listed notification appliances (bells).

6-1.4 These requirements are intended to address the reception of a notification signal and not its information content.

The requirements in this chapter do not presume to address the *content* of a textual message or a message contained in a coded audible or visible signal. Rather, the requirements address the ability of notification appliances to deliver a message.

6-1.5 Interconnection of Appliances. The interconnection of appliances, the control configurations, the power supply, and the use of the information provided by notification appliances for fire alarm systems are described in Chapter 1 and Chapter 3.

6-2   General.

6-2.1 Definitions.

Classification of Notification Signals.   For the purpose of this chapter, notification signals for fire alarm systems are classified as listed below:

Coded.   An audible or visible signal conveying several discrete bits or units of information. Notification signal examples are numbered strokes of an impact-type appliance and numbered flashes of a visible appliance.

Appendix subsection A-1-5.4.2.1 describes the recommended assignment for coded signals.

Noncoded.   An audible or visible signal conveying one discrete bit of information.

Noncoded Perceptually Constant.   The continuous operation of a notification appliance (for example, a bell, horn, siren, or light) that is energized continuously.

Noncoded Perceptually Repetitious.   The interrupted operation of a notification appliance (for example, a bell, horn, siren, or light) that is energized at a continuous uniform rate.

The American National Standard Emergency Evacuation Signal, as described in 3-7.2, is considered a noncoded perceptually repetitious signal because the temporal pattern does not convey more than one discrete bit of information.

Noncoded Single Event.   One stroke of an impact-type appliance or one flash of a strobe flash appliance. This should not be used for fire alarm purposes.

Textual. An audible or visible signal conveying a stream of information. An example of an audible textual signal is a voice message.

An example of a visible textual signal is an LCD unit that conveys phrases or sentences that make up a textual message.

General Audible. Labeled ratings are in accordance with ANSI S12.31, *Precision Methods for the Determination of Sound Power Levels of Broad Band Noise Sources in Reverberation Rooms*, and ANSI S12.32, *Precision Methods for the Determination of Sound Power Levels of Discrete Frequency and Narrow Band Noise Sources in Reverberation Rooms*, unless otherwise noted.

General/Notification. Audible or visible signals used for alerting the general public or specific individuals responsible for implementation and direction of emergency action.

General Visible. Definitions are in accordance with IES RP-16, *Nomenclature and Definitions for Illuminating Engineering*, unless otherwise noted.

Operating Mode, Private. Audible or visible signaling only to those persons directly concerned with the implementation and direction of emergency action initiation and procedure in the area protected by the fire alarm system.

Some locations where there may be individuals who receive signals in the private operating mode are a room containing PBX switchboard operators, a building engineer's office, a plant manager's office, a boiler room office, a fire brigade leader's office, a nurse's station in a health care facility, and a guard station in either the lobby of an office building or at the main gate of an industrial facility.

Operating Mode, Public. Audible or visible signaling to occupants or inhabitants of the area protected by the fire alarm system.

6-2.2 Nameplates.

6-2.2.1 The notification appliances shall include on their nameplates reference to electrical requirements and rated audible or visible performance, or both, as defined by the listing authority.

6-2.2.2 The audible appliances shall include on their nameplates reference to their parameters or reference to installation documents (supplied with the appliance) that include the parameters in accordance with 6-3.1. The visible appliances shall include on their nameplates reference to their parameters or reference to installation documents (supplied with the appliance) that include the parameters in accordance with 6-4.2.1.

In order to guide designers and installers of fire alarm systems so that the system will deliver audible and visible information with appropriate intensity, the nameplate must state the capabilities of the appliance as determined through tests conducted by the listing organization.

6-2.3 **Physical Construction.** All material for audible, textual, and visible appliances shall be moisture-, fire-, and climate-resistant in accordance with the stated purpose and shall be designed and fabricated to render them damage- and tamper-resistant.

It is essential to maintain the operational integrity of audible and visible notification appliances despite their possible location in relatively hostile environments.

6-2.4 Where subject to obvious mechanical damage, appliances shall be suitably protected.

This protection is usually provided by an enclosure that protects the actual audible or visible mechanism. In the case of speakers, a mechanical baffle protects the cone from being punctured by a sharp object.

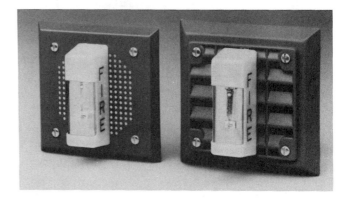

Figure 6.2 *Notification appliances showing mechanical baffles.*

6-2.5 Appliances shall be supported, in all cases, independently of their attachments to the circuit conductors.

It is not permitted to physically support the appliance by means of the wires that connect the appliance to the notification appliance circuit of the fire alarm system. See Figure 6.3.

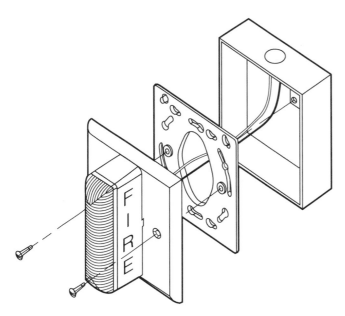

Figure 6.3 Notification appliance with independent support. (See 6-2.5.)

6-3 Audible Characteristics.

6-3.1* Public Mode.

A-6-3.1 The typical average ambient sound level for the following occupancies are intended only for design guidance purposes:

Locations	Average Ambient Sound Level
Business occupancies	55 dBA
Educational occupancies	45 dBA
Industrial occupancies	80 dBA
Institutional occupancies	50 dBA
Mercantile occupancies	40 dBA
Piers and water-surrounded structures	40 dBA
Places of assembly	55 dBA
Residential occupancies	35 dBA
Storage occupancies	30 dBA
Thoroughfares, high density urban	70 dBA
Thoroughfares, medium density urban	55 dBA
Thoroughfares, rural and suburban	40 dBA
Tower occupancies	35 dBA
Underground structures and windowless buildings	40 dBA
Vehicles and vessels	50 dBA

The typical average ambient sound levels noted should not be used in lieu of actual sound level measurements.

These sound levels, previously known as "equivalent sound levels," are defined as the mean square A-weighted sound pressure measured over a 24-hour period. This value can be experimentally determined for a particular occupancy by using a recording sound pressure level meter. The technician taking the measurement obtains the mean square result for the area under the curve by using integral calculus. The levels listed in A-6-3.1 were derived from acoustic reference literature and were originally based on this kind of calculation. See also 6-3.1.3.

It should be noted that measurements taken by a major manufacturer of fire alarm equipment in a large sampling of hotel rooms with through-the-wall air conditioning units determined that the average ambient sound level in those rooms was 55 dBA.

6-3.1.1 Audible signals intended for operation in the public mode shall have a sound level of not less than 75 dBA at 10 ft (3 m) or more than 130 dBA at the minimum hearing distance from the audible appliance.

There is a rule-of-thumb that states that the output of an audible notification appliance is reduced by 6 dBA if the distance between the appliance and the listener is doubled. Naturally, the accuracy of this rule-of-thumb depends on many intervening variables, particularly the acoustic properties of the materials in the listening space.

6-3.1.2 To ensure that audible public mode signals are clearly heard, it shall be required that their sound level be at least 15 dBA above the average ambient sound level or 5 dBA above the maximum sound level having a duration of at least 60 seconds (whichever is greater), measured 5 ft (1.5 m) above the floor in the occupiable area.

As a practical matter, most authorities having jurisdiction will measure the sound level of audible notification appliances to ensure that it is at least 5 dBA above the maximum sound level. It is more difficult to determine the average ambient sound level than it is to determine the maximum sound level that lasts at least 60 seconds. Care must be exercised in selecting the source of the maximum sound level for each occupancy.

6-3.1.3 Temporary sound sources not normally found continuously in the occupied area need not be considered in measuring maximum sound level. The average ambient sound level is the root mean square, A-weighted sound pressure measured over a 24-hour period.

Whether or not a sound source is really temporary must be determined with care. In a college dormitory room, for example, a student's radio may be portable, yet because it is usually part of the permanent equipment in the room, many authorities having jurisdiction do not believe it should be considered a "temporary" sound source.

6-3.1.4 An average sound level greater than 115 dBA shall require the use of a visible signal appliance(s) in accordance with Section 6-4.

There are a number of occupancies where the normal sound level is so high that it would be impractical to rely solely on audible notification appliances. A drop forge shop, a large casino, a rock music dance hall, or a newspaper press room are all candidates for the addition of visible signal appliances to help ensure the signals will be perceived by the occupants.

6-3.1.5 Each section of a floor divided by a required 2-hour rated fire wall shall be considered as a separate area.

NOTE: The typical average ambient sound level should be considered.

The presumption here is that the construction materials sufficient for the wall to achieve a 2-hour rating would also significantly limit the transmission of sound.

6-3.2 **Private Mode.** Audible signals intended for operation in the private mode shall have a sound level of not less than 45 dBA at 10 ft (3 m) or more than 130 dBA at the minimum hearing distance from the audible appliance. An average sound level greater than 115 dBA requires the use of a visible signal appliance(s) in accordance with Section 6-4.

Areas that use private mode signaling are expected to have a less intense average ambient sound level and a lower maximum sound level. It is important that the sound level of the audible notification appliance be adequate, but in delivering private mode signals does not unduly startle the occupants.

Figure 6.4 *Visible notification appliance. (See 6-3.1.4.)*

Figure 6.5 *Audible notification appliance for high ambient noise areas. (See 6-3.1.4.)*

6-3.3 **Audibility.** The sound level of an installed audible signal shall be adequate to perform its intended function and shall meet the requirements of the authority having jurisdiction or other applicable standards.

While additional requirements for textual audible appliances are covered in Section 6-8, this subsection can be used by an authority having jurisdiction to enforce the intelligibility of textual signals. If a textual audible notification appliance produced a signal of adequate sound level, but the message was not intelligible, then such a signal would not be adequate.

6-3.4 **Mechanical Equipment Rooms.** Where audible appliances are installed in mechanical equipment rooms, the average ambient sound level that shall be used for design guidance is at least 85 dBA for all occupancies.

Note that the required value of 85 dBA is only a minimum value. Some mechanical equipment rooms may have an average ambient sound level that exceeds this value. Based on the average ambient sound level of 85 dBA, the audible notification appliances would need to deliver 100 dBA throughout the room.

6-3.5 **Sleeping Areas.**

6-3.5.1 Where audible appliances are installed to signal sleeping areas, the maximum of 15 dBA above the average ambient sound or a minimum of 70 dBA shall be provided.

6-3.5.2 Sound level measurements at any point within the sleeping areas shall be the maximum of 15 dbA above the average ambient sound or a minimum of 70 dbA.

Subsection 6-3.5 indicates that in rooms where people sleep, the sound level delivered by the audible notification appliance must be no less than 70 dBA nor more than 15 dBA above the average ambient sound level. If the average ambient sound level in the sleeping area is 40 dBA, then the audible notification appliances must deliver 70 dBA, but not more than 70 dBA. If the average ambient sound level in the sleeping area is 60 dBA, then the audible notification appliances must deliver 75 dBA, but not more than 75 dBA.

6-3.6 **Noncoded Audible Signal Appliances.** The purpose and scope of 6-3.6 is to provide requirements for location and spacing of noncoded audible appliances.

6-3.7 **Location of Audible Signal Appliances.** Where ceiling heights permit, wall-mounted appliances shall have their tops at heights above the finished floors of not less than 90 in. (2.30 m) and below the finished ceilings of not less than 6 in. (0.15 m). This shall not preclude ceiling-mounted or recessed appliances.

Exception: Combination audible/visible appliances installed in sleeping areas shall comply with 6-4.4.3.

In rooms with a ceiling height of at least 8 feet, the top of a wall-mounted audible notification appliance must be at least 7½ feet above the floor and at least 6 inches down from the ceiling. The exception points to 6-4.4.3 for location of combination units within sleeping areas. Where the linear dimension of a sleeping room exceeds 16 feet, 6-4.4.3.2 requires a combination audible/visible notification appliance to be located within 16 feet of the pillow.

6-3.7.1 Where combination audible/visible appliances are installed, the location of the installed appliance shall be determined by the requirements of 6-4.4.

Exception: Where the combination audible/visible appliance serves as an integral part of a smoke detector, the mounting location shall be in accordance with Chapter 2.

Subsection 6-3.7.1 requires that the location of a combination audible/visible notification appliance follow the requirements of 6-4.4. The bottom of a wall-mounted appliance must be no less than 6 feet 8 inches above the floor, nor greater than 8 feet above the floor. For additional information relating to the exception, see 6-4.4.3.1, which specifies that smoke detectors in sleeping areas be installed in accordance with Chapter 2 and Chapter 5.

6-4 Visible Characteristics, Public Mode.

6-4.1 There are two methods of visible signaling. These are methods in which the message of notification of an emergency condition is conveyed by direct viewing of the illuminating appliance or by means of illumination of the surrounding area.

NOTE: One method of determining compliance with Section 6-4 is that the product be listed in accordance with UL 1971, *Signaling Applications for the Hearing Impaired.*

Private mode visible signaling is almost always directly viewed. Public mode visible signaling may be either directly

Table 6.1 Visible Signaling Appliance Requirements — Comparison of Standards

Requirement	ADA	ANSI 117.1	NFPA 72-1993 Chap. 6	UL 1971[1]	UL 1971[1] Proposed Revisions
Min. intensity (non-sleeping)	75 cd (50′ spacing)[7]	15 cd (20′ spacing)	15 cd (20′ spacing)	15 cd	15 cd
Min. intensity (sleeping) Note: See Mounting.	No requirement[11]	110 cd (signaling appliance) 177 cd (combination unit[9])	110 cd 177 cd	110 cd 177 cd	110 cd 177 cd
Min. intensity (corridors)	75 cd (50′ spacing)	15 cd (100′ spacing[6])[10]	15 cd (100′ spacing[8])[10]	15 cd	15 cd
Flash rate	1–3 Hz	1/3–3 Hz	1/3–3 Hz	1–3 Hz[2]	1/3–3 Hz
Mounting (nonsleeping and corridors)	Lower of 80″⇑; 6″⇓	Wall: 80-96″⇑[5,6]	Wall: 80-96″⇑ (5″ min⇓) Ceiling <30′⇑	No requirement	No requirement
Mounting (sleeping)	No requirement[11]	Appliance: Wall: 80-96″⇑[5,6] Comb. Unit[9]: Wall 4-12″⇓	>24″⇓ if 110 cd[3] <24″⇓ if 177 cd	<6′⇑ if 110 cd[12] >6′⇑ if 177 cd	>24″⇓ if 110 cd[4] <24″⇓ if 177 cd

⇑ = Above floor
⇓ = Below ceiling
[1] UL 1971, *Signaling Devices for the Hearing Impaired*, requires polar distribution of the light.
[2] May change to 1/3 − 3 Hz.
[3] If room is larger than 16 feet x 6 feet, appliance shall be located within 16 feet of the pillow, measured horizontally.
[4] Detector and signaling device in same room. No requirement if in separate rooms. Distance to ceiling is measured from the top of the signal light lens.
[5] ANSI 117.1 does not mention ceiling mounting or minimum distance from ceiling except by reference to NFPA 72, Chap. 6, in an appendix note.

[6] NFPA 72, Chap. 6, measures from bottom of device.
[7] In large rooms >100 feet across, without obstructions 6 feet above finish floor, devices may be placed around perimeter spaced <100 feet apart in lieu of suspending from ceiling.
[8] Maximum of 15 feet from end of corridor.
[9] Combination single-station smoke detector and visible signaling appliance.
[10] Requirement assumes direct view in a confined corridor.
[11] ADA refers to UL's requirement of 110 cd for sleeping rooms.
[12] Original specification assumed 8-foot ceilings. Height above floor does not accommodate higher ceilings.

viewed (e.g., a combination audible/visible notification appliance such as a horn/strobe) or indirectly viewed (e.g., a stroboscopic lamp used for notifying the hearing impaired).

With the passage of the Americans with Disabilities Act (ADA), a great deal of controversy has erupted over the requirements of various codes and standards with respect to visible notification appliances. Table 6.1 summarizes various requirements.

6-4.2 Light Pulse Characteristics. The flash rate shall not exceed three flashes per second nor be less than one flash every three seconds.

6-4.2.1 A maximum pulse duration shall be 0.2 sec with a maximum duty cycle of 40 percent. The pulse duration is defined as the time interval between initial and final points of 10 percent of maximum signal.

The light intensity of a pulsed source may be graphed as a bell-shaped curve. The duration of the pulse is measured beginning at the point the upward side of the curve exceeds 10 percent of the maximum intensity to the point where the downward side of the curve drops below 10 percent of the maximum intensity.

6-4.2.2 The light source color shall be clear or nominal white and shall not exceed 1000 candela (cd) (effective intensity).

6-4.3 **Appliance Photometrics.** Visible notification appliances used in the public mode shall be located so that the operating effect of the appliance can be seen by the intended viewers and shall be of a type, size, intensity, and number so that the viewer can discern when they have been illuminated, regardless of the viewer's orientation.

In the same manner that signals produced by audible notification appliances must be clearly heard, signals produced by visible notification appliances must also be clearly seen without regard to the viewer's position.

6-4.4 **Appliance Location.** Wall-mounted appliances shall have their bottoms at heights above the finished floor of not less than 80 in. (2 m) and no greater than 96 in. (2.4 m). Ceiling-mounted appliances shall be installed per Table 6-4.4.1(b).

Exception: Appliances installed in sleeping areas shall comply with 6-4.4.3.

In rooms with sufficient ceiling height, the bottom of a wall-mounted visible notification appliance must be at least 6 feet 8 inches above the floor, but not more than 8 feet above the floor. The exception refers to 6-4.4.3 for location of combination units within sleeping areas. Where the linear dimension of a sleeping room exceeds 16 feet, 6-4.4.3.2 requires a combination audible/visible notification appliance to be located within 16 feet of the pillow.

6-4.4.1* **Spacing Allocation for Rooms.**

A-6-4.4.1 Areas so large that they exceed the rectangular dimensions given in Figures A-6-4.4.1(a), (b), and (c) require additional appliances. Often, proper placement of appliances can be facilitated by breaking down the area into multiple squares and dimensions that fit most appropriately. [*See Figures A-6-4.4.1(a), (b), (c), and (d).*] An area 40 ft (12.2 m) wide and 74 ft (22.6 m) long can be covered with two 60-cd appliances. Irregular areas will take more careful planning to make sure that at least one 15-cd appliance is installed per 20 ft by 20 ft (6.09 m by 6.09 m) room.

These figures were added to avoid misinterpretation of the text. Figure A-6-4.4.1(a) demonstrates how a nonsquare

Table 6-4.4.1(a) Room Spacing Allocation for Wall-Mounted Visible Appliances

	Minimum Required Light Output, Candela (cd) (Effective Intensity)		
Maximum Room Size	One Light Per Room (cd)	Two Lights per Room (Located on Opposite Walls) (cd)	Four Lights per Room One Light per Wall) (cd)
20' × 20'	15	-	-
30' × 30'	30	15	-
40' × 40'	60	30	15
50' × 50'	95	60	30
60' × 60'	135	95	30
70' × 70'	185	110	60
80' × 80'	-	140	60
90' × 90'	-	180	95
100' × 100'	-	-	95
110' × 110'	-	-	135
120' × 120'	-	-	160
130' × 130'	-	-	185

Table 6-4.4.1(b) Room Spacing Allocation for Ceiling-Mounted Visible Appliances

	Minimum Required Light Output, Candels (cd) (Effective Intensity)	
Maximum Room Size	Maximum Ceiling Height	One Light (cd)
20' × 20'	10'	15
30' × 30'	10'	30
40' × 40'	10'	60
50' × 50'	10'	95
20' × 20'	20'	30
30' × 30'	20'	45
40' × 40'	20'	80
50' × 50'	20'	115
20' × 20'	30'	55
30' × 30'	30'	75
40' × 40'	30'	115
50' × 50'	30'	150

NOTE 1: Where ceiling heights exceed 30 ft, visible signaling appliances shall be suspended at or below 30 ft or wall-mounted in accordance with Table 6-4.4.1(a).

NOTE 2: The above is based on locating the visible signaling appliance at the center of the room. Where it is not located at the center of the room, the effective intensity (cd) shall be determined by doubling the distance from the appliance to the farthest wall to obtain the maximum room size.

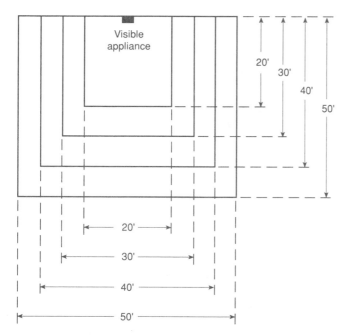

Figure 6-4.4.1 *Room spacing allocation for wall-mounted visible appliances.*

Note: The above is based on locating the visible signaling appliance at the halfway distance of the longest wall. In square rooms with appliances not centered or nonsquare rooms, the effective intensity (cd) from one visible signaling appliance shall be determined by maximum room size dimensions obtained either by the distance to the farthest wall or by double the distance to the farthest adjacent wall, whichever is greater, as shown in Table 6-4.4.1(a).

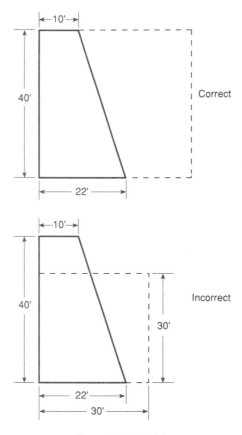

Figure A-6-4.4.1(a)

or nonrectangular room can be fitted into the spacing allocation of Tables 6-4.4.1(a) and (b). Figure A-6-4.4.1(b) demonstrates how to divide a room or area into smaller areas to enable the use of lower intensity lights. Figures A-6-4.4.1(c) and (d) show the correct and incorrect placement of multiple visible notification appliances in a room.

6-4.4.1.1 Spacing shall be in accordance with Figure 6-4.4.1 and Tables 6-4.4.1(a) and (b). A maximum separation between appliances shall not exceed 100 ft (30 m).

6-4.4.1.2 If a room configuration is not square, the square room size that will entirely encompass the room or subdivide the room into multiple squares shall be used.

Figure 6-4.4.1 and Tables 6-4.4.1(a) and (b) help ensure that a sufficient number of properly-sized visible notification appliances are installed in each protected space to afford complete coverage. The key to proper coverage in irregu-

lar spaces is to break down the space into a series of squares and provide proper coverage for each square as if it were an independent space.

6-4.4.2* Spacing Allocation for Corridors.

6-4.4.2.1 Table 6-4.4.2 applies to corridors not exceeding 20 ft (6.1 m) wide. For corridors greater than 20 ft (6.1 m) wide, refer to Figure 6-4.4.1 and Tables 6-4.4.1(a) and (b). In a corridor application, visible appliances shall be rated not less than 15 cd.

6-4.4.2.2 The visible appliances shall be located no more than 15 ft (4.57 m) from the end of the corridor with a separation no greater than 100 ft (30.4 m) between appliances. Where there is an interruption of the concentrated viewing path, such as a fire door, an elevation change, or any other obstruction, the area shall be considered as a separate corridor.

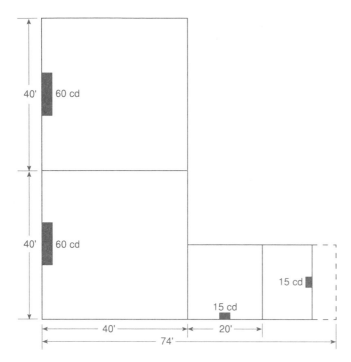

NOTE: Dashed lines represent imaginary walls.

Figure A-6-4.4.1(b)   *Room spacing allocation for ceiling-mounted visible appliances.*

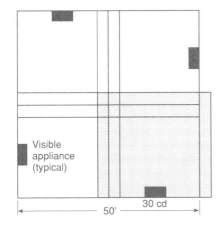

Figure A-6-4.4.1(c)   *Room spacing allocation — correct.*

6-4.4.3* Sleeping Areas.

A-6-4.4.3   Effective intensity is the conventional method of equating the brightness of a flashing light to that of a steady burning light as seen by a human observer. The units of effective intensity are expressed in candelas. For example, a flashing light that

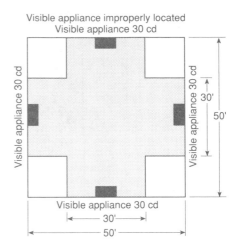

NOTE: See Table 6-4.4.1(a) for correction.

Figure A-6-4.4.1(d)   *Room spacing allocation — incorrect.*

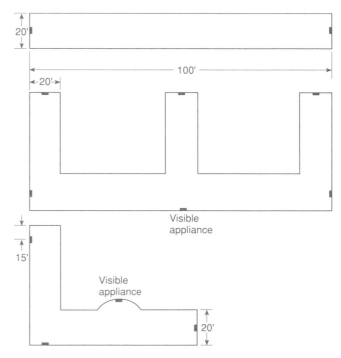

Figure A-6-4.4.2   *Corridor and elevator area spacing allocation.*

has an effective intensity of 15 candelas has the same apparent brightness to an observer as a 15-candela steady burning light source.

6-4.4.3.1   Smoke detectors shall be installed in accordance with the applicable requirements of Chapter 2 and Chapter 5.

Table 6-4.4.2 Corridor Spacing Allocation for Wall-Mounted Visible Appliances

Corridor Length (ft)	Minimum Number of 15-cd Visible Appliances Required
0- 30	1
31-130	2
131-230	3
231-330	4
331-430	5
431-530	6

Table 6-4.4.3 Effective Intensity Requirements for Sleeping Area

Visible Notification Appliance Distance from Ceiling to Top of Lens	Intensity
greater than or equal to 24″	110 cd
less than 24″	177 cd

This requirement reinforces the detector coverage requirements of Chapters 2 and 5. While not explicitly mentioned in 6-4.4.3.1, smoke detectors do have integral notification appliances, and these appliances should conform to the requirements of this chapter.

6-4.4.3.2 Table 6-4.4.3 applies to sleeping areas having no linear dimension greater than 16 ft (4.87 m). For larger rooms, the visible notification appliance shall be located within 16 ft (4.87 m) of the pillow.

The use of visible notification appliances in sleeping areas was not addressed in the earlier version of this chapter (formerly part of NFPA 72G, *Guide for the Installation, Maintenance, and Use of Notification Appliances for Protective Signaling Systems*). With the completion of the UL research report covering emergency signaling devices for the hearing impaired, NFPA was able to set minimum requirements for light intensity of visible notification appliances in sleeping areas. In order to offset the obscuration effect of smoke near the ceiling, it was decided that a value of 177 cd would be required when the notification appliance is located within 24 inches of the ceiling, and a value of 110 cd would be used when the distance is greater than 24 inches.

6-4.4.4 Where visible appliances are required, a minimum of one appliance shall be installed in the concentrated viewing path such as might be experienced in such areas as classrooms, theater stages, etc.

6-5 Visible Characteristics, Private Mode.

Visible signals used in the private mode shall be adequate for their intended purpose.

Visible notification appliances in the private mode are almost always used in conjunction with an audible notification appliance to call the viewer's attention to the visible appliance. Many visible appliances in the private mode provide annunciated information that helps the viewer locate the source of an alarm, supervisory, or trouble signal.

6-6 Supplementary Visible Signaling Method.

A supplementary visible appliance is intended to augment an audible or visible signal.

A supplementary visible notification appliance is not intended to serve as one of the required visible notification appliances, but rather is intended to supplement an audible notification appliance when visible notification appliances are not required or to supplement the required visible notification appliances.

6-6.1 A supplementary visible appliance shall comply with its marked rated performance.

Recognizing that this appliance is not satisfying a requirement, but is providing a supplemental function, this subsection makes it mandatory that the appliance function as marked and rated. This requirement discourages manufacturers from overrating the marking, which might not be detected since the appliances are supplementary, and gives the authority having jurisdiction a basis for verifying the performance of such appliances.

6-6.2 Supplementary visible notification appliances shall be permitted to be located less than 80 in. (2 m) above the floor.

Because such an appliance is supplementary, it need not meet the mandatory height requirement for visible appliances.

6-7 Coded Appliance Characteristics.

All requirements for noncoded appliances shall be met. In addition, the appliances shall differentiate several bits or units of information from all other information conveyed by that appliance.

Section 6-7 requires that coded appliances be designed so that they can properly convey coded information in a fashion that permits prompt decoding.

6-8 Textual Audible Appliances.

6-8.1 Performance. The textual appliance shall reproduce normal voice frequencies.

Textual audible appliances convey voice information. While the actual range of normal voice frequencies is not specified here, a textual appliance having bandpass similar for telephonic communications would normally serve as an appropriate measure of the suitability of the frequency response.

6-8.2 Loudspeaker Appliance. The sound level in dBA of the loudspeaker appliance evacuation tone signals of the particular mode installed shall comply with all the requirements in 6-3.1.

In addition to conveying textual information, textual audible appliances are also used to produce tones used to warn occupants to evacuate the protected premises. This requirement ensures that the sound level requirements of 6-3.1 will be met by the textual audible appliances.

6-8.3 Location of Loudspeaker Appliances. Where ceiling heights permit, wall-mounted loudspeaker appliances shall have their tops at heights above the finished floors of not less than 90 in. (2.30 m) and below the finished ceilings of not less than

Figure 6.6 Typical loudspeakers used as textual notification appliances.

6 in. (0.15 m). This does not preclude ceiling-mounted or recessed appliances.

This subsection requires textual audible appliances to be mounted at least 6 inches down from the finished ceiling and at least 7 feet 6 inches above the finished floor when the ceiling height of the protected space permits. This required location helps ensure that the output of the appliances will be appropriately distributed throughout the protected space.

6-8.3.1 Where loudspeaker/visible appliances are installed, the height of the installed appliance shall comply with 6-4.4.

Exception: Combination loudspeaker/visible appliances installed in sleeping areas shall comply with 6-4.4.3.

As with other combination audible/visible appliances, 6-8.3.1 and its exception give preference to the visible appliance with respect to location.

6-8.4 Telephone Appliance. The telephone appliance shall be in accordance with EIA Tr 41.3, *Telephones*.

The referenced Electronic Industries Association standard helps ensure the quality and technical suitability of a telephone handset.

6-8.5 Location of Telephone Appliances. Wall-mounted telephone appliances or related jacks shall be of convenient heights not to exceed 66 in. (1.7 m), except that where accessible to the general public, one telephone appliance per location should be no higher than 54 in. (1.37 m) with clear access to the wall at least 30 in. (0.76 m) wide.

The word "accessible" in this context means "available to and intended to be used by the general public." This includes use by floor or section fire wardens who might be required to communicate with the building emergency communication center by means of the fire alarm system telephones. However, fire alarm system telephones reserved exclusively for the use of fire fighters or other emergency services personnel would not need to meet the mounting height requirement of 6-8.5.

6-9 Textual Visible Appliances.

6-9.1 The temporary textual visible appliance shall be a nonstorage display that produces either visible alphanumerics subtending a character angle to the observing eye of not less than 10 minutes of arc or visible pictorial images.

A temporary textual visible appliance has no memory that would permit scrolling backward and forward to display information that has been received over a given period of time. The viewing angle helps define the types of display equipment that would be suitable for fire alarm use, ensuring that a reasonable viewing angle is provided.

6-9.1.1 The alphanumeric display shall have an equivalent minimum 7 by 5 matrix character definition, a minimum grey scale contrast as defined by 10 shades of grey, and a character retentivity from ½ minute to 5 minutes.

6-9.1.2 The pictorial display shall have a minimum of 250 line scan per frame, a minimum of 250 points per line scan, each arranged on a scale of 10 shades of grey, and shall have 30 frames per second. The display shall have an aspect ratio of 1:1.33.

Figure 6.7 Temporary textual visible appliance (annunciator).

Subsections 6-9.1.1 and 6-9.1.2 help ensure the overall quality of the display and its ability to accurately reproduce the information it is intended to convey.

6-9.1.3 The permanent textual visible appliance shall be a storage display that produces retrieved alphanumerics or retrieved pictorial images defined in accordance with 6-9.1. The retrieval time for the permanent textual visible appliance shall be not less than 1 year.

A permanent textual visible display has a memory that allows the retrieval of information. The memory must permit retrieval of at least 1 year's worth of information.

6-9.2 **Location.** All textual visible appliances in the private mode shall be located in rooms accessible only to those persons directly concerned with the implementation and direction of emergency action initiation and procedure in the areas protected by the fire alarm system.

Exception: In the lobby of a building where required by the authority having jurisdiction.

This subsection intends to limit access to private mode textual visible displays to only those persons authorized to obtain such information. The exception permits the authority having jurisdiction to specify a more public location of the textual visible appliance, presumably for use by responding emergency forces.

7

Inspection, Testing, and Maintenance

Contents, Chapter 7

7-1 General

    7-1.5 Special Hazards Systems and Equipment

    7-1.6 System Reacceptance Testing

7-2 Test Methods

    7-2.1 Central Stations

    Table 7-2.2 Test Methods

      1. Control Equipment

      2. Engine-Driven Generator

      3. Secondary (Standby) Power Supply

      4. Uninterrupted Power Supply (UPS)

      5. Batteries — General Tests

      6. Battery Tests (Specific Types)

      7. Public Reporting System Tests

      8. Transient Suppressors

      9. Control Panel Trouble Signals

      10. Remote Annunciators

      11. Conductors/Metallic

      12. Conductors/Non-Metallic

      13. Initiating Devices

      14. Alarm Notification Appliances

      15. Special Hazard Equipment

      16. Transmission and Receiving Equipment, Off Premises

      17. Emergency Communication Equipment

      18. Interface Equipment

      19. Guard's Tour Equipment

      20. Special Procedures

7-3 Inspection and Testing Frequency

    7-3.1 Visual Inspection

    Table 7-3.1 Visual Inspection Frequencies

      1. Alarm Indicating Appliances — Supervised

      2. Batteries

      3. Control Equipment: Fire Alarm Systems Monitored for Alarm, Supervisory, Trouble Signals

      4. Control Equipment: Fire Alarm Systems Unmonitored for Alarm, Supervisory, Trouble Signals

      5. Control Panel Trouble Signals

      6. Emergency Voice/Alarm Communications Equipment

      7. Fiber Optic Cable Connections

      8. Guard's Tour Equipment

      9. Initiating Devices

      10. Interface Equipment

      11. Remote Annunciators

12. Special Procedures
13. Transient Suppressors
14. Transmission and Receiving Equipment — Off
    Premises

7-3.2 Testing

Table 7-3.2 Testing Frequencies

1. Alarm Notification Appliances
2. Batteries — Central Station Facilities
3. Batteries — Fire Alarm Systems
4. Batteries — Public Fire Alarm Reporting Systems
5. Conductors/Metallic
6. Conductors/Nonmetallic
7. Control Equipment: Fire Alarm Systems Monitored
   for Alarm, Supervisory, Trouble Signals
8. Control Equipment: Fire Alarm Systems Unmonitored
   for Alarm, Supervisory, Trouble Signals
9. Control Unit Trouble Signals
10. Emergency Voice/Alarm Communications Equipment
11. Engine-Driven Generator
12. Fiber Optic Cable Power
13. Guard's Tour Equipment
14. Initiating Devices
15. Interface Equipment
16. Off-Premises Transmission Equipment
17. Remote Annunciators
18. Retransmission Equipment
19. Special Hazard Equipment
20. Special Procedures
21. System and Receiving Equipment — Off-Premises

7-4 Maintenance

7-4.4.1 Digital Alarm Communicator Transmitter
        (DACT)
7-4.4.2 Digital Alarm Communicator Receiver (DACR)
7-4.4.3 Digital Alarm Radio System (DARS)
    7-4.4.3.1 Digital Alarm Radio Receiver (DARR)
    7-4.4.3.2 McCulloh Systems

7-5 Records

7-5.1 Record of Inspection
7-5.2 Permanent Records

7-1 General.

7-1.1 This chapter covers the requirements for the inspection, testing, and maintenance of the fire alarm systems described in Chapters 3 and 4 and for their initiation and notification components described in Chapters 5 and 6. The testing and maintenance requirements for household fire warning equipment are located in Chapter 2.

This chapter centralizes all testing and maintenance requirements for protected premises fire alarm systems and for supervising station fire alarm systems. It does not include the testing and maintenance requirements for household fire warning equipment, which remain in Chapter 2 in order for that chapter to remain self-contained.

7-1.1.1 Inspection, testing, and maintenance programs shall satisfy the requirements of this code and the equipment manufacturer's instructions.

This subsection incorporates any manufacturer's instructions into the requirements of this code, which means these instructions should be enforced as are the requirements of this code.

7-1.1.2 Nothing in this chapter is intended to prevent the use of other test methods or testing devices, provided these other methods or devices are equivalent in effectiveness and safety and meet the intent of the requirements of this chapter.

The authority having jurisdiction has the responsibility of ensuring that the alternative methods are indeed equivalent.

7-1.2 The owner or his designated representative shall be responsible for inspection, testing, and maintenance of the system and alterations or additions to this system. Delegation of responsibility shall be in writing, with a copy of such delegation made available to the authority having jurisdiction.

The owner of the system is responsible for testing and maintaining the system. If the owner intends to delegate this responsibility to others, the delegation must be in writing. This may take the form of a testing and maintenance contract with a qualified contractor, or the delegation may be to a staff specialist. (*See 7-1.2.1.*)

Figure 7.1 *Technician testing fire alarm system.*

7-1.2.1 Inspection, testing, or maintenance shall be permitted to be done by a person or organization other than the owner when conducted under a written contract. Delegation of responsibility shall be in writing, with a copy of such delegation made available to the authority having jurisdiction.

7-1.2.2 Service personnel shall be qualified and experienced in the inspection, testing, and maintenance of fire alarm systems. Examples of qualified personnel shall be permitted to include but are not limited to:

(a) Factory trained and certified

(b) National Institute for Certification in Engineering Technologies Fire Alarm certified

(c) International Municipal Signal Association Fire Alarm certified

(d) Certified by state or local authority

(e) Trained and qualified personnel employed by an organization listed by a national testing laboratory for the servicing of fire alarm systems.

The International Municipal Signal Association is the professional association of those who oversee public fire service communication systems and traffic signals.

Any of the methods of qualification stated in 7-1.2.2 will help ensure that the persons testing and maintaining fire alarm systems are truly able to do so with an appropriate level of knowledge, skill, and understanding.

7-1.3 Before proceeding with any testing, all persons and facilities who would receive an alarm, supervisory, or trouble signal, and building occupants, shall be notified to prevent unnecessary response. At the conclusion of testing, those previously notified (and others necessary) shall be further notified that testing has been concluded.

It is very important that everyone who may be affected by the testing be notified that testing will take place. This notification should include, but not be limited to, the building owner, building manager, switchboard operator, building engineer, building or floor fire wardens, building maintenance personnel, public fire service communication center, alarm company supervising station, and building occupants. It is also necessary to establish a fire emergency plan that makes provision for notifying occupants, the public fire service communication center, and the supervising station of the alarm company in case a fire occurs while testing is being conducted.

7-1.3.1 The owner or his designated representative and service personnel shall coordinate system testing to prevent interruption of critical building systems or equipment.

Where the fire alarm system has interface connections to other building systems, it is important that such interfaces be managed so the testing does not cause disruption to building systems or equipment that may be critical to the continuity of building operations.

7-1.4 Prior to system maintenance or testing, the system certificate and the information regarding the system and system alterations including specifications, wiring diagrams, and floor plans shall be made available by the owner or designated representative to the service personnel.

Service personnel cannot effectively maintain or test a system without full access to the fire alarm system documentation. It would be difficult to properly maintain a system without this vital information.

7-1.5 **Special Hazards Systems and Equipment.** Special hazards systems and equipment shall include but not be limited to preaction and deluge sprinkler systems, Halon systems, carbon dioxide systems, dry chemical systems, foam systems, and fire pump controllers.

7-1.5.1 Where a special hazards system has its own control unit that is connected to and monitored by a protected premises fire alarm system, testing shall be limited to the point of interface.

Testing of special hazards fire protection systems that are equipped with their own fire alarm control unit should be conducted as a separate series of tests. Only the interface functions between the separate control unit and the building fire alarm system should be tested as part of the building fire alarm system testing.

7-1.5.2 Where the special hazards system does not have its own control unit and the protected premises fire alarm system is used to provide complete control of the special hazards equipment, testing shall include verification of the simulated release of the extinguishing agent or activation of the fire pump controls.

Where the building fire alarm system also controls the release of a special hazard fire protection system, the special hazard system operation must be tested as part of the building fire alarm system testing procedures. Care must be taken to ensure that the special hazards system is not inadvertently actuated.

7-1.5.3 Only qualified service personnel familiar with the special hazards system and equipment used shall be permitted to perform the required tests.

7-1.6 **System Reacceptance Testing.** Reacceptance test shall be performed after system components are added or deleted; after any modification, repair, or adjustment to system hardware or wiring; or after any change to software. All components, circuits, system operations, or software functions known to be affected by the change or identified by a means that indicates the system operational changes shall be 100 percent tested. In addition, 10 percent of initiating devices that are not directly affected by the change, up to a maximum of 50 devices, shall also be tested and proper system operation verified.

Whenever a fire alarm system is modified, the modification, no matter how simplistic it may seem, may affect the overall operation of the modified portion of the system. This requirement ensures that the affected portion will be completely reacceptance tested, and that a random sampling of other portions of the fire alarm system will be tested to help ensure that some other seemingly unrelated portion of the system has not been adversely affected by the modification.

Formal Interpretation 93-1
Reference: 7-1.6

Question: Is it the intent of the committee to mandate a complete retesting of an entire system including all devices, circuits, and connections when only a single device or circuit has been modified?
Answer: No.

Issue Edition: NFPA 72-1993
Reference: 7-1.6
Issue Date: September 9, 1993 ■

7-2 Test Methods.

7-2.1* **Central Stations.** The installation shall be inspected at the request of the authority having jurisdiction for complete information regarding the system, including specifications, wiring diagrams, and floor plans having been submitted for approval prior to installation of equipment and wiring.

A-7-2.1 Where the authority having jurisdiction strongly suspects significant deterioration or otherwise improper operation by a central station, a surprise inspection to test the operation of the central station may be made but requires extreme precaution. This test will be conducted without advising the central station, but the public fire service communication center must definitely be contacted when manual, waterflow alarms, or automatic fire detection systems are tested so that the fire department will not respond. In addition, persons normally receiving calls for supervisory alarms should be notified when gate valves, pump power, etc., are tested. Confirmation of the authenticity of the test procedure is recommended and should be a matter for resolution between plant management and the central station.

Installation of fire alarm systems for central station service is traditionally supervised by a highly skilled and knowledgeable field construction foreman. In many cases, field drawings will not have been developed prior to installation. Thus, some of the normally expected pre-installation documentation may be missing. The authority having jurisdiction has the right to request that such information be supplied, particularly if the ability of the system provider is in question. Certainly, upon completion of the installation, as-built drawings would be a necessary part of the documentation needed to maintain and test the system properly.

7-2.1.1 The installation shall be inspected to ensure all devices, combinations of devices, and equipment constructed and installed shall be approved for the purpose for which they are intended.

7-2.2* Fire alarm systems and other systems and equipment that may be associated with fire alarm systems and accessory equipment shall be tested according to Table 7-2.2.

Table 7-2.2 gives largely self-explanatory procedures for how tests are to be performed on various components, portions, and subsystems of the fire alarm system. Careful review of the table shows that the acceptance test is expected to include 100 percent of all devices, appliances, interfaced systems, and control unit functions. Under no circumstances does the code allow less than a 100 percent acceptance test.

A-7-2.2 **Test Methods.** The following wiring diagrams are representative of typical circuits encountered in the field and are not intended to be all-inclusive.

The noted styles are as indicated in Table 3-5.1, 3-6.1, 3-7.1, and 4-2.3.2.2.2.3.

The noted systems are as indicated in NFPA 170, *Standard for Firesafety Symbols.*

Since ground-fault detection is not required for all circuits, tests for ground-fault detection should be limited to those circuits equipped with ground-fault detection.

An individual point-identifying (addressable) initiating device operates on a signaling line circuit and not on a Style A, B, C, D, or E (Class B and Class A) initiating device circuit.

All of the following initiating device circuits are illustrative of either alarm or supervisory signaling. Alarm and supervisory initiating devices are not permitted on the same initiating device circuit.

In addition to losing its ability to receive an alarm from an initiating device located beyond an open fault, a Style A (Class B) initiating device circuit also loses its ability to receive an alarm when a single ground fault is present.

Style C and Style E (Class B and Class A) initiating device circuits can discriminate between an alarm condition and a wire-to-wire short. In these circuits, a wire-to-wire short provides a trouble indication. However, a wire-to-wire short will prevent alarm operation. Shorting-type initiating devices cannot be used without an additional current or voltage limiting element.

Directly connected system smoke detectors, commonly referred to as two-wire detectors, should be listed as being electrically and functionally compatible with the control unit and the specific subunit or module to which they are connected. If the detectors and the units or modules are not compatible, it is possible that, during an alarm condition, the detector's visible indicator will illuminate, but no change of state to the alarm condition will occur at the control unit. Incompatibility can also prevent proper system operation at extremes of operating voltage, temperature, and other environmental conditions.

If two or more two-wire detectors with integral relays are connected to a single initiating device circuit and their relay contacts are used to control essential building functions (e.g., fan shutdown, elevator recall, etc.), it should be clearly noted that the circuit may

Table 7-2.2 Test Methods

DEVICE	METHOD
1. Control Equipment:	
a. Functions	All functions of the system, including operation of the system in various alarm and trouble modes for which it is designed (e.g., open circuit, grounded circuits, power outage, etc.), shall be tested in accordance with the manufacturer's instructions.
	The functions provided with the control panel and designed to be accomplished by the fire alarm system must be tested to ensure proper operation.
b. Fuses	Remove fuse and verify rating and supervision.
c. Interfaced Equipment	Integrity of single or multiple circuits providing interface between two or more control panels shall be verified.
	Interfaced equipment connections shall be tested by operating or simulating operation of the equipment being supervised. Signals required to be transmitted shall be verified at the control panel.
	The wiring connections must be tested by simulating a single open and again with a single ground to verify proper indications for the monitoring of the interfaced equipment wiring integrity. In addition, the interfaced equipment must be placed in a simulated trouble condition to test for proper supervisory signal receipt and reaction at the main control unit.
d. Lamps and LEDs	Lamps and LEDs shall be illuminated.
e. Primary (Main) Power Supply	All secondary (standby) power shall be disconnected and tested under maximum load, including all alarm appliances requiring simultaneous operation. All secondary (standby) power shall be reconnected at end of test. For redundant power supplies, each shall be tested separately.
2. Engine-Driven Generator	If an engine-driven generator dedicated to the fire alarm system is used as a required power source, operation of the generator shall be verified in accordance with NFPA 110, *Standard for Emergency and Standby Power Systems,* by the building owner.
3. Secondary (Standby) Power Supply	Disconnect all primary (main) power supplies and verify that required trouble indication for loss of primary power occurs. Measure or verify system's standby and alarm current demand and, using manufacturer's data, verify whether batteries are adequate to meet standby and alarm requirements. Operate general alarm systems for a minimum of five minutes and emergency voice communication systems for a minimum of fifteen minutes. Reconnect primary (main) power supply at end of test.
4. Uninterrupted Power Supply (UPS)	If a UPS system dedicated to the fire alarm system is used as a required power source, verify the operation of the UPS system in accordance with NFPA 111, *Standard on Stored Electrical Energy Emergency and Standby Power Systems,* by the building owner.
5. Batteries — General Tests:	
a. Visual Inspection	Inspect batteries for corrosion or leakage. Check and ensure tightness of connections. If necessary, clean and coat the battery terminals or connections. Visually inspect electrolyte level in lead acid batteries.
b. Battery Replacement	Batteries shall be replaced in accordance with the recommendations of the alarm equipment manufacturer, or when the recharged battery voltage or current falls below the manufacturer's recommendations.
c. Charger Test	Check operation of battery charger in accordance with charger test for the specific type of battery.
d. Discharge Test	With the battery charger disconnected, load test the batteries following the manufacturer's recommendations. The voltage level shall not fall below the levels specified.
	Exception: An artificial load equal to the full fire alarm load connected to the battery shall be permitted to be utilized in conducting this test.

Table 7-2.2   Test Methods (cont.)

DEVICE	METHOD
Batteries — General Tests (cont.)	
e. Load Voltage Test	With the battery charger disconnected, measure the terminal voltage while supplying the maximum load required by its application.
	The voltage level shall not fall below the levels specified for the specific type of battery. If the voltage falls below the level specified, corrective action shall be taken and the batteries retested.
	Exception: An artificial load equal to the full fire alarm load connected to the battery shall be permitted to be utilized in conducting this test.
f. Open Circuit Voltage	With the battery charger disconnected, measure the open circuit voltage of the battery.
6. Battery Tests (Specific Types):	
a. Primary Batteries:	
1. Load Voltage Test*	The maximum load for a No. 6 primary battery shall not be more than 2 amperes per cell. An individual (1.5-volt) cell shall be replaced when a load of 1 ohm reduces the voltage below 1 volt. A 6-volt assembly shall be replaced where a test load of 4 ohms reduces the voltage below 4 volts.
b. Lead-Acid Type:	
1. Charger Test	With the batteries fully charged and connected to the charger, measure the voltage across the batteries with a voltmeter. The voltage shall be 2.30 volts per cell $+/-$ .02 volts (at 25°C) or as specified by the equipment manufacturer.
2. Load Voltage Test*	Under load, the battery shall not fall below 2.05 volts per cell.
3. Specific Gravity	The specific gravity of the liquid in the pilot cell or all of the cells shall be measured as required. The specific gravity shall be within the range specified by the manufacturer. Although the specified specific gravity may vary from manufacturer to manufacturer, a range of 1.205 – 1.220 is typical for regular lead acid batteries, while 1.240 – 1.260 is typical for high performance batteries. A hydrometer that only shows a pass or fail condition of the battery and does not indicate the specific gravity shall not be used since such a reading does not give a true indication of the battery condition.
c. Nickel-Cadmium Type:	
1. Charger Test	With the batteries fully charged and connected to the charger, place an amp meter in series with the battery under charge. The charging current shall be in accordance with the manufacturer's recommendations for the type of battery used. In the absence of specific information, this usually is 1/30 to 1/25 of the battery rating. (Example: 4000mAh $\times$ 1/25 = 160ma charging current at 25°C.)
2. Load Voltage Test*	Under load, the float voltage for the entire battery shall be 1.42 volts per cell nominal. If possible, cells shall be measured individually.
d. Sealed Lead-Acid Type:	
1. Charger Test	With the batteries fully charged and connected to the charger, measure the voltage across the batteries with a voltmeter. The voltage should be 2.30 volts per cell $+/-$ .02 volts (at 25°C) or as specified by the equipment manufacturer.
2. Load Voltage Test*	Under load, the float voltage shall not fall below 2.05 volts per cell.
7. Public Reporting System Tests	In addition to the tests and inspection required above, the following requirements shall apply.
	Manual tests of the power supply for public reporting circuits shall be made and recorded atleast once during each 24-hour period. Such tests shall include:
	(a) Current strength of each circuit. Changes in current of any circuit, amounting to 10 percent of normal current, shall be investigated immediately.
	(b) Voltage across terminals of each circuit, inside of terminals of protective devices. Changes in voltage of any circuit, amounting to 10 percent of normal voltage, shall be investigated immediately.
	(c) Voltage between ground and circuits. Where this test shows a reading in excess of 50 percent of that shown in test (b) above, the trouble shall be immediately located and cleared; readings in excess of 25 percent shall be given early attention. These readings shall be taken with a voltmeter of not more than 100-ohms resistance per volt.

Table 7-2.2   Test Methods (cont.)

DEVICE	METHOD
Public Reporting System Tests (cont.)	

NOTE 1:   The voltmeter sensitivity has been changed from 1000 ohms per volt to 100 ohms per volt so that false ground readings (caused by induced voltages) will be minimized.

NOTE 2:   Systems in which each circuit is supplied by an independent current source (Forms 3 and 4) will require tests between ground and each side of each circuit. Common current source systems (Form 2) will require voltage tests between ground and each terminal of each battery and other current source.

(d) A ground current reading shall be acceptable in lieu of (c) above. When this method of testing is used, all grounds showing a current reading in excess of 5 percent of the normal line current shall given immediate attention.

(e) Voltage across terminals of common battery, on switchboard side of fuses.

(f) Voltage between common battery terminals and ground. Abnormal ground readings shall be investigated immediately.

NOTE: Tests (e) and (f) apply only to those systems using a common battery. If more than one common battery is used, each common battery is to be tested.

8. Transient Suppressors

Lightning protection equipment shall be inspected and maintained per manufacturer's specifications.

Additional inspections shall be required after any lightning strikes.

Equipment located in moderate to severe areas outlined in NFPA 780, *Lightning Protection Code*, Appendix I, shall be inspected semi-annually and after any lightning strikes.

In areas prone to lightning storms, the owner should be advised to notify the service company when a storm has occurred so that all original lightning protection can be checked.

9. Control Panel Trouble Signals:

a. Audible and Visual

Verify operation of panel trouble signals and ring back feature for systems using a trouble silencing switch that requires resetting.

b. Disconnect Switches

When control unit (panel) has disconnect or isolating switches, verify that each switch performs its intended function and a trouble signal is received when a supervised function is disconnected.

c. Ground-Fault Monitoring Circuit

When system has ground detection feature, verify that a ground fault indication is given whenever any installation conductor is grounded.

Each conductor should be grounded temporarily to ensure proper ground detection circuit operation. This information should be recorded on the acceptance test report for future troubleshooting information.

d. Transmission of Signals to Off-Premises Location

Actuate an appropriate initiating device and verify that alarm signal is received at the off-premises location.

Create a trouble condition and verify that a trouble signal is received at the off-premises location.

Actuate a supervisory device and verify that a supervisory signal is received at the off-premises location. If transmission carrier is capable of operation under a single or multiple fault condition, activate an initiating device during such fault condition and verify that a trouble signal is received at the off-premises location in addition to the alarm signal.

10. Remote Annunciators

Verify for proper operation and confirm proper identification. Where provided, verify proper operation under a fault condition.

Remote annunciation is very important to the fire department personnel responding to the alarm. Its intent is to reduce the time in finding the source of alarm by providing clear and accurate information to the fire service.

11. Conductors/Metallic:

a. Stray Voltage

All installation conductors shall be tested with a volt/ohm meter to verify that there are no stray (unwanted) voltages between installation conductors or between installation conductors and ground. Unless a different threshold is specified in the system manufacturer's documentation, the maximum allowable stray voltage shall not exceed 1 volt ac/dc.

b. Ground Faults

All installation conductors other than those intentionally and permanently grounded shall be tested for isolation from ground per the manufacturer's recommendations.

Table 7-2.2 Test Methods (cont.)

DEVICE	METHOD
Conductors/Metallic (cont.)	
c. Short Circuit Faults	All installation conductors other than those intentionally connected together shall be tested for conductor-to-conductor isolation per the manufacturer's recommendations. These same circuits shall be tested conductor-to-ground, also.
d. Loop Resistance	With each initiating and indicating circuit installation conductor pair short-circuited at the far end, measure and record the resistance of each circuit. Verify that the loop resistance does not exceed the manufacturer's specified limits.
	The assumption here is that if the loop resistance exceeds the manufacturer's specified limits, the wiring will be changed as it should be.
12. Conductors/Non-Metallic:	
a. Circuits' Integrity	Test each initiating device, indicating appliance, and signaling line circuit to confirm that the integrity of installation conductors are being properly supervised.
b. Fiber Optics	The fiber optic transmission line shall be tested in accordance with the manufacturer's instructions or by the use of an optical power meter, or an optical time domain reflectometer to measure the relative power lost of the line. This relative figure for each fiber optic line shall be recorded in the fire alarm control panel. If the power level drops 2 percent or more from the figure recorded during the initial acceptance test, the transmission line, section thereof, or connectors shall be repaired and/or replaced by a qualified technician to bring the line back into compliance with an accepted transmission level per manufacturer's recommendations.
c. Supervision	Introduction of a fault in any supervised circuit shall result in a suitable trouble indication at the control unit. One connection shall be opened at no less than 10 percent of the initiating device, indicating appliance, and signaling line circuits.
	Test each initiating device, indicating appliance, and signaling line circuit for proper alarm response.
	The word "supervision" here means the monitoring of the circuit conductor's integrity.
13. Initiating Devices:	
	NOTE: See Table 3-6.1 for description of circuit performance and capacity.
a. Electromechanical Releasing Device:	
1. Nonrestorable-Type Link	Remove the fusible link and operate the associated device to ensure proper operation. Lubricate any moving parts as necessary.
2. Restorable-Type Link	Remove the fusible link and operate the associated device to ensure proper operation. Lubricate any moving parts as necessary.
	NOTE: Fusible thermal link detectors are commonly used to close fire doors and fire dampers. They can be actuated by the presence of external heat, which causes a solder element in the link to fuse, and by an electric thermal device which, when energized, generates heat within the body of the link, causing the link to fuse and separate.
b. Extinguishing System Alarm Switch	Mechanically or electrically operate the switch and verify receipt of signal by the control panel.
c. Fire-Gas and Other Detectors	Fire-gas detectors and other fire detectors shall be tested as prescribed by the manufacturer and as necessary for the application.
d. Heat Detectors:	
1. Fixed-Temperature and/or Rate-of-Rise or Rate-of-Compensation, Restorable Line or Spot Type (Except Pneumatic Tube)	Heat test with a heat source per manufacturer's recommendations for response within 1 minute. Precaution should be taken to avoid damage to the nonrestorable fixed-temperature element of a combination rate-of-rise/fixed-temperature element.
2. Fixed-Temperature, Non-restorable Line Type	Do not heat test. Test mechanically and electrically for function. Measure and record loop resistance. Investigate changes from acceptance test.
3. Fixed-Temperature, Non-restorable Spot Type	After 15 years, replace all devices or laboratory test two detectors per 100. Replace the two detectors with new devices. If a failure occurs on any of the detectors removed, additional detectors shall be removed and tested to determine either a general problem involving faulty detectors or a localized problem involving one or two defective detectors.
	The laboratory test referenced here is conducted by a testing laboratory engaged in the listing or approval of heat detectors.

Table 7-2.2 Test Methods (cont.)

DEVICE	METHOD
Initiating Devices (cont.)	
4. Nonrestorable (General)	Do not heat test. Test mechanically and electrically for function.
5. Restorable Line Type, Pneumatic Tube Only	Heat source (where test chambers are in circuit) or pressure pump.
e. Fire Alarm Boxes	Operate per manufacturer's instruction. For key operated pre-signal fire alarm boxes, test both pre-signal and general alarm circuit.
f. Radiant Energy Fire Detectors	Flame detectors and spark/ember detectors shall be tested in accordance with the manufacturer's instructions to determine that each detector is operative.
	Flame detector and spark/ember detector sensitivity shall be determined using either:
	(a) A calibrated test method, or
	(b) The manufacturer's calibrated sensitivity test instrument, or
	(c) Listed control panel arranged for the purpose, or
	(d) Other calibrated sensitivity test method acceptable to the authority having jurisdiction that is directly proportional to the input signal from a fire consistent with the detector listing or approval.
	Detectors found to be outside of the approved range of sensitivity shall be replaced or adjusted to bring them into the approved range if designed to be field adjustable.
	Flame detector and spark/ember detector sensitivity shall not be determined using a light source that administers an unmeasured quantity of radiation at an undefined distance from the detector.
g. Smoke Detectors:	
1. All Types	The detectors shall be tested in place to ensure smoke entry into the sensing chamber and an alarm response. Testing with smoke or listed aerosol acceptable to the manufacturer, or other means acceptable to the detector manufacturer shall be permitted as one acceptable test method.
	This is a "go, no-go" type of test and does not test the detector's sensitivity. This section requires more than just a visual test.
	Ensure that each smoke detector is within its listed and marked sensitivity range by testing using either:
	(a) A calibrated test method, or
	(b) The manufacturer's calibrated sensitivity test instrument, or
	(c) Listed control equipment arranged for the purpose, or
	(d) Other calibrated sensitivity test method acceptable to the authority having jurisdiction.
	NOTE: The detector sensitivity cannot be tested or measured using any spray device that administers an unmeasured concentration of aerosol into the detector.
2. Air Sampling:	
Wilson Cloud Chamber	Per manufacturer's recommended test methods, including verification of sampling from each method.
Photoelectric-Type	Verify detector alarm response through the end sampling port on each pipe run, as well as verifying air flow through all other ports.
3. Duct-Type	Air duct detectors shall be tested or inspected to ensure that the device will sample the air stream. The test shall be made in accordance with the manufacturer's instructions.
4. Projected Beam-Type	The detector shall be tested by introducing smoke, other aerosol, or an optical filter into the beam path.
5. Smoke Detector with Built-in Thermal Element	Operate both portions of the detector independently as described for the respective devices.
	This is a new requirement to test both portions of a combination unit wherever possible. The code is silent on the issue of one feature failing, but it is assumed that if a combination smoke/heat detector is being used, and one feature fails a test, the entire unit is removed and replaced.
6. Smoke Detectors with Control Output Functions	When individual fire detectors are used to control the operation of equipment as permitted by 3-7.1, the control capability shall remain operable even if all of the initiating devices connected to the sameinitiating circuit are in an alarm state.
	This requirement disallows the use of two-wire smoke detectors for controlling operations on an initiating device circuit when other devices are installed on the same circuit. If, for instance, the smoke detector tries to alarm after a manual fire alarm box has been activated, the smoke detector will not alarm.

Table 7-2.2   Test Methods (cont.)

DEVICE	METHOD
Initiating Devices (cont.)	
h. Initiating Devices, Supervisory:	
1. Control Valve Switch	Operate valve and verify signal receipt within the first two revolutions of the hand wheel or within one-fifth of the travel distance, or manufacturer's specifications.
2. High or Low Air Pressure Switch	Operate switch and verify that receipt of signal is obtained where the required pressure is increased or decreased 10 psi from the required pressure level.
3. Room Temperature Switch	Operate switch and verify receipt of signal to indicate the decrease in room temperature to 40°F (4.4°C) and its restoration to above 40°F (4.4°C).
4. Water Level Switch	Operate switch and verify the receipt of signal indicating the water level raised or lowered 3 in. (76.2 mm) from the required level within a pressure tank, or 12 in. (305 mm) from the required level of a nonpressure tank, and its restoral to required level.
5. Water Temperature Switch	Operate switch and verify receipt of signal to indicate the decrease in water temperature to 40°F (4.4°C) and its restoration to above 40°F (4.4°C).
i. Waterflow Device:	
1. Mechanical, Electrosonic, or Pressure Type	Flow water through an inspector's test connection indicating the flow of water equal to that from a single sprinkler of the smallest orifice size installed in the system for wet-pipe systems, or an alarm test bypass connection for dry-pipe, pre-action, or deluge systems in accordance with NFPA 25, *Standard for the Inspection, Testing, and Maintenance of Water-Based Fire Protection Systems*.
14. Alarm Notification Appliances:	
a. Audible	Measure sound pressure level with sound level meter meeting ANSI S-1.4a, *Sound Level Meters*, Type 2 requirements. Measure and record levels throughout protected area.
	It makes sense to choose areas physically remote from the audible notification appliances to measure and record sound levels. If these areas comply with code requirements, then further measurements may be deemed unnecessary by the authority having jurisdiction.
b. Speakers	Measure sound pressure level with sound level meter meeting ANSI S-1.4a, *Sound Level Meters*, Type 2 requirements. Measure and record levels throughout protected area. Verify voice clarity.
	In order to comply with the sound level requirements of the code, many installers will attempt to tap speakers at a higher wattage rather than increase the number of speakers in an area. This incorrect approach to sound level compliance leads to distortion of voice messages through the speakers. Therefore, in addition to meeting the sound level requirements of the code, clarity or intelligibility must be checked when speakers are used for voice communications.
c. Visible	Test in accordance with manufacturer's instructions. Verify device locations are per approved layout and confirm that no floor plan changes affect the approved layout.
	If the area being tested is occupied, it is advisable to ensure that none of the visible notification appliances are blocked by shelving, furniture, ceiling-mounted light fixtures, or movable partitions. Also, the owner should be advised to keep all viewing paths to visible notification appliances clear.
15. Special Hazard Equipment:	Caution should be used when testing the interfaced special hazard equipment to avoid unnecessary actuation. Never assume that the previous tests were conducted or conducted properly when testing these interconnections. After all equipment has been tested independently, ensure all previous connections or test switches are placed in their normal positions.
a. Abort Switch (IRI-Type)	Operate abort switch. Verify correct sequence and operation.
b. Abort Switch (Recycle-Type)	Operate abort switch. Verify correct matrix develops with each sensor operated.
c. Abort Switch (Special-Type)	Operate abort switch. Verify correct sequence and operation in accordance with authority having jurisdiction. Note sequence on as-built drawings or in owner's manual.
d. Cross Zone Detection Circuit	Operate one sensor or detector on each zone. Verify that correct sequence occurs with operation of first zone and then with operation of second zone.

Table 7-2.2 Test Methods (cont.)

DEVICE	METHOD
Special Hazard Equipment (cont.)	
e. Matrix Type Circuit	Operate all sensors in system. Verify correct matrix develops with each sensor operated.
f. Release Solenoid Circuit	Use solenoid with equal current requirements. Verify operation of solenoid.
g. Squibb Release Circuit	Use AGI flashbulb or other test light acceptable to the manufacturer. Verify operation of flashbulb or light.
h. Verified, Sequential, or Counting Zone Circuit	Operate required sensors at a minimum of four locations in circuit. Verify correct sequence with both the first and second detector in alarm.
i. All Above Devices and/or Circuits	Verify supervision of circuits by creating an open circuit. Note specific trouble indications.
16. Transmission and Receiving Equipment, Off Premises:	
a. All Equipment	Verify all system functions and features in accordance with manufacturer's instructions.
	Remove primary power, actuate an initiating device, and verify that the initiating device signal is received at the monitoring station. On completion of test, restore system to normal.
	Where test jacks are used, the first and last tests shall be made without the use of the test jack.
b. Transmitters—Digital Alarm Communicator Systems (DACS)	Verify that the failure of the primary transmission path at the protected premises shall result in a trouble signal being transmitted via the secondary path to the monitoring station within 4 minutes of the detection of the fault.
	Check that all timing devices are synchronized to ensure the proper compliance to the time limits for reporting.
1. DACT	See 7-4.4.1.
	Verify the DACT is connected to two separate lines (numbers) at the protected premises.
	In turn, disconnect each telephone line at the protected premises and verify local annunciation and that trouble signals are transmitted to the monitoring station over the other line (number). Transmission shall be initiated within 4 minutes of the detection of the fault.
2. DACR	Verify that the DACR equipment is connected to a minimum of two separate incoming telephone lines (numbers). Verify that if the lines (numbers) are in a single hunt group, they are individually accessible.
	Use 7-4.4.2(b).
	Use 7-4.4.2(c).
	Use 7-4.4.2(d).
3. DARR	Verify supervision of the following conditions at the monitoring station:
	(a) Failure of ac power supplying the radio equipment
	(b) Receiver malfunction
	(c) Antenna and interconnecting cable malfunction
	(d) Indication of automatic switchover of the DARR
	(e) Data transmission line between the DARR and the monitoring station.
4. McCulloh Systems	Verify that signals are received during one of the following signaling line fault conditions:
	(a) Open
	(b) Ground
	(c) Wire-to-wire short
	(d) Open and ground.
17. Emergency Communication Equipment:	
a. Amplifier/Tone Generators	Verification of proper switching and operation of backup equipment.
b. Call-in Signal Silence	Operate function and verify receipt of proper visual and audible signals at control panel.

Table 7-2.2   Test Methods (cont.)

DEVICE	METHOD
Emergency Communication Equipment (cont.)	
c. Off-hook Indicator (Ring Down)	Install phone set or remove phone from hook and verify receipt of signal at control panel.
d. Phone Jacks	Visual inspection and initiate communication path through jack.
	During an acceptance test, it is expected that all phone jacks on each floor or zone will be checked for proper operation.
e. Phone Set	Activate each phone set and verify proper operation.
f. System Performance	Operate system with a minimum of any five handsets simultaneously. Verify acceptable voice quality and clarity.
18. Interface Equipment	Interface equipment connections shall be tested by operating or simulating the equipment being supervised. Signals required to be transmitted shall be verified at the control panel. Test frequency for interface equipment shall be the same as the frequency required by the applicable NFPA standard(s) for the equipment being supervised.
	The signals being verified include the status (alarm, trouble, or supervisory conditions) of the interfaced equipment. The main fire alarm control unit will indicate a supervisory signal for any trouble or supervisory conditions at the interfaced equipment.
19. Guard's Tour Equipment	Test the device in accordance with manufacturer's specifications.
20. Special Procedures:	
a. Alarm Verification	Verify time delay and alarm response for smoke detector circuits identified as having alarm verification.
b. Multiplex Systems	Verify communication between sending and receiving units under both normal and standby power.
	Verify communication between sending and receiving units under open circuit and short-circuit trouble conditions.
	Verify communication between sending and receiving units in all directions when multiple communication pathways are provided.
	When redundant central control equipment is provided, verify switchover and all required functions and operations of secondary control equipment.
	Verify all system functions and features in accordance with manufacturer's instructions.
	System functions and features should be verified in accordance with the circuit styles as designed as well as based on the manufacturer's specifications.

be capable of supplying only enough energy to support one detector/relay combination in an alarm mode. If control of more than one building function is required, each detector/relay combination used to control separate functions should be connected to separate initiating device circuits, or they should be connected to an initiating device circuit that will provide adequate power to permit all the detectors connected to the circuit to be in the alarm mode simultaneously. During acceptance and reacceptance testing, this feature should always be tested and verified.

A speaker is an alarm indicating appliance, and, when used in the following diagrams, the principle of operation and supervision is the same as for other audible alarm indicating appliances (e.g., bells, horns, etc.).

Wiring Diagrams.

NOTE: Where testing circuits, verify the correct wiring size, insulation type, and conductor fill in accordance with the requirements in NFPA 70, *National Electrical Code.*

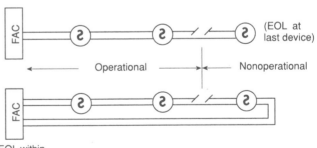

EOL - End-of-line device
FAC - Fire alarm control unit

Disconnect conductor at device or control unit, then reconnect. Temporarily connect a ground to either leg of conductors, then remove ground. Both operations should indicate audible and visual trouble with subsequent restoral at control unit. Conductor-to-conductor short should initiate alarm, Style A and Style B (Class B) indicate trouble Style C (Class B). Style A (Class B) will not initiate alarm while in trouble condition.

Figure A-7-2.2(a)

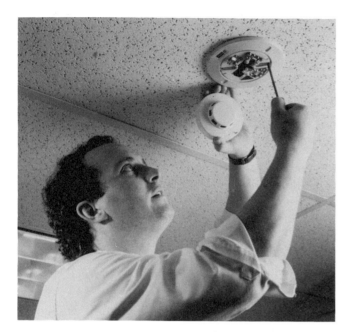

Figure 7.2 *Technician disconnecting conductor at device.*

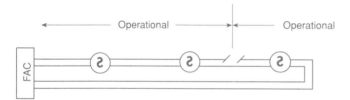

Disconnect a conductor at a device mid-point in the circuit. Operate a device on either side of device with disconnected conductor. Reset control unit and reconnect conductor. Repeat test with a ground applied to either conductor in place of the disconnected conductor. Both operations should indicate audible and visual trouble, then alarm or supervisory indication with subsequent restoral.

Figure A-7-2.2(b)

Testing of supervised remote relays to be conducted in same manner as indicating appliances.

Circuit Styles.

NOTE: Some testing laboratories and authorities having jurisdiction permit systems to be classified as a Style 7 (Class A) by the application of two circuits of the same style operating in parallel. An example of this is to take two series circuits, either Style 0.5 or Style 1.0 (Class B), and operate them in parallel. The logic being that should a condition occur on one of the circuits, the remaining parallel circuit would be operative.

In order to understand the principles of the circuit, perform alarm receipt capability on a single circuit and indicate on the certificate of completion the style type based on the performance.

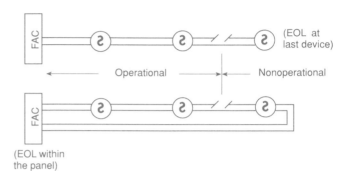

Remove smoke detector if installed with plug-in base or disconnect conductor beyond first device from control unit. Activate smoke detector per manufacturer's recommendations between control unit and circuit break. Restore detector and/or circuit. Control unit should indicate trouble where fault occurs and alarm where detectors are activated between the break and the control unit.

Figure A-7-2.2(c)

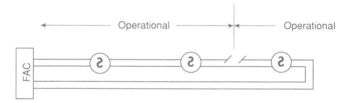

Disconnect conductor at a smoke detector or remove if installed with a plug-in base mid-point in the circuit. Operate a device on either side of device with the fault. Reset control unit and reconnect conductor or detector. Repeat test with a ground applied to either conductor in place of the disconnected conductor or removed device. Both operations should indicate audible and visual trouble, then alarm indication with subsequent restoral.

Figure A-7-2.2(d)

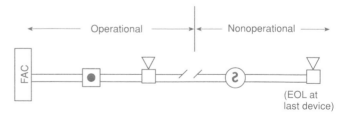

Disconnect a conductor either at indicating or initiating device. Activate initiating device between fault and control unit. Activate additional smoke detectors between device first activated and control unit. Restore circuit, initiating devices, and control unit. Confirm that all indicating appliances on the circuit operate from the control unit up to the fault and that all smoke detectors tested and their associated ancillary functions, if any, operated.

Figure A-7-2.2(e)

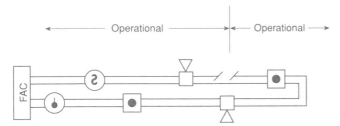

Testing of the circuit is similar to that described above. Confirm all indicating appliances operate on either side of fault.

Figure A-7-2.2(f)

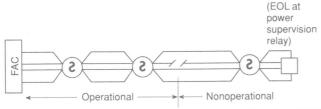

Testing of the circuit is similar to that described in A-7-2.2(c) and A-7-2.2(d). Disconnect a leg of the power supply circuit beyond the first device on the circuit. Activate initiating device between fault and control unit. Restore circuits, initiating devices, and control unit. Audible and visual trouble should indicate at the control unit where either initiating or power circuit is faulted. All initiating devices between the circuit fault and the control unit should activate. In addition, removal of a smoke detector from a plug-in type base can also break the power supply circuit. When circuits contain various powered and nonpowered devices on the same initiating circuit, verify that the nonpowered devices beyond the power circuit fault can still initiate an alarm. A return loop should be brought back to the last powered device and the power supervisory relay to incorporate into the end-of-line device.

Figure A-7-2.2(g)

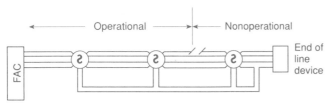

Testing of the circuit is similar to that described in A-7-2.2(c) with the addition of a power circuit.

Figure A-7-2.2(h)

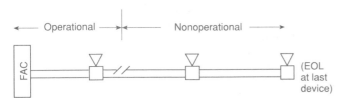

Testing of the indicating appliances connected to Style W and Style Y (Class B) is similar to that described in A-7-2.2(c).

Figure A-7-2.2(i)

Style 0.5. This signaling circuit operates as a series circuit in performance. This is identical to the historical series audible signaling circuits. Any type of break or ground in one of the conductors or the internal of the multiple interface device, and the total circuit is rendered operative.

To test and verify this type of circuit, either lift a conductor or place an earth ground on a conductor or a terminal point where the signaling circuit attaches to the multiplex interface device.

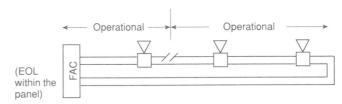

Testing of the indicating appliances connected to Style X and Style Z (Class B and Class A) is similar to that described in A-7-2.2(d).

Figure A-7-2.2(j)

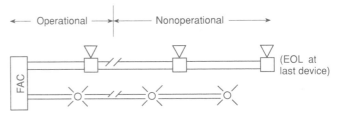

Testing of the indicating appliances connected to Style X and Style Z (Class B and Class A) is similar to that described in A-7-2.2(d).

Figure A-7-2.2(k)

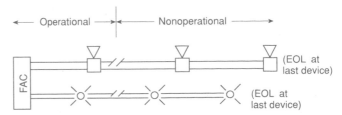

Testing of the indicating appliances connected to Style X and Style Z (Class B and Class A) is similar to that described in A-7-2.2(d).

Figure A-7-2.2(l)

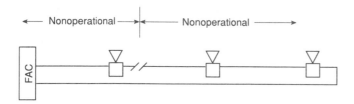

An open fault in the circuit wiring should cause a trouble condition.

Figure A-7-2.2(m)

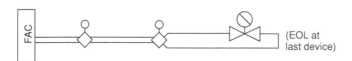

An open fault in the circuit wiring of operation of the valve switch (or any supervisory signal device) should cause a trouble condition.

Figure A-7-2.2(n)

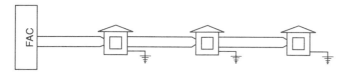

An open fault in the circuit wiring or operation of the valve switch should cause a trouble signal.

Figure A-7-2.2(o)

Disconnect a leg of municipal circuit at master box. Verify alarm sent to public communication center. Disconnect leg of auxiliary circuit. Verify trouble condition on control unit. Restore circuits. Activate control unit and send alarm signal to communication center. Verify control unit in trouble condition until master box reset.

Figure A-7-2.2(p)

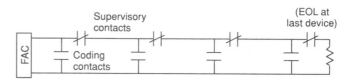

Figure A-7-2.2(q) Self-explanatory test.

Style 0.5(a) functions so that when a box is operated, the supervisory contacts open, making the succeeding devices nonoperative while the operating box sends a coded signal. Any alarms occurring in any successive devices will not be received at the receiving station during this period.

Figure A-7-2.2(r) Style 0.5(a) (Class B) series.

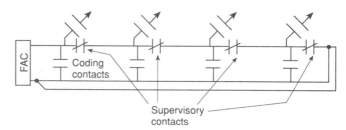

The contact closes when the device is operated and remains closed to shut out the remainder of the system until the code is complete.

Figure A-7-2.2(s) Style 0.5(b) (Class B) shunt.

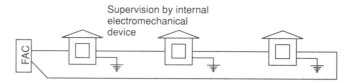

An open or ground fault on the circuit should cause a trouble condition at the control unit.

Figure A-7-2.2(t) Style 0.5(c) (Class B) positive supervised successive.

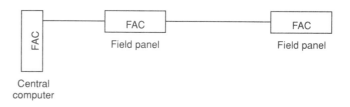

This is a series circuit identical to diagram for Style 0.5, except that the fire alarm system hardware has enhanced performance. A single earth ground can be placed on a conductor or multiplex interface device, and the circuit and hardware still have alarm operability.

If a conductor break or an internal fault occurs in the pathway of the circuit conductors, the entire circuit becomes inoperative.

To verify alarm receipt capability and the resulting trouble signal, place an earth ground on one of the conductors or at the point where the signaling circuit attaches to the multiplex interface device. Then place one of the transmitters or an initiating devices into alarm.

Figure A-7-2.2(u) Style 1.0 (Class B).

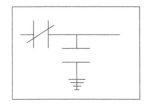

Figure A-7-2.2(v)   *Typical transmitter layout.*

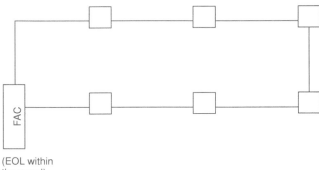

(EOL within
the panel)

This is the central station McCulloh redundant type circuit and has alarm receipt capability on either side of a single break.

(a) To test, lift one of the conductors and operate a transmitter or initiating device on each side of the break. This activity should be repeated for each conductor.

(b) Place an earth ground on a conductor and operate a single transmitter or initiating device to verify alarm receipt capability and trouble condition for each conductor.

(c) Repeat the instructions of (a) and (b) at the same time and verify alarm receipt capability and that a trouble condition results.

Figure A-7-2.2(w)   *Typical McCulloh loop.*

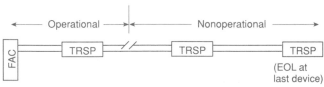

TRSP = Transponder

This is a parallel circuit whose multiplex interface devices transmit signal and operating power over the same conductors. The multiplex interface devices may be operable up to the point of a single break. Verify by lifting a conductor and causing an alarm condition on one of the units between the central alarm unit and the break. Either lift a conductor to verify the trouble condition or place an earth ground on the conductors. Test for all the valuations shown on the signaling table.

On ground fault testing verify alarm receipt capability by actuating a multiplex interface initiating device or a transmitter.

Figure A-7-2.2(x)   *Style 3.0 (Class B).*

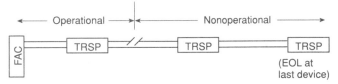

(EOL at
last device)

Repeat the instructions for Style 3.0 (Class B) and verify the trouble conditions by either lifting a conductor or placing a ground on the conductor.

Figure A-7-2.2(y)   *Style 3.5 (Class B).*

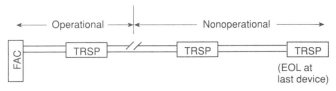

(EOL at
last device)

Repeat the instructions for Style 3.0 (Class B) and include a loss of carrier if the signal is being used.

Figure A-7-2.2(z)   *Style 4.0 (Class B).*

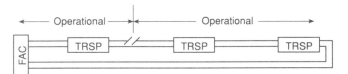

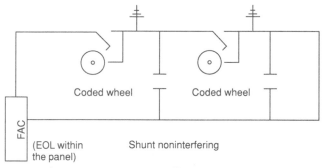

Coded wheel   Coded wheel

(EOL within   Shunt noninterfering
the panel)

Repeat the instructions for Style 3.5 (Class B). Verify alarm receipt capability while lifting a conductor by actuating a multiple interface device or transmitter on each side of the break.

Figure A-7-2.2(aa)   *Style 4.5 (Class B)*

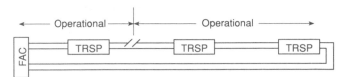

Verify the alarm receipt capability and trouble annunciation by lifting a conductor and actuating a multiplex interfacing device or a transmitter on each side of the break. For the earth ground verification, place an earth ground and certify alarm receipt capability and trouble annunciation by actuating a single multiplex interfacing device or a transmitter.

Figure A-7-2.2(bb)   *Style 5.0 (Class A).*

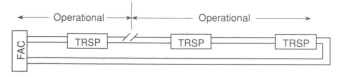

Repeat the instructions for Style 2.0 (Class A) [(a) through (c)]. Verify the remaining steps for trouble annunciation for the various combinations.

Figure A-7-2.2(cc)   *Style 6.0 (Class A).*

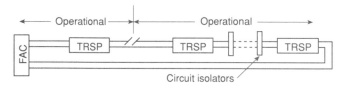

For the portions of the circuits electrically located between the monitoring points of circuit isolators, follow the instructions for a Style 7.0 (Class A) circuit. It should be clearly noted that the alarm receipt capability for remaining portions of the circuit protection isolators is not the capability of the circuit, but permissible with enhanced system capabilities.

Figure A-7-2.2(dd)   *Style 6.0 (with circuit isolators) (Class A).*

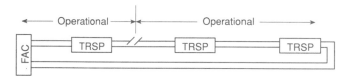

Repeat the instructions for testing of Style 6.0 (Class A) for alarm receipt capability and trouble annunciation.

NOTE 1:   A portion of the circuit between the alarm processor or central supervising station and the first circuit isolator does not have alarm receipt capability in the presence of a wire-to-wire short. The same is true for the portion of the circuit from the last isolator to the alarm processor or the central supervising station.

NOTE 2:   Some manufacturers of this type of equipment have isolators as part of the base assembly. Therefore in the field this component may not be readily observable without the assistance of the manufacturer's representative.

Figure A-7-2.2(ee)   *Style 7.0 (Class A).*

7-3   Inspection and Testing Frequency.

7-3.1   Visual Inspection.

7-3.1.1   Visual inspection shall be performed in accordance with the schedules in this chapter or more frequently where required by the authority having jurisdiction. The visual inspection shall be made to ensure that there are no changes that would affect equipment performance, such as building modifications, occupancy hazards, and environmental effects.

Where the authority having jurisdiction suspects that building conditions are changing more rapidly than normal and that these changes are likely to affect the performance of the fire alarm system, the authority having jurisdiction may require more frequent visual inspections.

Exception: Items in areas that are inaccessible for safety considerations due to continuous process operations, energized electrical equipment, etc., shall be inspected during each scheduled shutdown but not more than every 18 months.

7-3.1.2   Where automatic testing is performed at a frequency of not less than weekly by a remotely monitored fire alarm control unit specifically listed for this application, the visual inspection frequency shall be permitted to be extended to annually. (*See Table 7-3.1.*)

Some fire alarm system control units monitor analog information for the alarm initiating devices and, thus, are able to detect changes in ambient conditions at or internal to the initiating device. Extending the visual inspection frequency to annually when such control units are employed recognizes that these units provide more complete monitoring for integrity.

Table 7-3.1 gives largely self-explanatory frequencies for how often visual inspections are to be performed on various components, portions, and subsystems of the fire alarm system.

7-3.2*   **Testing.** Testing shall be performed in accordance with the schedules in this chapter or more frequently where required by the authority having jurisdiction. Where automatic testing is performed at least weekly by a remotely monitored fire alarm control unit specifically listed for the application, the manual testing frequency shall be permitted to be extended to annually. (*See Table 7-3.2.*)

Table 7-3.2 gives largely self-explanatory frequencies for how often tests are to be performed on various components, portions, and subsystems of the fire alarm system.

Exception: Devices in areas that are inaccessible for safety considerations, such as continuous process operations, shall be tested during scheduled shutdowns at intervals approved by the authority having jurisdiction.

Table 7-3.1   Visual Inspection Frequencies

	Init./Reaccpt.	Monthly	Quarterly	Semiann.	Ann.
1. **Alarm Indicating Appliances — Supervised**	X			X	
2. **Batteries**					
a. Lead-Acid	X	X			
b. Nickel-Cadmium	X			X	
c. Primary (Dry Cell)	X	X			
d. Sealed Lead-Acid	X			X	
3. **Control Equipment: Fire Alarm Systems Monitored for Alarm, Supervisory, Trouble Signals**					
a. Fuses	X				X
b. Interfaced Equipment	X				X
c. Lamps and LEDs	X				X
d. Primary (Main) Power Supply	X				X
4. **Control Equipment: Fire Alarm Systems Unmonitored for Alarm, Supervisory, Trouble Signals**					
a. Fuses	X				X
b. Interfaced Equipment	X				X
c. Lamps and LEDs	X				X
d. Primary (Main) Power Supply	X				X
5. **Control Panel Trouble Signals**	X			X	
6. **Emergency Voice/Alarm Communications Equipment**	X			X	
7. **Fiber Optic Cable Connections**	X				X
8. **Guard's Tour Equipment**	X			X	
9. **Initiating Devices**					
a. Air Sampling	X			X	
b. Duct Detectors	X			X	
c. Electromechanical Releasing Device	X			X	
d. Extinguishing System Switches	X			X	
e. Fire Alarm Boxes	X			X	
f. Heat Detectors	X			X	
g. Radiant Energy Fire Detectors	X			X	
h. Smoke Detectors	X			X	
i. Supervisory Signal Devices	X		X		
j. Waterflow Devices	X		X		
10. **Interface Equipment**	X			X	
11. **Remote Annunciators**	X			X	
12. **Special Procedures**	X			X	
13. **Transient Suppressors**	X			X	
14. **Transmission and Receiving Equipment — Off Premises**					
a. All Equipment	X			X	
b. DACT — Telephone Line	X			X	
c. DACR — Telephone Line	X	X			
d. DACR — Signal Receipt	X (DAILY)				

Table 7-3.2 Testing Frequencies

	Init./Reaccpt.	Monthly	Quarterly	Semiann.	Ann.	Table 7-2.2 Reference
1. Alarm Notification Appliances						14
a. Audible Devices	X				X	
b. Speakers	X				X	
c. Visible Devices	X				X	
2. Batteries — Central Station Facilities						
a. Lead-Acid Type						6b
1. Charger Test	X				X	
(Replace battery as needed.)						
2. Discharge Test (30 min.)	X	X				
3. Load Voltage Test	X	X				
4. Specific Gravity	X			X		
b. Nickel-Cadmium Type						6c
1. Charger Test	X		X			
(Replace battery as needed.)						
2. Discharge Test (30 min.)	X				X	
3. Load Voltage Test	X				X	
c. Sealed Lead-Acid Type	X	X				6d
1. Charger Test		X	X			
(Replace battery as needed.)						
2. Discharge Test (30 min.)	X	X				
3. Load Voltage Test	X	X				
3. Batteries — Fire Alarm Systems						
a. Lead-Acid Type						6b
1. Charger Test	X				X	
(Replace battery as needed.)						
2. Discharge Test (30 min.)	X			X		
3. Load Voltage Test	X			X		
4. Specific Gravity	X			X		
b. Nickel-Cadmium Type						6c
1. Charger Test	X				X	
(Replace battery as needed.)						
2. Discharge Test (30 min.)	X				X	
3. Load Voltage Test	X			X		
c. Primary Type (Dry Cell)						6a
1. Load Voltage Test	X	X				
d. Sealed Lead-Acid Type						6d
1. Charger Test	X				X	
(Replace battery every 4 years.)						
2. Discharge Test (30 min.)	X				X	
3. Load Voltage Test	X			X		
4. Batteries — Public Fire Alarm Reporting Systems X (DAILY)						

Voltage tests in accordance with Table 7-2.2, Public Reporting System Tests, paragraphs (a) - (f).

	Init./Reaccpt.	Monthly	Quarterly	Semiann.	Ann.	Table 7-2.2 Reference
a. Lead-Acid Type						6b
1. Charger Test	X				X	
(Replace battery as needed.)						
2. Discharge Test (2 hours)	X		X			
3. Load Voltage Test	X		X			
4. Specific Gravity	X			X		
b. Nickel-Cadmium Type						6c
1. Charger Test	X				X	
(Replace battery as needed.)						
2. Discharge Test (2 hours)	X				X	
3. Load Voltage Test	X		X			

Table 7-3.2   Testing Frequencies (cont.)

		Init./Reaccpt.	Monthly	Quarterly	Semiann.	Ann.	Table 7-2.2 Reference
	c. Sealed Lead-Acid Type						6d
	1. Charger Test (Replace battery as needed.)	X				X	
	2. Discharge Test (2 hours)	X				X	
	3. Load Voltage Test	X		X			
5.	**Conductors/Metallic**	X					11
6.	**Conductors/Nonmetallic**	X					12
7.	**Control Equipment: Fire Alarm Systems Monitored for Alarm, Supervisory, Trouble Signals**						1, 7 and 16
	a. Functions	X				X	
	b. Fuses	X				X	
	c. Interfaced Equipment	X				X	
	d. Lamps and LEDs	X				X	
	e. Primary (Main) Power Supply	X				X	
	f. Transponders	X				X	
8.	**Control Equipment: Fire Alarm Systems Unmonitored for Alarm, Supervisory, Trouble Signals**						1
	a. Functions	X		X			
	b. Fuses	X		X			
	c. Interfaced Equipment	X		X			
	d. Lamps and LEDs	X		X			
	e. Primary (Main) Power Supply	X		X			
	f. Transponders	X		X			
9.	**Control Unit Trouble Signals**	X				X	9
10.	**Emergency Voice/Alarm Communications Equipment**	X				X	17
11.	**Engine-Driven Generator**	X (WEEKLY)					
12.	**Fiber Optic Cable Power**	X				X	19
13.	**Guard's Tour Equipment**	X				X	
14.	**Initiating Devices**						13
	a. Duct Detectors	X				X	
	b. Electromechanical Releasing Device	X				X	
	c. Extinguishing System Switches	X				X	
	d. Fire-Gas and Other Detectors	X				X	
	e. Heat Detectors	X				X	
	f. Fire Alarm Boxes	X				X	
	g. Radiant Energy Fire Detectors	X				X	
	h. Smoke Detectors - Functional	X				X	
	i. Smoke Detectors - Sensitivity (See 7-3.2.1.)						
	j. Supervisory Signal Devices	X		X			
	k. Waterflow Devices	X		X			
15.	**Interface Equipment**	X				X	18
16.	**Off-Premises Transmission Equipment**	X		X			
17.	**Remote Annunciators**	X				X	10
18.	**Retransmission Equipment**	X (See 7-3.4.)					
19.	**Special Hazard Equipment**	X				X	15
20.	**Special Procedures**	X				X	20

Table 7-3.2   Testing Frequencies (cont.)

	Init./Reaccpt.	Monthly	Quarterly	Semiann.	Ann.	Table 7-2.2 Reference
21. System and Receiving Equipment — Off-Premises						16
a. Operational						
1. Functional — All	X				X	
2. Transmitters — WF & Supervisory	X		X			
3. Transmitters — All Others	X				X	
4. Receivers	X	X				
b. Standby Loading — All Receivers	X	X				
c. Standby Power						
1. Receivers — All	X	X				
2. Transmitters — All	X				X	
d. Telephone Line — All Receivers	X	X				
e. Telephone Line — All Transmitters	X				X	

NOTE: For testing addressable and analog described devices, which are normally affixed to either a single molded assembly or twist lock type affixed to a base, TESTING SHALL BE DONE UTILIZING THE SIGNALING STYLE CIRCUITS (Styles 0.5 through 7). The addressable term was determined by the Technical Committee in Formal Interpretation 79-8 on NFPA 72D and Formal Interpretation 87-1 on NFPA 72A. Analog type detectors shall be tested with the same criteria.

A-7-3.2   Batteries. To maximize battery life, nickel-cadmium batteries should be charged as follows:

Float Voltage	1.42 Volts/Cell + .01 volts
High Rate Voltage	1.58 Volts/Cell + .07 − 0.00 volts

NOTE: High and low gravity voltages are (+) 0.07 volts and (−) 0.03 volts respectively.

To maximize battery life, the battery voltage for lead-acid cells should be maintained within the limits shown in the following table:

Float Voltage	High Gravity Battery (Lead Calcium)	Low Gravity Battery (Lead Antimony)
Maximum	2.25 Volts/Cell	2.17 Volts/Cell
Minimum	2.20 Volts/Cell	2.13 Volts/Cell
High Rate Voltage		2.33 Volts/Cell

The following procedure is recommended for checking state of charge for nickel-cadmium batteries:

(a) Switch the battery charger from float to high-rate mode.

(b) The current, as indicated on the charger ammeter, will immediately rise to the maximum output of the charger, and the battery voltage, as shown on the charger voltmeter, will start to rise at the same time.

(c) The actual value of the voltage rise is unimportant since it depends on many variables; the length of time it takes for the voltage to rise is the important factor.

(d) If, for example, the voltage rises rapidly in a few minutes, then holds steady at the new value, the battery was fully charged.

At the same time, the current will drop to slightly above its original value.

(e) In contrast, if the voltage rises slowly and the output current remains high, the high-rate charge should be continued until the voltage remains constant. Such a condition is an indication that the battery was not fully charged, and the float voltage should be increased slightly.

7-3.2.1* Detector sensitivity shall be checked within 1 year after installation and every alternate year thereafter. After the second required calibration test, if sensitivity tests indicate that the detector has remained within its listed and marked sensitivity range, the length of time between calibration tests shall be permitted to be extended not to exceed 5 years. If the frequency is extended, records of detector-caused unwanted alarms and subsequent trends of these alarms shall be maintained. In zone or in areas where unwanted alarms show any increase over the previous year, calibration tests shall be performed.

To ensure that each smoke detector is within its listed and marked sensitivity range, it shall be tested using either:

(a) A calibrated test method, or

(b) The manufacturer's calibrated sensitivity test instrument, or

(c) Listed control equipment arranged for the purpose, or

(d) A smoke detector/control unit arrangement whereby the detector causes a signal at the control unit where its sensitivity is outside its acceptable sensitivity range, or

(e) Other calibrated sensitivity test method acceptable to the authority having jurisdiction.

Detectors found to have a sensitivity outside the listed and marked sensitivity range shall be cleaned and recalibrated or replaced.

Exception: Detectors listed as field adjustable may be either adjusted within the listed and marked sensitivity range, cleaned, and recalibrated, or replaced.

The detector sensitivity shall not be tested or measured using any device that administers an unmeasured concentration of smoke or other aerosol into the detector.

> **Formal Interpretation 87-1**
> Reference: 7-2.2
>
> *Background:* There is a large population of installed smoke detectors which were manufactured in compliance with testing laboratory standards which are no longer used. It isnot possible to test the listed sensitivity range of some of these detectors.
>
> *Question 1:* Is it the code's intent that these detectors need not be tested for sensitivity?
> *Answer:* No.
>
> *Question 2:* It is the code's intent that these detectors be replaced with new detectors whose sensitivity can be tested?
> *Answer:* No.
>
> *Issue Edition:* NFPA 72E-1987
> *Reference:* 8-2.4.2
> *Issue Date:* July 11, 1989 ■

A-7-3.2.1 It is suggested that the annual test can be conducted in segments so that all devices are tested annually.

The requirements of this subsection give the most possible options in testing the sensitivity of smoke detectors. It is important to note that each of the five options provides a measured means of ensuring sensitivity. Further, after two successful tests where sensitivity has remained stable, extending sensitivity testing to 5-year intervals recognizes the apparent stability of the environment in which the detector is installed and the apparent stability of the detector itself.

When frequency of sensitivity testing is extended, the required records of detector operation will help warn of changes in the environment or changes in the stability of the detector. Any such changes may warrant more frequent testing.

> **Formal Interpretation 84-2**
> Reference: 7-3.2.1
>
> *Question 1:* Is the intent of this test to determine that the detector sensitivity is within the UL requirements for this detector?
> *Answer:* Yes.
>
> *Question 2:* Would the manufacturer have to supply the fire officials with the exact sensitivity of the detector?
> *Answer:* No. However, the authority having jurisdiction has the option to require this information.
>
> *Question 3:* Would it be acceptable to supply the fire officials with the information that the detector sensitivity falls within its UL listed range?
> *Answer:* Yes.
>
> *Issue Edition:* NFPA 72E-1984
> *Reference:* 8-2.2.2
> *Issue Date:* March 1985 ■

The acceptable test methods include:

(a) The use of a calibrated test method, such as the operation of the test button where a calibrated wire is raised into the reflective path of a photoelectric detector. The wire provides a test of the outside sensitivity limit of the detector.

(b) The use of a manufacturer's sensitivity instrument to check the sensitivity window of the detector.

(c) A system control unit designed to connect internal sensitivity test instruments to each detector upon manual command from the control unit. This test method in effect provides remote sensitivity testing.

(d) A system control unit/detector combination where the control unit receives analog information from the detector and stores it in memory. Thus, the control unit contains a history of both the stability of the environment in which the detector is installed and the stability of the detector itself. By comparing the information stored in memory, the control unit can detect changes in the environment or in the detector's stability that might affect system operation.

(e) Other calibrated test methods such as one that delivers a precisely measured amount of a test aerosol product acceptable to the detector manufacturer. By comparing the minimum and maximum quantities of aerosol product delivered to the detector, the window of sensitivity can be determined.

7-3.2.2   Test frequency of interfaced equipment shall be the same as specified by the applicable NFPA standards for the equipment being supervised.

For example, the test frequency for a carbon dioxide special hazard fire extinguishing system shall be the frequency specified in NFPA 12, *Standard on Carbon Dioxide Extinguishing Systems.*

7-3.3   Single-station smoke detectors shall be inspected, tested, and maintained as specified by Chapter 2.

7-3.4   Test of all circuits extending from the central station shall be made at intervals of not more than 24 hours.

Operators at the central station initiate these tests to verify that all circuits are operational.

7-4   Maintenance.

7-4.1   Fire alarm system equipment shall be periodically maintained in accordance with manufacturer's instructions. The frequency of maintenance will depend on the type of equipment and the local ambient conditions.

7-4.2   Any accumulation of dust and dirt may adversely effect device and appliance performance. The frequency of cleaning will depend on the type of equipment and the local ambient conditions.

7-4.3   All apparatus requiring rewinding or resetting to maintain normal operation shall be restored to normal as promptly as possible after each test and alarm and kept in normal condition for operation. All test signals received shall be recorded to indicate date, time, and type.

Subsections 7-4.2 and 7-4.3 require that periodic maintenance be performed. The emphasis is on cleaning, which should be done in strict accordance with the manufactur-

er's instructions as often as ambient conditions necessitate cleaning.

Subsections 7-4.3 through 7-4.4.3.2 give specific directions on how to perform tests of various types of off-premises transmission technologies, as well as operational procedures following tests, such as the requirements of 7-4.3. Further, 7-4.4 gives the testing frequency of retransmission means between the supervising station and the public fire service communication center.

7-4.4   The retransmission means as defined in Section 4-3 shall be tested at intervals of not more than 12 hours. The retransmission signal and the time and date of the retransmission shall be recorded in the central station.

Exception: Where the retransmission means is the public switched telephone network, it need only be tested weekly to confirm its operation to each public fire service communications center.

7-4.4.1   Digital Alarm Communicator Transmitter (DACT).

(a) Verify the DACT is capable of seizing the telephone line (going off-hook) at the protected premises, disconnecting an outgoing or incoming telephone call, and preventing its use for outgoing calls until signal transmission is completed.

Figure 7.3   Digital alarm communicator with programmer.

(b) Verify the DACT has the means to satisfactorily obtain an available dial tone, dial the number(s) of the DACR, obtain verification that the DACR is ready to receive signals, transmit the signal, and receive acknowledgment that the DACR has accepted the signal. In no event shall the time from going off-hook to on-hook exceed 90 seconds per attempt.

(c) Verify the DACT has a suitable means to reset and retry if the first attempt to complete a signal transmission sequence is unsuccessful. A failure to complete connection shall not prevent subsequent attempts to transmit an alarm if such alarm is generated from any other initiating device circuit. Additional attempts shall be made until the signal transmission sequence has been completed to a minimum of five and a maximum of ten attempts.

(d) If the maximum number of attempts to complete the sequence is reached, an indication of the failure shall be made at the premises.

(e) Verify the DACT is connected to two separate lines (numbers) at the protected premise by disconnecting the primary phone line of the DACT. The DACT trouble signal shall be transmitted within the time specified in accordance with 7-4.4.1(a). Operate an initiating device to test the secondary transmission of the DACT. The DACT shall be capable of selecting the operable line (number) in the event of failure in either (line number).

(f) Failure of either of the telephone lines (numbers) at the protected premises shall be annunciated at the protected premises, and a trouble signal shall be transmitted to the central station over the other line (number). Transmission shall be initiated within 4 minutes of the detection of the fault.

7-4.4.2 Digital Alarm Communicator Receiver (DACR).

(a) Verify that at least two separate incoming telephone lines are in a hunt group and are individually accessible. These lines shall be used for no other purpose than receiving signals from DACTs. These lines (numbers) shall be unlisted.

(b) The failure of any telephone line (number) connected to the DACR due to loss of line voltage shall be annunciated visually and audibly in the central station.

(c) The loading capacity for hunt group shall be in accordance to Table 4-2.3.2.2.2.3 or be capable of demonstrating a 90 percent probability of immediately answering the incoming call.

(d) Verify a signal is received on each individual incoming DARC line at least once every 24 hours.

(e) The verification of the 24-hour DARC line test should be done early enough in the day to allow repairs to be made by the telephone company.

7-4.4.3 Digital Alarm Radio System (DARS).

(a) When DARS is used, verify that when any DACT signal transmission is unsuccessful, the information is transmitted by means of the DART. The DACT shall continue its normal transmission as required.

(b) The failure of the telephone line at the protected premises shall result in a trouble signal being transmitted to the central station by means of the DART within 4 minutes of the detection of the fault.

(c) Each DART shall automatically initiate and complete a test transmission sequence to its associated DARR at least once every 24 hours. A successful DART signal transmission sequence of any other type shall be considered sufficient to fulfill the requirement to test integrity of the reporting system, if signal processing is automated so that 24-hour delinquencies must be acknowledged by central station personnel.

7-4.4.3.1 Digital Alarm Radio Receiver (DARR).

Verify supervision in the central station of the following conditions:

(a) Failure of ac power supplying the radio equipment
(b) Receiver malfunction
(c) Antenna and interconnecting cable malfunction
(d) Indication of automatic switchover of the DARR
(e) Data transmission line between the DARR and the central station.

7-4.4.4 McCulloh Systems.

Verify that when end-to-end metallic continuity is present, proper signals shall be received From other points under one of the following signaling line fault conditions at one point in the line:

(a) Open
(b) Ground
(c) Wire-to-wire short
(d) Open group.

7-5 Records.

7-5.1 Record of Inspection. A permanent record of all inspections, testing, and maintenance shall be provided, which includes

the following information of periodic tests and all the applicable information requested in Figure 7-5.1.

(a) Date

(b) Test frequency

(c) Name of property

(d) Address

(e) Name of person performing inspection, maintenance, and/or tests, affiliation, business address, and telephone number

(f) Approving agency's(ies') name, address, and representative

(g) Designation of the detector(s) tested ("Tests performed in accordance with Section _____.")

(h) Functional test of detectors

(i) Check of all smoke detectors

(j) Loop resistance for all fixed temperature line type heat detectors

(k) Other tests as required by equipment manufacturers

(l) Other tests as required by the authority having jurisdiction

(m) Signatures of tester and approved authority representative.

7-5.2 Permanent Records. After successful completion of acceptance tests satisfactory to theauthority having jurisdiction, a set of reproducible as-built installation drawings, operation and maintenance manuals, and a written sequence of operation shall be provided to the building owner or his designated represen-

tative. In addition, inspection, testing, and maintenance reports shall be provided for the owner or a designated representative. It shall be the responsibility of the owner to maintain these records for the life of system and to keep them available for examination by any authority having jurisdiction. Paper or electronic media shall be acceptable.

A historic record of the system installation that includes the information required by 7-5.2 gives troubleshooters valuable assistance in promptly diagnosing and repairing system faults.

7-5.3 Where off-premise monitoring is provided, records of signals, tests, and operations recorded at the monitoring center shall be maintained for not less than 12 months. Upon request, a hardcopy record shall be available for examination by the authority having jurisdiction. Paper or electronic media shall be acceptable.

7-5.4 Where the operation of a device, circuit, control panel function, or special hazard system interface is simulated, it shall be noted on the certificate that the operation was simulated and who it was simulated by.

For future reference in determining overall system reliability, it is important to know when interfaced system operation is fully tested by actual operation of the interfaced system or when its operation is merely simulated.

INSPECTION AND TESTING FORM

DATE: _____

TIME: _____

SERVICE ORGANIZATION

NAME: _____

ADDRESS: _____

REPRESENTATIVE: _____

LICENSE NO.: _____

TELEPHONE: _____

PROPERTY NAME (USER)

NAME: _____

ADDRESS: _____

OWNER CONTRACT: _____

TELEPHONE: _____

MONITORING ENTITY

CONTACT: _____

TELEPHONE: _____

MONITORING ACCOUNT REF. NO.: _____

APPROVING AGENCY

CONTACT: _____

TELEPHONE: _____

TYPE TRANSMISSION

[ ] - McCulloh
[ ] - Multiplex
[ ] - Digital
[ ] - Reverse Priority
[ ] - RF
[ ] - Other (Specify)

SERVICE

[ ] - Weekly
[ ] - Monthly
[ ] - Quarterly
[ ] - Semi-Annually
[ ] - Annually
[ ] - Other (Specify)

_____

_____

_____

PANEL MANUFACTURE: _____   MODEL NO.: _____

CIRCUIT STYLES: _____

NO. OF CIRCUITS: _____

SOFTWARE REV.: _____

LAST DATE SYSTEM HAD ANY SERVICE PERFORMED: _____

LAST DATE THAT ANY SOFTWARE OR CONFIGURATION WAS REVISED: _____

Figure 7-5.1   Inspection and Testing Form.

ALARM INITIATING DEVICES AND CIRCUIT INFORMATION

QTY OF	CIRCUIT STYLE	
_____	_____	MANUAL STATIONS
_____	_____	ION DETECTORS
_____	_____	PHOTO DETECTORS
_____	_____	DUCT DETECTORS
_____	_____	HEAT DETECTORS
_____	_____	WATERFLOW SWITCHES
_____	_____	SUPERVISORY SWITCHES
_____	_____	OTHER: (SPECIFY) _____
		_____
		_____
		_____

ALARM INDICATING APPLIANCES AND CIRCUIT INFORMATION

QTY OF	CIRCUIT STYLE	
_____	_____	BELLS
_____	_____	HORNS
_____	_____	CHIMES
_____	_____	STROBES
_____	_____	SPEAKERS
_____	_____	OTHER: (SPECIFY) _____
		_____
		_____
_____	_____	_____

NO. OF ALARM INDICATING CIRCUITS: _____

ARE CIRCUITS SUPERVISED?   [ ] YES   [ ] NO

Figure 7-5.1 Inspection and Testing Form. (cont.)

SUPERVISORY SIGNAL INITIATING DEVICES AND CIRCUIT INFORMATION

QTY OF　　　　　　　　　　**CIRCUIT STYLE**

QTY OF	CIRCUIT STYLE	
_____	_____	BUILDING TEMP.
_____	_____	SITE WATER TEMP.
_____	_____	SITE WATER LEVEL
_____	_____	FIRE PUMP POWER
_____	_____	FIRE PUMP RUNNING
_____	_____	FIRE PUMP AUTO POSITION
_____	_____	FIRE PUMP OR PUMP CONTROLLER TROUBLE
_____	_____	FIRE PUMP RUNNING
_____	_____	GENERATOR IN AUTO POSITION
_____	_____	GENERATOR OR CONTROLLER TROUBLE
_____	_____	SWITCH TRANSFER
_____	_____	GENERATOR ENGINE RUNNING
_____	_____	OTHER: _____
		_____

SIGNALING LINE CIRCUITS

Quantity and style (See NFPA 72, Table 3-6.1) of signaling line circuits connected to system:

Quantity _____　　　Style(s) _____

SYSTEM POWER SUPPLIES

a. Primary (Main): Nominal Voltage _____ , Amps _____

Overcurrent Protection: Type _____ , Amps _____

Location (Panel Number): _____

Disconnecting Means Location: _____

b. Secondary (Standby):

_____ Storage Battery: Amp-Hr. Rating _____

Calculated capacity to operate system, in hours: _____ 24 _____ 60 _____

_____ Engine-driven generator dedicated to fire alarm system:

Location of fuel storage: _____

Figure 7-5.1 Inspection and Testing Form. (cont.)

TYPE BATTERY

[ ] Dry Cell

[ ] Nickel Cadmium

[ ] Sealed Lead-Acid

[ ] Lead-Acid

[ ] Other (Specify) _____

   c. Emergency or standby system used as a backup to primary power supply, instead of using a secondary power supply:

      _____ Emergency system described in NFPA 70, Article 700

      _____ Legally required standby described in NFPA 70, Article 701

      _____ Optional standby system described in NFPA 70, Article 702, which also meets the performance requirements of Article 700 or 701.

PRIOR TO ANY TESTING

NOTIFICATIONS ARE MADE:	YES	NO	WHO	TIME
MONITORING ENTITY	[ ]	[ ]	_____	_____
BUILDING OCCUPANTS	[ ]	[ ]	_____	_____
BUILDING MANAGEMENT	[ ]	[ ]	_____	_____
OTHER (SPECIFY)	[ ]	[ ]	_____	_____
AHJ (NOTIFIED) OF ANY IMPAIRMENTS	[ ]	[ ]		

SYSTEM TESTS AND INSPECTIONS

TYPE	VISUAL	FUNCTIONAL	COMMENTS
CONTROL PANEL	[ ]	[ ]	_____
INTERFACE EQ.	[ ]	[ ]	_____
LAMPS/LEDS	[ ]	[ ]	_____
FUSES	[ ]	[ ]	_____
PRIMARY POWER SUPPLY	[ ]	[ ]	_____
TROUBLE SIGNALS	[ ]	[ ]	_____
DISCONNECT SWITCHES	[ ]	[ ]	_____
GROUND FAULT MONITORING	[ ]	[ ]	_____

Figure 7-5.1 *Inspection and Testing Form. (cont.)*

SECONDARY POWER

TYPE	VISUAL	FUNCTIONAL	COMMENTS
BATTERY CONDITION	[ ]		_____
LOAD VOLTAGE		[ ]	_____
DISCHARGE TEST		[ ]	_____
CHARGER TEST		[ ]	_____
SPECIFIC GRAVITY		[ ]	_____
TRANSIENT SUPPRESSORS	[ ]		_____
REMOTE ANNUNCIATORS	[ ]	[ ]	_____
NOTIFICATION APPLIANCES			
AUDIBLE	[ ]	[ ]	_____
VISUAL	[ ]	[ ]	_____
SPEAKERS	[ ]	[ ]	_____
VOICE CLARITY	[ ]		_____

INITIATING AND SUPERVISORY DEVICE TESTS AND INSPECTIONS

LOC. & S/N	DEVICE TYPE	VISUAL CHECK	FUNCTIONAL TEST	FACTORY SETTING	MEAS. SETTING	PASS	FAIL
_____	_____	[ ]	[ ]	_____	_____	[ ]	[ ]
_____	_____	[ ]	[ ]	_____	_____	[ ]	[ ]
_____	_____	[ ]	[ ]	_____	_____	[ ]	[ ]
_____	_____	[ ]	[ ]	_____	_____	[ ]	[ ]
_____	_____	[ ]	[ ]	_____	_____	[ ]	[ ]
_____	_____	[ ]	[ ]	_____	_____	[ ]	[ ]

COMMENTS: _____

_____

_____

_____

_____

_____

_____

Figure 7-5.1 *Inspection and Testing Form. (cont.)*

	VISUAL	FUNCTIONAL	COMMENTS
EMERGENCY COMMUNICATIONS EQUIPMENT			
PHONE SET	[ ]	[ ]	_____
PHONE JACKS	[ ]	[ ]	_____
OFF-HOOK INDICATOR	[ ]	[ ]	_____
AMPLIFIER(S)	[ ]	[ ]	_____
TONE GENERATOR(S)	[ ]	[ ]	_____
CALL IN SIGNAL	[ ]	[ ]	_____
SYSTEM PERFORMANCE	[ ]	[ ]	_____

	VISUAL	DEVICE OPERATION	SIMULATED OPERATION
INTERFACE EQUIPMENT			
(SPECIFY) _____	[ ]	[ ]	[ ]
(SPECIFY) _____	[ ]	[ ]	[ ]
(SPECIFY) _____	[ ]	[ ]	[ ]
SPECIAL HAZARD SYSTEMS			
(SPECIFY) _____	[ ]	[ ]	[ ]
(SPECIFY) _____	[ ]	[ ]	[ ]
(SPECIFY) _____	[ ]	[ ]	[ ]

SPECIAL PROCEDURES: _____

_____

_____

_____

_____

_____

_____

COMMENTS: _____

_____

_____

_____

_____

_____

_____

Figure 7-5.1 Inspection and Testing Form. (cont.)

ON/OFF PREMISES MONITORING:

	YES	NO	TIME	COMMENTS
ALARM SIGNAL	[ ]	[ ]	_____	_____
ALARM RESTORAL	[ ]	[ ]	_____	_____
TROUBLE SIGNAL	[ ]	[ ]	_____	_____
SUPERVISORY SIGNAL	[ ]	[ ]	_____	_____
SUPERVISORY RESTORAL	[ ]	[ ]	_____	_____

NOTIFICATIONS THAT TESTING IS COMPLETE:

	YES	NO	WHO	TIME
BUILDING MANAGEMENT	[ ]	[ ]	_____	_____
MONITORING AGENCY	[ ]	[ ]	_____	_____
BUILDING OCCUPANTS	[ ]	[ ]	_____	_____
OTHER (SPECIFY)	[ ]	[ ]	_____	_____

THE FOLLOWING DID NOT OPERATE CORRECTLY: _____

_____

_____

_____

SYSTEM RESTORED TO NORMAL OPERATION: DATE _____ TIME _____

THIS TESTING WAS PERFORMED IN ACCORDANCE WITH APPLICABLE NFPA STANDARDS.

NAME OF INSPECTOR: _____

DATE: _____   TIME: _____

SIGNATURE: _____

NAME OF OWNER OR REPRESENTATIVE: _____

DATE: _____   TIME: _____

SIGNATURE: _____

Figure 7-5.1 *Inspection and Testing Form. (cont.)*

8

Referenced
Publications

The following documents or portions thereof are referenced within this code and should be considered part of the requirements of this document. The edition indicated for each reference is the current edition as of the date of the NFPA issuance of this document.

8-1.1 **NFPA Publications.** National Fire Protection Association, 1 Batterymarch Park, P.O. Box 9101, Quincy, MA 02269-9101.

NFPA 10, *Standard for Portable Fire Extinguishers,* 1990 edition.

NFPA 13, *Standard for the Installation of Sprinkler Systems,* 1991 edition.

NFPA 13D, *Standard for the Installation of Sprinkler Systems in One- and Two-Family Dwellings and Mobile Homes,* 1991 edition.

NFPA 13R, *Standard for the Installation of Sprinkler Systems in Residential Occupancies Up to and Including Four Stories in Height,* 1991 edition.

NFPA 20, *Standard for the Installation of Centrifugal Fire Pumps,* 1993 edition.

NFPA 25, *Standard for the Inspection, Testing, and Maintenance of Water-Based Fire Protection Systems,* 1992 edition.

NFPA 37, *Standard for the Installation and Use of Stationary Combustion Engines and Gas Turbines,* 1990 edition.

NFPA 54, *National Fuel Gas Code,* 1992 edition.

NFPA 58, *Standard for the Storage and Handling of Liquefied Petroleum Gases,* 1992 edition.

NFPA 70, *National Electrical Code,* 1993 edition.

NFPA 90A, *Standard for the Installation of Air Conditioning and Ventilating Systems,* 1993 edition.

NFPA 110, *Standard for Emergency and Standby Power Systems,* 1993 edition.

NFPA 601, *Standard on Guard Service in Fire Loss Prevention,* 1992 edition.

NFPA 780, *Lightning Protection Code,* 1992 edition.

NFPA 1221, *Standard for the Installation, Maintenance, and Use of Public Fire Service Communication Systems,* 1991 edition.

8-1.2 **ANSI Publications.** American National Standards Institute, 1430 Broadway, New York, NY 10036.

ANSI A-58.1-1982, *Building Code Requirements for Minimum Design Loads in Buildings and Other Structures.*

ANSI S-1.4a-1985, *Specifications for Sound Level Meters.*

ANSI S3.41-1990, *Audible Emergency Evacuation Signals.*

ANSI S12.31-1980, *Precision Methods for the Determination of Sound Power Levels of Broad Band Noise Sources in Reverberation Rooms.*

ANSI S12.32-1980, *Precision Methods for the Determination of Sound Power Levels of Discrete Frequency and Narrow Band Noise Sources in Reverberation Rooms.*

ANSI/ASME A17.1, *Safety Code for Elevators and Escalators.*

ANSI/IEEE C2, *The National Electrical Safety Code.*

ANSI/UL 217, *Single and Multiple Station Smoke Detectors, Third Edition.*

ANSI/UL 268, *Smoke Detectors for Fire Protective Signaling Systems, Second Edition.*

ANSI/UL 827-1988, *Central Stations for Watchman, Fire Alarm and Supervisory Service.*

8-1.3 **EIA Publication.** Electronic Industries Association, 2001 I Street NW, Washington, DC 20006.

EIA Tr 41.3, *Telephones.*

8-1.4 **IES Publication.** Illuminating Engineering Society of North America, 345 East 47th Street, New York, NY 10017.

IES RP-16-1987, *Nomenclature and Definitions for Illuminating Engineering.*

A

Explanatory Material

This Appendix is not a part of the requirements of this NFPA document, but is included for information purposes only.

The material contained in Appendix A of this standard is included within the text of this handbook and, therefore, is not repeated here.

B

Engineering Guide for Automatic Fire Detector Spacing

Contents, Appendix B

B-1 Introduction
  B-1.2 Purpose
    B-1.2.1 Design
    B-1.2.2 Analysis
  B-1.3 Relationship to Listed Spacings
  B-1.4 Required Data
    B-1.4.1 Analysis
    B-1.4.2 Design
B-2 Fire Development and Ceiling Height Considerations
  B-2.1 General
    B-2.1.4 High Ceilings
  B-2.2 Fire Development
    B-2.2.2 Fire Size
      B-2.2.2.2 Example
    B-2.2.3 Fire Growth
    B-2.2.4 Selection of Fire Size
  B-2.3 Ceiling Height
B-3 Heat Detector Spacing
  B-3.1 General
  B-3.2 Fixed-Temperature Heat Detector Spacing
    Table B-3.2.2 Time Constants for any Listed Heat Detector
    Table B-3.2.4 Design Tables Index
    B-3.2.6 Example
  B-3.3 Rate-of-Rise Heat Detector Spacing
    B-3.3.5 Example
  B-3.4 Design Curves
    B-3.4.1.1 Fixed-Temperature Heat Detectors
    B-3.4.1.2 Rate-of-Rise Heat Detectors
    B-3.4.3.1 Example 1
    B-3.4.3.2 Example 2
B-4 Analysis of Existing Heat Detection Systems
  Table B-4.1 Analysis Tables Index
    B-4.1.1 Example
B-5 Smoke Detector Spacing for Flaming Fires
  B-5.6.3 Example 1
  B-5.6.4 Example 2
B-6 Theoretical Considerations
  B-6.1 Introduction
  B-6.2 Temperature and Velocity Correlations
  B-6.3 Heat Detector Model
  B-6.4 Ambient Temperature Considerations
  B-6.5 Heat and Smoke Analogy — Smoke Detector Model

This Appendix is not a part of the requirements of this NFPA document, but is included for information purposes only.

Appendix B was formerly Appendix C in the 1990 edition of NFPA 72E. This engineering guide is a result of intensive work on the part of the original Appendix C subcommittee members to put the research sponsored by the Fire Detection Institute into a useful design guide. The work began in 1974 with Factory Mutual Research Corporation conducting a research and testing program under the direction of Gunnar Heskestad. In 1977 the National Institute of Standards and Technology (formerly the National Bureau of Standards) performed an analysis of this research and published its findings. In 1979 the Fire Detection Institute also published a summary analysis of the Factory Mutual research. The original Appendix C subcommittee was formed sometime in early 1981 and, at the direction of NFPA 72E chairman Patrick Phillips, was instructed to review the material published to date and develop information that could be used in NFPA 72E. This subcommittee, chaired by Wayne D. Moore, included Jack Abbot, Craig Beyler, Harold Cutler, Robert P. Schifiliti, Walter Schuchard, Larry Stanley, and Ralph Transue. In 1984, seven years after the initial research information was published, Appendix C of NFPA 72E appeared for the first time. In that same edition of the standard, a new table appeared in the heat detector section — Table 3-5.1.2 (now Table 5-2.7.1.2) — that required the adjustment of detector spacing based on ceiling height. Since that time, under the direction of Robert Schifiliti, the task group has revised the guide to include faster fire growth rates, fire data on common furnishings, and tables for design and analysis of detector response with examples.

B-1 Introduction.

B-1.1 **Scope.** This appendix provides information intended to supplement Chapter 5 and includes a procedure for determining heat detector spacing based on the size and rate of growth of fire to be detected, various ceiling heights, and ambient temperature. The effects of ceiling height and the size and rate of growth of a flaming fire on smoke detector spacing are also treated. A procedure for analyzing the response of existing heat detection systems is also presented.

B-1.1.1 This appendix utilizes the results of fire research funded by the Fire Detection Institute to provide test data and analysis to the NFPA Technical Committee on Detection Devices. (*See reference 10 in Appendix C.*)

B-1.1.2 This appendix is based on full-scale fire tests in which all fires were geometrically growing flaming fires.

B-1.1.3 The tables and graphs in this appendix were produced using test data and data correlations for wood fuels having a total heat of combustion of about 20,900 kJ/kg and a convective heat release rate fraction equal to 75 percent of the total heat release rate. Users should refer to references 12 and 13 in Appendix C for fuels or burning conditions substantially different from these conditions.

B-1.1.4 The guidance applicable to smoke detectors is limited to a theoretical analysis based on the flaming fire test data and is not intended to address the detection of smoldering fires.

B-1.2 **Purpose.** The purpose of this appendix is to assist fire alarm system engineers concerned with spacing and response of heat or smoke detectors.

B-1.2.1 **Design.** This appendix provides a method for modifying the listed spacing of both rate-of-rise and fixed-temperature heat detectors required to achieve detector response to a geometrically growing flaming fire at a specific fire size, taking into account the height of the ceiling on which the detectors are mounted and the fire safety objectives for the space. This procedure also permits modification of listed spacing of fixed temperature heat detectors to account for variation of ambient temperature (T_o) from standard test conditions.

B-1.2.2 **Analysis.** This appendix may be used to estimate the fire size that can be detected by an existing array of listed heat detectors installed at a given spacing for a given ceiling height in known ambient conditions.

B-1.2.3 This appendix is also intended to explain the effect of rate of fire growth and fire size of a flaming fire, as well as the effect of ceiling height on the spacing and response of smoke detectors.

B-1.2.4 This methodology utilizes theories of fire development, fire plume dynamics, and detector performance, which are the major factors influencing detector response. However, it does not consider several lesser phenomena that, in general, are unlikely to have significant influence. A discussion of ceiling drag, heat loss to the ceiling, radiation to the detector from a fire, re-radiation of heat from a detector to its surroundings, and the heat of fusion of eutectic materials in fusible elements of heat detectors and their possible limitations on the design method are provided in references 4, 11, and 14 in Appendix C.

B-1.3 **Relationship to Listed Spacings.** Listed spacings for heat detectors are based on relatively large fires (approximately 1200 Btu/sec), burning at a constant rate. [The listed spacing is based on the distance from a fire at which an ordinary degree heat detector actuates prior to operation of a 160°F (71°C) sprinkler installed at a 10-ft (3-m) spacing.] [*See Figure A-5-2.7.1(a).*]

Design spacing for this type of fire can be determined using the material in Chapter 5.

When smaller or larger fires and varying growth rates must be considered, the designer may use the material presented by this appendix.

B-1.4 **Required Data.** The following data are required to use the methods in this appendix for either analysis or design.

B-1.4.1 **Analysis.**

T_O	Ambient temperature
H	Ceiling height or clearance above fuel
T_S	Detector operating temperature (heat detectors only)
ΔT_S/min	Rate of temperature change set point for rate-of-rise heat detectors
RTI	Response time index for the detector (heat detectors only) or its listed spacing
α or t_g	Fuel fire intensity coefficient or t_g, the fire growth time
S	The actual installed spacing of the existing detectors

B-1.4.2 **Design.**

T_O	Ambient temperature
H	Ceiling height or clearance above fuel
T_S	Detector operating temperature (heat detectors only)
ΔT_S/min	Rate of temperature change set point for rate-of-rise heat detectors
RTI	Response time index for the detector (heat detectors only) or its listed spacing
α or t_g	Fuel fire intensity coefficient or t_g, the fire growth time
Q_d or t_d	The threshold fire size at which response must occur or the time to detector response

B-1.4.3 The terms and data listed above are defined in more detail in the following sections.

B-2 Fire Development and Ceiling Height Considerations.

B-2.1 **General.** The purpose of this section is to discuss the effects of ceiling height and the selection of a threshold fire size that may be used as the basis for determination of type and spacing of automatic fire detectors in a specific situation.

B-2.1.1 A detector will ordinarily operate sooner in detecting the fire if it is nearer the fire.

B-2.1.2 Generally, height is the most important single dimension where ceiling heights exceed 16 ft (4.9 m).

B-2.1.3 As smoke and heat rise from a fire, they tend to spread in the general form of an inverted cone. Therefore, the concentration within the cone varies inversely as a variable exponential function of the distance from the source. This effect is very significant in the early stages of a fire, because the angle of the cone is wide. As a fire intensifies, the angle of the cone narrows and the significance of the effect of height is lessened.

B-2.1.4 **High Ceilings.** As the ceiling height increases, a larger-size fire is required to actuate the same detector in the same time. In view of this, it is mandatory that the designer of a fire detection system calling for heat detectors consider the size of the fire and rate of heat release that may be permitted to develop before detection is ultimately obtained.

B-2.1.5 The most sensitive detectors suitable for the maximum ambient temperature at heights above 30 ft (9.1 m) should be employed.

B-2.1.6 Spacing recommended by testing laboratories for the location of detectors is an indication of their relative sensitivity. This applies with each detection principle; however, detectors operating on various physical principles have different inherent sensitivities to different types of fires and fuels.

B-2.1.7 Reduction of listed spacing may be required for any of the following purposes:

(a) Faster response of the device to a fire

(b) Response of the device to a smaller fire

(c) Accommodation to room geometry

(d) Other special considerations, such as air movement, or ceiling or other obstructions.

B-2.2 Fire Development.

B-2.2.1 Fire development will vary depending on the combustion characteristics of the fuels involved and the physical configuration of the fuels. After ignition, most fires grow in an accelerating pattern.

B-2.2.2 Fire Size.

B-2.2.2.1 Fires can be characterized by their rate of heat release, measured in terms of the number of Btus per second (kW) generated. Typical maximum heat release rates, Q_m, for a number of different fuels and fuel configurations are provided in Tables B-2.2.2.1(a) and (b).

In Table B-2.2.2.1(a):

$$Q_m = q_A$$

Where:

Q_m = the maximum or peak heat release rate in Btu/sec

q = the heat release rate density per unit floor area in Btu/sec/ft^2

A = the floor area of the fuel in ft^2.

B-2.2.2.2 Example. A particular hazard analysis is to be based on a fire scenario involving a 10-ft by 10-ft stack of wood pallets 5 ft high. Approximately what peak heat release rate can be expected?

From Table B-2.2.2.1(a), the heat release rate density (q) for 5-ft high wood pallets is about 330 Btu/sec/ft^2.

The area is 10 ft by 10 ft = 100 ft^2.

$Q_m = q_A = 330 \times 100 = 33,000$ Btu/sec.

The fire would have a medium to fast fire growth rate reaching 1000 Btu/sec in about 90 to 190 seconds.

B-2.2.2.3 The National Institute of Standards and Technology (former National Bureau of Standards) has developed a large-

scale calorimeter for measuring the heat release rates of burning furniture. Two reports issued by NIST (*see references 3 and 13 in Appendix C*) describe the apparatus and data collected during two test series.

Test data from 40 furniture calorimeter tests have been used to independently verify the power-law fire growth model, $Q = \alpha t^2$. (*See reference 14 in Appendix C.*) Here Q is the instantaneous heat release rate, α is the fire intensity coefficient, and t is time. The fire growth time, t_g, is arbitrarily defined as the time after established burning when the fire would reach a burning rate of 1000 Btu/sec. In terms of t_g:

$\alpha = 1000/t_g^2$ Btu/sec^3

$\alpha = 1055/t_g^2$ kW/sec^2

and

$Q = (1000/t_g^2)t^2$ Btu/sec

$Q = (1055/t_g^2)t^2$ kW.

Graphs of heat release data from the 40 furniture calorimeter tests can be found in reference 8. Best fit power-law fire growth curves have been superimposed on the graphs. Data from the best fit curves can be used with this appendix to design or analyze fire detection systems that must respond to similar items burning under a flat ceiling. Table B-2.2.2.3 is a summary of that data.

For reference, the table contains the test numbers used in the original NIST reports. The virtual time of origin, t_v, is the time at which the fires began to obey the power-law fire growth model. Prior to t_v, the fuels may have smoldered but did not burn vigorously with an open flame. The model curves are then predicted by:

$Q = \alpha(t - t_v)^2$ Btu/sec or kW

$Q = (1000/t_g^2)(t - t_v)^2$ Btu/sec

$Q = (1055/t_g^2)(t - t_v)^2$ kW.

Figure B-2.2.2.3 is an example of actual test data with a power-law curve superimposed. This shows how the model may be used to approximate the growth phase of the fire.

For tests 19, 21, 29, 42, and 67, different power-law curves were used to model the initial and the latter realms of burning. In examples such as these, engineers must choose the fire growth parameter that best describes the realm of burning that the detection system is being designed to respond to.

In addition to heat release rate data, the original NIST reports contain data on particulate conversion and radiation from the test

Table B-2.2.2.1(a)   Maximum Heat Release Rates

Warehouse Materials	Growth Time (tg) (sec)	Heat Release Density (q)	Classification (s = slow, m = medium, f = fast)
1. Wood pallets, stack, 1^1/2 ft high (6-12% moisture)	150-310	110	f-m
2. Wood pallets, stack, 5 ft high (6-12% moisture)	90-190	330	f-m
3. Wood pallets, stack, 10 ft high (6-12% moisture)	80-110	600	f
4. Wood pallets, stack, 16 ft high (6-12% moisture)	75-105	900	f
5. Mail bags, filled, stored 5 ft high	190	35	m
6. Cartons, compartmented, stacked 15 ft high	60	200	f
7. Paper, vertical rolls, stacked 20 ft high	15-28	—	†
8. Cotton (also PE, PE/Cot, Acrylic/Nylon/PE), garments in 12-ft high rack	20-42	—	†
9. Cartons on pallets, rack storage, 15-30 ft high	40-280	—	f-m
10. Paper products, densely packed in cartons, rack storage, 20 ft high	470	—	s
11. PE letter trays, filled, stacked 5 ft high on cart	190	750	m
12. PE trash barrels in cartons, stacked 15 ft high	55	250	f
13. FRP shower stalls in cartons, stacked 15 ft high	85	110	f
14. PE bottles, packed in Item 6	85	550	f
15. PE bottles in cartons, stacked 15 ft high	75	170	f
16. PE pallets, stacked 3 ft high	130	—	f
17. PE pallets, stacked 6-8 ft high	30-55	—	f
18. PU mattress, single, horizontal	110	—	f
19. PE insulation board, rigid foam, stacked 15 ft high	8	170	†
20. PS jars, packed in Item 6	55	1200	f
21. PS tubs nested in cartons, stacked 14 ft high	105	450	f
22. PS toy parts in cartons, stacked 15 ft high	110	180	f
23. PS insulation board, rigid, stacked 14 ft high	7	290	†
24. PVC bottles, packed in Item 6	9	300	†
25. PP tubs, packed in Item 6	10	390	†
26. PP and PE film in rolls, stacked 14 ft high	40	350	†
27. Distilled spirits in barrels, stacked 20 ft high	23-40	—	†
28. Methyl alcohol	—	65	—
29. Gasoline	—	200	—
30. Kerosene	—	200	—
31. Diesel oil	—	180	—

For SI Units: 1 ft = 0.305 m.

NOTE: The heat release rates per unit floor area are for fully involved combustibles, assuming 100 percent combustion efficiency. The growth times shown are those required to exceed 1000 Btu/sec heat release rate for developing fires assuming 100 percent combustion efficiency.

(PE = polyethylene; PS = polystyrene; PVC = polyvinyl chloride; PP = polypropylene; PU = polyurethane; FRP = fiberglass-reinforced polyester.)

†Fire growth rate exceeds design data.

Table B-2.2.2.1(b)   Maximum Heat Release Rates from Fire Detection Institute Analysis

Materials	Approximate Values Btu/sec
Medium wastebasket with milk cartons	100
Large barrel with milk cartons	140
Upholstered chair with polyurethane foam	350
Latex foam mattress (heat at room door)	1200
Furnished living room (heat at open door)	4000-8000

specimens. These data can be used to determine the threshold fire size (heat release rate) at which tenability becomes endangered or when additional fuel packages might become involved in the fire.

B-2.2.2.4 A fire detection system can be designed to detect a fire at a certain size in terms of its heat release rate. This is called the threshold fire size (Q_d). The threshold size is the rate of heat release at which detection is desired.

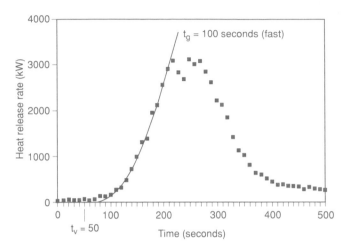

Figure B-2.2.2.3   Test 38, foam sofa.

B-2.2.3   Fire Growth.

B-2.2.3.1   A second important consideration concerning fire development is the time (t_g) it takes for fire to reach a given heat release rate. Table B-2.2.1(a) and Table B-2.2.3 provide the times required to reach a heat release rate of 1000Btu/sec (1055 kW) for a variety of materials in various configurations.

B-2.2.3.2   For purposes of this appendix, fires are classified as being either slow-, medium-, or fast-developing. (*See Figure B-2.2.3.2.*)

B-2.2.3.2.1   The slow-developing fire is defined as one that would take 400 or more seconds (6 minutes, 40 seconds) from the time

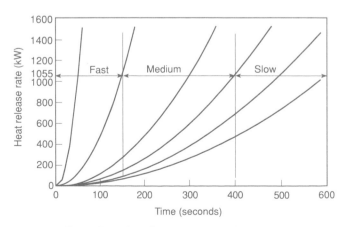

Figure B-2.2.3.2   Power-law heat release rates.

that established burning occurs until the fire reaches a heat release rate of 1000 Btu/sec (1055 kW). Using the relationships discussed in B-2.2.2.3, this corresponds to an α of 0.0062 Btu/sec³ or less (0.0066 kW/sec² or less).

B-2.2.3.2.2   The medium-developing fire is one that would take 150 seconds (2 minutes, 30 seconds) or more and less than 400 seconds (6 minutes, 40 seconds) from the time that established burning occursuntil the fire reaches a heat release rate of 1000 Btu/sec (1055 kW). Using the relationships discussed in B-2.2.2.3, this corresponds to $0.0444 \leq \alpha < 0.0062$ Btu/sec³ ($0.0469 \leq \alpha < 0.0066$ kW/sec²).

B-2.2.3.2.3   The fast-developing fire is one that would take less than 150 seconds (2 minutes, 30 seconds) from the time that established burning occurs until the fire reaches a heat release rate of 1000 Btu/sec (1055 kW). Using the relationships discussed in B-2.2.2.3, this corresponds to an α greater than 0.0444 Btu/sec³ (0.0469 kW/sec²).

B-2.2.3.3   The design fires used in this guide grow according to the following equation: $Q = (1000/t_g^2)t^2$ [where Q is the heat release rate in Btu/sec; t_g is the fire growth time (149 sec = fast, 150–399 sec = medium, 400 sec = slow); and t_g is the time, in seconds, after established burning occurs].

B-2.2.4   **Selection of Fire Size.**   The selection of threshold fire size, Q_d, should be based on an understanding of the characteristics of a specified space and fire safety objectives for that space.

For example, in a particular installation it may be desirable to detect a typical wastebasket fire. Table B-2.2.2.1(b) includes data for a fire involving a comparable array of combustibles, specifically milk cartons in a wastebasket. Such a fire is indicated to produce a peak heat release rate of 100 Btu/sec.

B-2.3   Ceiling Height.

B-2.3.1   The Fire Detection Institute data are based on the height of the ceiling above the fire. In this guide, it is recommended that the designer use the actual distance from floor to ceiling, since the ceiling height will thereby be more conservative and actual detector response will improve when the potential fuel in a room is above floor level.

Table B-2.2.2.3   Furniture Heat Release Rates

Test No.	Item/Mass/Description	Growth Time (tg) (sec)	Classification (s = slow, m = medium, f = fast)	Alpha (α) (kW/sec²)	Virtual time (tv) (sec)	Maximum Heat Release Rates (kW)
Test 15	Metal wardrobe, 41.4 kg (total)	50	f	0.4220	10	750
Test 18	Chair F33 (trial loveseat), 39.2 kg	400	s	0.0066	140	950
Test 19	Chair F21, 28.15 kg (initial)	175	m	0.0344	110	350
Test 19	Chair F21, 28.15 kg (later)	50	f	0.4220	190	2000
Test 21	Metal wardrobe, 40.8 kg (total) (initial)	250	m	0.0169	10	250
Test 21	Metal wardrobe, 40.8 kg (total) (average)	120	f	0.0733	60	250
Test 21	Metal wardrobe, 40.8 kg (total) (later)	100	f	0.1055	30	140
Test 22	Chair F24, 28.3 kg	350	m	0.0086	400	700
Test 23	Chair F23, 31.2 kg	400	s	0.0066	100	700
Test 24	Chair F22, 31.9 kg	2000	s	0.0003	150	300
Test 25	Chair F26, 19.2 kg	200	m	0.0264	90	800
Test 26	Chair F27, 29.0 kg	200	m	0.0264	360	900
Test 27	Chair F29, 14.0 kg	100	f	0.1055	70	1850
Test 28	Chair F28, 29.2 kg	425	s	0.0058	90	700
Test 29	Chair F25, 27.8 kg (later)	60	f	0.2931	175	700
Test 29	Chair F25, 27.8 kg (initial)	100	f	0.1055	100	2000
Test 30	Chair F30, 25.2 kg	60	f	0.2931	70	950
Test 31	Chair F31 (loveseat), 39.6 kg	60	F	0.2931	145	2600
Test 37	Chair F31 (loveseat), 40.4 kg	80	f	0.1648	100	2750
Test 38	Chair F32 (sofa), 51.5 kg	100	f	0.1055	50	3000
Test 39	¹/₂-in. plywood wardrobe with fabrics, 68.5 kg	35	†	0.8612	20	3250
Test 40	¹/₂-in. plywood wardrobe with fabrics, 68.32 kg	35	†	0.8612	40	3500
Test 41	¹/₈-in. plywood wardrobe with fabrics, 36.0 kg	40	†	0.6594	40	6000
Test 42	¹/₈-in. plywood wardrobe with fire-retardant int. fin. (initial growth)	70	f	0.2153	50	2000
Test 42	¹/₈-in. plywood wardrobe with fire-retardant int. fin. (later growth)	30	†	1.1722	100	5000
Test 43	Repeat of ¹/₂-in. plywood wardrobe, 67.62 kg	30	†	1.1722	50	3000
Test 44	¹/₈-in. plywood wardrobe with fire-retardant latex paint, 37.26 kg	90	f	0.1302	30	2900
Test 45	Chair F21, 28.34 kg	100	f	0.1055	120	2100
Test 46	Chair F21, 28.34 kg	45	†	0.5210	130	2600
Test 47	Chair, adj. back metal frame, foam cushions, 20.82 kg	170	m	0.0365	30	250
Test 48	Easy chair C07, 11.52 kg	175	m	0.0344	90	950
Test 49	Easy chair F-34, 15.68 kg	200	m	0.0264	50	200
Test 50	Chair, metal frame, minimum cushion, 16.52 kg	200	m	0.0264	120	3000
Test 51	Chair, molded fiberglass, no cushion, 5.28 kg	120	f	0.0733	20	35
Test 52	Molded plastic patient chair, 11.26 kg	275	m	0.0140	2090	700
Test 53	Chair, metal frame, padded seat and back, 15.54 kg	350	m	0.0086	50	280
Test 54	Loveseat, metal frame, foam cushions, 27.26 kg	500	s	0.0042	210	300
Test 56	Chair, wood frame, latex foam cushions, 11.2 kg	500	s	0.0042	50	85
Test 57	Loveseat, wood frame, foam cushions, 54.6 kg	350	m	0.0086	500	1000
Test 61	Wardrobe, ³/₄-in. particleboard, 120.33 kg	150	m	0.0469	0	1200
Test 62	Bookcase, plywood with aluminum frame, 30.39 kg	65	f	0.2497	40	25
Test 64	Easy chair, molded flexible urethane frame, 15.98 kg	1000	s	0.0011	750	450
Test 66	Easy chair, 23.02 kg	76	f	0.1827	3700	600
Test 67	Mattress and boxspring, 62.36 kg (later)	350	m	0.0086	400	500
Test 67	Mattress and boxspring, 62.36 kg (initial)	1100	s	0.0009	90	400

For SI Units: 1 ft = 0.305 m; 1000 Btu/sec = 1055 kW; 1 lb = 0.456kg.

†Fire growth exceeds design data.

B-2.3.2 Where the designer desires to consider the height of the potential fuel in the room, the distance between the fuel and the ceiling should be used in place of the ceiling height in the tables and graphs. This should be considered only where the minimum height of the potential fuel is always constant, and where the concept is acceptable to the authority having jurisdiction.

B-2.3.3 The procedures presented in this appendix are based on an analysis of test data for ceiling heights up to 30 ft (9.1 m). No data was analyzed for ceilings greater than 30 ft (9.1 m); therefore, in such installations, engineering judgment and manufacturer's recommendations should be used.

B-3   Heat Detector Spacing.

B-3.1   General.

B-3.1.1 This section discusses procedures for determination of installed spacing of listed heat detectors used to detect flaming fires.

B-3.1.2 The determination of the installed spacing of heat detectors using these procedures adjusts the listed spacing to reflect the effects of ceiling height, threshold fire size, rate of fire development, and, for fixed temperature detectors, the ambient temperature and the temperature rating of the detector.

B-3.1.3 Other factors that will affect detector response, such as beams and joists, are treated in Chapter 5.

B-3.1.4 The difference between the rated temperature of a fixed temperature detector (T_s) and the maximum ambient temperature (T_o) at the ceiling should be as small as possible. To reduce unwanted alarms, the difference between operating temperature and ambient temperature should be not less than 20°F (11°C).

B-3.1.5 Listed rate-of-rise heat detectors are designed to activate at a nominal rate of temperature rise of 15°F (8°C) per minute.

B-3.1.6 The listed spacing of a detector is an indicator of the detector's sensitivity. Given the same temperature rating, a detector listed for a 50-ft (15.2-m) spacing is more sensitive than one listed for a 20-ft (6.1-m) spacing.

B-3.1.7 Where using combination detectors incorporating both fixed temperature and rate-of-rise heat detection principles to detect a geometrically growing fire, the data herein for rate-of-rise detectors should be used in selecting an installed spacing because the rate-of-rise principle controls the response.

B-3.1.8 Rate-compensated detectors are not specifically covered by this guide. However, a conservative approach to predicting their performance is to use the fixed temperature heat detector guidance contained herein.

B-3.2   Fixed-Temperature Heat Detector Spacing.

B-3.2.1 Tables B-3.2.2 and B-3.2.4(a) through (y) are to be used to determine the installed spacing of fixed-temperature heat detectors. The analytical basis for the tables is presented in a later section of this appendix. This section describes how the tables are to be used.

B-3.2.1.1 Except for ceiling height, the nearest value shown in the tables will provide sufficient accuracy for these calculations. Interpolation is allowable but not necessary except for ceiling height.

B-3.2.2 Given the detector's listed spacing and the detector's rated temperature (T_s), use Table B-3.2.2 to find the detector time constant (τ_o). The time constant is a measure of the detector's sensitivity. (See B-3.3.)

B-3.2.2.1 Response time index (RTI) can also be used to describe the sensitivity of a fixed temperature heat detector. (See Section B-4.)

B-3.2.3 Estimate the minimum ambient temperature (T_o) expected at the ceiling of the space to be protected. Calculate the temperature change (ΔT) of the detector required for detection ($\Delta T = T_s - T_o$).

B-3.2.3.1 Selection of the minimum ambient temperature requires engineering judgment. Use of the absolute minimum ambient temperature will result in the most conservative designs. This is true because it is then assumed that the detector must absorb enough energy to raise its temperature from the low ambient value up to its operating temperature. A review of historical data may show very low ambient temperatures that occur relatively infrequently, such as every one hundred years or so.

Depending on actual design considerations, it may be more prudent to use an average minimum ambient temperature. In any

Table B-3.2.2 Time Constants (τ_o) for Any Listed Heat Detector*

Listed Spacing (ft)	128°	135°	ULI 145°	160°	170°	196°	FMRC All Temps.
10	400	330	262	195	160	97	196
15	250	190	156	110	89	45	110
20	165	135	105	70	52	17	70
25	124	100	78	48	32		48
30	95	80	61	36	22		36
40	71	57	41	18			
50	59	44	30				
70	36	24	9				

NOTE 1: These time constants are based on an analysis of the Underwriters Laboratories Inc. and Factory Mutual listing test procedures. Plunge test (*see reference 8 in Appendix C*) results performed on the detector to be used will give a more accurate time constant. See Section B-5 for a further discussion of detector time constants.

NOTE 2: These time constants can be converted to response time index (RTI) values by multiplying by $\sqrt{5}$ ft/sec. (*See B-3-3.*)

*At a reference velocity of 5 ft/sec.

case, a sensitivity analysis should be performed to determine the effect of changing the ambient temperature on the design results.

B-3.2.4 Having determined the detector's sensitivity (time constant or RTI) (*see B-3.2.2*), the temperature change of the detector required for detection (*see B-3.2.3*), the threshold fire size (*see B-3.2.2*), the fire growth rate (*see B-3.2.3*), and the ceiling height, use Tables B-3.2.4(a) through (y) to determine the required installed spacing. Table B-3.2.4 is an index to the tables.

B-3.2.5 Installed spacings listed as zero in the tables indicate that the detector chosen will not respond within the design objectives.

B-3.2.6 Example.

Input:

Ceiling height: 8 ft (2.4 m)

Detector type: Fixed temperature 135°F (57°C)

UL listed spacing: 30 ft (9.1 m)

Fire:

Q_d: 500 Btu/sec (527 kW)

Fire growth rate: Slow

t_g: 600 sec

α: 0.003 Btu/sec³

Environmental conditions:

T_o: 55°F (12.8°C)

Table B-3.2.4 Design Tables Index

	Threshold Fire Size (Btu/sec) Q_d	Fire Growth Period (sec) t_g	Alpha (Btu/sec³) α
Table B-3.2.4(a)	250	50	0.400
Table B-3.2.4(b)	250	150	0.044
Table B-3.2.4(c)	250	300	0.011
Table B-3.2.4(d)	250	500	0.004
Table B-3.2.4(e)	250	600	0.003
Table B-3.2.4(f)	500	50	0.400
Table B-3.2.4(g)	500	150	0.044
Table B-3.2.4(h)	500	300	0.011
Table B-3.2.4(i)	500	500	0.004
Table B-3.2.4(j)	500	600	0.003
Table B-3.2.4(k)	750	50	0.400
Table B-3.2.4(l)	750	150	0.044
Table B-3.2.4(m)	750	300	0.011
Table B-3.2.4(n)	750	500	0.004
Table B-3.2.4(o)	750	600	0.003
Table B-3.2.4(p)	1000	50	0.400
Table B-3.2.4(q)	1000	150	0.044
Table B-3.2.4(r)	1000	300	0.011
Table B-3.2.4(s)	1000	500	0.004
Table B-3.2.4(t)	1000	600	0.003
Table B-3.2.4(u)	2000	50	0.400
Table B-3.2.4(v)	2000	150	0.044
Table B-3.2.4(w)	2000	300	0.011
Table B-3.2.4(x)	2000	500	0.004
Table B-3.2.4(y)	2000	600	0.003

Required installed spacing:

From Table B-3.2.2, the detector time constant (τ_o) is 80 seconds.

(RTI = $80 \sqrt{5}$ = 180 ft$^{1/2}$ sec$^{1/2}$)

$\Delta T = T_s - T_o = 135 - 55 = 80°F$

From Table B-3.2.4(j):

For τ_o = 75 sec — spacing = 17 ft

For τ_o = 100 sec — spacing = 16 ft

By interpolation:

Spacing = 17 − [(17−16/80−75/100−75)] = 16.8, round to 17.0 ft

NOTE: Interpolation for τ_o = 80 seconds was not required, but was included for demonstration. If the ceiling height is 16 ft, the required spacing would be 8.8 ft. Using the detector in the above example, at a ceiling height of 28 ft, no practical spacing would ensure detection of the fire at the threshold fire size of 500 Btu/sec. A more sensitive detector would need to be used. Alternatively, the design objectives could be changed to accept a larger fire. These results clearly illustrate the need to consider ceiling height in the design of a detection system.

For SI Units: 1 ft = 0.305 m.

Table B-3.2.4(a) Q_d, Threshold Fire Size at Response: 250 Btu/sec; t_g: 50 seconds to 1000 Btu/sec; α: 0.400 Btu/sec^3

τ	RTI	ΔT	CEILING HEIGHT IN FEET							τ	RTI	ΔT	CEILING HEIGHT IN FEET						
			4.0	8.0	12.0	16.0	20.0	24.0	28.0				4.0	8.0	12.0	16.0	20.0	24.0	28.0
			INSTALLED SPACING OF DETECTORS										INSTALLED SPACING OF DETECTORS						
25	56	40	7	5	2	0	0	0	0	225	503	40	2	0	0	0	0	0	0
25	56	60	6	3	1	0	0	0	0	225	503	60	1	0	0	0	0	0	0
25	56	80	5	2	0	0	0	0	0	225	503	80	0	0	0	0	0	0	0
25	56	100	4	2	0	0	0	0	0	225	503	100	0	0	0	0	0	0	0
25	56	120	4	1	0	0	0	0	0	225	503	120	0	0	0	0	0	0	0
25	56	140	3	1	0	0	0	0	0	225	503	140	0	0	0	0	0	0	0
50	112	40	5	3	1	0	0	0	0	250	559	40	2	0	0	0	0	0	0
50	112	60	4	2	0	0	0	0	0	250	559	60	0	0	0	0	0	0	0
50	112	80	3	1	0	0	0	0	0	250	559	80	0	0	0	0	0	0	0
50	112	100	3	0	0	0	0	0	0	250	559	100	0	0	0	0	0	0	0
50	112	120	2	0	0	0	0	0	0	250	559	120	0	0	0	0	0	0	0
50	112	140	2	0	0	0	0	0	0	250	559	140	0	0	0	0	0	0	0
75	168	40	4	2	0	0	0	0	0	275	615	40	1	0	0	0	0	0	0
75	168	60	3	1	0	0	0	0	0	275	615	60	0	0	0	0	0	0	0
75	168	80	2	0	0	0	0	0	0	275	615	80	0	0	0	0	0	0	0
75	168	100	2	0	0	0	0	0	0	275	615	100	0	0	0	0	0	0	0
75	168	120	2	0	0	0	0	0	0	275	615	120	0	0	0	0	0	0	0
75	168	140	1	0	0	0	0	0	0	275	615	140	0	0	0	0	0	0	0
100	224	40	3	1	0	0	0	0	0	300	671	40	1	0	0	0	0	0	0
100	224	60	2	0	0	0	0	0	0	300	671	60	0	0	0	0	0	0	0
100	224	80	2	0	0	0	0	0	0	300	671	80	0	0	0	0	0	0	0
100	224	100	1	0	0	0	0	0	0	300	671	100	0	0	0	0	0	0	0
100	224	120	1	0	0	0	0	0	0	300	671	120	0	0	0	0	0	0	0
100	224	140	1	0	0	0	0	0	0	300	671	140	0	0	0	0	0	0	0
125	280	40	3	0	0	0	0	0	0	325	727	40	1	0	0	0	0	0	0
125	280	60	2	0	0	0	0	0	0	325	727	60	0	0	0	0	0	0	0
125	280	80	1	0	0	0	0	0	0	325	727	80	0	0	0	0	0	0	0
125	280	100	1	0	0	0	0	0	0	325	727	100	0	0	0	0	0	0	0
125	280	120	0	0	0	0	0	0	0	325	727	120	0	0	0	0	0	0	0
125	280	140	0	0	0	0	0	0	0	325	727	140	0	0	0	0	0	0	0
150	335	40	2	0	0	0	0	0	0	350	783	40	1	0	0	0	0	0	0
150	335	60	2	0	0	0	0	0	0	350	783	60	0	0	0	0	0	0	0
150	335	80	1	0	0	0	0	0	0	350	783	80	0	0	0	0	0	0	0
150	335	100	0	0	0	0	0	0	0	350	783	100	0	0	0	0	0	0	0
150	335	120	0	0	0	0	0	0	0	350	783	120	0	0	0	0	0	0	0
150	335	140	0	0	0	0	0	0	0	350	783	140	0	0	0	0	0	0	0
175	391	40	2	0	0	0	0	0	0	375	839	40	0	0	0	0	0	0	0
175	391	60	1	0	0	0	0	0	0	375	839	60	0	0	0	0	0	0	0
175	391	80	1	0	0	0	0	0	0	375	839	80	0	0	0	0	0	0	0
175	391	100	0	0	0	0	0	0	0	375	839	100	0	0	0	0	0	0	0
175	391	120	0	0	0	0	0	0	0	375	839	120	0	0	0	0	0	0	0
175	391	140	0	0	0	0	0	0	0	375	839	140	0	0	0	0	0	0	0
200	447	40	2	0	0	0	0	0	0	400	894	40	0	0	0	0	0	0	0
200	447	60	1	0	0	0	0	0	0	400	894	60	0	0	0	0	0	0	0
200	447	80	0	0	0	0	0	0	0	400	894	80	0	0	0	0	0	0	0
200	447	100	0	0	0	0	0	0	0	400	894	100	0	0	0	0	0	0	0
200	447	120	0	0	0	0	0	0	0	400	894	120	0	0	0	0	0	0	0
200	447	140	0	0	0	0	0	0	0	400	894	140	0	0	0	0	0	0	0

NOTE: Detector time constant at a reference velocity of 5 ft/sec.

For SI Units: 1 ft = 0.305 m

1000 BTU/sec = 1055 kW

Table B-3.2.4(b) Q_d, Threshold Fire Size at Response: 250 Btu/sec; t_g: 150 Seconds to 1000 Btu/sec; α: 0.044 Btu/sec³

τ	RTI	ΔT	CEILING HEIGHT IN FEET							τ	RTI	ΔT	CEILING HEIGHT IN FEET						
			4.0	8.0	12.0	16.0	20.0	24.0	28.0				4.0	8.0	12.0	16.0	20.0	24.0	28.0
			INSTALLED SPACING OF DETECTORS										INSTALLED SPACING OF DETECTORS						
25	56	40	15	12	9	6	3	0	0	225	503	40	5	3	1	0	0	0	0
25	56	60	12	9	6	3	0	0	0	225	503	60	4	2	0	0	0	0	0
25	56	80	10	7	4	1	0	0	0	225	503	80	3	1	0	0	0	0	0
25	56	100	9	6	2	0	0	0	0	225	503	100	2	0	0	0	0	0	0
25	56	120	8	4	1	0	0	0	0	225	503	120	2	0	0	0	0	0	0
25	56	140	7	4	1	0	0	0	0	225	503	140	2	0	0	0	0	0	0
50	112	40	11	9	6	3	1	0	0	250	559	40	5	2	0	0	0	0	0
50	112	60	9	6	3	1	0	0	0	250	559	60	3	1	0	0	0	0	0
50	112	80	7	5	2	0	0	0	0	250	559	80	3	0	0	0	0	0	0
50	112	100	6	4	1	0	0	0	0	250	559	100	2	0	0	0	0	0	0
50	112	120	6	3	1	0	0	0	0	250	559	120	2	0	0	0	0	0	0
50	112	140	5	2	0	0	0	0	0	250	559	140	1	0	0	0	0	0	0
75	168	40	9	7	4	2	0	0	0	275	615	40	4	2	0	0	0	0	0
75	168	60	7	5	2	0	0	0	0	275	615	60	3	1	0	0	0	0	0
75	168	80	6	3	1	0	0	0	0	275	615	80	2	0	0	0	0	0	0
75	168	100	5	3	0	0	0	0	0	275	615	100	2	0	0	0	0	0	0
75	168	120	4	2	0	0	0	0	0	275	615	120	2	0	0	0	0	0	0
75	168	140	4	1	0	0	0	0	0	275	615	140	1	0	0	0	0	0	0
100	224	40	8	6	3	1	0	0	0	300	671	40	4	2	0	0	0	0	0
100	224	60	6	4	2	0	0	0	0	300	671	60	3	1	0	0	0	0	0
100	224	80	5	3	1	0	0	0	0	300	671	80	2	0	0	0	0	0	0
100	224	100	4	2	0	0	0	0	0	300	671	100	2	0	0	0	0	0	0
100	224	120	4	1	0	0	0	0	0	300	671	120	1	0	0	0	0	0	0
100	224	140	3	1	0	0	0	0	0	300	671	140	1	0	0	0	0	0	0
125	280	40	7	5	2	1	0	0	0	325	727	40	4	2	0	0	0	0	0
125	280	60	5	3	1	0	0	0	0	325	727	60	3	1	0	0	0	0	0
125	280	80	4	2	0	0	0	0	0	325	727	80	2	0	0	0	0	0	0
125	280	100	4	1	0	0	0	0	0	325	727	100	2	0	0	0	0	0	0
125	280	120	3	1	0	0	0	0	0	325	727	120	1	0	0	0	0	0	0
125	280	140	3	0	0	0	0	0	0	325	727	140	1	0	0	0	0	0	0
150	335	40	6	4	2	0	0	0	0	350	783	40	4	2	0	0	0	0	0
150	335	60	5	2	1	0	0	0	0	350	783	60	3	0	0	0	0	0	0
150	335	80	4	2	0	0	0	0	0	350	783	80	2	0	0	0	0	0	0
150	335	100	3	1	0	0	0	0	0	350	783	100	2	0	0	0	0	0	0
150	335	120	3	0	0	0	0	0	0	350	783	120	1	0	0	0	0	0	0
150	335	140	2	0	0	0	0	0	0	350	783	140	1	0	0	0	0	0	0
175	391	40	6	3	1	0	0	0	0	375	839	40	3	1	0	0	0	0	0
175	391	60	4	2	0	0	0	0	0	375	839	60	2	0	0	0	0	0	0
175	391	80	3	1	0	0	0	0	0	375	839	80	2	0	0	0	0	0	0
175	391	100	3	1	0	0	0	0	0	375	839	100	1	0	0	0	0	0	0
175	391	120	2	0	0	0	0	0	0	375	839	120	1	0	0	0	0	0	0
175	391	140	2	0	0	0	0	0	0	375	839	140	0	0	0	0	0	0	0
200	447	40	5	3	1	0	0	0	0	400	894	40	3	1	0	0	0	0	0
200	447	60	4	2	0	0	0	0	0	400	894	60	2	0	0	0	0	0	0
200	447	80	3	1	0	0	0	0	0	400	894	80	2	0	0	0	0	0	0
200	447	100	3	0	0	0	0	0	0	400	894	100	1	0	0	0	0	0	0
200	447	120	2	0	0	0	0	0	0	400	894	120	1	0	0	0	0	0	0
200	447	140	2	0	0	0	0	0	0	400	894	140	0	0	0	0	0	0	0

NOTE: Detector time constant at a reference velocity of 5 ft/sec.

For SI Units: 1 ft = 0.305 m

1000 BTU/sec = 1055 kW

Table B-3.2.4(c) Q_d, Threshold Fire Size at Response: 250 Btu/sec; t_g: 300 Seconds to 1000 Btu/sec; α: 0.011 Btu/sec^3

τ	RTI	ΔT	4.0	8.0	12.0	16.0	20.0	24.0	28.0
			INSTALLED SPACING OF DETECTORS						
25	56	40	21	18	14	10	6	3	0
25	56	60	17	13	9	5	2	0	0
25	56	80	14	10	6	3	0	0	0
25	56	100	12	8	4	1	0	0	0
25	56	120	11	7	3	0	0	0	0
25	56	140	10	6	2	0	0	0	0
50	112	40	17	14	11	7	4	2	0
50	112	60	13	10	7	4	1	0	0
50	112	80	11	8	5	2	0	0	0
50	112	100	10	6	3	0	0	0	0
50	112	120	8	5	2	0	0	0	0
50	112	140	8	4	1	0	0	0	0
75	168	40	14	11	8	6	3	1	0
75	168	60	11	8	5	3	1	0	0
75	168	80	9	6	3	1	0	0	0
75	168	100	8	5	2	0	0	0	0
75	168	120	7	4	1	0	0	0	0
75	168	140	6	3	1	0	0	0	0
100	224	40	12	10	7	4	2	0	0
100	224	60	10	7	4	2	0	0	0
100	224	80	8	5	3	1	0	0	0
100	224	100	7	4	2	0	0	0	0
100	224	120	6	3	1	0	0	0	0
100	224	140	5	3	0	0	0	0	0
125	280	40	11	9	6	3	1	0	0
125	280	60	9	6	3	1	0	0	0
125	280	80	7	4	2	0	0	0	0
125	280	100	6	3	1	0	0	0	0
125	280	120	5	3	1	0	0	0	0
125	280	140	5	2	0	0	0	0	0
150	335	40	10	8	5	3	1	0	0
150	335	60	8	5	3	1	0	0	0
150	335	80	6	4	2	0	0	0	0
150	335	100	6	3	1	0	0	0	0
150	335	120	5	2	0	0	0	0	0
150	335	140	4	2	0	0	0	0	0
175	391	40	9	7	4	2	1	0	0
175	391	60	7	5	2	1	0	0	0
175	391	80	6	3	1	0	0	0	0
175	391	100	5	3	1	0	0	0	0
175	391	120	4	2	0	0	0	0	0
175	391	140	4	1	0	0	0	0	0
200	447	40	9	6	4	2	0	0	0
200	447	60	7	4	2	0	0	0	0
200	447	80	5	3	1	0	0	0	0
200	447	100	5	2	0	0	0	0	0
200	447	120	4	2	0	0	0	0	0
200	447	140	3	1	0	0	0	0	0
225	503	40	8	6	3	2	0	0	0
225	503	60	6	4	2	0	0	0	0
225	503	80	5	3	1	0	0	0	0
225	503	100	4	2	0	0	0	0	0
225	503	120	4	1	0	0	0	0	0
225	503	140	3	1	0	0	0	0	0
250	559	40	8	5	3	1	0	0	0
250	559	60	6	3	1	0	0	0	0
250	559	80	5	2	0	0	0	0	0
250	559	100	4	2	0	0	0	0	0
250	559	120	3	1	0	0	0	0	0
250	559	140	3	1	0	0	0	0	0
275	615	40	7	5	3	1	0	0	0
275	615	60	6	3	1	0	0	0	0
275	615	80	4	2	0	0	0	0	0
275	615	100	4	1	0	0	0	0	0
275	615	120	3	1	0	0	0	0	0
275	615	140	3	0	0	0	0	0	0
300	671	40	7	5	2	1	0	0	0
300	671	60	5	3	1	0	0	0	0
300	671	80	4	2	0	0	0	0	0
300	671	100	3	1	0	0	0	0	0
300	671	120	3	1	0	0	0	0	0
300	671	140	3	0	0	0	0	0	0
325	727	40	7	4	2	0	0	0	0
325	727	60	5	3	1	0	0	0	0
325	727	80	4	2	0	0	0	0	0
325	727	100	3	1	0	0	0	0	0
325	727	120	3	1	0	0	0	0	0
325	727	140	2	0	0	0	0	0	0
350	783	40	6	4	2	0	0	0	0
350	783	60	5	2	1	0	0	0	0
350	783	80	4	2	0	0	0	0	0
350	783	100	3	1	0	0	0	0	0
350	783	120	3	0	0	0	0	0	0
350	783	140	2	0	0	0	0	0	0
375	839	40	6	4	2	0	0	0	0
375	839	60	4	2	0	0	0	0	0
375	839	80	4	1	0	0	0	0	0
375	839	100	3	1	0	0	0	0	0
375	839	120	2	0	0	0	0	0	0
375	839	140	2	0	0	0	0	0	0
400	894	40	6	3	2	0	0	0	0
400	894	60	4	2	0	0	0	0	0
400	894	80	3	1	0	0	0	0	0
400	894	100	3	1	0	0	0	0	0
400	894	120	2	0	0	0	0	0	0
400	894	140	2	0	0	0	0	0	0

NOTE: Detector time constant at a reference velocity of 5 ft/sec.

For SI Units: 1 ft = 0.305 m
1000 BTU/sec = 1055 kW

Table B-3.2.4(d) Q_d, Threshold Fire Size at Response: 250 Btu/sec; t_g: 500 Seconds to 1000 Btu/sec; α: 0.004 Btu/sec³

τ	RTI	ΔT	CEILING HEIGHT IN FEET							τ	RTI	ΔT	CEILING HEIGHT IN FEET						
			4.0	8.0	12.0	16.0	20.0	24.0	28.0				4.0	8.0	12.0	16.0	20.0	24.0	28.0
			INSTALLED SPACING OF DETECTORS										INSTALLED SPACING OF DETECTORS						
25	56	40	26	22	17	13	9	5	1	225	503	40	11	9	6	4	2	0	0
25	56	60	20	16	11	7	3	0	0	225	503	60	9	6	4	1	0	0	0
25	56	80	17	12	8	4	0	0	0	225	503	80	7	5	2	0	0	0	0
25	56	100	15	10	5	1	0	0	0	225	503	100	6	3	1	0	0	0	0
25	56	120	13	8	4	0	0	0	0	225	503	120	5	3	1	0	0	0	0
25	56	140	11	7	2	0	0	0	0	225	503	140	5	2	0	0	0	0	0
50	112	40	21	18	14	11	7	4	1	250	559	40	11	8	6	3	1	0	0
50	112	60	17	13	9	6	2	0	0	250	559	60	8	6	3	1	0	0	0
50	112	80	14	10	6	3	0	0	0	250	559	80	7	4	2	0	0	0	0
50	112	100	12	8	4	1	0	0	0	250	559	100	6	3	1	0	0	0	0
50	112	120	11	7	3	0	0	0	0	250	559	120	5	2	0	0	0	0	0
50	112	140	9	6	2	0	0	0	0	250	559	140	4	2	0	0	0	0	0
75	168	40	18	15	12	9	6	3	0	275	615	40	10	8	5	3	1	0	0
75	168	60	14	11	8	5	2	0	0	275	615	60	8	5	3	1	0	0	0
75	168	80	12	9	5	2	0	0	0	275	615	80	6	4	2	0	0	0	0
75	168	100	10	7	4	1	0	0	0	275	615	100	5	3	1	0	0	0	0
75	168	120	9	6	2	0	0	0	0	275	615	120	5	2	0	0	0	0	0
75	168	140	8	5	1	0	0	0	0	275	615	140	4	2	0	0	0	0	0
100	224	40	16	14	10	7	4	0	0	300	671	40	10	1	0	0	7	5	3
100	224	60	13	10	7	4	1	0	0	300	671	60	7	0	0	0	5	3	1
100	224	80	11	8	4	2	0	0	0	300	671	80	6	0	0	0	4	1	0
100	224	100	9	6	3	0	0	0	0	300	671	100	5	0	0	0	3	1	0
100	224	120	8	5	2	0	0	0	0	300	671	120	4	0	0	0	2	0	0
100	224	140	7	4	1	0	0	0	0	300	671	140	4	0	0	0	2	0	0
125	280	40	15	12	9	6	4	1	0	325	727	40	9	1	0	0	7	4	2
125	280	60	12	9	6	3	1	0	0	325	727	60	7	0	0	0	5	2	1
125	280	80	10	7	4	1	0	0	0	325	727	80	6	0	0	0	3	1	0
125	280	100	8	5	2	0	0	0	0	325	727	100	5	0	0	0	2	0	0
125	280	120	7	4	1	0	0	0	0	325	727	120	4	0	0	0	2	0	0
125	280	140	6	3	1	0	0	0	0	325	727	140	4	0	0	0	1	0	0
150	335	40	14	11	8	5	3	1	0	350	783	40	9	1	0	0	6	4	2
150	335	60	11	8	5	3	1	0	0	350	783	60	7	0	0	0	4	2	0
150	335	80	9	6	3	1	0	0	0	350	783	80	6	0	0	0	3	1	0
150	335	100	8	5	2	0	0	0	0	350	783	100	5	0	0	0	2	0	0
150	335	120	7	4	1	0	0	0	0	350	783	120	4	0	0	0	2	0	0
150	335	140	6	3	1	0	0	0	0	350	783	140	3	0	0	0	1	0	0
175	391	40	13	10	7	5	2	1	0	375	839	40	9	0	0	0	6	4	2
175	391	60	10	7	4	2	0	0	0	375	839	60	6	0	0	0	4	2	0
175	391	80	8	5	3	1	0	0	0	375	839	80	5	0	0	0	3	1	0
175	391	100	7	4	2	0	0	0	0	375	839	100	4	0	0	0	2	0	0
175	391	120	6	3	1	0	0	0	0	375	839	120	4	0	0	0	2	0	0
175	391	140	5	3	0	0	0	0	0	375	839	140	3	0	0	0	1	0	0
200	447	40	12	9	7	4	2	1	0	400	894	40	8	0	0	0	6	4	2
200	447	60	9	7	4	2	0	0	0	400	894	60	6	0	0	0	4	2	0
200	447	80	8	5	2	1	0	0	0	400	894	80	5	0	0	0	3	1	0
200	447	100	6	4	1	0	0	0	0	400	894	100	4	0	0	0	2	0	0
200	447	120	6	3	1	0	0	0	0	400	894	120	4	0	0	0	1	0	0
200	447	140	5	2	0	0	0	0	0	400	894	140	3	0	0	0	1	0	0

NOTE: Detector time constant at a reference velocity of 5 ft/sec.

For SI Units: 1 ft = 0.305 m

1000 BTU/sec = 1055 kW

Table B-3.2.4(e) Q_d, Threshold Fire Size at Response: 250 Btu/sec; t_g: 600 Seconds to 1000 Btu/sec; α: 0.003 Btu/sec^3

τ	RTI	ΔT	CEILING HEIGHT IN FEET — INSTALLED SPACING OF DETECTORS						
			4.0	8.0	12.0	16.0	20.0	24.0	28.0
25	56	40	28	23	18	14	9	5	2
25	56	60	22	17	12	8	4	0	0
25	56	80	18	13	8	4	0	0	0
25	56	100	15	10	6	2	0	0	0
25	56	120	13	8	4	0	0	0	0
25	56	140	12	7	3	0	0	0	0
50	112	40	23	19	15	12	8	4	1
50	112	60	18	14	10	6	3	0	0
50	112	80	15	11	7	3	0	0	0
50	112	100	13	9	5	1	0	0	0
50	112	120	11	7	3	0	0	0	0
50	112	140	10	6	2	0	0	0	0
75	168	40	20	17	13	10	7	3	1
75	168	60	16	12	9	5	2	0	0
75	168	80	13	10	6	3	0	0	0
75	168	100	11	8	4	1	0	0	0
75	168	120	10	6	3	0	0	0	0
75	168	140	9	5	2	0	0	0	0
100	224	40	18	15	12	9	5	3	0
100	224	60	14	11	8	4	2	0	0
100	224	80	12	8	5	2	0	0	0
100	224	100	10	7	4	1	0	0	0
100	224	120	9	5	2	0	0	0	0
100	224	140	8	5	1	0	0	0	0
125	280	40	16	14	10	7	5	2	0
125	280	60	13	10	7	4	1	0	0
125	280	80	11	8	4	2	0	0	0
125	280	100	9	6	3	0	0	0	0
125	280	120	8	5	2	0	0	0	0
125	280	140	7	4	1	0	0	0	0
150	335	40	15	12	9	7	4	2	0
150	335	60	12	9	6	3	1	0	0
150	335	80	10	7	4	1	0	0	0
150	335	100	8	5	3	0	0	0	0
150	335	120	7	4	2	0	0	0	0
150	335	140	7	4	1	0	0	0	0
175	391	40	14	11	9	6	3	1	0
175	391	60	11	8	5	3	1	0	0
175	391	80	9	6	3	1	0	0	0
175	391	100	8	5	2	0	0	0	0
175	391	120	7	4	1	0	0	0	0
175	391	140	6	3	1	0	0	0	0
200	447	40	13	11	8	5	3	1	0
200	447	60	10	8	5	2	1	0	0
200	447	80	8	6	3	1	0	0	0
200	447	100	7	4	2	0	0	0	0
200	447	120	6	4	1	0	0	0	0
200	447	140	6	3	1	0	0	0	0
225	503	40	12	10	7	5	2	1	0
225	503	60	10	7	4	2	0	0	0
225	503	80	8	5	3	1	0	0	0
225	503	100	7	4	2	0	0	0	0
225	503	120	6	3	1	0	0	0	0
225	503	140	5	3	0	0	0	0	0
250	559	40	12	9	7	4	2	1	0
250	559	60	9	7	4	2	0	0	0
250	559	80	8	5	2	1	0	0	0
250	559	100	6	4	1	0	0	0	0
250	559	120	6	3	1	0	0	0	0
250	559	140	5	2	0	0	0	0	0
275	615	40	11	9	6	4	2	0	0
275	615	60	9	6	4	1	0	0	0
275	615	80	7	5	2	0	0	0	0
275	615	100	6	3	1	0	0	0	0
275	615	120	5	3	1	0	0	0	0
275	615	140	5	2	0	0	0	0	0
300	671	40	11	8	6	3	1	0	0
300	671	60	8	6	3	1	0	0	0
300	671	80	7	4	2	0	0	0	0
300	671	100	6	3	1	0	0	0	0
300	671	120	5	2	0	0	0	0	0
300	671	140	4	2	0	0	0	0	0
325	727	40	10	8	5	3	1	0	0
325	727	60	8	5	3	1	0	0	0
325	727	80	6	4	2	0	0	0	0
325	727	100	6	3	1	0	0	0	0
325	727	120	5	2	0	0	0	0	0
325	727	140	4	2	0	0	0	0	0
350	783	40	10	7	5	3	1	0	0
350	783	60	8	5	3	1	0	0	0
350	783	80	6	4	2	0	0	0	0
350	783	100	5	3	1	0	0	0	0
350	783	120	5	2	0	0	0	0	0
350	783	140	4	2	0	0	0	0	0
375	839	40	10	7	5	3	1	0	0
375	839	60	7	5	3	1	0	0	0
375	839	80	6	3	1	0	0	0	0
375	839	100	5	3	1	0	0	0	0
375	839	120	4	2	0	0	0	0	0
375	839	140	4	1	0	0	0	0	0
400	894	40	9	7	4	2	1	0	0
400	894	60	7	5	2	1	0	0	0
400	894	80	6	3	1	0	0	0	0
400	894	100	5	2	0	0	0	0	0
400	894	120	4	2	0	0	0	0	0
400	894	140	4	1	0	0	0	0	0

NOTE: Detector time constant at a reference velocity of 5 ft/sec.

For SI Units: 1 ft = 0.305 m
1000 BTU/sec = 1055 kW

Table B-3.2.4(f) Q_d, Threshold Fire Size at Response: 500 Btu/sec; t_g: 50 Seconds to 1000 Btu/sec; α: 0.400 Btu/sec^3

τ	RTI	ΔT	4.0	8.0	12.0	16.0	20.0	24.0	28.0
			\multicolumn{7}{c}{INSTALLED SPACING OF DETECTORS}						
25	56	40	13	11	8	5	2	1	0
25	56	60	11	8	5	3	1	0	0
25	56	80	9	6	4	1	0	0	0
25	56	100	8	5	3	1	0	0	0
25	56	120	7	4	2	0	0	0	0
25	56	140	7	4	1	0	0	0	0
50	112	40	10	7	5	2	1	0	0
50	112	60	8	5	3	1	0	0	0
50	112	80	7	4	2	0	0	0	0
50	112	100	6	3	1	0	0	0	0
50	112	120	5	3	0	0	0	0	0
50	112	140	5	2	0	0	0	0	0
75	168	40	8	6	3	1	0	0	0
75	168	60	6	4	2	0	0	0	0
75	168	80	5	3	1	0	0	0	0
75	168	100	4	2	0	0	0	0	0
75	168	120	4	2	0	0	0	0	0
75	168	140	3	1	0	0	0	0	0
100	224	40	7	4	2	0	0	0	0
100	224	60	5	3	1	0	0	0	0
100	224	80	4	2	0	0	0	0	0
100	224	100	4	1	0	0	0	0	0
100	224	120	3	1	0	0	0	0	0
100	224	140	3	0	0	0	0	0	0
125	280	40	6	4	2	0	0	0	0
125	280	60	5	2	0	0	0	0	0
125	280	80	4	2	0	0	0	0	0
125	280	100	3	1	0	0	0	0	0
125	280	120	3	0	0	0	0	0	0
125	280	140	2	0	0	0	0	0	0
150	335	40	5	3	1	0	0	0	0
150	335	60	4	2	0	0	0	0	0
150	335	80	3	1	0	0	0	0	0
150	335	100	3	0	0	0	0	0	0
150	335	120	2	0	0	0	0	0	0
150	335	140	2	0	0	0	0	0	0
175	391	40	5	3	1	0	0	0	0
175	391	60	4	2	0	0	0	0	0
175	391	80	3	1	0	0	0	0	0
175	391	100	2	0	0	0	0	0	0
175	391	120	2	0	0	0	0	0	0
175	391	140	2	0	0	0	0	0	0
200	447	40	5	2	0	0	0	0	0
200	447	60	3	1	0	0	0	0	0
200	447	80	3	0	0	0	0	0	0
200	447	100	2	0	0	0	0	0	0
200	447	120	2	0	0	0	0	0	0
200	447	140	1	0	0	0	0	0	0

τ	RTI	ΔT	4.0	8.0	12.0	16.0	20.0	24.0	28.0
			\multicolumn{7}{c}{INSTALLED SPACING OF DETECTORS}						
225	503	40	4	2	0	0	0	0	0
225	503	60	3	1	0	0	0	0	0
225	503	80	2	0	0	0	0	0	0
225	503	100	2	0	0	0	0	0	0
225	503	120	2	0	0	0	0	0	0
225	503	140	1	0	0	0	0	0	0
250	559	40	4	2	0	0	0	0	0
250	559	60	3	1	0	0	0	0	0
250	559	80	2	0	0	0	0	0	0
250	559	100	2	0	0	0	0	0	0
250	559	120	1	0	0	0	0	0	0
250	559	140	1	0	0	0	0	0	0
275	615	40	4	2	0	0	0	0	0
275	615	60	3	0	0	0	0	0	0
275	615	80	2	0	0	0	0	0	0
275	615	100	2	0	0	0	0	0	0
275	615	120	1	0	0	0	0	0	0
275	615	140	1	0	0	0	0	0	0
300	671	40	3	1	0	0	0	0	0
300	671	60	2	0	0	0	0	0	0
300	671	80	2	0	0	0	0	0	0
300	671	100	1	0	0	0	0	0	0
300	671	120	1	0	0	0	0	0	0
300	671	140	0	0	0	0	0	0	0
325	727	40	3	1	0	0	0	0	0
325	727	60	2	0	0	0	0	0	0
325	727	80	2	0	0	0	0	0	0
325	727	100	1	0	0	0	0	0	0
325	727	120	1	0	0	0	0	0	0
325	727	140	0	0	0	0	0	0	0
350	783	40	3	1	0	0	0	0	0
350	783	60	2	0	0	0	0	0	0
350	783	80	2	0	0	0	0	0	0
350	783	100	1	0	0	0	0	0	0
350	783	120	0	0	0	0	0	0	0
350	783	140	0	0	0	0	0	0	0
375	839	40	3	1	0	0	0	0	0
375	839	60	2	0	0	0	0	0	0
375	839	80	1	0	0	0	0	0	0
375	839	100	1	0	0	0	0	0	0
375	839	120	0	0	0	0	0	0	0
375	839	140	0	0	0	0'	0	0	0
400	894	40	3	0	0	0	0	0	0
400	894	60	2	0	0	0	0	0	0
400	894	80	1	0	0	0	0	0	0
400	894	100	1	0	0	0	0	0	0
400	894	120	0	0	0	0	0	0	0
400	894	140	0	0	0	0	0	0	0

NOTE: Detector time constant at a reference velocity of 5 ft/sec.

For SI Units: 1 ft = 0.305 m

1000 BTU/sec = 1055 kW

Table B-3.2.4(g) Q_d, Threshold Fire Size at Response: 500 Btu/sec; t_g: 150 Seconds to 1000 Btu/sec; α: 0.044 Btu/sec³

τ	RTI	ΔT	CEILING HEIGHT IN FEET INSTALLED SPACING OF DETECTORS							τ	RTI	ΔT	CEILING HEIGHT IN FEET INSTALLED SPACING OF DETECTORS						
			4.0	8.0	12.0	16.0	20.0	24.0	28.0				4.0	8.0	12.0	16.0	20.0	24.0	28.0
25	56	40	24	22	18	15	11	8	5	225	503	40	9	7	5	3	1	0	0
25	56	60	20	17	13	10	6	3	0	225	503	60	7	5	3	1	0	0	0
25	56	80	17	14	10	6	3	0	0	225	503	80	6	4	2	0	0	0	0
25	56	100	15	11	8	4	1	0	0	225	503	100	5	3	1	0	0	0	0
25	56	120	13	10	6	3	0	0	0	225	503	120	5	2	0	0	0	0	0
25	56	140	12	8	5	1	0	0	0	225	503	140	4	2	0	0	0	0	0
50	112	40	19	16	14	11	8	5	2	250	559	40	9	7	4	2	1	0	0
50	112	60	15	13	10	7	4	1	0	250	559	60	7	5	2	1	0	0	0
50	112	80	13	10	7	4	2	0	0	250	559	80	6	3	1	0	0	0	0
50	112	100	11	9	5	3	0	0	0	250	559	100	5	3	1	0	0	0	0
50	112	120	10	7	4	1	0	0	0	250	559	120	4	2	1	0	0	0	0
50	112	140	9	6	3	1	0	0	0	250	559	140	4	2	0	0	0	0	0
75	168	40	16	14	11	8	5	3	1	275	615	40	8	6	4	2	0	0	0
75	168	60	13	10	8	5	2	1	0	275	615	60	7	4	2	0	0	0	0
75	168	80	11	8	5	3	1	0	0	275	615	80	5	3	1	0	0	0	0
75	168	100	10	7	4	2	0	0	0	275	615	100	5	2	0	0	0	0	0
75	168	120	8	6	3	1	0	0	0	275	615	120	4	2	0	0	0	0	0
75	168	140	8	5	2	0	0	0	0	275	615	140	3	1	0	0	0	0	0
100	224	40	14	12	9	6	4	2	1	300	671	40	8	6	3	2	0	0	0
100	224	60	11	9	6	4	2	0	0	300	671	60	6	4	2	0	0	0	0
100	224	80	10	7	4	2	0	0	0	300	671	80	5	3	1	0	0	0	0
100	224	100	8	6	3	1	0	0	0	300	671	100	4	2	0	0	0	0	0
100	224	120	7	5	2	0	0	0	0	300	671	120	4	2	0	0	0	0	0
100	224	140	7	4	2	0	0	0	0	300	671	140	3	1	0	0	0	0	0
125	280	40	13	10	8	5	3	1	0	325	727	40	8	5	3	1	0	0	0
125	280	60	10	8	5	3	1	0	0	325	727	60	6	4	2	0	0	0	0
125	280	80	8	6	3	1	0	0	0	325	727	80	5	3	1	0	0	0	0
125	280	100	7	5	2	1	0	0	0	325	727	100	4	2	0	0	0	0	0
125	280	120	6	4	2	0	0	0	0	325	727	120	3	1	0	0	0	0	0
125	280	140	6	3	1	0	0	0	0	325	727	140	3	1	0	0	0	0	0
150	335	40	12	9	7	4	2	1	0	350	783	40	7	5	3	1	0	0	0
150	335	60	9	7	4	2	1	0	0	350	783	60	6	3	1	0	0	0	0
150	335	80	8	5	3	1	0	0	0	350	783	80	5	2	0	0	0	0	0
150	335	100	7	4	2	0	0	0	0	350	783	100	4	2	0	0	0	0	0
150	335	120	6	3	1	0	0	0	0	350	783	120	3	1	0	0	0	0	0
150	335	140	5	3	1	0	0	0	0	350	783	140	3	1	0	0	0	0	0
175	391	40	11	8	6	4	2	0	0	375	839	40	7	5	3	1	0	0	0
175	391	60	8	6	4	2	0	0	0	375	839	60	5	3	1	0	0	0	0
175	391	80	7	5	2	1	0	0	0	375	839	80	4	2	0	0	0	0	0
175	391	100	6	4	2	0	0	0	0	375	839	100	4	2	0	0	0	0	0
175	391	120	5	3	1	0	0	0	0	375	839	120	3	1	0	0	0	0	0
175	391	140	5	2	0	0	0	0	0	375	839	140	3	0	0	0	0	0	0
200	447	40	10	8	5	3	1	0	0	400	894	40	7	4	2	1	0	0	0
200	447	60	8	5	3	1	0	0	0	400	894	60	5	3	1	0	0	0	0
200	447	80	7	4	2	0	0	0	0	400	894	80	4	2	0	0	0	0	0
200	447	100	6	3	1	0	0	0	0	400	894	100	3	1	0	0	0	0	0
200	447	120	5	2	1	0	0	0	0	400	894	120	3	1	0	0	0	0	0
200	447	140	4	2	0	0	0	0	0	400	894	140	3	0	0	0	0	0	0

NOTE: Detector time constant at a reference velocity of 5 ft/sec.

For SI Units: 1 ft = 0.305 m

1000 BTU/sec = 1055 kW

Table B-3.2.4(h) Q_d, Threshold Fire Size at Response: 500 Btu/sec; t_g: 300 Seconds to 1000 Btu/sec; α: 0.011 Btu/sec³

τ	RTI	ΔT	CEILING HEIGHT IN FEET / INSTALLED SPACING OF DETECTORS						
			4.0	8.0	12.0	16.0	20.0	24.0	28.0
25	56	40	34	30	25	21	17	13	9
25	56	60	27	23	18	14	10	6	2
25	56	80	23	18	14	9	5	2	0
25	56	100	20	15	11	7	3	0	0
25	56	120	18	13	8	4	1	0	0
25	56	140	16	11	7	3	0	0	0
50	112	40	27	24	21	17	14	10	7
50	112	60	22	18	15	11	8	4	1
50	112	80	18	15	11	8	4	1	0
50	112	100	16	12	9	5	2	0	0
50	112	120	14	11	7	3	0	0	0
50	112	140	13	9	5	2	0	0	0
75	168	40	23	21	18	14	11	8	5
75	168	60	19	16	13	9	6	3	1
75	168	80	16	13	9	6	3	1	0
75	168	100	14	11	7	4	1	0	0
75	168	120	12	9	6	3	0	0	0
75	168	140	11	8	4	1	0	0	0
100	224	40	21	18	15	12	9	6	4
100	224	60	17	14	11	8	5	2	0
100	224	80	14	11	8	5	2	0	0
100	224	100	12	9	6	3	1	0	0
100	224	120	11	8	5	2	0	0	0
100	224	140	10	7	4	1	0	0	0
125	280	40	19	16	14	11	8	5	3
125	280	60	15	12	10	7	4	2	0
125	280	80	13	10	7	4	2	0	0
125	280	100	11	8	5	3	1	0	0
125	280	120	10	7	4	2	0	0	0
125	280	140	9	6	3	1	0	0	0
150	335	40	17	15	12	10	7	4	2
150	335	60	14	11	8	6	3	1	0
150	335	80	12	9	6	4	1	0	0
150	335	100	10	7	5	2	0	0	0
150	335	120	9	6	3	1	0	0	0
150	335	140	8	5	3	1	0	0	0
175	391	40	16	14	11	9	6	4	2
175	391	60	13	10	8	5	3	1	0
175	391	80	11	8	5	3	1	0	0
175	391	100	9	7	4	2	0	0	0
175	391	120	8	6	3	1	0	0	0
175	391	140	7	5	2	0	0	0	0
200	447	40	15	13	10	8	5	3	1
200	447	60	12	10	7	4	2	1	0
200	447	80	10	8	5	3	1	0	0
200	447	100	9	6	4	1	0	0	0
200	447	120	8	5	3	1	0	0	0
200	447	140	7	4	2	0	0	0	0
225	503	40	14	12	10	7	5	3	1
225	503	60	11	9	6	4	2	0	0
225	503	80	10	7	4	2	1	0	0
225	503	100	8	6	3	1	0	0	0
225	503	120	7	5	2	1	0	0	0
225	503	140	6	4	2	0	0	0	0
250	559	40	14	11	9	6	4	2	1
250	559	60	11	8	6	3	2	0	0
250	559	80	9	6	4	2	0	0	0
250	559	100	8	5	3	1	0	0	0
250	559	120	7	4	2	0	0	0	0
250	559	140	6	4	1	0	0	0	0
275	615	40	13	11	8	6	4	2	1
275	615	60	10	8	5	3	1	0	0
275	615	80	9	6	4	2	0	0	0
275	615	100	7	5	3	1	0	0	0
275	615	120	6	4	2	0	0	0	0
275	615	140	6	3	1	0	0	0	0
300	671	40	12	10	8	5	3	2	0
300	671	60	10	7	5	3	1	0	0
300	671	80	8	6	3	1	0	0	0
300	671	100	7	5	2	1	0	0	0
300	671	120	6	4	2	0	0	0	0
300	671	140	6	3	1	0	0	0	0
325	727	40	12	10	7	5	3	1	0
325	727	60	9	7	5	2	1	0	0
325	727	80	8	5	3	1	0	0	0
325	727	100	7	4	2	0	0	0	0
325	727	120	6	3	1	0	0	0	0
325	727	140	5	3	1	0	0	0	0
350	783	40	12	9	7	4	3	1	0
350	783	60	9	7	4	2	1	0	0
350	783	80	7	5	3	1	0	0	0
350	783	100	6	4	2	0	0	0	0
350	783	120	6	3	1	0	0	0	0
350	783	140	5	3	1	0	0	0	0
375	839	40	11	9	6	4	2	1	0
375	839	60	9	6	4	2	0	0	0
375	839	80	7	5	3	1	0	0	0
375	839	100	6	4	2	0	0	0	0
375	839	120	5	3	1	0	0	0	0
375	839	140	5	2	0	0	0	0	0
400	894	40	11	8	6	4	2	1	0
400	894	60	8	6	4	2	0	0	0
400	894	80	7	4	2	1	0	0	0
400	894	100	6	3	1	0	0	0	0
400	894	120	5	3	1	0	0	0	0
400	894	140	5	2	0	0	0	0	0

NOTE: Detector time constant at a reference velocity of 5 ft/sec.

For SI Units: 1 ft = 0.305 m
1000 BTU/sec = 1055 kW

Table B-3.2.4(i) Q_d, Threshold Fire Size at Response: 500 Btu/sec; t_g: 500 Seconds to 1000 Btu/sec; α: 0.004 Btu/sec^3

τ	RTI	ΔT	CEILING HEIGHT IN FEET						
			4.0	8.0	12.0	16.0	20.0	24.0	28.0
			INSTALLED SPACING OF DETECTORS						
25	56	40	41	35	30	25	20	16	11
25	56	60	32	26	21	16	12	7	3
25	56	80	27	21	16	11	7	3	0
25	56	100	23	17	12	8	4	0	0
25	56	120	20	15	10	5	1	0	0
25	56	140	18	13	8	3	0	0	0
50	112	40	34	30	26	22	18	14	10
50	112	60	27	23	18	14	10	6	3
50	112	80	23	18	14	10	6	2	0
50	112	100	20	15	11	7	3	0	0
50	112	120	17	13	9	5	1	0	0
50	112	140	16	11	7	3	0	0	0
75	168	40	30	26	23	19	15	12	8
75	168	60	24	20	16	13	9	5	2
75	168	80	20	16	12	9	5	2	0
75	168	100	17	14	10	6	2	0	0
75	168	120	15	11	8	4	1	0	0
75	168	140	14	10	6	2	0	0	0
100	224	40	27	24	20	17	14	10	7
100	224	60	21	18	15	11	8	4	2
100	224	80	18	15	11	8	4	1	0
100	224	100	16	12	9	5	2	0	0
100	224	120	14	10	7	5	3	0	0
100	224	140	13	9	5	2	0	0	0
125	280	40	25	22	19	15	12	9	6
125	280	60	20	17	13	10	7	4	1
125	280	80	16	13	10	7	4	1	0
125	280	100	14	11	8	5	2	0	0
125	280	120	13	9	6	3	0	0	0
125	280	140	11	8	5	2	0	0	0
150	335	40	23	20	17	14	11	8	5
150	335	60	18	15	12	9	6	3	1
150	335	80	15	12	9	6	3	1	0
150	335	100	13	10	7	4	1	0	0
150	335	120	12	9	5	3	0	0	0
150	335	140	11	7	4	1	0	0	0
175	391	40	21	19	16	13	10	7	4
175	391	60	17	14	11	8	5	3	1
175	391	80	14	11	8	5	3	1	0
175	391	100	12	9	6	3	1	0	0
175	391	120	11	8	5	2	0	0	0
175	391	140	10	7	4	1	0	0	0
200	447	40	20	18	15	12	9	6	4
200	447	60	16	13	10	7	5	2	1
200	447	80	13	11	8	5	2	0	0
200	447	100	12	9	6	3	1	0	0
200	447	120	10	7	4	2	0	0	0
200	447	140	9	6	3	1	0	0	0
225	503	40	19	17	14	11	8	6	3
225	503	60	15	13	10	7	4	2	0
225	503	80	13	10	7	4	2	0	0
225	503	100	11	8	5	3	1	0	0
225	503	120	10	7	4	2	0	0	0
225	503	140	9	6	3	1	0	0	0
250	559	40	18	16	13	10	8	5	3
250	559	60	14	12	9	6	4	2	0
250	559	80	12	9	7	4	2	0	0
250	559	100	10	8	5	2	1	0	0
250	559	120	9	6	4	1	0	0	0
250	559	140	8	6	3	1	0	0	0
275	615	40	17	15	12	10	7	5	2
275	615	60	14	11	8	6	3	1	0
275	615	80	12	9	6	3	1	0	0
275	615	100	10	7	5	2	0	0	0
275	615	120	9	6	3	1	0	0	0
275	615	140	8	5	3	1	0	0	0
300	671	40	17	14	12	9	6	4	2
300	671	60	13	11	8	5	3	1	0
300	671	80	11	8	6	3	1	0	0
300	671	100	10	7	4	2	0	0	0
300	671	120	8	6	3	1	0	0	0
300	671	140	8	5	2	0	0	0	0
325	727	40	16	14	11	9	6	4	2
325	727	60	13	10	8	5	3	1	0
325	727	80	11	8	5	3	1	0	0
325	727	100	9	7	4	2	0	0	0
325	727	120	8	5	3	1	0	0	0
325	727	140	7	5	2	0	0	0	0
350	783	40	16	13	11	8	6	3	2
350	783	60	12	10	7	5	2	1	0
350	783	80	10	8	5	3	1	0	0
350	783	100	9	6	4	2	0	0	0
350	783	120	8	5	3	1	0	0	0
350	783	140	7	4	2	0	0	0	0
375	839	40	15	13	10	8	5	3	1
375	839	60	12	9	7	4	2	1	0
375	839	80	10	7	5	2	1	0	0
375	839	100	8	6	3	1	0	0	0
375	839	120	7	5	2	1	0	0	0
375	839	140	7	4	2	0	0	0	0
400	894	40	14	12	10	7	5	3	1
400	894	60	11	9	6	4	2	1	0
400	894	80	9	7	4	2	1	0	0
400	894	100	8	6	3	1	0	0	0
400	894	120	7	5	2	1	0	0	0
400	894	140	6	4	2	0	0	0	0

NOTE: Detector time constant at a reference velocity of 5 ft/sec.

For SI Units: 1 ft = 0.305 m

1000 BTU/sec = 1055 kW

Table B-3.2.4(j) Q_d, Threshold Fire Size at Response: 500 Btu/sec; t_g: 600 Seconds to 1000 Btu/sec; α: 0.003 Btu/sec³

τ	RTI	ΔT	\multicolumn{7}{c}{CEILING HEIGHT IN FEET}
			4.0	8.0	12.0	16.0	20.0	24.0	28.0
			\multicolumn{7}{c}{INSTALLED SPACING OF DETECTORS}
25	56	40	43	37	31	26	21	17	12
25	56	60	34	27	22	17	12	8	4
25	56	80	28	22	16	12	7	3	0
25	56	100	24	18	13	8	4	0	0
25	56	120	21	15	10	6	1	0	0
25	56	140	19	13	8	4	0	0	0
50	112	40	36	32	27	23	19	15	11
50	112	60	29	24	20	15	11	7	3
50	112	80	24	19	15	10	6	2	0
50	112	100	21	16	11	7	3	0	0
50	112	120	18	14	9	5	1	0	0
50	112	140	17	12	7	3	0	0	0
75	168	40	32	29	25	21	17	13	9
75	168	60	26	22	18	14	10	6	3
75	168	80	21	17	13	9	6	2	0
75	168	100	19	14	10	6	3	0	0
75	168	120	17	12	8	4	1	0	0
75	168	140	15	11	7	3	0	0	0
100	224	40	29	26	22	19	15	12	8
100	224	60	23	20	16	12	9	5	2
100	224	80	19	16	12	8	5	2	0
100	224	100	17	13	9	6	2	0	0
100	224	120	15	11	7	4	1	0	0
100	224	140	14	10	6	2	0	0	0
125	280	40	27	24	20	17	14	10	7
125	280	60	21	18	15	11	8	5	2
125	280	80	18	15	11	8	4	1	0
125	280	100	16	12	9	5	2	0	0
125	280	120	14	10	7	3	1	0	0
125	280	140	12	9	5	2	0	0	0
150	335	40	25	22	19	16	13	9	6
150	335	60	20	17	14	10	7	4	1
150	335	80	17	14	10	7	4	1	0
150	335	100	15	11	8	5	2	0	0
150	335	120	13	10	6	3	0	0	0
150	335	140	12	8	5	2	0	0	0
175	391	40	23	21	18	15	12	8	6
175	391	60	19	16	13	9	6	3	1
175	391	80	16	13	9	6	3	1	0
175	391	100	14	10	7	4	1	0	0
175	391	120	12	9	6	3	0	0	0
175	391	140	11	8	4	2	0	0	0
200	447	40	22	19	17	14	11	8	5
200	447	60	18	15	12	9	6	3	1
200	447	80	15	12	9	6	3	1	0
200	447	100	13	10	7	4	1	0	0
200	447	120	11	8	5	2	0	0	0
200	447	140	10	7	4	1	0	0	0
225	503	40	21	18	16	13	10	7	4
225	503	60	17	14	11	8	5	3	1
225	503	80	14	11	8	5	3	1	0
225	503	100	12	9	6	3	1	0	0
225	503	120	11	8	5	2	0	0	0
225	503	140	10	7	4	1	0	0	0
250	559	40	20	18	15	12	9	6	4
250	559	60	16	13	10	7	5	2	1
250	559	80	13	11	8	5	2	0	0
250	559	100	12	9	6	3	1	0	0
250	559	120	10	7	4	2	0	0	0
250	559	140	9	6	3	1	0	0	0
275	615	40	19	17	14	11	8	6	3
275	615	60	15	13	10	7	4	2	0
275	615	80	13	10	7	4	2	0	0
275	615	100	11	8	5	3	1	0	0
275	615	120	10	7	4	2	0	0	0
275	615	140	9	6	3	1	0	0	0
300	671	40	18	16	13	11	8	5	3
300	671	60	15	12	9	6	4	2	0
300	671	80	12	10	7	4	2	0	0
300	671	100	11	8	5	2	1	0	0
300	671	120	9	7	4	1	0	0	0
300	671	140	8	6	3	1	0	0	0
325	727	40	18	15	13	10	7	5	3
325	727	60	14	11	9	6	4	2	0
325	727	80	12	9	6	4	2	0	0
325	727	100	10	7	5	2	1	0	0
325	727	120	9	6	4	1	0	0	0
325	727	140	8	5	3	1	0	0	0
350	783	40	17	15	12	9	7	4	2
350	783	60	13	11	8	6	3	1	0
350	783	80	11	9	6	3	1	0	0
350	783	100	10	7	4	2	0	0	0
350	783	120	9	6	3	1	0	0	0
350	783	140	8	5	2	1	0	0	0
375	839	40	17	14	12	9	6	4	2
375	839	60	13	11	8	5	3	1	0
375	839	80	11	8	6	3	1	0	0
375	839	100	9	7	4	2	0	0	0
375	839	120	8	6	3	1	0	0	0
375	839	140	7	5	2	0	0	0	0
400	894	40	16	14	11	9	6	4	2
400	894	60	13	10	7	5	3	1	0
400	894	80	11	8	5	3	1	0	0
400	894	100	9	6	4	2	0	0	0
400	894	120	8	5	3	1	0	0	0
400	894	140	7	5	2	0	0	0	0

NOTE: Detector time constant at a reference velocity of 5 ft/sec.

For SI Units: 1 ft = 0.305 m

1000 BTU/sec = 1055 kW

Table B-3.2.4(k) Q_d, Threshold Fire Size at Response: 750 Btu/sec; t_g: 50 Seconds to 1000 Btu/sec; α: 0.400 Btu/sec^3

τ	RTI	ΔT	CEILING HEIGHT IN FEET / INSTALLED SPACING OF DETECTORS						
			4.0	8.0	12.0	16.0	20.0	24.0	28.0
25	56	40	18	15	13	10	7	4	2
25	56	60	15	12	9	6	4	1	0
25	56	80	13	10	7	4	2	0	0
25	56	100	11	9	6	3	1	0	0
25	56	120	10	7	4	2	0	0	0
25	56	140	9	6	4	1	0	0	0
50	112	40	14	11	9	6	3	2	0
50	112	60	11	9	6	3	1	0	0
50	112	80	9	7	4	2	0	0	0
50	112	100	8	6	3	1	0	0	0
50	112	120	7	5	2	0	0	0	0
50	112	140	7	4	2	0	0	0	0
75	168	40	11	9	6	4	2	0	0
75	168	60	9	7	4	2	0	0	0
75	168	80	8	5	3	1	0	0	0
75	168	100	7	4	2	0	0	0	0
75	168	120	6	3	1	0	0	0	0
75	168	140	5	3	1	0	0	0	0
100	224	40	10	7	5	3	1	0	0
100	224	60	8	5	3	1	0	0	0
100	224	80	7	4	2	0	0	0	0
100	224	100	6	3	1	0	0	0	0
100	224	120	5	3	1	0	0	0	0
100	224	140	4	2	0	0	0	0	0
125	280	40	9	6	4	2	0	0	0
125	280	60	7	5	2	1	0	0	0
125	280	80	6	3	1	0	0	0	0
125	280	100	5	3	1	0	0	0	0
125	280	120	4	2	0	0	0	0	0
125	280	140	4	2	0	0	0	0	0
150	335	40	8	6	3	1	0	0	0
150	335	60	6	4	2	0	0	0	0
150	335	80	5	3	1	0	0	0	0
150	335	100	4	2	0	0	0	0	0
150	335	120	4	2	0	0	0	0	0
150	335	140	3	1	0	0	0	0	0
175	391	40	7	5	3	1	0	0	0
175	391	60	6	3	1	0	0	0	0
175	391	80	5	2	0	0	0	0	0
175	391	100	4	2	0	0	0	0	0
175	391	120	3	1	0	0	0	0	0
175	391	140	3	1	0	0	0	0	0
200	447	40	7	4	2	1	0	0	0
200	447	60	5	3	1	0	0	0	0
200	447	80	4	2	0	0	0	0	0
200	447	100	4	1	0	0	0	0	0
200	447	120	3	1	0	0	0	0	0
200	447	140	3	0	0	0	0	0	0
225	503	40	6	4	2	0	0	0	0
225	503	60	5	3	1	0	0	0	0
225	503	80	4	2	0	0	0	0	0
225	503	100	3	1	0	0	0	0	0
225	503	120	3	1	0	0	0	0	0
225	503	140	2	0	0	0	0	0	0
250	559	40	6	4	2	0	0	0	0
250	559	60	4	2	0	0	0	0	0
250	559	80	4	2	0	0	0	0	0
250	559	100	3	1	0	0	0	0	0
250	559	120	3	0	0	0	0	0	0
250	559	140	2	0	0	0	0	0	0
275	615	40	6	3	1	0	0	0	0
275	615	60	4	2	0	0	0	0	0
275	615	80	3	1	0	0	0	0	0
275	615	100	3	1	0	0	0	0	0
275	615	120	2	0	0	0	0	0	0
275	615	140	2	0	0	0	0	0	0
300	671	40	5	3	1	0	0	0	0
300	671	60	4	2	0	0	0	0	0
300	671	80	3	1	0	0	0	0	0
300	671	100	3	0	0	0	0	0	0
300	671	120	2	0	0	0	0	0	0
300	671	140	2	0	0	0	0	0	0
325	727	40	5	3	1	0	0	0	0
325	727	60	4	2	0	0	0	0	0
325	727	80	3	1	0	0	0	0	0
325	727	100	2	0	0	0	0	0	0
325	727	120	2	0	0	0	0	0	0
325	727	140	2	0	0	0	0	0	0
350	783	40	5	3	1	0	0	0	0
350	783	60	4	1	0	0	0	0	0
350	783	80	3	1	0	0	0	0	0
350	783	100	2	0	0	0	0	0	0
350	783	120	2	0	0	0	0	0	0
350	783	140	2	0	0	0	0	0	0
375	839	40	5	2	0	0	0	0	0
375	839	60	3	1	0	0	0	0	0
375	839	80	3	0	0	0	0	0	0
375	839	100	2	0	0	0	0	0	0
375	839	120	2	0	0	0	0	0	0
375	839	140	1	0	0	0	0	0	0
400	894	40	4	2	0	0	0	0	0
400	894	60	3	1	0	0	0	0	0
400	894	80	2	0	0	0	0	0	0
400	894	100	2	0	0	0	0	0	0
400	894	120	2	0	0	0	0	0	0
400	894	140	1	0	0	0	0	0	0

NOTE: Detector time constant at a reference velocity of 5 ft/sec.

For SI Units: 1 ft = 0.305 m
1000 BTU/sec = 1055 kW

Table B-3.2.4(l) Q_d, Threshold Fire Size at Response: 750 Btu/sec; t_g: 150 Seconds to 1000 Btu/sec; α: 0.044 Btu/sec³

τ	RTI	ΔT	CEILING HEIGHT IN FEET							τ	RTI	ΔT	CEILING HEIGHT IN FEET						
			4.0	8.0	12.0	16.0	20.0	24.0	28.0				4.0	8.0	12.0	16.0	20.0	24.0	28.0
			INSTALLED SPACING OF DETECTORS										INSTALLED SPACING OF DETECTORS						
25	56	40	32	29	26	22	18	15	11	225	503	40	13	11	8	6	4	2	1
25	56	60	26	23	19	15	12	8	4	225	503	60	10	8	6	3	2	0	0
25	56	80	23	19	15	11	8	4	1	225	503	80	9	6	4	2	0	0	0
25	56	100	20	16	12	8	5	1	0	225	503	100	8	5	3	1	0	0	0
25	56	120	18	14	10	6	3	0	0	225	503	120	7	4	2	0	0	0	0
25	56	140	16	12	8	5	1	0	0	225	503	140	6	4	1	0	0	0	0
50	112	40	25	23	20	17	14	11	8	250	559	40	12	10	8	5	3	2	0
50	112	60	21	18	15	12	8	5	3	250	559	60	10	7	5	3	1	0	0
50	112	80	18	15	12	8	5	2	0	250	559	80	8	6	4	2	0	0	0
50	112	100	16	13	9	6	3	1	0	250	559	100	7	5	2	1	0	0	0
50	112	120	14	11	8	4	2	0	0	250	559	120	6	4	2	0	0	0	0
50	112	140	13	10	6	3	1	0	0	250	559	140	6	3	1	0	0	0	0
75	168	40	22	19	17	14	11	8	5	275	615	40	12	10	7	5	3	1	0
75	168	60	18	15	12	9	6	4	1	275	615	60	9	7	5	3	1	0	0
75	168	80	15	12	9	6	4	1	0	275	615	80	8	5	3	1	0	0	0
75	168	100	13	10	7	5	2	0	0	275	615	100	7	4	2	1	0	0	0
75	168	120	12	9	6	3	1	0	0	275	615	120	6	4	2	0	0	0	0
75	168	140	11	8	5	2	0	0	0	275	615	140	5	3	1	0	0	0	0
100	224	40	19	17	14	12	9	6	4	300	671	40	11	9	7	4	3	1	0
100	224	60	16	13	10	8	5	3	1	300	671	60	9	7	4	2	1	0	0
100	224	80	13	11	8	5	3	1	0	300	671	80	7	5	3	1	0	0	0
100	224	100	12	9	6	3	1	0	0	300	671	100	6	4	2	0	0	0	0
100	224	120	10	8	5	2	1	0	0	300	671	120	6	3	1	0	0	0	0
100	224	140	9	7	4	1	0	0	0	300	671	140	5	3	1	0	0	0	0
125	280	40	17	15	13	10	7	5	3	325	727	40	11	9	6	4	2	1	0
125	280	60	14	12	9	6	4	2	1	325	727	60	9	6	4	2	1	0	0
125	280	80	12	9	7	4	2	0	0	325	727	80	7	5	3	1	0	0	0
125	280	100	10	8	5	3	1	0	0	325	727	100	6	4	2	0	0	0	0
125	280	120	9	7	4	2	0	0	0	325	727	120	5	3	1	0	0	0	0
125	280	140	8	6	3	1	0	0	0	325	727	140	5	2	1	0	0	0	0
150	335	40	16	14	11	9	6	4	2	350	783	40	10	8	6	4	2	1	0
150	335	60	13	10	8	5	3	1	0	350	783	60	8	6	4	2	0	0	0
150	335	80	11	8	6	3	1	0	0	350	783	80	7	4	2	1	0	0	0
150	335	100	9	7	4	2	1	0	0	350	783	100	6	3	2	0	0	0	0
150	335	120	8	6	3	1	0	0	0	350	783	120	5	3	1	0	0	0	0
150	335	140	8	5	3	1	0	0	0	350	783	140	5	2	0	0	0	0	0
175	391	40	15	13	10	8	5	3	2	375	839	40	10	8	5	3	2	0	0
175	391	60	12	9	7	5	2	1	0	375	839	60	8	6	3	2	0	0	0
175	391	80	10	8	5	3	1	0	0	375	839	80	6	4	2	0	0	0	0
175	391	100	9	6	4	2	0	0	0	375	839	100	6	3	1	0	0	0	0
175	391	120	8	5	3	1	0	0	0	375	839	120	5	3	1	0	0	0	0
175	391	140	7	4	2	0	0	0	0	375	839	140	4	2	0	0	0	0	0
200	447	40	14	12	9	7	4	3	1	400	894	40	10	7	5	3	2	0	0
200	447	60	11	9	6	4	2	1	0	400	894	60	8	5	3	1	0	0	0
200	447	80	9	7	4	2	1	0	0	400	894	80	6	4	2	0	0	0	0
200	447	100	8	6	3	1	0	0	0	400	894	100	5	3	1	0	0	0	0
200	447	120	7	5	2	1	0	0	0	400	894	120	5	2	1	0	0	0	0
200	447	140	6	4	2	0	0	0	0	400	894	140	4	2	0	0	0	0	0

NOTE: Detector time constant at a reference velocity of 5 ft/sec.

For SI Units: 1 ft = 0.305 m
1000 BTU/sec = 1055 kW

Table B-3.2.4(m) Q_d, Threshold Fire Size at Response: 750 Btu/sec; t_g: 300 Seconds to 1000 Btu/sec; α: 0.011 Btu/sec³

τ	RTI	ΔT	CEILING HEIGHT IN FEET — INSTALLED SPACING OF DETECTORS						
			4.0	8.0	12.0	16.0	20.0	24.0	28.0
25	56	40	43	39	34	30	25	21	17
25	56	60	35	30	25	21	16	12	8
25	56	80	30	24	20	15	11	6	3
25	56	100	26	21	16	11	7	3	0
25	56	120	23	18	13	9	4	1	0
25	56	140	21	15	11	6	2	0	0
50	112	40	36	32	29	25	21	17	14
50	112	60	29	25	21	17	14	10	6
50	112	80	24	21	17	13	9	5	2
50	112	100	21	17	13	10	6	2	0
50	112	120	19	15	11	7	3	0	0
50	112	140	17	13	9	5	2	0	0
75	168	40	31	28	25	22	18	15	11
75	168	60	25	22	18	15	12	8	5
75	168	80	21	18	14	11	7	4	1
75	168	100	19	15	12	8	5	2	0
75	168	120	17	13	10	6	3	0	0
75	168	140	15	12	8	4	1	0	0
100	224	40	28	25	22	19	16	13	10
100	224	60	22	19	16	13	10	7	4
100	224	80	19	16	13	10	6	3	1
100	224	100	17	14	10	7	4	1	0
100	224	120	15	12	8	5	2	0	0
100	224	140	14	10	7	4	1	0	0
125	280	40	25	23	20	17	14	11	8
125	280	60	20	18	15	12	9	6	3
125	280	80	17	14	11	8	5	3	1
125	280	100	15	12	9	6	3	1	0
125	280	120	14	11	7	4	2	0	0
125	280	140	12	9	6	3	1	0	0
150	335	40	23	21	18	15	13	10	7
150	335	60	19	16	13	10	8	5	2
150	335	80	16	13	10	7	5	2	0
150	335	100	14	11	8	5	3	1	0
150	335	120	13	10	7	4	1	0	0
150	335	140	11	8	5	3	0	0	0
175	391	40	22	20	17	14	11	9	6
175	391	60	18	15	12	9	7	4	2
175	391	80	15	12	9	7	4	2	0
175	391	100	13	10	7	5	2	0	0
175	391	120	12	9	6	3	1	0	0
175	391	140	11	8	5	2	0	0	0
200	447	40	21	18	16	13	10	8	5
200	447	60	17	14	11	9	6	4	2
200	447	80	14	11	9	6	3	1	0
200	447	100	12	10	7	4	2	0	0
200	447	120	11	8	5	3	1	0	0
200	447	140	10	7	4	2	0	0	0

τ	RTI	ΔT	CEILING HEIGHT IN FEET — INSTALLED SPACING OF DETECTORS						
			4.0	8.0	12.0	16.0	20.0	24.0	28.0
225	503	40	20	17	15	12	9	7	5
225	503	60	16	13	11	8	5	3	1
225	503	80	13	11	8	5	3	1	0
225	503	100	12	9	6	4	2	0	0
225	503	120	10	8	5	2	1	0	0
225	503	140	9	7	4	2	0	0	0
250	559	40	19	16	14	11	9	6	4
250	559	60	15	12	10	7	5	3	1
250	559	80	13	10	7	5	3	1	0
250	559	100	11	8	6	3	1	0	0
250	559	120	10	7	4	2	1	0	0
250	559	140	9	6	4	1	0	0	0
275	615	40	18	16	13	10	8	6	3
275	615	60	14	12	9	7	4	2	1
275	615	80	12	9	7	4	2	1	0
275	615	100	10	8	5	3	1	0	0
275	615	120	9	7	4	2	0	0	0
275	615	140	8	6	3	1	0	0	0
300	671	40	17	15	12	10	7	5	3
300	671	60	14	11	9	6	4	2	1
300	671	80	11	9	6	4	2	1	0
300	671	100	10	7	5	3	1	0	0
300	671	120	9	6	4	2	0	0	0
300	671	140	8	5	3	1	0	0	0
325	727	40	16	14	12	9	7	5	3
325	727	60	13	11	8	6	3	2	0
325	727	80	11	9	6	4	2	0	0
325	727	100	10	7	5	2	1	0	0
325	727	120	8	6	3	1	0	0	0
325	727	140	8	5	3	1	0	0	0
350	783	40	16	14	11	9	6	4	2
350	783	60	13	10	8	5	3	1	0
350	783	80	11	8	6	3	1	0	0
350	783	100	9	7	4	2	1	0	0
350	783	120	8	6	3	1	0	0	0
350	783	140	7	5	2	1	0	0	0
375	839	40	15	13	11	8	6	4	2
375	839	60	12	10	7	5	3	1	0
375	839	80	10	8	5	3	1	0	0
375	839	100	9	6	4	2	0	0	0
375	839	120	8	5	3	1	0	0	0
375	839	140	7	5	2	1	0	0	0
400	894	40	15	13	10	8	5	3	2
400	894	60	12	9	7	5	3	1	0
400	894	80	10	7	5	3	1	0	0
400	894	100	8	6	4	2	0	0	0
400	894	120	7	5	3	1	0	0	0
400	894	140	7	4	2	0	0	0	0

NOTE: Detector time constant at a reference velocity of 5 ft/sec.

For SI Units: 1 ft = 0.305 m

1000 BTU/sec = 1055 kW

Table B-3.2.4(n) Q_d, Threshold Fire Size at Response: 750 Btu/sec; t_g: 500 Seconds to 1000 Btu/sec; α: 0.004 Btu/sec³

τ	RTI	ΔT	CEILING HEIGHT IN FEET						
			4.0	8.0	12.0	16.0	20.0	24.0	28.0
			INSTALLED SPACING OF DETECTORS						
25	56	40	52	45	39	34	29	24	20
25	56	60	41	34	28	23	18	14	9
25	56	80	34	28	22	17	12	8	4
25	56	100	29	23	18	13	8	4	0
25	56	120	26	20	14	10	5	1	0
25	56	140	23	17	12	7	3	0	0
50	112	40	44	40	35	30	26	22	18
50	112	60	35	30	26	21	17	12	8
50	112	80	30	25	20	15	11	7	3
50	112	100	26	21	16	12	7	3	0
50	112	120	23	18	13	9	5	1	0
50	112	140	21	16	11	7	3	0	0
75	168	40	39	35	31	27	24	20	16
75	168	60	31	27	23	19	15	11	7
75	168	80	26	22	18	14	10	6	3
75	168	100	23	19	15	10	7	3	0
75	168	120	20	16	12	8	4	1	0
75	168	140	18	14	10	6	2	0	0
100	224	40	35	32	29	25	21	18	14
100	224	60	28	25	21	17	14	10	6
100	224	80	24	20	16	13	9	5	2
100	224	100	21	17	13	10	6	2	0
100	224	120	19	15	11	7	4	0	0
100	224	140	17	13	9	5	2	0	0
125	280	40	32	30	26	23	20	16	13
125	280	60	26	23	19	16	12	9	6
125	280	80	22	19	15	12	8	5	2
125	280	100	19	16	12	9	5	2	0
125	280	120	17	14	10	7	3	0	0
125	280	140	16	12	8	5	2	0	0
150	335	40	30	28	25	21	18	15	12
150	335	60	24	21	18	15	11	8	5
150	335	80	21	17	14	11	7	4	1
150	335	100	18	15	11	8	5	2	0
150	335	120	16	13	9	6	3	0	0
150	335	140	15	11	8	4	1	0	0
175	391	40	28	26	23	20	17	14	10
175	391	60	23	20	17	14	10	7	4
175	391	80	19	16	13	10	7	4	1
175	391	100	17	14	11	7	4	1	0
175	391	120	15	12	9	5	2	0	0
175	391	140	14	10	7	4	1	0	0
200	447	40	27	24	22	18	16	12	10
200	447	60	22	19	16	13	10	7	4
200	447	80	18	15	12	9	6	3	1
200	447	100	16	13	10	7	4	1	0
200	447	120	14	11	8	5	2	0	0
200	447	140	13	10	7	4	1	0	0
225	503	40	26	23	20	17	14	12	9
225	503	60	20	18	15	12	9	6	3
225	503	80	17	14	11	8	6	3	1
225	503	100	15	12	9	6	3	1	0
225	503	120	13	11	7	4	2	0	0
225	503	140	12	9	6	3	1	0	0
250	559	40	24	22	19	16	14	11	8
250	559	60	19	17	14	11	8	5	3
250	559	80	16	14	11	8	5	3	1
250	559	100	14	12	9	6	3	1	0
250	559	120	13	10	7	4	2	0	0
250	559	140	12	9	6	3	1	0	0
275	615	40	23	21	18	16	13	10	7
275	615	60	19	16	13	10	8	5	3
275	615	80	16	13	10	7	5	2	1
275	615	100	14	11	8	5	3	1	0
275	615	120	12	9	7	4	1	0	0
275	615	140	11	8	5	3	1	0	0
300	671	40	22	20	18	15	12	9	7
300	671	60	18	15	13	10	7	5	2
300	671	80	15	13	10	7	4	2	0
300	671	100	13	11	8	5	2	1	0
300	671	120	12	9	6	3	1	0	0
300	671	140	11	8	5	2	0	0	0
325	727	40	22	19	17	14	11	9	6
325	727	60	17	15	12	9	7	4	2
325	727	80	15	12	9	6	4	2	0
325	727	100	13	10	7	5	2	1	0
325	727	120	11	9	6	3	1	0	0
325	727	140	10	7	5	2	0	0	0
350	783	40	21	19	16	13	11	8	6
350	783	60	17	14	12	9	6	4	2
350	783	80	14	12	9	6	4	2	0
350	783	100	12	10	7	4	2	0	0
350	783	120	11	8	5	3	1	0	0
350	783	140	10	7	4	2	0	0	0
375	839	40	20	18	16	13	10	8	5
375	839	60	16	14	11	8	6	3	2
375	839	80	14	11	8	6	3	1	0
375	839	100	12	9	7	4	2	0	0
375	839	120	11	8	5	3	1	0	0
375	839	140	10	7	4	2	0	0	0
400	894	40	20	17	15	12	10	7	5
400	894	60	16	13	11	8	5	3	1
400	894	80	13	11	8	5	3	1	0
400	894	100	11	9	6	4	2	0	0
400	894	120	10	8	5	3	1	0	0
400	894	140	9	7	4	2	0	0	0

NOTE: Detector time constant at a reference velocity of 5 ft/sec.

For SI Units: 1 ft = 0.305 m
1000 BTU/sec = 1055 kW

Table B-3.2.4(o) Q_d, Threshold Fire Size at Response: 750 Btu/sec; t_g: 600 Seconds to 1000 Btu/sec; α: 0.003 Btu/sec³

τ	RTI	ΔT	CEILING HEIGHT IN FEET						
			4.0	8.0	12.0	16.0	20.0	24.0	28.0
			INSTALLED SPACING OF DETECTORS						
25	56	40	55	47	41	35	30	25	21
25	56	60	43	36	29	24	19	14	10
25	56	80	36	28	23	18	13	8	4
25	56	100	31	24	18	13	9	4	0
25	56	120	27	20	15	10	6	1	0
25	56	140	24	18	12	8	3	0	0
50	112	40	47	42	37	32	28	23	19
50	112	60	37	32	27	22	18	13	9
50	112	80	31	26	21	16	12	8	4
50	112	100	27	22	17	12	8	4	0
50	112	120	24	19	14	9	5	1	0
50	112	140	22	16	11	7	3	0	0
75	168	40	42	38	34	29	25	21	17
75	168	60	33	29	25	20	16	12	8
75	168	80	28	24	19	15	11	7	3
75	168	100	24	20	15	11	7	3	0
75	168	120	22	17	13	9	5	1	0
75	168	140	20	15	11	6	3	0	0
100	224	40	38	35	31	27	23	19	16
100	224	60	30	27	23	19	15	11	7
100	224	80	26	22	18	14	10	6	3
100	224	100	22	18	14	10	7	3	0
100	224	120	20	16	12	8	4	1	0
100	224	140	18	14	10	6	2	0	0
125	280	40	35	32	29	25	22	18	14
125	280	60	28	25	21	17	14	10	7
125	280	80	24	20	16	13	9	6	2
125	280	100	21	17	13	10	6	3	0
125	280	120	19	15	11	7	4	1	0
125	280	140	17	13	9	5	2	0	0
150	335	40	33	30	27	23	20	17	13
150	335	60	26	23	20	16	13	9	6
150	335	80	22	19	15	12	8	5	2
150	335	100	19	16	12	9	5	2	0
150	335	120	17	14	10	7	3	0	0
150	335	140	16	12	8	5	2	0	0
175	391	40	31	28	25	22	19	15	12
175	391	60	25	22	18	15	12	9	5
175	391	80	21	18	14	11	8	4	2
175	391	100	18	15	12	8	5	2	0
175	391	120	16	13	10	6	3	0	0
175	391	140	15	11	8	5	1	0	0
200	447	40	29	27	24	21	17	14	11
200	447	60	23	21	17	14	11	8	5
200	447	80	20	17	14	10	7	4	1
200	447	100	17	14	11	8	4	2	0
200	447	120	15	12	9	6	3	0	0
200	447	140	14	11	7	4	1	0	0
225	503	40	28	26	23	19	16	13	10
225	503	60	22	20	17	13	10	7	4
225	503	80	19	16	13	10	7	4	1
225	503	100	16	13	10	7	4	1	0
225	503	120	15	12	8	5	2	0	0
225	503	140	13	10	7	4	1	0	0
250	559	40	27	24	21	18	15	12	10
250	559	60	21	19	16	13	10	7	4
250	559	80	18	15	12	9	6	3	1
250	559	100	16	13	10	7	4	1	0
250	559	120	14	11	8	5	2	0	0
250	559	140	13	10	7	4	1	0	0
275	615	40	26	23	20	18	15	12	9
275	615	60	21	18	15	12	9	6	4
275	615	80	17	15	12	9	6	3	1
275	615	100	15	12	9	6	3	1	0
275	615	120	13	11	7	5	2	0	0
275	615	140	12	9	6	3	1	0	0
300	671	40	25	22	20	17	14	11	8
300	671	60	20	17	14	11	8	6	3
300	671	80	17	14	11	8	5	3	1
300	671	100	15	12	9	6	3	1	0
300	671	120	13	10	7	4	2	0	0
300	671	140	12	9	6	3	1	0	0
325	727	40	24	22	19	16	13	10	8
325	727	60	19	16	14	11	8	5	3
325	727	80	16	13	10	8	5	2	1
325	727	100	14	11	8	5	3	1	0
325	727	120	12	10	7	4	2	0	0
325	727	140	11	8	5	3	1	0	0
350	783	40	23	21	18	15	13	10	7
350	783	60	18	16	13	10	7	5	3
350	783	80	15	13	10	7	5	2	1
350	783	100	13	11	8	5	3	1	0
350	783	120	12	9	6	4	1	0	0
350	783	140	11	8	5	3	1	0	0
375	839	40	22	20	17	15	12	9	7
375	839	60	18	15	13	10	7	5	2
375	839	80	15	12	10	7	4	2	0
375	839	100	13	10	8	5	0	1	0
375	839	120	12	9	6	3	1	0	0
375	839	140	11	8	5	2	0	0	0
400	894	40	22	19	17	14	11	9	6
400	894	60	17	15	12	9	7	4	2
400	894	80	15	12	9	6	4	2	0
400	894	100	13	10	7	5	2	1	0
400	894	120	11	9	6	3	1	0	0
400	894	140	10	7	5	2	0	0	0

NOTE: Detector time constant at a reference velocity of 5 ft/sec.

For SI Units: 1 ft = 0.305 m
1000 BTU/sec = 1055 kW

Table B-3.2.4(p) Q_d, Threshold Fire Size at Response: 1000 Btu/sec; t_g: 50 Seconds to 1000 Btu/sec; α: 0.400 Btu/sec^3

τ	RTI	ΔT	CEILING HEIGHT IN FEET / INSTALLED SPACING OF DETECTORS 4.0	8.0	12.0	16.0	20.0	24.0	28.0
25	56	40	22	20	17	14	11	8	5
25	56	60	18	16	13	10	7	4	2
25	56	80	16	13	10	7	4	2	0
25	56	100	14	11	8	5	3	0	0
25	56	120	13	10	7	4	1	0	0
25	56	140	12	9	6	3	1	0	0
50	112	40	17	15	12	9	7	4	2
50	112	60	14	11	9	6	4	2	0
50	112	80	12	9	7	4	2	0	0
50	112	100	11	8	5	3	1	0	0
50	112	120	10	7	4	2	0	0	0
50	112	140	9	6	3	1	0	0	0
75	168	40	14	12	9	7	4	2	1
75	168	60	12	9	7	4	2	1	0
75	168	80	10	7	5	3	1	0	0
75	168	100	9	6	4	2	0	0	0
75	168	120	8	5	3	1	0	0	0
75	168	140	7	4	2	0	0	0	0
100	224	40	12	10	8	5	3	1	0
100	224	60	10	8	5	3	1	0	0
100	224	80	8	6	4	2	0	0	0
100	224	100	7	5	3	1	0	0	0
100	224	120	7	4	2	0	0	0	0
100	224	140	6	4	1	0	0	0	0
125	280	40	11	9	6	4	2	1	0
125	280	60	9	7	4	2	1	0	0
125	280	80	8	5	3	1	0	0	0
125	280	100	7	4	2	0	0	0	0
125	280	120	6	3	1	0	0	0	0
125	280	140	5	3	1	0	0	0	0
150	335	40	0	8	5	3	2	0	0
150	335	60	8	6	3	2	0	0	0
150	335	80	7	4	2	0	0	0	0
150	335	100	6	3	2	0	0	0	0
150	335	120	5	3	1	0	0	0	0
150	335	140	5	2	0	0	0	0	0
175	391	40	9	7	5	3	1	0	0
175	391	60	7	5	3	1	0	0	0
175	391	80	6	4	2	0	0	0	0
175	391	100	5	3	1	0	0	0	0
175	391	120	5	2	0	0	0	0	0
175	391	140	4	2	0	0	0	0	0
200	447	40	9	6	4	2	1	0	0
200	447	60	7	5	2	1	0	0	0
200	447	80	6	3	1	0	0	0	0
200	447	100	5	3	1	0	0	0	0
200	447	120	4	2	0	0	0	0	0
200	447	140	4	2	0	0	0	0	0
225	503	40	8	6	4	2	0	0	0
225	503	60	6	4	2	0	0	0	0
225	503	80	5	3	1	0	0	0	0
225	503	100	5	2	0	0	0	0	0
225	503	120	4	2	0	0	0	0	0
225	503	140	3	1	0	0	0	0	0
250	559	40	8	5	3	1	0	0	0
250	559	60	6	4	2	0	0	0	0
250	559	80	5	3	1	0	0	0	0
250	559	100	4	2	0	0	0	0	0
250	559	120	4	2	0	0	0	0	0
250	559	140	3	1	0	0	0	0	0
275	615	40	7	5	3	1	0	0	0
275	615	60	6	3	1	0	0	0	0
275	615	80	5	2	0	0	0	0	0
275	615	100	4	2	0	0	0	0	0
275	615	120	3	1	0	0	0	0	0
275	615	140	3	1	0	0	0	0	0
300	671	40	7	5	3	1	0	0	0
300	671	60	5	3	1	0	0	0	0
300	671	80	4	2	0	0	0	0	0
300	671	100	4	2	0	0	0	0	0
300	671	120	3	1	0	0	0	0	0
300	671	140	3	1	0	0	0	0	0
325	727	40	7	4	2	1	0	0	0
325	727	60	5	3	1	0	0	0	0
325	727	80	4	2	0	0	0	0	0
325	727	100	3	1	0	0	0	0	0
325	727	120	3	1	0	0	0	0	0
325	727	140	3	0	0	0	0	0	0
350	783	40	6	4	2	0	0	0	0
350	783	60	5	3	1	0	0	0	0
350	783	80	4	2	0	0	0	0	0
350	783	100	3	1	0	0	0	0	0
350	783	120	3	1	0	0	0	0	0
350	783	140	2	0	0	0	0	0	0
375	839	40	6	4	2	0	0	0	0
375	839	60	5	2	0	0	0	0	0
375	839	80	4	2	0	0	0	0	0
375	839	100	3	1	0	0	0	0	0
375	839	120	3	0	0	0	0	0	0
375	839	140	2	0	0	0	0	0	0
400	894	40	6	4	2	0	0	0	0
400	894	60	4	2	0	0	0	0	0
400	894	80	3	1	0	0	0	0	0
400	894	100	3	1	0	0	0	0	0
400	894	120	2	0	0	0	0	0	0
400	894	140	2	0	0	0	0	0	0

NOTE: Detector time constant at a reference velocity of 5 ft/sec.

For SI Units: 1 ft = 0.305 m
1000 BTU/sec = 1055 kW

Table B-3.2.4(q) Q_d, Threshold Fire Size at Response: 1000 Btu/sec; t_g: 150 Seconds to 1000 Btu/sec; α: 0.044 Btu/sec^3

τ	RTI	ΔT	CEILING HEIGHT IN FEET 4.0	8.0	12.0	16.0	20.0	24.0	28.0
			INSTALLED SPACING OF DETECTORS						
25	56	40	39	36	32	28	25	21	17
25	56	60	32	28	24	21	17	13	9
25	56	80	27	24	20	16	12	8	4
25	56	100	24	20	16	12	8	4	1
25	56	120	22	18	14	10	6	2	0
25	56	140	20	16	12	8	4	0	0
50	112	40	31	29	26	22	19	16	13
50	112	60	25	23	19	16	13	9	6
50	112	80	22	19	15	12	9	6	3
50	112	100	19	16	13	9	6	3	0
50	112	120	17	14	11	7	4	1	0
50	112	140	16	13	9	6	2	0	0
75	168	40	27	24	22	19	16	13	10
75	168	60	22	19	16	13	10	7	4
75	168	80	19	16	13	10	7	4	2
75	168	100	16	14	11	7	4	2	0
75	168	120	15	12	9	6	3	1	0
75	168	140	13	10	7	4	2	0	0
100	224	40	24	22	19	16	13	10	8
100	224	60	19	17	14	11	8	6	3
100	224	80	16	14	11	8	5	3	1
100	224	100	14	12	9	6	3	1	0
100	224	120	13	10	7	5	2	0	0
100	224	140	12	9	6	3	1	0	0
125	280	40	21	19	17	14	11	9	6
125	280	60	17	15	12	10	7	4	2
125	280	80	15	12	10	7	4	2	1
125	280	100	13	10	8	5	3	1	0
125	280	120	12	9	6	4	1	0	0
125	280	140	11	8	5	3	1	0	0
150	335	40	20	18	15	13	10	7	5
150	335	60	16	14	11	8	6	4	2
150	335	80	14	11	9	6	3	2	0
150	335	100	12	9	7	4	2	1	0
150	335	120	11	8	5	3	1	0	0
150	335	140	10	7	4	2	1	0	0
175	391	40	18	16	14	11	9	6	4
175	391	60	15	13	10	7	5	3	1
175	391	80	13	10	8	5	3	1	0
175	391	100	11	9	6	4	2	0	0
175	391	120	10	7	5	2	1	0	0
175	391	140	9	6	4	2	0	0	0
200	447	40	17	15	13	10	8	5	3
200	447	60	14	12	9	7	4	2	1
200	447	80	12	9	7	4	2	1	0
200	447	100	10	8	5	3	1	0	0
200	447	120	9	7	4	2	1	0	0
200	447	140	8	6	3	1	0	0	0

τ	RTI	ΔT	CEILING HEIGHT IN FEET 4.0	8.0	12.0	16.0	20.0	24.0	28.0
			INSTALLED SPACING OF DETECTORS						
225	503	40	16	14	12	9	7	5	3
225	503	60	13	11	8	6	4	2	1
225	503	80	11	9	6	4	2	1	0
225	503	100	10	7	5	3	1	0	0
225	503	120	9	6	4	2	0	0	0
225	503	140	8	5	3	1	0	0	0
250	559	40	16	13	11	9	6	4	2
250	559	60	12	10	8	5	3	2	0
250	559	80	11	8	6	3	2	0	0
250	559	100	9	7	4	2	1	0	0
250	559	120	8	6	3	1	0	0	0
250	559	140	7	5	3	1	0	0	0
275	615	40	15	13	10	8	6	4	2
275	615	60	12	10	7	5	3	1	0
275	615	80	10	8	5	3	1	0	0
275	615	100	9	6	4	2	0	0	0
275	615	120	8	5	3	1	0	0	0
275	615	140	7	5	2	1	0	0	0
300	671	40	14	12	10	7	5	3	2
300	671	60	11	9	7	4	2	1	0
300	671	80	10	7	5	3	1	0	0
300	671	100	8	6	4	2	0	0	0
300	671	120	7	5	3	1	0	0	0
300	671	140	7	4	2	0	0	0	0
325	727	40	14	12	9	7	5	3	1
325	727	60	11	9	6	4	2	1	0
325	727	80	9	7	4	2	1	0	0
325	727	100	8	6	3	1	0	0	0
325	727	120	7	5	2	1	0	0	0
325	727	140	6	4	2	0	0	0	0
350	783	40	13	11	9	6	4	2	1
350	783	60	10	8	6	4	2	1	0
350	783	80	9	6	4	2	1	0	0
350	783	100	8	5	3	1	0	0	0
350	783	120	7	4	2	1	0	0	0
350	783	140	6	4	2	0	0	0	0
375	839	40	13	11	8	6	4	2	1
375	839	60	10	8	5	3	2	0	0
375	839	80	8	6	4	2	0	0	0
375	839	100	7	5	3	1	0	0	0
375	839	120	6	4	2	0	0	0	0
375	839	140	6	3	1	0	0	0	0
400	894	40	12	10	8	5	3	2	1
400	894	60	10	7	5	3	1	0	0
400	894	80	8	6	4	2	0	0	0
400	894	100	7	5	3	1	0	0	0
400	894	120	6	4	2	0	0	0	0
400	894	140	6	3	1	0	0	0	0

NOTE: Detector time constant at a reference velocity of 5 ft/sec.

For SI Units: 1 ft = 0.305 m

1000 BTU/sec = 1055 kW

Table B-3.2.4(r) Q_d, Threshold Fire Size at Response: 1000 Btu/sec; t_g: 300 Seconds to 1000 Btu/sec; α: 0.011 Btu/sec³

τ	RTI	ΔT	4.0	8.0	12.0	16.0	20.0	24.0	28.0
			CEILING HEIGHT IN FEET / INSTALLED SPACING OF DETECTORS						
25	56	40	52	47	42	37	32	28	23
25	56	60	42	36	31	26	22	17	13
25	56	80	35	30	25	20	15	11	7
25	56	100	31	25	20	15	11	7	3
25	56	120	28	22	17	12	8	4	0
25	56	140	25	19	14	10	5	1	0
50	112	40	43	40	36	32	28	24	20
50	112	60	35	31	27	23	19	15	11
50	112	80	30	25	21	17	13	9	5
50	112	100	26	22	18	13	9	6	2
50	112	120	23	19	15	11	7	3	0
50	112	140	21	17	12	8	4	1	0
75	168	40	37	35	31	28	24	21	17
75	168	60	30	27	24	20	16	13	9
75	168	80	26	22	19	15	11	8	4
75	168	100	23	19	15	12	8	5	1
75	168	120	20	17	13	9	6	2	0
75	168	140	19	15	11	7	4	1	0
100	224	40	34	31	28	25	22	18	15
100	224	60	27	24	21	18	14	11	8
100	224	80	23	20	17	13	10	7	4
100	224	100	20	17	14	10	7	4	1
100	224	120	18	15	12	8	5	2	0
100	224	140	17	13	10	6	3	0	0
125	280	40	31	29	26	23	19	16	13
125	280	60	25	22	19	16	13	10	7
125	280	80	21	18	15	12	9	6	3
125	280	100	19	16	13	9	6	3	1
125	280	120	17	14	10	7	4	1	0
125	280	140	15	12	9	6	3	0	0
150	335	40	29	27	24	21	18	15	12
150	335	60	23	21	18	15	12	9	6
150	335	80	20	17	14	11	8	5	2
150	335	100	17	14	11	8	5	3	1
150	335	120	16	13	10	6	4	1	0
150	335	140	14	11	8	5	2	0	0
175	391	40	27	25	22	19	16	13	11
175	391	60	22	19	16	13	10	8	5
175	391	80	18	16	13	10	7	4	2
175	391	100	16	13	10	8	5	2	0
175	391	120	15	12	9	6	3	1	0
175	391	140	13	10	7	4	2	0	0
200	447	40	25	23	21	18	15	12	9
200	447	60	20	18	15	12	10	7	4
200	447	80	17	15	12	9	6	4	2
200	447	100	15	13	10	7	4	2	0
200	447	120	14	11	8	5	3	1	0
200	447	140	12	10	7	4	2	0	0
225	503	40	24	22	19	17	14	11	9
225	503	60	19	17	14	11	9	6	4
225	503	80	16	14	11	8	6	3	1
225	503	100	14	12	9	6	4	2	0
225	503	120	13	10	7	5	2	1	0
225	503	140	12	9	6	3	1	0	0
250	559	40	23	21	18	16	13	10	8
250	559	60	18	16	13	11	8	5	3
250	559	80	16	13	10	8	5	3	1
250	559	100	14	11	8	6	3	1	0
250	559	120	12	10	7	4	2	0	0
250	559	140	11	8	6	3	1	0	0
275	615	40	22	20	17	15	12	9	7
275	615	60	18	15	13	10	7	5	3
275	615	80	15	12	10	7	5	2	1
275	615	100	13	11	8	5	3	1	0
275	615	120	12	9	6	4	2	0	0
275	615	140	11	8	5	3	1	0	0
300	671	40	21	19	16	14	11	9	6
300	671	60	17	15	12	9	7	4	2
300	671	80	14	12	9	7	4	2	1
300	671	100	13	10	7	5	3	1	0
300	671	120	11	9	6	3	1	0	0
300	671	140	10	8	5	3	1	0	0
325	727	40	20	18	16	13	11	8	6
325	727	60	16	14	11	9	6	4	2
325	727	80	14	11	9	6	4	2	1
325	727	100	12	10	7	4	2	1	0
325	727	120	11	8	6	3	1	0	0
325	727	140	10	7	5	2	1	0	0
350	783	40	20	18	15	13	10	8	5
350	783	60	16	13	11	8	6	4	2
350	783	80	13	11	8	6	3	2	0
350	783	100	12	9	7	4	2	1	0
350	783	120	10	8	5	3	1	0	0
350	783	140	9	7	4	2	1	0	0
375	839	40	19	17	14	12	9	7	5
375	839	60	15	13	10	8	5	3	2
375	839	80	13	10	8	5	3	1	0
375	839	100	11	9	6	4	2	0	0
375	839	120	10	7	5	3	1	0	0
375	839	140	9	6	4	2	0	0	0
400	894	40	18	16	14	11	9	7	4
400	894	60	15	12	10	7	5	3	1
400	894	80	12	10	7	5	3	1	0
400	894	100	11	8	6	3	2	0	0
400	894	120	10	7	5	2	1	0	0
400	894	140	9	6	4	2	0	0	0

NOTE: Detector time constant at a reference velocity of 5 ft/sec.

For SI Units: 1 ft = 0.305 m
1000 BTU/sec = 1055 kW

Table B-3.2.4(s) Q_d, Threshold Fire Size at Response: 1000 Btu/sec; t_g: 500 Seconds to 1000 Btu/sec; α: 0.004 Btu/sec³

τ	RTI	ΔT	CEILING HEIGHT IN FEET							τ	RTI	ΔT	CEILING HEIGHT IN FEET						
			4.0	8.0	12.0	16.0	20.0	24.0	28.0				4.0	8.0	12.0	16.0	20.0	24.0	28.0
			INSTALLED SPACING OF DETECTORS										INSTALLED SPACING OF DETECTORS						
25	56	40	62	54	48	42	37	31	27	225	503	40	31	29	26	23	20	17	14
25	56	60	49	41	35	29	24	19	15	225	503	60	25	22	19	16	13	10	7
25	56	80	41	33	27	22	17	12	8	225	503	80	21	18	15	12	9	6	3
25	56	100	35	28	22	17	12	8	4	225	503	100	19	16	13	9	6	3	1
25	56	120	31	24	18	13	9	4	0	225	503	120	17	14	10	7	4	2	0
25	56	140	28	21	16	11	6	2	0	225	503	140	15	12	9	6	3	0	0
50	112	40	53	48	43	38	33	29	24	250	559	40	30	28	25	22	19	16	13
50	112	60	42	37	32	27	22	18	14	250	559	60	24	21	18	15	12	9	7
50	112	80	35	30	25	20	16	11	7	250	559	80	20	18	15	11	8	6	3
50	112	100	31	25	20	16	11	7	3	250	559	100	18	15	12	9	6	3	1
50	112	120	27	22	17	12	8	4	0	250	559	120	16	13	10	7	4	1	0
50	112	140	25	19	14	10	6	2	0	250	559	140	14	11	8	5	2	0	0
75	168	40	47	43	39	35	31	26	22	275	615	40	29	27	24	21	18	15	12
75	168	60	38	33	29	25	20	16	12	275	615	60	23	20	18	15	12	9	6
75	168	80	32	27	23	19	14	10	6	275	615	80	20	17	14	11	8	5	3
75	168	100	28	23	19	14	10	6	3	275	615	100	17	14	11	8	5	3	1
75	168	120	25	20	16	11	7	3	0	275	615	120	15	12	9	6	4	1	0
75	168	140	22	18	13	9	5	1	0	275	615	140	14	11	8	5	2	0	0
100	224	40	43	40	36	32	28	24	20	300	671	40	28	25	23	20	17	14	11
100	224	60	34	31	27	23	19	15	11	300	671	60	22	20	17	14	11	8	5
100	224	80	29	25	21	17	13	9	6	300	671	80	19	16	13	10	7	5	2
100	224	100	25	21	17	13	9	6	2	300	671	100	16	14	11	8	5	2	1
100	224	120	23	19	15	11	7	3	0	300	671	120	15	12	9	6	3	1	0
100	224	140	21	16	12	8	5	1	0	300	671	140	13	10	7	5	2	0	0
125	280	40	39	37	33	30	26	22	19	325	727	40	27	25	22	19	16	13	11
125	280	60	32	28	25	21	17	14	10	325	727	60	21	19	16	13	10	8	5
125	280	80	27	23	20	16	12	9	5	325	727	80	18	16	13	10	7	4	2
125	280	100	24	20	16	12	9	5	2	325	727	100	16	13	10	7	5	2	0
125	280	120	21	17	14	10	6	3	0	325	727	120	14	11	8	6	3	1	0
125	280	140	19	15	11	8	4	1	0	325	727	140	13	10	7	4	2	0	0
150	335	40	37	34	31	28	24	21	17	350	783	40	26	24	21	18	15	13	10
150	335	60	30	27	23	20	16	13	9	350	783	60	21	18	15	13	10	7	5
150	335	80	25	22	18	15	11	8	5	350	783	80	18	15	12	9	7	4	2
150	335	100	22	19	15	12	8	5	2	350	783	100	15	13	10	7	4	2	0
150	335	120	20	16	13	9	6	2	0	350	783	120	14	11	8	5	3	1	0
150	335	140	18	14	11	7	4	1	0	350	783	140	12	10	7	4	2	0	0
175	391	40	35	32	29	26	23	19	16	375	839	40	25	23	20	18	15	12	9
175	391	60	28	25	22	18	15	12	8	375	839	60	20	18	15	12	9	7	4
175	391	80	24	20	17	14	10	7	4	375	839	80	17	14	12	9	6	4	2
175	391	100	21	18	14	11	7	4	1	375	839	100	15	12	9	7	4	2	0
175	391	120	19	15	12	8	5	2	0	375	839	120	13	11	8	5	3	1	0
175	391	140	17	13	10	7	3	1	0	375	839	140	12	9	6	4	1	0	0
200	447	40	33	30	28	24	21	18	15	400	894	40	24	22	20	17	14	11	9
200	447	60	26	24	20	17	14	11	8	400	894	60	19	17	14	12	9	6	4
200	447	80	22	19	16	13	10	7	4	400	894	80	16	14	11	8	6	3	1
200	447	100	20	17	13	10	7	4	1	400	894	100	14	12	9	6	4	2	0
200	447	120	18	14	11	8	5	2	0	400	894	120	13	10	7	5	2	1	0
200	447	140	16	13	9	6	3	1	0	400	894	140	12	9	6	4	1	0	0

NOTE: Detector time constant at a reference velocity of 5 ft/sec.

For SI Units: 1 ft = 0.305 m

1000 BTU/sec = 1055 kW

Table B-3.2.4(t) Q_d, Threshold Fire Size at Response: 1000 Btu/sec; t_g: 600 Seconds to 1000 Btu/sec; α: 0.003 Btu/sec³

τ	RTI	ΔT	CEILING HEIGHT IN FEET							τ	RTI	ΔT	CEILING HEIGHT IN FEET						
			4.0	8.0	12.0	16.0	20.0	24.0	28.0				4.0	8.0	12.0	16.0	20.0	24.0	28.0
			INSTALLED SPACING OF DETECTORS										INSTALLED SPACING OF DETECTORS						
25	56	40	65	56	50	43	38	33	28	225	503	40	34	32	29	26	22	19	16
25	56	60	51	43	36	30	25	20	15	225	503	60	27	25	21	18	15	12	8
25	56	80	42	34	28	23	18	13	8	225	503	80	23	20	17	14	10	7	4
25	56	100	36	29	23	17	13	8	4	225	503	100	20	17	14	11	7	4	1
25	56	120	32	25	19	14	9	5	1	225	503	120	18	15	12	8	5	2	0
25	56	140	29	21	16	11	6	2	0	225	503	140	17	13	10	6	3	1	0
50	112	40	56	51	45	40	35	30	26	250	559	40	33	30	27	24	21	18	15
50	112	60	45	39	33	28	23	19	14	250	559	60	26	23	20	17	14	11	8
50	112	80	37	31	26	21	16	12	8	250	559	80	22	19	16	13	10	7	4
50	112	100	33	26	21	16	12	7	3	250	559	100	19	16	13	10	7	4	1
50	112	120	29	23	18	13	8	4	0	250	559	120	17	14	11	8	5	2	0
50	112	140	26	20	15	10	6	2	0	250	559	140	16	13	9	6	3	1	0
75	168	40	50	46	42	37	32	28	24	275	615	40	31	29	26	23	20	17	14
75	168	60	40	35	31	26	22	17	13	275	615	60	25	23	19	16	13	10	7
75	168	80	34	29	24	20	15	11	7	275	615	80	21	18	15	12	9	6	3
75	168	100	30	25	20	15	11	7	3	275	615	100	19	16	13	9	6	3	1
75	168	120	26	21	17	12	8	4	0	275	615	120	17	14	11	7	4	2	0
75	168	140	24	19	14	10	5	2	0	275	615	140	15	12	9	6	3	0	0
100	224	40	46	43	38	34	30	26	22	300	671	40	30	28	25	22	19	16	13
100	224	60	37	33	28	24	20	16	12	300	671	60	24	22	19	16	13	10	7
100	224	80	31	27	23	18	14	10	6	300	671	80	21	18	15	12	9	6	3
100	224	100	27	23	18	14	10	6	3	300	671	100	18	15	12	9	6	3	1
100	224	120	24	20	15	11	7	3	0	300	671	120	16	13	10	7	4	1	0
100	224	140	22	17	13	9	5	1	0	300	671	140	15	12	8	5	3	0	0
125	280	40	43	40	36	32	28	24	21	325	727	40	29	27	24	21	18	15	13
125	280	60	34	31	27	23	19	15	11	325	727	60	23	21	18	15	12	9	6
125	280	80	29	25	21	17	13	10	6	325	727	80	20	17	14	11	8	5	3
125	280	100	25	21	17	13	10	6	2	325	727	100	17	15	12	8	6	3	1
125	280	120	23	19	15	11	7	3	0	325	727	120	16	13	10	7	4	1	0
125	280	140	21	16	12	8	5	1	0	325	727	140	14	11	8	5	2	0	0
150	335	40	40	37	34	30	26	23	19	350	783	40	28	26	23	21	18	15	12
150	335	60	32	29	25	21	18	14	11	350	783	60	23	20	17	14	11	9	6
150	335	80	27	24	20	16	12	9	5	350	783	80	19	17	14	11	8	5	3
150	335	100	24	20	16	13	9	5	2	350	783	100	17	14	11	8	5	3	1
150	335	120	21	17	14	10	6	3	0	350	783	120	15	12	9	6	3	1	0
150	335	140	19	15	12	8	4	1	0	350	783	140	14	11	8	5	2	0	0
175	391	40	38	35	32	28	25	21	18	375	839	40	28	25	23	20	17	14	11
175	391	60	30	27	24	20	17	13	10	375	839	60	22	19	17	14	11	8	5
175	391	80	26	22	19	15	12	8	5	375	839	80	19	16	13	10	7	5	2
175	391	100	22	19	15	12	8	5	2	375	839	100	16	14	11	8	5	2	1
175	391	120	20	17	13	9	6	3	0	375	839	120	15	12	9	6	3	1	0
175	391	140	18	15	11	7	4	1	0	375	839	140	13	10	7	5	2	0	0
200	447	40	36	33	30	27	24	20	17	400	894	40	27	25	22	19	16	13	11
200	447	60	29	26	22	19	16	12	9	400	894	60	21	19	16	13	10	8	5
200	447	80	24	21	18	14	11	8	4	400	894	80	18	15	13	10	7	4	2
200	447	100	21	18	15	11	8	4	2	400	894	100	16	13	10	7	5	2	1
200	447	120	19	16	12	9	5	2	0	400	894	120	14	11	8	6	3	1	0
200	447	140	17	14	10	7	4	1	0	400	894	140	13	10	7	4	2	0	0

NOTE: Detector time constant at a reference velocity of 5 ft/sec.

For SI Units: 1 ft = 0.305 m
1000 BTU/sec = 1055 kW

Table B-3.2.4(u) Q_d, Threshold Fire Size at Response: 2000 Btu/sec; t_g: 50 Seconds to 1000 Btu/sec; α: 0.400 Btu/sec^3

τ	RTI	ΔT	CEILING HEIGHT IN FEET — INSTALLED SPACING OF DETECTORS 4.0	8.0	12.0	16.0	20.0	24.0	28.0
25	56	40	35	33	31	28	25	21	18
25	56	60	30	27	24	21	18	15	11
25	56	80	26	23	20	17	14	10	7
25	56	100	23	21	17	14	11	7	4
25	56	120	21	18	15	12	8	5	2
25	56	140	20	17	13	10	7	3	1
50	112	40	28	26	23	21	18	15	12
50	112	60	23	21	18	15	13	10	7
50	112	80	20	18	15	12	9	6	4
50	112	100	18	15	13	10	7	4	2
50	112	120	16	14	11	8	5	3	1
50	112	140	15	12	10	7	4	2	0
75	168	40	24	22	19	17	14	11	9
75	168	60	20	17	15	12	9	7	4
75	168	80	17	15	12	9	7	4	2
75	168	100	15	13	10	7	5	3	1
75	168	120	14	11	8	6	3	1	0
75	168	140	13	10	7	5	2	1	0
100	224	40	21	19	17	14	11	9	6
100	224	60	17	15	13	10	7	5	3
100	224	80	15	13	10	7	5	3	1
100	224	100	13	11	8	6	3	2	0
100	224	120	12	10	7	4	2	1	0
100	224	140	11	8	6	3	2	0	0
125	280	40	19	17	15	12	10	7	5
125	280	60	16	13	11	8	6	4	2
125	280	80	13	11	9	6	4	2	1
125	280	100	12	10	7	5	2	1	0
125	280	120	11	8	6	3	2	0	0
125	280	140	10	7	5	3	1	0	0
150	335	40	18	16	13	11	8	6	4
150	335	60	14	12	10	7	5	3	1
150	335	80	12	10	7	5	3	1	0
150	335	100	11	8	6	4	2	0	0
150	335	120	10	7	5	3	1	0	0
150	335	140	9	6	4	2	0	0	0
175	391	40	16	14	12	9	7	5	3
175	391	60	13	11	9	6	4	2	1
175	391	80	11	9	7	4	2	1	0
175	391	100	10	8	5	3	1	0	0
175	391	120	9	7	4	2	1	0	0
175	391	140	8	6	3	2	0	0	0
200	447	40	15	13	11	8	6	4	2
200	447	60	12	10	8	5	3	2	0
200	447	80	11	8	6	4	2	1	0
200	447	100	9	7	5	3	1	0	0
200	447	120	8	6	4	2	0	0	0
200	447	140	8	5	3	1	0	0	0
225	503	40	15	12	10	8	5	3	2
225	503	60	12	9	7	5	3	1	0
225	503	80	10	8	5	3	2	0	0
225	503	100	9	6	4	2	1	0	0
225	503	120	8	5	3	1	0	0	0
225	503	140	7	5	3	1	0	0	0
250	559	40	14	12	9	7	5	3	2
250	559	60	11	9	6	4	2	1	0
250	559	80	9	7	5	3	1	0	0
250	559	100	8	6	4	2	0	0	0
250	559	120	7	5	3	1	0	0	0
250	559	140	7	4	2	1	0	0	0
275	615	40	13	11	9	6	4	2	1
275	615	60	10	8	6	4	2	1	0
275	615	80	9	7	4	2	1	0	0
275	615	100	8	5	3	2	0	0	0
275	615	120	7	5	2	1	0	0	0
275	615	140	6	4	2	0	0	0	0
300	671	40	13	10	8	6	4	2	1
300	671	60	10	8	5	3	2	0	0
300	671	80	8	6	4	2	1	0	0
300	671	100	7	5	3	1	0	0	0
300	671	120	7	4	2	1	0	0	0
300	671	140	6	4	2	0	0	0	0
325	727	40	12	10	8	5	3	2	1
325	727	60	10	7	5	3	1	0	0
325	727	80	8	6	4	2	0	0	0
325	727	100	7	5	3	1	0	0	0
325	727	120	6	4	2	0	0	0	0
325	727	140	6	3	1	0	0	0	0
350	783	40	12	9	7	5	3	2	0
350	783	60	9	7	5	3	1	0	0
350	783	80	8	5	3	2	0	0	0
350	783	100	7	4	2	1	0	0	0
350	783	120	6	4	2	0	0	0	0
350	783	140	5	3	1	0	0	0	0
375	839	40	11	9	7	4	3	1	0
375	839	60	9	7	4	2	1	0	0
375	839	80	7	5	3	1	0	0	0
375	839	100	6	4	2	0	0	0	0
375	839	120	6	3	2	0	0	0	0
375	839	140	5	3	1	0	0	0	0
400	894	40	11	9	6	4	2	1	0
400	894	60	9	6	4	2	1	0	0
400	894	80	7	5	3	1	0	0	0
400	894	100	6	4	2	0	0	0	0
400	894	120	5	3	1	0	0	0	0
400	894	140	5	3	1	0	0	0	0

NOTE: Detector time constant at a reference velocity of 5 ft/sec.

For SI Units: 1 ft = 0.305 m

1000 BTU/sec = 1055 kW

Table B-3.2.4(v) Q_d, Threshold Fire Size at Response: 2000 Btu/sec; t_g: 150 Seconds to 1000 Btu/sec; α: 0.044 Btu/sec^3

τ	RTI	ΔT	CEILING HEIGHT IN FEET 4.0	8.0	12.0	16.0	20.0	24.0	28.0	τ	RTI	ΔT	CEILING HEIGHT IN FEET 4.0	8.0	12.0	16.0	20.0	24.0	28.0
			INSTALLED SPACING OF DETECTORS										INSTALLED SPACING OF DETECTORS						
25	56	40	60	57	53	49	44	40	36	225	503	40	27	26	23	21	18	15	13
25	56	60	50	46	41	37	32	28	23	225	503	60	22	20	18	15	12	10	7
25	56	80	43	38	34	29	25	20	16	225	503	80	19	17	14	11	9	6	4
25	56	100	38	33	28	24	19	15	11	225	503	100	17	14	12	9	7	4	2
25	56	120	34	29	25	20	15	11	7	225	503	120	15	13	10	7	5	3	1
25	56	140	31	26	21	17	13	8	4	225	503	140	14	11	9	6	4	2	0
50	112	40	49	47	44	40	37	33	30	250	559	40	26	24	22	19	17	14	12
50	112	60	40	38	34	31	27	23	19	250	559	60	21	19	16	14	11	9	6
50	112	80	35	32	28	24	21	17	13	250	559	80	18	16	13	11	8	6	4
50	112	100	31	28	24	20	16	12	9	250	559	100	16	14	11	8	6	4	2
50	112	120	28	24	21	17	13	9	6	250	559	120	14	12	9	7	4	2	1
50	112	140	26	22	18	14	10	7	3	250	559	140	13	11	8	5	3	1	0
75	168	40	43	41	38	35	32	28	25	275	615	40	25	23	21	18	16	13	11
75	168	60	35	33	30	26	23	20	16	275	615	60	20	18	16	13	11	8	6
75	168	80	30	28	24	21	18	14	11	275	615	80	17	15	13	10	7	5	3
75	168	100	27	24	21	17	14	10	7	275	615	100	15	13	10	8	5	3	2
75	168	120	24	21	18	14	11	8	4	275	615	120	14	11	9	6	4	2	1
75	168	140	22	19	16	12	9	5	2	275	615	140	12	10	7	5	3	1	0
100	224	40	38	36	34	31	28	25	22	300	671	40	24	22	20	17	15	12	10
100	224	60	31	29	26	23	20	17	14	300	671	60	19	17	15	12	10	7	5
100	224	80	27	25	22	18	15	12	9	300	671	80	17	14	12	9	7	5	3
100	224	100	24	21	18	15	12	9	6	300	671	100	15	12	10	7	5	3	1
100	224	120	22	19	16	13	9	6	3	300	671	120	13	11	8	6	3	2	0
100	224	140	20	17	14	11	7	4	2	300	671	140	12	10	7	5	2	1	0
125	280	40	35	33	31	28	25	22	19	325	727	40	23	21	19	16	14	11	9
125	280	60	29	27	24	21	18	15	12	325	727	60	19	17	14	12	9	7	5
125	280	80	25	22	19	16	13	11	8	325	727	80	16	14	11	9	6	4	2
125	280	100	22	19	16	13	10	7	5	325	727	100	14	12	9	7	4	2	1
125	280	120	20	17	14	11	8	5	3	325	727	120	13	10	8	5	3	1	0
125	280	140	18	15	12	9	6	4	1	325	727	140	11	9	7	4	2	1	0
150	335	40	32	31	28	26	23	20	17	350	783	40	22	20	18	16	13	11	8
150	335	60	27	24	22	19	16	13	10	350	783	60	18	16	13	11	9	6	4
150	335	80	23	20	18	15	12	9	6	350	783	80	15	13	11	8	6	4	2
150	335	100	20	18	15	12	9	6	4	350	783	100	13	11	9	6	4	2	1
150	335	120	18	16	13	10	7	4	2	350	783	120	12	10	7	5	3	1	0
150	335	140	17	14	11	8	5	3	1	350	783	140	11	9	6	4	2	1	0
175	391	40	30	29	26	24	21	18	15	375	839	40	22	20	17	15	12	10	8
175	391	60	25	23	20	17	15	12	9	375	839	60	17	15	13	10	8	6	4
175	391	80	21	19	16	14	11	8	6	375	839	80	15	13	10	8	5	3	2
175	391	100	19	16	14	11	8	6	3	375	839	100	13	11	8	6	4	2	1
175	391	120	17	15	12	9	6	4	2	375	839	120	12	9	7	5	3	1	0
175	391	140	16	13	10	7	5	2	1	375	839	140	11	8	6	3	2	0	0
200	447	40	29	27	25	22	19	17	14	400	894	40	21	19	17	14	12	9	7
200	447	60	23	21	19	16	13	11	8	400	894	60	17	15	12	10	8	5	3
200	447	80	20	18	15	12	10	7	5	400	894	80	14	12	10	7	5	3	2
200	447	100	18	15	13	10	7	5	3	400	894	100	13	10	8	5	3	2	1
200	447	120	16	14	11	8	5	3	1	400	894	120	11	9	6	4	2	1	0
200	447	140	15	12	9	7	4	2	1	400	894	140	10	8	5	3	2	0	0

NOTE: Detector time constant at a reference velocity of 5 ft/sec.

For SI Units: 1 ft = 0.305 m

1000 BTU/sec = 1055 kW

Appendix B: Engineering Guide for Automatic Fire Detector Spacing

Table B-3.2.4(w) Q_d, Threshold Fire Size at Response: 2000 Btu/sec; t_g: 300 Seconds to 1000 Btu/sec; α: 0.011 Btu/sec³

τ	RTI	ΔT	CEILING HEIGHT IN FEET						
			4.0	8.0	12.0	16.0	20.0	24.0	28.0
			INSTALLED SPACING OF DETECTORS						
25	56	40	79	73	66	60	55	49	44
25	56	60	64	56	50	44	39	33	29
25	56	80	54	46	40	34	29	24	20
25	56	100	47	40	33	28	23	18	14
25	56	120	42	35	29	23	18	14	9
25	56	140	38	31	25	20	15	10	6
50	112	40	67	63	58	54	49	44	40
50	112	60	54	49	45	40	35	30	26
50	112	80	46	41	36	31	27	22	18
50	112	100	41	35	30	26	21	17	12
50	112	120	36	31	26	21	17	13	8
50	112	140	33	28	23	18	14	9	5
75	168	40	59	56	52	48	44	40	36
75	168	60	48	44	40	36	32	28	24
75	168	80	41	37	33	29	24	20	16
75	168	100	36	32	28	23	19	15	11
75	168	120	33	28	24	20	15	11	7
75	168	140	30	25	21	17	12	9	5
100	224	40	53	51	48	44	41	37	33
100	224	60	43	40	37	33	29	25	22
100	224	80	37	34	30	26	22	19	15
100	224	100	33	29	25	22	18	14	10
100	224	120	30	26	22	18	14	10	7
100	224	140	27	23	19	15	11	8	4
125	280	40	49	47	44	41	37	34	31
125	280	60	40	37	34	31	27	23	20
125	280	80	34	31	28	24	21	17	14
125	280	100	30	27	23	20	16	13	9
125	280	120	27	24	20	17	13	9	6
125	280	140	25	22	18	14	11	7	4
150	335	40	46	44	41	38	35	32	28
150	335	60	37	35	32	28	25	22	18
150	335	80	32	29	26	23	19	16	12
150	335	100	28	25	22	19	15	12	8
150	335	120	26	22	19	15	12	9	5
150	335	140	23	20	17	13	10	6	3
175	391	40	43	41	39	36	33	29	26
175	391	60	35	33	30	27	23	20	17
175	391	80	30	28	24	21	18	15	11
175	391	100	27	24	21	17	14	11	8
175	391	120	24	21	18	14	11	8	5
175	391	140	22	19	16	12	9	6	3
200	447	40	41	39	37	34	31	28	25
200	447	60	33	31	28	25	22	19	16
200	447	80	29	26	23	20	17	14	10
200	447	100	25	22	19	16	13	10	7
200	447	120	23	20	17	14	10	7	4
200	447	140	21	18	15	11	8	5	2

τ	RTI	ΔT	CEILING HEIGHT IN FEET						
			4.0	8.0	12.0	16.0	20.0	24.0	28.0
			INSTALLED SPACING OF DETECTORS						
225	503	40	39	37	35	32	29	26	23
225	503	60	32	30	27	24	21	18	15
225	503	80	27	25	22	19	16	13	10
225	503	100	24	21	18	15	12	9	6
225	503	120	22	19	16	13	10	7	4
225	503	140	20	17	14	11	8	5	2
250	559	40	37	36	33	30	28	25	22
250	559	60	30	28	25	23	19	17	14
250	559	80	26	24	21	18	15	12	9
250	559	100	23	20	17	14	11	8	6
250	559	120	21	18	15	12	9	6	3
250	559	140	19	16	13	10	7	4	2
275	615	40	36	34	32	29	26	23	20
275	615	60	29	27	24	21	18	16	13
275	615	80	25	23	20	17	14	11	8
275	615	100	22	19	17	14	11	8	5
275	615	120	20	17	14	11	8	6	3
275	615	140	18	15	12	9	7	4	2
300	671	40	35	33	30	28	25	22	19
300	671	60	28	26	23	20	18	15	13
300	671	80	24	22	19	16	13	10	8
300	671	100	21	19	16	13	10	7	5
300	671	120	19	16	14	11	8	5	3
300	671	140	17	15	12	9	6	4	1
325	727	40	33	32	29	27	24	21	18
325	727	60	27	25	22	20	17	14	11
325	727	80	23	21	18	15	12	10	7
325	727	100	20	18	15	12	10	7	4
325	727	120	18	16	13	10	7	5	2
325	727	140	17	14	11	8	6	3	1
350	783	40	32	31	28	26	23	20	17
350	783	60	26	24	21	19	16	13	11
350	783	80	22	20	17	15	12	9	7
350	783	100	20	17	15	12	9	6	4
350	783	120	18	15	12	10	7	4	2
350	783	140	16	14	11	8	5	3	1
375	839	40	31	30	27	25	22	19	16
375	839	60	25	23	21	18	15	13	10
375	839	80	22	19	17	14	11	9	6
375	839	100	19	17	14	11	9	6	4
375	839	120	17	15	12	9	6	4	2
375	839	140	16	13	10	8	5	3	1
400	894	40	30	29	26	24	21	18	16
400	894	60	25	23	20	17	15	12	9
400	894	80	21	19	16	13	11	8	5
400	894	100	19	16	13	11	8	6	3
400	894	120	17	14	11	9	6	4	2
400	894	140	15	13	10	7	5	2	1

NOTE: Detector time constant at a reference velocity of 5 ft/sec.

For SI Units: 1 ft = 0.305 m
                1000 BTU/sec = 1055 kW

Table B-3.2.4(x) Q_d, Threshold Fire Size at Response: 2000 Btu/sec; t_g: 500 Seconds to 1000 Btu/sec; α: 0.004 Btu/sec³

τ	RTI	ΔT	4.0	8.0	12.0	16.0	20.0	24.0	28.0
			\multicolumn{7}{c}{INSTALLED SPACING OF DETECTORS — CEILING HEIGHT IN FEET}
25	56	40	92	82	74	67	60	54	49
25	56	60	72	62	55	48	42	36	31
25	56	80	61	51	43	37	32	26	22
25	56	100	52	43	36	30	25	20	15
25	56	120	46	37	31	25	20	15	10
25	56	140	42	33	27	21	16	11	7
50	112	40	81	74	68	62	56	51	46
50	112	60	64	57	51	45	40	35	30
50	112	80	54	47	41	35	30	25	20
50	112	100	47	40	34	29	23	19	14
50	112	120	42	35	29	24	19	14	10
50	112	140	38	31	25	20	15	11	6
75	168	40	73	68	63	58	53	48	43
75	168	60	58	53	47	42	37	33	28
75	168	80	49	44	38	33	28	24	19
75	168	100	43	37	32	27	22	18	13
75	168	120	39	33	27	22	18	13	9
75	168	140	35	29	24	19	14	10	6
100	224	40	67	63	58	54	50	45	41
100	224	60	54	49	45	40	35	31	27
100	224	80	46	41	36	31	27	23	18
100	224	100	40	35	30	26	21	17	13
100	224	120	36	31	26	21	17	13	9
100	224	140	33	27	23	18	14	10	6
125	280	40	62	59	55	51	47	43	38
125	280	60	50	46	42	38	33	29	25
125	280	80	43	38	34	30	25	21	17
125	280	100	37	33	29	24	20	16	12
125	280	120	34	29	25	20	16	12	8
125	280	140	31	26	22	17	13	9	5
150	335	40	58	55	52	48	44	40	36
150	335	60	47	44	40	36	32	28	24
150	335	80	40	36	32	28	24	20	16
150	335	100	35	31	27	23	19	15	11
150	335	120	32	27	23	19	15	11	8
150	335	140	29	25	20	16	12	9	5
175	391	40	55	53	49	46	42	38	35
175	391	60	44	41	38	34	30	26	23
175	391	80	38	34	31	27	23	19	15
175	391	100	33	30	26	22	18	14	11
175	391	120	30	26	22	18	15	11	7
175	391	140	27	23	20	16	12	8	4
200	447	40	52	50	47	44	40	36	33
200	447	60	42	39	36	32	29	25	21
200	447	80	36	33	29	26	22	18	15
200	447	100	32	28	25	21	17	14	10
200	447	120	29	25	21	18	14	10	7
200	447	140	26	22	19	15	11	8	4
225	503	40	50	48	45	42	38	35	31
225	503	60	40	38	34	31	27	24	20
225	503	80	35	31	28	24	21	17	14
225	503	100	30	27	24	20	16	13	9
225	503	120	27	24	20	17	13	10	6
225	503	140	25	21	18	14	11	7	4
250	559	40	48	46	43	40	37	33	30
250	559	60	39	36	33	30	26	23	19
250	559	80	33	30	27	23	20	17	13
250	559	100	29	26	23	19	16	12	9
250	559	120	26	23	20	16	13	9	6
250	559	140	24	21	17	14	10	7	4
275	615	40	46	44	41	38	35	32	29
275	615	60	37	35	32	28	25	22	19
275	615	80	32	29	26	22	19	16	13
275	615	100	28	25	22	18	15	12	8
275	615	120	25	22	19	15	12	9	5
275	615	140	23	20	16	13	10	6	3
300	671	40	45	43	40	37	34	31	27
300	671	60	36	34	31	27	24	21	18
300	671	80	31	28	25	22	18	15	12
300	671	100	27	24	21	18	14	11	8
300	671	120	24	21	18	15	11	8	5
300	671	140	22	19	16	13	9	6	3
325	727	40	43	41	39	36	33	30	26
325	727	60	35	32	30	26	23	20	17
325	727	80	30	27	24	21	18	14	11
325	727	100	26	23	20	17	14	11	8
325	727	120	24	21	17	14	11	8	5
325	727	140	22	19	15	12	9	6	3
350	783	40	42	40	37	35	31	28	25
350	783	60	34	31	29	26	22	19	16
350	783	80	29	26	23	20	17	14	11
350	783	100	25	23	19	16	13	10	7
350	783	120	23	20	17	14	11	7	5
350	783	140	21	18	15	12	8	5	3
375	839	40	41	39	36	33	30	27	24
375	839	60	33	31	28	25	22	19	16
375	839	80	28	25	22	19	16	13	10
375	839	100	25	22	19	16	13	10	7
375	839	120	22	19	16	13	10	7	4
375	839	140	20	17	14	11	8	5	2
400	894	40	40	38	35	32	29	27	24
400	894	60	32	30	27	24	21	18	15
400	894	80	27	25	22	19	16	13	10
400	894	100	24	21	18	15	12	9	6
400	894	120	22	19	16	13	10	7	4
400	894	140	20	17	14	11	8	5	2

NOTE: Detector time constant at a reference velocity of 5 ft/sec.

For SI Units: 1 ft = 0.305 m

1000 BTU/sec = 1055 kW

Table B-3.2.4(y) Q_d, Threshold Fire Size at Response: 2000 Btu/sec; t_g: 600 Seconds to 1000 Btu/sec; α: 0.003 Btu/sec^3

| τ | RTI | ΔT | \multicolumn{7}{c}{CEILING HEIGHT IN FEET} |
			4.0	8.0	12.0	16.0	20.0	24.0	28.0
			\multicolumn{7}{c}{INSTALLED SPACING OF DETECTORS}
25	56	40	96	85	78	68	62	56	50
25	56	60	75	64	56	49	43	37	32
25	56	80	63	52	44	38	32	27	22
25	56	100	54	44	37	31	25	20	15
25	56	120	48	38	31	25	20	15	11
25	56	140	43	34	27	21	16	12	7
50	112	40	86	78	71	64	58	53	48
50	112	60	68	60	53	47	41	36	31
50	112	80	57	49	42	36	31	26	21
50	112	100	49	41	35	29	24	19	15
50	112	120	44	36	30	24	19	15	10
50	112	140	40	32	26	21	16	11	7
75	168	40	78	72	66	61	55	50	45
75	168	60	62	56	50	44	39	34	29
75	168	80	52	46	40	34	30	25	20
75	168	100	46	39	33	28	23	19	14
75	168	120	41	34	28	23	19	14	10
75	168	140	37	30	25	20	15	11	6
100	224	40	72	67	62	57	52	48	43
100	224	60	57	52	47	42	37	32	28
100	224	80	49	43	38	33	28	24	19
100	224	100	43	37	32	27	22	18	13
100	224	120	38	32	27	22	18	13	9
100	224	140	35	29	24	19	14	10	6
125	280	40	67	63	59	54	50	45	41
125	280	60	54	49	45	40	35	31	27
125	280	80	46	41	36	31	27	23	18
125	280	100	40	35	30	26	21	17	13
125	280	120	36	31	26	21	17	13	9
125	280	140	33	27	23	18	14	10	6
150	335	40	63	60	56	52	47	43	39
150	335	60	51	47	42	38	34	30	26
150	335	80	43	39	34	30	26	22	18
150	335	100	38	33	29	25	20	16	12
150	335	120	34	29	25	21	16	12	8
150	335	140	31	26	22	17	13	9	5
175	391	40	60	57	53	49	45	41	37
175	391	60	48	44	41	36	32	28	24
175	391	80	41	37	33	29	25	21	17
175	391	100	36	32	28	24	19	15	12
175	391	120	32	28	24	20	16	12	8
175	391	140	29	25	21	17	13	9	5
200	447	40	57	54	51	47	43	40	36
200	447	60	46	43	39	35	31	27	23
200	447	80	39	35	32	28	24	20	16
200	447	100	34	30	26	23	19	15	11
200	447	120	31	27	23	19	15	11	7
200	447	140	28	24	20	16	12	8	5
225	503	40	54	52	49	45	42	38	34
225	503	60	44	41	37	34	30	26	22
225	503	80	37	34	30	26	23	19	15
225	503	100	33	29	25	22	18	14	11
225	503	120	30	26	22	18	14	11	7
225	503	140	27	23	19	15	12	8	4
250	559	40	52	50	47	43	40	36	33
250	559	60	42	39	36	32	29	25	21
250	559	80	36	33	29	25	22	18	15
250	559	100	32	28	24	21	17	14	10
250	559	120	28	25	21	17	14	10	7
250	559	140	26	22	19	15	11	8	4
275	615	40	50	48	45	42	39	35	32
275	615	60	41	38	35	31	28	24	21
275	615	80	35	32	28	25	21	18	14
275	615	100	30	27	24	20	17	13	10
275	615	120	27	24	20	17	13	10	6
275	615	140	25	22	18	14	11	7	4
300	671	40	49	47	44	40	37	34	30
300	671	60	39	37	33	30	27	23	20
300	671	80	33	31	27	24	20	17	13
300	671	100	29	26	23	19	16	12	9
300	671	120	26	23	20	16	13	9	6
300	671	140	24	21	17	14	10	7	4
325	727	40	47	45	42	39	36	33	29
325	727	60	38	35	32	29	26	22	19
325	727	80	32	30	26	23	20	16	13
325	727	100	29	25	22	19	15	12	9
325	727	120	26	22	19	16	12	9	6
325	727	140	23	20	17	13	10	7	3
350	783	40	46	44	41	38	35	32	28
350	783	60	37	34	31	28	25	22	18
350	783	80	31	29	25	22	19	16	12
350	783	100	28	25	21	18	15	12	8
350	783	120	25	22	18	15	12	9	5
350	783	140	23	20	16	13	9	6	3
375	839	40	44	42	40	37	34	31	27
375	839	60	36	33	30	27	24	21	18
375	839	80	31	28	25	21	18	15	12
375	839	100	27	24	21	18	14	11	8
375	839	120	24	21	18	15	11	8	5
375	839	140	22	19	16	12	9	6	3
400	894	40	43	41	39	36	33	30	26
400	894	60	35	32	30	26	23	20	17
400	894	80	30	27	24	21	18	14	11
400	894	100	26	23	20	17	14	11	8
400	894	120	23	21	17	14	11	8	5
400	894	140	21	18	15	12	9	6	3

NOTE: Detector time constant at a reference velocity of 5 ft/sec.

For SI Units: 1 ft = 0.305 m

1000 BTU/sec = 1055 kW

B-3.3 Rate-of-Rise Heat Detector Spacing.

B-3.3.1 Tables B-3.3.2(a) and B-3.3.2(b) are to be used to determine the installed spacing of rate-of-rise heat detectors. The analytical basis for the tables is presented in Section B-6. This section shows how the tables are to be used.

B-3.3.2 Table B-3.3.2(a) provides installed spacings for rate-of-rise heat detectors required to achieve detection for a specific threshold for size, fire growth rate, and ceiling height. This table may be used directly to determine installed spacings for 50-ft (15.2-m) listed spacing detectors.

Table B-3.3.2(b)   Spacing Modifiers for Rate-of-Rise Heat Detectors

| Listed Spacing (ft) | Fire Growth Rate | | |
	Slow	Medium	Fast
15	0.57	0.55	0.45
20	0.72	0.63	0.62
25	0.84	0.78	0.76
30	0.92	0.86	0.85
40	0.98	0.96	0.95
50	1.00	1.00	1.00
70	1.00	1.01	1.02

For SI Units: 1 ft = 0.305 m.

B-3.3.2(a)   Installed Spacings for Rate-of-Rise Heat Detectors (Threshold Fire Size and Growth Rate)

Ceiling Height (ft)	Qd = 1000 Btu/sec			Qd = 750 Btu/sec			Qd = 500 Btu/sec			Qd = 250 Btu/sec			Qd = 100 Btu/sec		
	s	m	f	s	m	f	s	m	f	s	m	f	s	m	f
4	28	32	32	26	28	27	22	24	23	16	17	16	11	11	10
5	27	31	31	25	27	27	21	23	22	15	16	15	10	10	9
6	26	30	31	24	26	27	20	22	22	15	15	15	9	9	9
7	25	29	30	23	26	26	19	21	21	14	14	14	9	9	8
8	24	29	30	22	25	26	18	21	21	13	13	14	8	8	8
9	23	28	29	21	24	25	17	20	20	12	13	13	7	7	7
10	22	27	29	20	23	25	16	19	20	12	12	13	7	7	7
11	21	27	28	18	23	24	15	19	19	11	12	12	6	6	6
12	20	26	26	17	22	24	15	18	19	10	11	12	5	5	5
13	19	25	27	16	22	23	14	18	18	9	11	11	5	5	5
14	18	24	27	15	21	22	13	17	18	9	10	11	4		
15	16	24	26	14	20	21	12	17	17	8	10	10			
16	15	23	25	13	19	21	11	16	16	7	9	10			
17	14	22	25	12	19	20	10	15	16	6	9	9			
18	13	22	24	11	18	20	9	14	15		8	8			
19	12	21	23	10	17	19	8	14	14		8	8			
20	11	20		9	16	19	7	13	14		7	7			
21	10	19		8	15	18		12	13		7				
22	9	19		7	15	17		12	13		6				
23	8	18			14	17		11	12		5				
24		17			13	16		11	11		5				
25		16			12	15		10	10		4				
26		15			12	15		9	10						
27		14			11	14		9							
28		13			11	13		8							
29		13			10			8							
30		12			10			7							

s = slow fire, m = medium fire, f = fast fire.

B-3.3.2.1 Tables B-3.3.2(a) and B-3.3.2(b) use the following values for t_g:

Fast fire growth rate, t_g = 150 seconds

Medium fire growth rate, t_g = 300 seconds

Slow fire growth rate, t_g = 600 seconds.

B-3.3.3 For rate-of-rise heat detectors with a listed spacing of other than 50 ft (15.2 m), installed spacing obtained from Table B-3.3.2(a) must be multiplied by the modifier shown in Table B-3.3.2(b) for the appropriate listed spacing and fire growth rate. This takes into account the difference in sensitivity between the detector and a 50-ft (15.2-m) listed detector.

B-3.3.4 Having determined the threshold fire size (*see B-2.2.2*), the fire growth rate (*see B-2.2.3*), the detector's listed spacing, and the ceiling height, use Table B-3.3.2(a) to determine the corrected spacing for 50-ft (15.2-m) listed detectors. Use Table B-3.3.2(b) to determine the spacing modifier. Find the required installed spacing by multiplying the corrected spacing by the spacing modifier.

B-3.3.5 Example.

Input:
   Ceiling height: 12 ft (3.7 m)
   Detector type: Combination rate-of-rise, fixed temperature 30-ft (9.1-m) listed spacing
   Q_d: 500 Btu/sec
   Fire growth rate: Medium
Spacing:
   From Table B-3.3.2(a), installed spacing = 18 ft (5.5 m)
   From Table B-3.3.2(b), spacing modifier = 0.86
   Installed spacing = 18 × 0.86 = 15.5 ft (4.7 m)

NOTE: This answer may be rounded to either 15 ft (4.6 m) or 16 ft (4.9 m). Use of 15 ft (4.6 m) would be slightly conservative. However, depending on field conditions, use of 16 ft (4.9 m) may fit the space better.

B-3.4 Design Curves.

B-3.4.1 The design curves [Figures B-3.4.1 (a) through (i)] may also be used to determine the installed spacings of heat detectors. However, they are not as comprehensive as the tables, because the tables include additional fire growth rates, fire sizes, and detector sensitivities.

B-3.4.1.1 Fixed-Temperature Heat Detectors. Figures B-3.4.1(a) through (f) can be used directly to determine the installed spacing for fixed-temperature heat detectors having listed spacings of 30 ft and 50 ft (9.1 m and 15.2 m), respectively, where the difference between the detectors' rated temperature (T_s) and the ambient temperature (T_o) is 65°F (36°C). When ΔT is not 65°F (36°C), tables previously discussed in B-3.3 should be used.

B-3.4.1.2 Rate-of-Rise Heat Detectors. Figures B-3.4.1(g), (h), and (i) can be used directly to determine the installed spacing for rate-of-rise heat detectors having a listed spacing of 50 ft (15.2 m).

B-3.4.2 To use the curves, the same format must be followed as with tables. The designer must first determine how large a fire can be tolerated before detection can occur. This is the threshold fire size, Q_d. Curves are presented, in most cases, for values of Q_d = 1000, 750, 500, 250, and 100 Btu/sec (1055, 791, 527, 264, and 105 kW). Interpolation between values of Q_d on a given graph is allowable. Table B-2.2.2.1(a) and Table B-2.2.2.3 also contain examples of various fuels and their fire growth rates under specified conditions.

B-3.4.3 Once a threshold size and expected fire growth rate have been selected, an installed detector spacing can be obtained from Figures B-3.4.1(a) through (i) for a certain detector's listed spacing, ambient temperature, and ceiling height. As in B-3.2.6, to determine the installed spacing of 135°F (57°C) fixed temperature heat detectors with a listed spacing of 30 ft (9.1 m) and to detect a slowly developing fire at a threshold fire size of 500 Btu/sec (527 kW) in a room 10 ft (3 m) high with an ambient temperature of 70°F (21°C), the following procedure is used.

B-3.4.3.1 Example 1.

Input:
   Ceiling height: 10 ft (3 m)
   Detector type: Fixed temperature 135°F (57°C)
      UL listed spacing: 30 ft (9.1 m)
   Fire:
      Q_d: 500 Btu/sec (527 kW)
      Fire growth rate: Slow
      t_g: 600 sec
   Environmental conditions:
      T_o: 70°F (21°C)
      ΔT = 135 − 70 = 65°F (36°C)
Required installed spacing:
   From Figure B-3.4.1(a), use an installed spacing of 18 ft (5.2 m) (17.5 ft rounded to 18 ft).

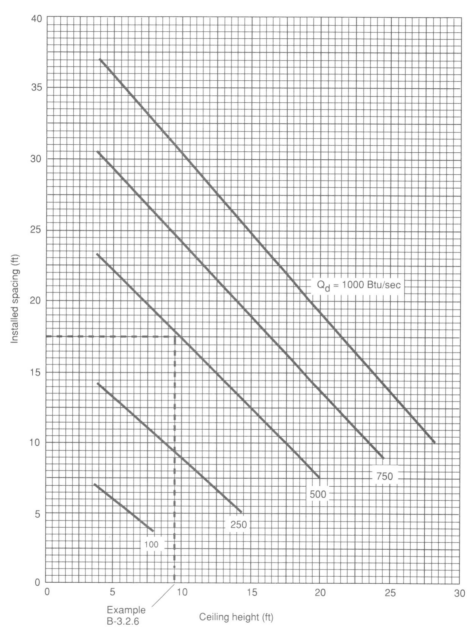

Figure B-3.4.1(a)   *Heat detector, fixed temperature, 30-ft (9.1-m) listed spacing, slow fire. ΔT = 65°F (36.1°C).*

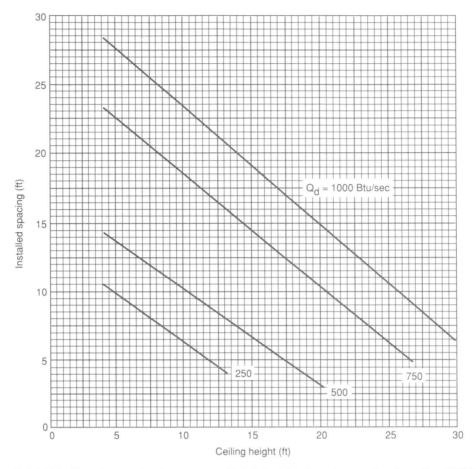

Figure B-3.4.1(b) *Heat detector, fixed temperature, 30-ft (9.1-m) listed spacing, medium fire. ΔT = 65°F (36°C).*

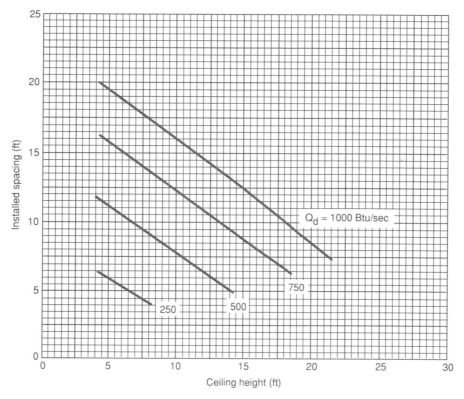

Figure B-3.4.1(c)   *Heat detector, fixed temperature, 30-ft (9.1-m) listed spacing, fast fire. $\Delta T = 65°F$ (36°C).*

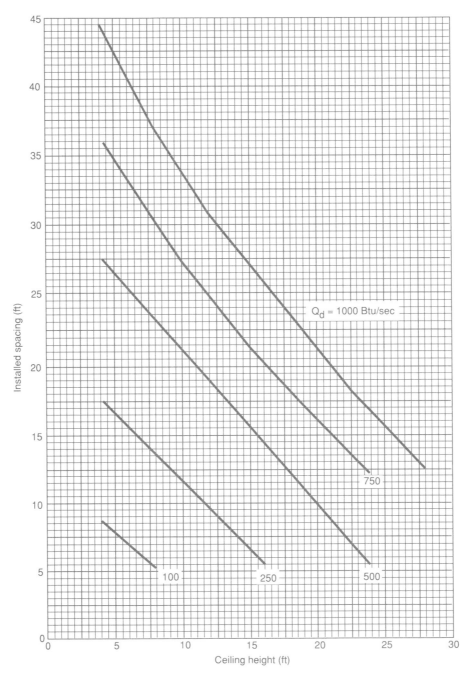

Figure B-3.4.1(d) *Heat detector, fixed temperature, 50-ft (15.2-m) listed spacing, slow fire. ΔT = 65°F (36°C).*

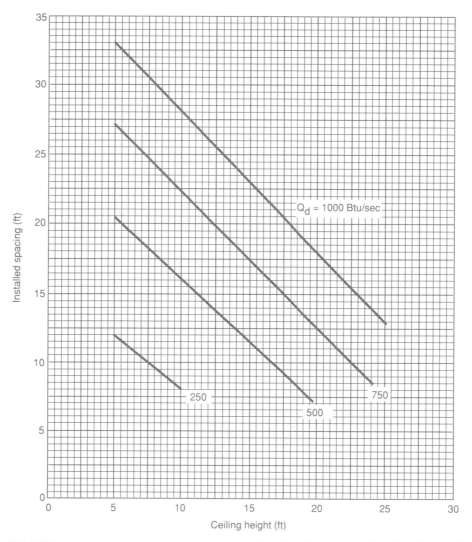

Figure B-3.4.1(e) *Heat detector, fixed temperature, 50-ft (15.2-m) listed spacing, medium fire. ΔT = 65°F (36°C).*

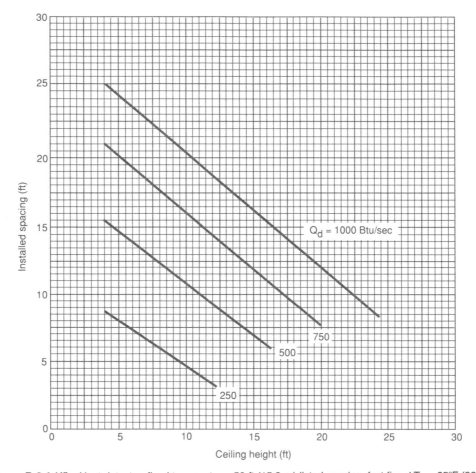

Figure B-3.4.1(f) *Heat detector, fixed temperature, 50-ft (15.2-m) listed spacing, fast fire. ΔT = 65°F (36°C).*

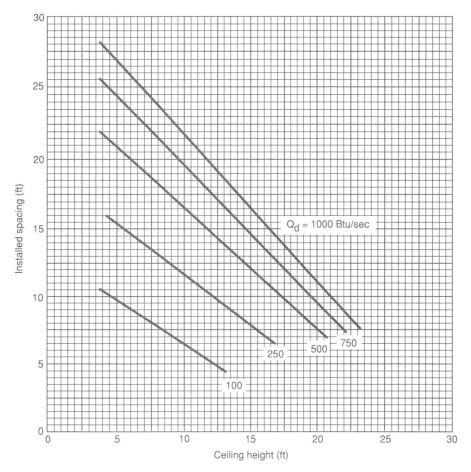

Figure B-3.4.1(g)   *Heat detector, rate-of-rise, 50-ft (15.2-m) listed spacing, slow fire.*

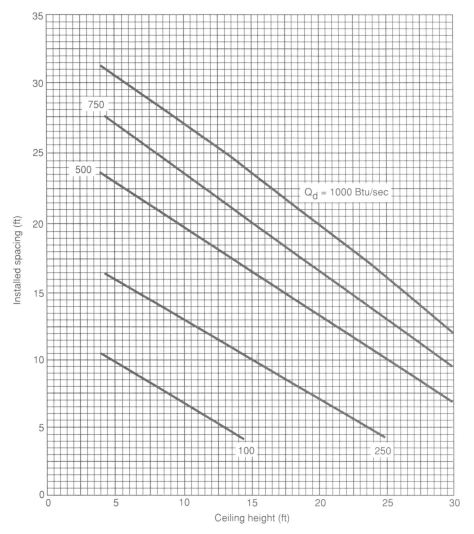

Figure B-3.4.1(h)   *Heat detector, rate-of-rise, 50-ft (15.2-m) listed spacing, medium fire.*

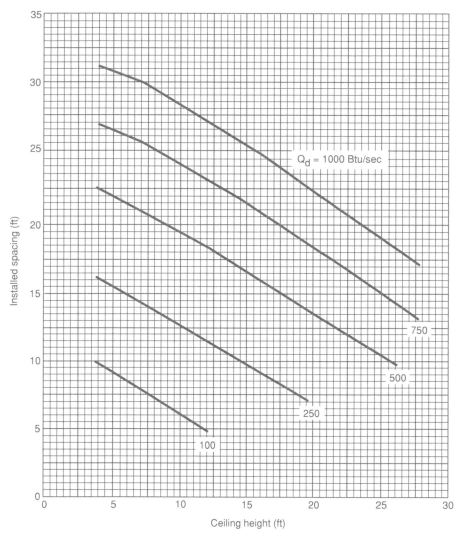

Figure B-3.4.1(i)   *Heat detector, rate-of-rise, 50-ft (15.2-m) listed spacing, fast fire.*

Note that if the ceiling height is 15 ft (4.6 m), the same graph gives an installed spacing of 12 ft (3.5 m). A ceiling height of 20 ft (6.1 m) would require a spacing of 8 ft (2.4 m). This change in spacing clearly illustrates the need to consider ceiling height in the design of a detection system.

B-3.4.3.2 Example 2.

Input:

Ceiling height: 10 ft (3 m)

Detector type: Combination rate-of-rise and fixed temperature

UL listed spacing: 50 ft (15.2 m)

Fire:

Q_d: 500 Btu/sec (527 kW)

Fire growth rate: Fast

t_g: 150 sec

Environmental conditions:

T_o: 70°F (21°C)

ΔT: 65°F (36°C)

Spacing:

From Figure B-3.4.1(i), use an installed spacing of 20 ft (5.8 m) (19.5 ft rounded to 20 ft).

A 30-ft (9.1-m) fixed temperature detector would require a 7.5-ft (2.5-m) spacing.

If the fire growth rate was slow, as in Example 1, the rate-of-rise detector would require an installed spacing of 16 ft (4.88 m).

B-4 Analysis of Existing Heat Detection Systems.

B-4.1 Tables B-4.1(a) through (nn) can be used to determine the size fire (heat release rate) that existing fixed-temperature heat detection systems will respond to. Table B-4.1 is an index to Tables B-4.1(a) through (nn).

The use of the analysis tables is similar to that described for new designs. The difference is that the spacing of the existing detectors must be known. An estimate of the fire intensity coefficient (α) or the fire growth time (t_g) must also be made for the fuel that is expected to burn.

Table B-4.1   Analysis Tables Index

	Installed Spacing (ft)	Fire Growth Rate (sec) τ_g	Alpha (Btu/sec^3) α
Table B-4.1(a)	8	50	0.400
Table B-4.1(b)	8	150	0.044
Table B-4.1(c)	8	300	0.011
Table B-4.1(d)	8	500	0.004
Table B-4.1(e)	8	600	0.003
Table B-4.1(f)	10	50	0.400
Table B-4.1(g)	10	150	0.044
Table B-4.1(h)	10	300	0.011
Table B-4.1(i)	10	500	0.004
Table B-4.1(j)	10	600	0.003
Table B-4.1(k)	12	50	0.400
Table B-4.1(l)	12	150	0.044
Table B-4.1(m)	12	300	0.011
Table B-4.1(n)	12	500	0.004
Table B-4.1(o)	12	600	0.003
Table B-4.1(p)	15	50	0.400
Table B-4.1(q)	15	150	0.044
Table B-4.1(r)	15	300	0.011
Table B-4.1(s)	15	500	0.004
Table B-4.1(t)	15	600	0.003
Table B-4.1(u)	20	50	0.400
Table B-4.1(v)	20	150	0.044
Table B-4.1(w)	20	300	0.011
Table B-4.1(x)	20	500	0.004
Table B-4.1(y)	20	600	0.003
Table B-4.1(z)	25	50	0.400
Table B-4.1(aa)	25	150	0.044
Table B-4.1(bb)	25	300	0.011
Table B-4.1(cc)	25	500	0.004
Table B-4.1(dd)	25	600	0.003
Table B-4.1(ee)	30	50	0.400
Table B-4.1(ff)	30	150	0.044
Table B-4.1(gg)	30	300	0.011
Table B-4.1(hh)	30	500	0.004
Table B-4.1(ii)	30	600	0.003
Table B-4.1(jj)	50	50	0.400
Table B-4.1(kk)	50	150	0.044
Table B-4.1(ll)	50	300	0.011
Table B-4.1(mm)	50	500	0.004
Table B-4.1(nn)	50	600	0.003

Table B-4.1(a) Installed Spacing of Heat Detector: 8 feet; t_g: 50 Seconds to 1000 Btu/sec; α: 0.400 Btu/sec^3

τ	RTI	ΔT	\multicolumn{7}{c}{CEILING HEIGHT IN FEET}						
			4.0	8.0	12.0	16.0	20.0	24.0	28.0
			\multicolumn{7}{c}{FIRE SIZE AT DETECTOR RESPONSE (BTU/SEC)}						
25	56	40	300	402	535	668	832	1016	1219
25	56	60	368	508	687	877	1106	1365	1657
25	56	80	450	618	838	1102	1381	1722	2110
25	56	100	512	716	985	1308	1661	2090	2585
25	56	120	573	815	1132	1517	1949	2473	3082
25	56	140	654	919	1282	1730	2265	2870	3601
50	112	40	422	571	755	926	1136	1366	1614
50	112	60	546	738	976	1211	1496	1811	2157
50	112	80	642	883	1181	1484	1846	2251	2699
50	112	100	754	1033	1383	1752	2194	2692	3248
50	112	120	865	1179	1582	2018	2542	3138	3810
50	112	140	928	1305	1773	2318	2895	3592	4386
75	168	40	542	722	908	1137	1389	1659	1948
75	168	60	702	932	1219	1492	1826	2193	2589
75	168	80	813	1111	1472	1824	2245	2710	3217
75	168	100	931	1289	1718	2146	2656	3221	3844
75	168	120	1016	1451	1955	2464	3063	3733	4475
75	168	140	1149	1629	2193	2778	3470	4247	5115
100	224	40	625	841	1101	1332	1614	1920	2246
100	224	60	802	1087	1427	1742	2122	2535	2978
100	224	80	944	1305	1728	2128	2604	3125	3687
100	224	100	1050	1503	2012	2501	3074	3703	4388
100	224	120	1222	1723	2298	2867	3537	4276	5088
100	224	140	1360	1925	2573	3226	3995	4849	5791
125	280	40	729	967	1208	1501	1820	2160	2519
125	280	60	912	1238	1622	1972	2394	2850	3337
125	280	80	1036	1472	1959	2409	2936	3508	4123
125	280	100	1233	1730	2294	2830	3461	4150	4895
125	280	120	1398	1968	2614	3240	3976	4782	5661
125	280	140	1561	2201	2926	3642	4484	5411	6246
150	335	40	793	1066	1340	1664	2013	2384	2775
150	335	60	979	1362	1797	2187	2649	3145	3674
150	335	80	1185	1656	2186	2673	3247	3868	4533
150	335	100	1378	1933	2554	3138	3825	4570	5373
150	335	120	1568	2201	2911	3590	4389	5259	6202
150	335	140	1757	2462	3257	4033	4944	5942	7027
175	391	40	882	1175	1468	1818	2195	2595	3016
175	391	60	1046	1483	1965	2391	2890	3425	3993
175	391	80	1301	1819	2397	2923	3542	4210	4923
175	391	100	1520	2127	2802	3431	4170	4970	5827
175	391	120	1734	2423	3193	3923	4782	5713	6718
175	391	140	1947	2712	3573	4405	5382	6447	7601
200	447	40	925	1257	1586	1964	2369	2797	3247
200	447	60	1168	1625	2136	2587	3121	3692	4298
200	447	80	1415	1977	2599	3162	3825	4537	5295
200	447	100	1658	2313	3040	3711	4501	5352	6262
200	447	120	1897	2637	3464	4242	5158	6148	7212
200	447	140	2133	2952	3875	4761	5802	6932	8152

τ	RTI	ΔT	\multicolumn{7}{c}{CEILING HEIGHT IN FEET}						
			4.0	8.0	12.0	16.0	20.0	24.0	28.0
			\multicolumn{7}{c}{FIRE SIZE AT DETECTOR RESPONSE (BTU/SEC)}						
225	503	40	968	1337	1754	2111	2537	2991	3468
225	503	60	1254	1747	2294	2774	3342	3949	4590
225	503	80	1527	2129	2794	3392	4096	4851	5653
225	503	100	1794	2494	3268	3980	4819	5720	6681
225	503	120	2057	2845	3724	4549	5520	6567	7689
225	503	140	2317	3185	4168	5104	6206	7400	8683
250	559	40	1011	1417	1865	2247	2698	3177	3681
250	559	60	1339	1866	2447	2955	3556	4197	4873
250	559	80	1637	2278	2982	3614	4358	5155	5999
250	559	100	1928	2669	3489	4241	5126	6076	7087
250	559	120	2215	3046	3890	4842	5870	6972	8150
250	559	140	2499	3412	4356	5431	6597	7852	9197
275	615	40	1093	1513	1981	2380	2854	3358	3887
275	615	60	1424	1982	2596	3131	3763	4437	5147
275	615	80	1746	2422	3165	3829	4612	5449	6334
275	615	100	2061	2840	3618	4488	5424	6421	7479
275	615	120	2371	3242	4128	5129	6209	7365	8597
275	615	140	2679	3633	4622	5753	6977	8291	9697
300	671	40	1151	1595	2089	2508	3005	3533	4087
300	671	60	1507	2096	2740	3301	3964	4670	5413
300	671	80	1853	2563	3259	4032	4859	5735	6661
300	671	100	2192	3007	3820	4734	5714	6756	7862
300	671	120	2526	3434	4359	5409	6540	7748	9033
300	671	140	2859	3849	4881	6066	7346	8718	10183
325	727	40	1208	1677	2194	2633	3152	3704	4282
325	727	60	1589	2207	2804	3461	4160	4898	5672
325	727	80	1959	2701	3428	4236	5100	6014	6978
325	727	100	2322	3171	4018	4973	5996	7084	8234
325	727	120	2680	3623	4585	5682	6862	8121	9457
325	727	140	3038	4061	5133	6371	7706	9135	10657
350	783	40	1265	1756	2297	2754	3296	3871	4472
350	783	60	1671	2315	2937	3623	4352	5119	5925
350	783	80	2064	2836	3592	4435	5335	6287	7289
350	783	100	2451	3331	4211	5207	6272	7403	8599
350	783	120	2834	3808	4805	5949	7177	8485	9872
350	783	140	3218	4270	5380	6669	8058	9542	11121
375	839	40	1321	1835	2398	2874	3437	4034	4658
375	839	60	1751	2422	3069	3782	4539	5336	6172
375	839	80	2169	2969	3753	4630	5565	6553	7592
375	839	100	2579	3489	4401	5436	6543	7716	8955
375	839	120	2987	3990	5021	6210	7486	8842	10279
375	839	140	3303	4445	5620	6961	8403	9941	11575
400	894	40	1377	1912	2423	2982	3574	4193	4840
400	894	60	1831	2527	3197	3937	4723	5549	6415
400	894	80	2272	3100	3911	4821	5791	6814	7890
400	894	100	2707	3645	4586	5660	6807	8023	9304
400	894	120	3141	4169	5233	6466	7788	9192	10677
400	894	140	3456	4640	5857	7247	8741	10332	12020

NOTE: Detector time constant at a reference velocity of 5 ft/sec.

For SI Units: 1 ft = 0.305 m

1000 BTU/sec = 1055 kW

Table B-4.1(b)   Installed Spacing of Heat Detector: 8 feet; t_g: 50 Seconds to 1000 Btu/sec; α: 0.044 Btu/sec^3

τ	RTI	ΔT	CEILING HEIGHT IN FEET FIRE SIZE AT DETECTOR RESPONSE (BTU/SEC) 4.0	8.0	12.0	16.0	20.0	24.0	28.0	τ	RTI	ΔT	CEILING HEIGHT IN FEET FIRE SIZE AT DETECTOR RESPONSE (BTU/SEC) 4.0	8.0	12.0	16.0	20.0	24.0	28.0
25	56	40	118	167	232	311	400	507	631	225	503	40	425	570	726	906	1102	1312	1538
25	56	60	154	226	322	440	584	752	952	225	503	60	558	759	992	1227	1500	1797	2119
25	56	80	194	286	415	579	781	1026	1309	225	503	80	693	939	1231	1533	1885	2271	2694
25	56	100	228	346	512	726	993	1319	1699	225	503	100	809	1108	1461	1833	2265	2744	3272
25	56	120	263	409	614	883	1221	1633	2118	225	503	120	926	1274	1686	2130	2645	3220	3859
25	56	140	299	473	721	1049	1462	1969	2573	225	503	140	1026	1434	1909	2425	3025	3700	4456
50	112	40	171	237	320	410	517	638	775	250	559	40	453	608	774	964	1170	1392	1628
50	112	60	224	317	435	574	728	913	1126	250	559	60	596	810	1041	1304	1591	1902	2239
50	112	80	281	397	550	735	949	1205	1504	250	559	80	738	1002	1311	1629	1997	2401	2841
50	112	100	329	474	666	901	1185	1516	1912	250	559	100	876	1185	1556	1946	2397	2896	3444
50	112	120	377	552	785	1074	1428	1846	2347	250	559	120	982	1358	1794	2258	2795	3392	4053
50	112	140	424	630	906	1254	1683	2202	2808	250	559	140	1107	1531	2029	2567	3192	3891	4671
75	168	40	216	296	395	498	620	756	906	275	615	40	480	645	820	1021	1237	1469	1716
75	168	60	283	395	533	683	861	1063	1291	275	615	60	646	863	1103	1379	1680	2005	2355
75	168	80	352	492	668	876	1107	1381	1696	275	615	80	783	1063	1389	1722	2107	2527	2984
75	168	100	413	585	803	1063	1360	1714	2125	275	615	100	921	1256	1647	2054	2525	3043	3611
75	168	120	472	678	939	1255	1622	2063	2578	275	615	120	1039	1440	1898	2382	2941	3560	4242
75	168	140	531	770	1076	1451	1901	2427	3055	275	615	140	1177	1624	2146	2706	3355	4078	4881
100	224	40	255	349	462	577	713	863	1027	300	671	40	507	681	865	1075	1301	1543	1801
100	224	60	343	467	622	788	983	1202	1446	300	671	60	680	911	1163	1452	1766	2105	2469
100	224	80	416	578	776	996	1254	1548	1880	300	671	80	827	1123	1445	1811	2213	2650	3123
100	224	100	488	685	929	1214	1530	1904	2333	300	671	100	967	1325	1736	2161	2650	3187	3774
100	224	120	559	792	1081	1424	1811	2273	2806	300	671	120	1109	1523	2000	2503	3083	3723	4427
100	224	140	636	898	1234	1637	2101	2656	3301	300	671	140	1246	1715	2260	2842	3514	4261	5087
125	280	40	291	397	523	650	799	962	1140	325	727	40	533	717	909	1129	1365	1616	1884
125	280	60	391	532	704	885	1097	1333	1593	325	727	60	714	959	1222	1524	1851	2203	2580
125	280	80	476	657	877	1114	1392	1705	2056	325	727	80	881	1184	1517	1899	2317	2770	3259
125	280	100	558	779	1046	1342	1690	2086	2534	325	727	100	1014	1393	1823	2264	2772	3328	3933
125	280	120	647	899	1214	1571	1992	2476	3029	325	727	120	1169	1601	2100	2622	3222	3884	4608
125	280	140	723	1017	1382	1813	2300	2878	3543	325	727	140	1314	1803	2371	2974	3670	4440	5288
150	335	40	325	443	581	719	880	1056	1246	350	783	40	559	751	952	1181	1426	1688	1965
150	335	60	435	593	781	976	1204	1456	1733	350	783	60	747	1005	1280	1594	1933	2298	2688
150	335	80	531	732	971	1226	1523	1855	2224	350	783	80	917	1239	1589	1986	2418	2887	3392
150	335	100	634	869	1157	1473	1842	2259	2728	350	783	100	1072	1462	1885	2365	2892	3466	4089
150	335	120	720	999	1340	1719	2164	2671	3245	350	783	120	1228	1679	2197	2737	3359	4041	4786
150	335	140	805	1128	1522	1967	2491	3093	3780	350	783	140	1380	1890	2480	3104	3822	4615	5486
175	391	40	357	486	637	784	957	1145	1347	375	839	40	584	785	994	1232	1486	1757	2045
175	391	60	478	650	854	1063	1307	1574	1866	375	839	60	780	1050	1336	1662	2014	2391	2795
175	391	80	584	803	1061	1332	1649	1999	2386	375	839	80	953	1294	1658	2070	2518	3002	3523
175	391	100	694	952	1262	1598	1989	2427	2915	375	839	100	1122	1528	1967	2464	3009	3601	4242
175	391	120	790	1094	1460	1861	2330	2860	3456	375	839	120	1286	1754	2292	2851	3492	4195	4960
175	391	140	892	1236	1656	2125	2675	3301	4011	375	839	140	1446	1975	2587	3231	3971	4787	5681
200	447	40	396	530	676	846	1031	1230	1444	400	894	40	609	818	1036	1282	1545	1826	2122
200	447	60	519	705	924	1146	1405	1687	1995	400	894	60	813	1094	1392	1729	2093	2483	2899
200	447	80	646	873	1148	1435	1769	2138	2543	400	894	80	989	1348	1726	2153	2615	3114	3651
200	447	100	752	1031	1363	1718	2129	2588	3096	400	894	100	1171	1593	2048	2562	3123	3733	4393
200	447	120	870	1188	1576	1998	2490	3042	3660	400	894	120	1343	1829	2359	2962	3623	4346	5132
200	447	140	959	1337	1785	2277	2852	3503	4236	400	894	140	1511	2058	2692	3356	4118	4956	5872

NOTE: Detector time constant at a reference velocity of 5 ft/sec.

For SI Units: 1 ft = 0.305 m

1000 BTU/sec = 1055 kW

Table B-4.1(c)   Installed Spacing of Heat Detector: 8 feet; t_g: 300 Seconds to 1000 Btu/sec; α: 0.011 Btu/sec^3

τ	RTI	ΔT	__CEILING HEIGHT IN FEET__ FIRE SIZE AT DETECTOR RESPONSE (BTU/SEC)							τ	RTI	ΔT	__CEILING HEIGHT IN FEET__ FIRE SIZE AT DETECTOR RESPONSE (BTU/SEC)						
			4.0	8.0	12.0	16.0	20.0	24.0	28.0				4.0	8.0	12.0	16.0	20.0	24.0	28.0
25	56	40	70	104	152	211	285	374	477	225	503	40	247	337	443	553	680	820	973
25	56	60	95	149	223	321	443	592	767	225	503	60	335	458	606	766	951	1158	1389
25	56	80	122	196	302	442	620	838	1099	225	503	80	413	572	763	976	1223	1503	1819
25	56	100	148	246	387	575	815	1110	1463	225	503	100	489	684	919	1186	1499	1858	2267
25	56	120	174	299	479	719	1027	1405	1858	225	503	120	569	794	1075	1406	1782	2225	2735
25	56	140	201	354	576	873	1253	1721	2283	225	503	140	639	902	1230	1621	2072	2605	3224
50	112	40	101	144	200	267	345	439	547	250	559	40	264	360	471	587	720	866	1025
50	112	60	136	200	284	389	517	668	849	250	559	60	357	488	644	811	1005	1220	1458
50	112	80	172	257	372	520	703	926	1187	250	559	80	441	610	811	1032	1289	1578	1902
50	112	100	205	315	465	661	906	1206	1560	250	559	100	522	728	975	1252	1576	1945	2363
50	112	120	239	374	563	811	1124	1508	1963	250	559	120	607	844	1138	1474	1868	2322	2842
50	112	140	273	437	666	970	1356	1830	2399	250	559	140	682	959	1301	1705	2167	2712	3341
75	168	40	127	178	242	318	402	502	616	275	615	40	280	382	499	620	759	911	1076
75	168	60	170	245	339	453	586	746	931	275	615	60	380	518	681	856	1057	1280	1525
75	168	80	215	311	438	595	786	1012	1280	275	615	80	469	646	856	1086	1352	1651	1984
75	168	100	255	378	540	745	998	1303	1661	275	615	100	555	771	1028	1316	1650	2029	2457
75	168	120	296	445	646	903	1223	1612	2072	275	615	120	643	893	1199	1546	1953	2418	2948
75	168	140	336	514	756	1069	1461	1942	2510	275	615	140	723	1014	1369	1779	2261	2817	3456
100	224	40	150	209	281	361	455	561	682	300	671	40	297	403	520	652	797	955	1126
100	224	60	201	286	390	514	654	821	1013	300	671	60	401	546	717	899	1108	1338	1591
100	224	80	253	361	500	667	864	1099	1374	300	671	80	496	682	901	1140	1415	1722	2064
100	224	100	300	436	611	827	1088	1397	1764	300	671	100	593	813	1081	1378	1723	2113	2550
100	224	120	347	511	725	993	1322	1714	2183	300	671	120	679	941	1259	1617	2035	2512	3052
100	224	140	393	587	843	1167	1568	2055	2628	300	671	140	763	1067	1436	1858	2352	2921	3571
125	280	40	171	237	317	403	504	618	745	325	727	40	317	425	546	684	834	998	1174
125	280	60	230	323	437	567	719	893	1093	325	727	60	422	574	753	941	1157	1395	1656
125	280	80	289	408	557	736	941	1184	1466	325	727	80	522	716	945	1192	1476	1792	2143
125	280	100	342	490	678	906	1173	1493	1867	325	727	100	623	854	1132	1439	1795	2194	2641
125	280	120	395	573	801	1081	1420	1819	2294	325	727	120	713	988	1317	1687	2116	2604	3155
125	280	140	447	656	926	1262	1674	2163	2748	325	727	140	802	1119	1502	1936	2442	3023	3684
150	335	40	192	264	350	443	551	671	805	350	783	40	332	445	571	714	871	1040	1222
150	335	60	261	360	482	619	780	963	1170	350	783	60	443	602	787	982	1206	1451	1719
150	335	80	322	451	612	801	1015	1267	1557	350	783	80	548	750	987	1242	1535	1861	2220
150	335	100	381	542	742	981	1258	1587	1969	350	783	100	653	894	1182	1499	1865	2274	2731
150	335	120	440	631	873	1166	1511	1923	2406	350	783	120	747	1033	1375	1755	2196	2694	3256
150	335	140	497	721	1006	1356	1778	2276	2867	350	783	140	841	1171	1566	2012	2531	3123	3795
175	391	40	211	289	383	481	596	723	863	375	839	40	347	465	596	744	906	1081	1269
175	391	60	287	394	525	670	839	1030	1245	375	839	60	463	629	821	1023	1253	1506	1781
175	391	80	353	493	664	859	1087	1348	1646	375	839	80	573	783	1029	1292	1594	1928	2296
175	391	100	419	591	803	1055	1341	1679	2070	375	839	100	682	933	1232	1558	1933	2353	2820
175	391	120	482	687	943	1248	1603	2025	2517	375	839	120	781	1078	1431	1822	2274	2784	3356
175	391	140	545	784	1083	1447	1875	2387	2987	375	839	140	885	1221	1629	2086	2618	3222	3906
200	447	40	229	314	413	518	639	772	919	400	894	40	362	485	620	774	941	1121	1314
200	447	60	311	426	566	719	896	1095	1318	400	894	60	483	655	846	1062	1300	1560	1842
200	447	80	384	533	715	918	1156	1427	1733	400	894	80	604	817	1070	1341	1651	1994	2371
200	447	100	454	638	862	1119	1421	1770	2169	400	894	100	710	971	1280	1615	2001	2430	2907
200	447	120	523	741	1010	1328	1694	2126	2626	400	894	120	814	1122	1486	1888	2351	2872	3454
200	447	140	596	844	1158	1535	1974	2497	3106	400	894	140	919	1270	1690	2160	2704	3320	4015

NOTE: Detector time constant at a reference velocity of 5 ft/sec.

For SI Units: 1 ft = 0.305 m

1000 BTU/sec = 1055 kW

Table B-4.1(d) Installed Spacing of Heat Detector: 8 feet; t_g: 500 Seconds to 1000 Btu/sec; α: 0.004 Btu/sec^3

τ	RTI	ΔT	CEILING HEIGHT IN FEET						
			4.0	8.0	12.0	16.0	20.0	24.0	28.0
			FIRE SIZE AT DETECTOR RESPONSE (BTU/SEC)						
25	56	40	49	78	118	171	237	318	413
25	56	60	70	116	183	272	385	523	689
25	56	80	91	158	256	386	553	759	1005
25	56	100	113	203	336	513	740	1020	1357
25	56	120	136	252	422	651	944	1306	1741
25	56	140	160	304	515	799	1163	1613	2154
50	112	40	70	104	148	204	273	355	453
50	112	60	96	149	220	312	427	568	735
50	112	80	124	196	297	431	601	809	1058
50	112	100	150	246	381	562	792	1075	1415
50	112	120	178	298	471	703	1000	1365	1803
50	112	140	206	353	567	855	1222	1676	2221
75	168	40	87	126	176	236	309	393	493
75	168	60	120	178	255	351	470	612	782
75	168	80	152	231	338	476	649	860	1109
75	168	100	184	286	427	612	845	1131	1471
75	168	120	215	343	521	757	1057	1425	1867
75	168	140	247	402	621	912	1283	1740	2289
100	224	40	103	147	201	266	341	430	533
100	224	60	141	205	288	389	512	658	830
100	224	80	179	264	378	521	698	912	1163
100	224	100	214	324	472	661	898	1188	1530
100	224	120	250	386	571	811	1114	1486	1928
100	224	140	286	449	674	970	1345	1805	2354
125	280	40	118	166	225	295	374	466	573
125	280	60	160	231	319	426	552	702	878
125	280	80	203	295	415	564	746	961	1218
125	280	100	243	360	515	711	952	1245	1590
125	280	120	282	427	619	865	1173	1548	1993
125	280	140	322	494	727	1028	1407	1871	2423
150	335	40	131	184	248	320	404	501	611
150	335	60	179	255	349	462	592	747	926
150	335	80	226	325	452	607	793	1013	1273
150	335	100	270	395	557	759	1005	1300	1650
150	335	120	313	466	666	918	1231	1610	2058
150	335	140	356	537	778	1085	1469	1937	2492
175	391	40	144	201	269	345	434	535	649
175	391	60	196	278	378	496	631	790	973
175	391	80	247	353	487	648	837	1063	1327
175	391	100	295	428	597	806	1058	1356	1711
175	391	120	342	503	711	971	1289	1669	2123
175	391	140	389	579	828	1142	1532	2004	2562
200	447	40	157	217	290	370	463	568	685
200	447	60	213	300	405	526	668	832	1020
200	447	80	268	380	520	688	882	1113	1381
200	447	100	320	460	637	852	1107	1413	1771
200	447	120	370	539	755	1022	1346	1731	2189
200	447	140	421	620	877	1199	1594	2067	2633
225	503	40	169	233	310	394	491	599	721
225	503	60	232	321	432	558	705	873	1065
225	503	80	288	406	553	727	926	1162	1435
225	503	100	343	490	675	897	1158	1469	1832
225	503	120	397	574	799	1073	1403	1792	2254
225	503	140	451	659	925	1254	1656	2134	2703
250	559	40	181	249	329	417	518	630	756
250	559	60	247	342	458	588	740	914	1110
250	559	80	307	432	584	765	970	1211	1488
250	559	100	366	520	712	941	1208	1524	1892
250	559	120	424	609	841	1122	1455	1853	2320
250	559	140	480	697	971	1309	1717	2200	2774
275	615	40	192	264	348	439	544	660	789
275	615	60	263	362	483	618	775	953	1154
275	615	80	326	456	615	798	1012	1258	1541
275	615	100	388	549	748	984	1257	1578	1951
275	615	120	449	642	882	1171	1510	1914	2386
275	615	140	509	734	1017	1362	1773	2266	2844
300	671	40	203	278	366	461	569	690	823
300	671	60	277	381	507	647	809	992	1198
300	671	80	345	481	645	834	1053	1305	1593
300	671	100	410	578	783	1026	1305	1632	2010
300	671	120	474	674	922	1219	1564	1974	2451
300	671	140	541	770	1062	1415	1833	2331	2915
325	727	40	214	292	384	482	594	719	855
325	727	60	292	400	531	676	843	1030	1240
325	727	80	362	504	675	869	1094	1351	1644
325	727	100	431	606	818	1063	1352	1685	2069
325	727	120	498	706	961	1266	1617	2033	2516
325	727	140	568	806	1106	1467	1892	2396	2985
350	783	40	224	306	401	503	619	747	887
350	783	60	306	419	555	704	875	1068	1282
350	783	80	380	527	704	903	1134	1397	1694
350	783	100	452	633	852	1104	1398	1738	2126
350	783	120	522	737	1000	1312	1670	2092	2580
350	783	140	594	841	1149	1518	1949	2460	3055
375	839	40	235	320	419	523	643	774	918
375	839	60	320	437	577	731	907	1104	1324
375	839	80	397	550	732	937	1173	1442	1744
375	839	100	472	659	885	1143	1444	1790	2184
375	839	120	545	767	1038	1357	1722	2150	2644
375	839	140	619	875	1191	1569	2007	2524	3125
400	894	40	245	333	435	543	666	801	949
400	894	60	333	455	600	758	939	1141	1364
400	894	80	414	572	760	970	1212	1486	1794
400	894	100	492	685	918	1182	1489	1841	2241
400	894	120	572	798	1075	1396	1773	2208	2707
400	894	140	645	908	1233	1619	2063	2588	3194

NOTE: Detector time constant at a reference velocity of 5 ft/sec.

For SI Units: 1 ft = 0.305 m

1000 BTU/sec = 1055 kW

Table B-4.1(e) Installed Spacing of Heat Detector: 8 feet; t_g: 600 Seconds to 1000 Btu/sec; α: 0.003 Btu/sec^3

τ	RTI	ΔT	CEILING HEIGHT IN FEET FIRE SIZE AT DETECTOR RESPONSE (BTU/SEC)						
			4.0	8.0	12.0	16.0	20.0	24.0	28.0
25	56	40	44	71	110	160	225	303	397
25	56	60	63	108	173	259	370	505	668
25	56	80	83	148	244	372	536	738	980
25	56	100	104	192	323	497	721	997	1328
25	56	120	126	240	408	633	922	1279	1709
25	56	140	149	291	499	780	1139	1584	2119
50	112	40	62	93	135	188	254	334	429
50	112	60	86	135	203	292	404	542	705
50	112	80	111	180	278	409	575	779	1024
50	112	100	136	228	360	537	763	1042	1376
50	112	120	161	279	448	676	968	1328	1761
50	112	140	188	332	542	826	1188	1636	2175
75	168	40	77	112	158	215	284	365	462
75	168	60	106	160	233	325	440	579	744
75	168	80	136	210	312	446	614	821	1066
75	168	100	165	262	398	578	807	1088	1425
75	168	120	194	316	489	720	1015	1378	1813
75	168	140	224	373	586	872	1238	1689	2231
100	224	40	91	130	180	241	312	396	495
100	224	60	124	184	261	357	475	616	784
100	224	80	158	238	346	483	654	863	1110
100	224	100	191	295	436	619	851	1134	1471
100	224	120	224	353	531	765	1062	1428	1866
100	224	140	257	413	631	920	1288	1742	2287
125	280	40	103	147	201	265	339	427	528
125	280	60	141	206	288	388	510	653	824
125	280	80	180	265	378	520	695	905	1155
125	280	100	216	326	472	660	895	1181	1520
125	280	120	252	388	571	810	1110	1478	1916
125	280	140	289	451	675	968	1340	1796	2341
150	335	40	115	162	220	288	366	456	561
150	335	60	158	227	313	419	542	690	863
150	335	80	200	291	409	556	734	947	1200
150	335	100	239	355	508	701	939	1229	1570
150	335	120	279	421	611	854	1158	1530	1970
150	335	140	318	488	718	1016	1391	1851	2398
175	391	40	127	177	239	309	391	485	593
175	391	60	173	247	338	448	575	727	903
175	391	80	218	315	439	591	774	989	1246
175	391	100	261	384	543	741	983	1274	1620
175	391	120	304	454	650	898	1206	1581	2024
175	391	140	347	524	761	1064	1443	1906	2456
200	447	40	138	192	257	330	416	513	624
200	447	60	187	266	362	476	607	763	942
200	447	80	237	339	468	625	810	1031	1291
200	447	100	283	412	576	780	1027	1321	1670
200	447	120	329	485	688	942	1255	1630	2078
200	447	140	374	559	802	1111	1495	1961	2513
225	503	40	148	205	274	351	440	541	654
225	503	60	202	284	385	504	639	798	981
225	503	80	254	362	496	658	848	1073	1336
225	503	100	303	438	609	818	1070	1368	1720
225	503	120	352	515	725	985	1302	1681	2132
225	503	140	401	593	843	1158	1546	2013	2572
250	559	40	158	219	291	371	463	567	684
250	559	60	215	302	407	528	670	832	1019
250	559	80	271	384	524	691	885	1114	1380
250	559	100	323	464	641	856	1110	1414	1770
250	559	120	375	545	761	1027	1350	1732	2187
250	559	140	426	626	884	1204	1598	2069	2630
275	615	40	168	232	307	390	486	593	713
275	615	60	231	320	429	554	700	866	1056
275	615	80	287	405	550	723	921	1155	1425
275	615	100	343	490	673	893	1152	1460	1820
275	615	120	397	574	797	1069	1397	1783	2241
275	615	140	451	658	923	1250	1649	2123	2689
300	671	40	178	244	323	409	508	619	742
300	671	60	244	337	451	579	729	900	1093
300	671	80	303	426	577	754	957	1195	1469
300	671	100	362	514	703	930	1193	1506	1869
300	671	120	419	602	831	1110	1440	1834	2296
300	671	140	475	690	962	1295	1699	2178	2747
325	727	40	187	257	339	428	530	644	770
325	727	60	256	353	471	604	758	933	1130
325	727	80	319	447	602	782	992	1234	1512
325	727	100	380	538	733	965	1234	1551	1919
325	727	120	440	630	865	1150	1485	1884	2350
325	727	140	499	721	999	1340	1746	2233	2806
350	783	40	196	269	354	446	551	668	797
350	783	60	269	370	492	628	786	965	1166
350	783	80	334	467	627	811	1026	1273	1555
350	783	100	398	562	763	1000	1274	1596	1968
350	783	120	461	657	899	1190	1530	1934	2404
350	783	140	523	751	1037	1384	1796	2287	2864
375	839	40	205	281	369	464	572	692	824
375	839	60	281	386	512	652	814	997	1201
375	839	80	349	486	652	841	1060	1312	1598
375	839	100	416	585	792	1035	1313	1640	2016
375	839	120	481	683	932	1229	1574	1983	2458
375	839	140	549	781	1073	1427	1844	2341	2922
400	894	40	214	292	383	481	592	715	851
400	894	60	292	401	531	675	841	1028	1236
400	894	80	364	506	676	869	1094	1350	1640
400	894	100	433	608	820	1065	1352	1684	2065
400	894	120	501	709	964	1268	1618	2032	2511
400	894	140	571	810	1110	1470	1893	2395	2981

NOTE: Detector time constant at a reference velocity of 5 ft/sec.

For SI Units: 1 ft = 0.305 m

1000 BTU/sec = 1055 kW

Table B-4.1(f)   Installed Spacing of Heat Detector: 10 feet; t_g: 50 Seconds to 1000 Btu/sec; α: 0.400 Btu/sec³

τ	RTI	ΔT	CEILING HEIGHT IN FEET — FIRE SIZE AT DETECTOR RESPONSE (BTU/SEC)							τ	RTI	ΔT	CEILING HEIGHT IN FEET — FIRE SIZE AT DETECTOR RESPONSE (BTU/SEC)						
			4.0	8.0	12.0	16.0	20.0	24.0	28.0				4.0	8.0	12.0	16.0	20.0	24.0	28.0
25	56	40	376	499	623	779	956	1152	1366	225	503	40	1234	1667	2041	2468	2925	3406	3911
25	56	60	486	641	811	1027	1273	1550	1860	225	503	60	1622	2186	2681	3249	3858	4503	5183
25	56	80	570	769	1013	1271	1591	1957	2371	225	503	80	1998	2671	3274	3976	4731	5535	6386
25	56	100	675	902	1193	1517	1916	2377	2906	225	503	100	2367	3132	3838	4670	5569	6529	7551
25	56	120	751	1024	1371	1766	2249	2814	3465	225	503	120	2731	3577	4380	5340	6382	7499	8692
25	56	140	827	1146	1551	2040	2593	3268	4050	225	503	140	3096	3911	4902	5994	7178	8452	9818
50	112	40	552	692	904	1084	1306	1550	1813	250	559	40	1317	1778	2175	2628	3111	3620	4152
50	112	60	712	895	1171	1418	1723	2059	2426	250	559	60	1740	2337	2860	3461	4105	4786	5503
50	112	80	828	1106	1391	1738	2129	2561	3038	250	559	80	2150	2859	3494	4237	5035	5882	6778
50	112	100	945	1282	1633	2055	2531	3064	3658	250	559	100	2553	3355	4096	4976	5925	6936	8011
50	112	120	1035	1446	1905	2371	2936	3575	4292	250	559	120	2953	3739	4669	5689	6787	7963	9215
50	112	140	1177	1626	2143	2685	3343	4094	4942	250	559	140	3267	4270	5234	6384	7631	8970	10402
75	168	40	673	890	1088	1331	1598	1885	2191	275	615	40	1398	1886	2305	2783	3291	3826	4386
75	168	60	885	1157	1418	1744	2104	2495	2916	275	615	60	1855	2484	3033	3667	4345	5061	5813
75	168	80	1000	1370	1725	2136	2590	3086	3625	275	615	80	2300	3042	3707	4489	5329	6219	7159
75	168	100	1174	1602	2024	2518	3066	3670	4333	275	615	100	2737	3485	4340	5272	6270	7331	8456
75	168	120	1330	1822	2315	2893	3538	4255	5047	275	615	120	3175	3975	4954	6027	7181	8413	9723
75	168	140	1484	2037	2602	3264	4010	4843	5769	275	615	140	3488	4440	5548	6761	8070	9472	10968
100	224	40	801	1051	1276	1554	1858	2183	2528	300	671	40	1478	1992	2432	2933	3467	4027	4613
100	224	60	985	1337	1662	2037	2446	2885	3356	300	671	60	1970	2628	3202	3867	4578	5328	6115
100	224	80	1195	1623	2026	2494	3005	3560	4158	300	671	80	2448	3221	3915	4735	5615	6547	7529
100	224	100	1389	1893	2375	2934	3550	4221	4951	300	671	100	2922	3687	4584	5560	6605	7716	8890
100	224	120	1581	2155	2714	3365	4086	4877	5742	300	671	120	3301	4199	5231	6356	7563	8850	10217
100	224	140	1771	2411	3046	3790	4618	5532	6537	300	671	140	3705	4696	5858	7129	8498	9962	11520
125	280	40	908	1193	1447	1759	2097	2457	2837	325	727	40	1557	2096	2555	3079	3637	4222	4834
125	280	60	1130	1528	1889	2307	2761	3246	3763	325	727	60	2084	2769	3367	4063	4805	5588	6409
125	280	80	1364	1853	2303	2823	3389	3998	4652	325	727	80	2596	3397	4117	4975	5894	6866	7889
125	280	100	1595	2164	2699	3319	3997	4732	5525	325	727	100	3106	3883	4821	5842	6933	8090	9313
125	280	120	1822	2466	3083	3803	4594	5456	6392	325	727	120	3489	4421	5501	6677	7937	9278	10699
125	280	140	2046	2759	3458	4277	5183	6175	7257	325	727	140	3919	4945	6160	7488	8915	10439	12059
150	335	40	971	1308	1605	1949	2320	2713	3127	350	783	40	1635	2198	2676	3222	3804	4413	5049
150	335	60	1257	1702	2101	2560	3055	3583	4144	350	783	60	2196	2908	3528	4253	5027	5842	6696
150	335	80	1528	2070	2564	3132	3749	4410	5117	350	783	80	2744	3476	4308	5209	6166	7178	8241
150	335	100	1794	2421	3004	3681	4418	5213	6067	350	783	100	3291	4076	5052	6117	7253	8456	9726
150	335	120	2055	2760	3430	4214	5072	6002	7006	350	783	120	3675	4639	5765	6991	8302	9695	11170
150	335	140	2314	3089	3845	4737	5716	6782	7939	350	783	140	4130	5189	6456	7839	9323	10905	12585
175	391	40	1035	1419	1755	2130	2531	2954	3400	375	839	40	1713	2298	2794	3362	3966	4600	5260
175	391	60	1381	1869	2303	2800	3335	3903	4505	375	839	60	2308	3044	3685	4440	5244	6090	6976
175	391	80	1688	2277	2811	3426	4091	4801	5558	375	839	80	2891	3636	4501	5438	6433	7483	8586
175	391	100	1988	2666	3294	4025	4818	5670	6582	375	839	100	3366	4257	5279	6386	7566	8814	10130
175	391	120	2284	3042	3760	4606	5527	6521	7591	375	839	120	3858	4853	6024	7298	8659	10104	11631
175	391	140	2577	3406	4214	5173	6223	7361	8590	375	839	140	4339	5428	6745	8183	9723	11362	13100
200	447	40	1151	1552	1902	2302	2732	3185	3661	400	894	40	1790	2396	2909	3499	4126	4782	5466
200	447	60	1503	2030	2495	3029	3601	4209	4851	400	894	60	2419	3178	3840	4622	5457	6334	7251
200	447	80	1844	2477	3047	3706	4417	5175	5980	400	894	80	3038	3793	4691	5662	6694	7782	8923
200	447	100	2179	2903	3571	4353	5201	6108	7076	400	894	100	3522	4440	5501	6649	7872	9165	10527
200	447	120	2509	3313	4076	4980	5962	7019	8151	400	894	120	4040	5062	6278	7599	9009	10504	12083
200	447	140	2837	3712	4567	5591	6709	7917	9215	400	894	140	4546	5662	7029	8519	10115	11810	13606

NOTE: Detector time constant at a reference velocity of 5 ft/sec.

For SI Units: 1 ft = 0.305 m

1000 BTU/sec = 1055 kW

Table B-4.1(g) Installed Spacing of Heat Detector: 10 feet; t_g: 150 Seconds to 1000 Btu/sec; α: 0.044 Btu/sec³

τ	RTI	ΔT	CEILING HEIGHT IN FEET 4.0	8.0	12.0	16.0	20.0	24.0	28.0		τ	RTI	ΔT	CEILING HEIGHT IN FEET 4.0	8.0	12.0	16.0	20.0	24.0	28.0
			FIRE SIZE AT DETECTOR RESPONSE (BTU/SEC)											FIRE SIZE AT DETECTOR RESPONSE (BTU/SEC)						
25	56	40	152	209	280	362	461	576	709		225	503	40	544	694	871	1064	1273	1497	1737
25	56	60	205	284	389	519	671	856	1071		225	503	60	727	931	1173	1441	1734	2052	2395
25	56	80	252	358	502	683	904	1164	1473		225	503	80	895	1170	1460	1803	2181	2595	3046
25	56	100	298	435	621	858	1151	1504	1912		225	503	100	1034	1378	1738	2157	2623	3136	3701
25	56	120	345	514	746	1044	1416	1864	2385		225	503	120	1195	1587	2011	2507	3063	3680	4634
25	56	140	392	595	877	1241	1696	2247	2890		225	503	140	1345	1790	2280	2856	3504	4230	5040
50	112	40	221	296	380	481	596	727	872		250	559	40	580	742	928	1132	1352	1588	1840
50	112	60	297	398	525	669	841	1040	1268		250	559	60	775	994	1250	1532	1840	2173	2531
50	112	80	362	496	664	860	1097	1374	1695		250	559	80	948	1246	1555	1915	2311	2743	3212
50	112	100	426	593	805	1064	1366	1730	2154		250	559	100	1116	1473	1849	2289	2775	3310	3895
50	112	120	490	691	950	1269	1649	2107	2645		250	559	120	1279	1694	2138	2658	3237	3878	4585
50	112	140	561	791	1098	1482	1952	2506	3166		250	559	140	1440	1909	2422	3024	3698	4450	5284
75	168	40	277	369	467	584	715	861	1021		275	615	40	616	787	984	1198	1429	1676	1939
75	168	60	372	495	635	803	995	1212	1456		275	615	60	822	1055	1324	1621	1943	2290	2663
75	168	80	454	614	807	1022	1280	1576	1913		275	615	80	1001	1322	1647	2024	2438	2888	3375
75	168	100	535	731	971	1245	1574	1957	2397		275	615	100	1187	1564	1958	2417	2924	3479	4085
75	168	120	622	848	1135	1472	1879	2355	2908		275	615	120	1362	1797	2261	2804	3406	4070	4800
75	168	140	697	963	1302	1713	2194	2772	3446		275	615	140	1534	2025	2560	3187	3887	4664	5523
100	224	40	328	435	545	677	823	983	1158		300	671	40	663	832	1038	1263	1504	1762	2036
100	224	60	439	582	738	926	1136	1371	1632		300	671	60	880	1115	1397	1707	2043	2405	2792
100	224	80	537	720	926	1171	1450	1766	2122		300	671	80	1065	1398	1736	2130	2561	3028	3533
100	224	100	641	857	1122	1418	1770	2174	2633		300	671	100	1256	1652	2063	2542	3069	3644	4270
100	224	120	730	990	1307	1667	2098	2596	3168		300	671	120	1443	1898	2382	2946	3571	4258	5010
100	224	140	819	1121	1492	1920	2434	3034	3726		300	671	140	1625	2138	2695	3346	4071	4873	5757
125	280	40	383	487	617	763	923	1097	1286		325	727	40	695	875	1091	1325	1577	1845	2130
125	280	60	501	662	835	1040	1268	1520	1799		325	727	60	918	1172	1467	1791	2141	2516	2918
125	280	80	625	822	1044	1310	1611	1947	2322		325	727	80	1123	1471	1823	2234	2681	3166	3687
125	280	100	729	973	1249	1580	1956	2382	2862		325	727	100	1325	1738	2166	2664	3210	3805	4451
125	280	120	831	1121	1467	1850	2307	2828	3421		325	727	120	1522	1996	2499	3085	3732	4441	5216
125	280	140	935	1268	1670	2123	2665	3288	4002		325	727	140	1716	2249	2827	3502	4252	5078	5986
150	335	40	426	543	685	844	1017	1204	1406		350	783	40	727	918	1142	1387	1648	1927	2222
150	335	60	559	738	925	1147	1392	1662	1957		350	783	60	957	1229	1537	1873	2236	2626	3041
150	335	80	694	914	1155	1442	1762	2119	2513		350	783	80	1179	1542	1909	2335	2799	3300	3839
150	335	100	812	1081	1380	1733	2133	2581	3082		350	783	100	1392	1822	2266	2783	3348	3963	4628
150	335	120	930	1246	1602	2025	2507	3052	3667		350	783	120	1600	2093	2614	3222	3890	4621	5418
150	335	140	1034	1406	1839	2317	2886	3535	4271		350	783	140	1805	2330	2955	3655	4428	5279	6211
175	391	40	467	595	750	920	1106	1306	1521		375	839	40	758	959	1193	1447	1718	2007	2312
175	391	60	628	798	1011	1249	1511	1797	2108		375	839	60	996	1304	1605	1954	2330	2733	3162
175	391	80	760	1002	1261	1567	1907	2283	2697		375	839	80	1234	1589	1992	2434	2914	3432	3987
175	391	100	899	1186	1504	1880	2302	2772	3295		375	839	100	1459	1880	2364	2900	3484	4118	4802
175	391	120	1013	1362	1744	2192	2699	3268	3906		375	839	120	1678	2161	2726	3355	4045	4798	5616
175	391	140	1146	1540	1981	2503	3099	3773	4534		375	839	140	1894	2434	3081	3804	4601	5476	6431
200	447	40	506	646	811	994	1191	1403	1631		400	894	40	789	1000	1242	1505	1787	2085	2400
200	447	60	678	866	1093	1347	1625	1927	2254		400	894	60	1035	1359	1671	2032	2422	2837	3280
200	447	80	825	1086	1362	1687	2047	2442	2874		400	894	80	1289	1657	2074	2532	3027	3561	4132
200	447	100	966	1284	1623	2021	2465	2957	3501		400	894	100	1524	1959	2461	3014	3617	4269	4973
200	447	120	1109	1477	1880	2352	2884	3478	4138		400	894	120	1754	2251	2837	3486	4197	4971	5810
200	447	140	1247	1667	2133	2682	3305	4005	4790		400	894	140	1982	2536	3205	3951	4771	5669	6648

NOTE: Detector time constant at a reference velocity of 5 ft/sec.

For SI Units: 1 ft = 0.305 m

1000 BTU/sec = 1055 kW

Table B-4.1(h)   Installed Spacing of Heat Detector: 10 feet; t_g: 300 Seconds to 1000 Btu/sec; α: 0.011 Btu/sec³

τ	RTI	ΔT	CEILING HEIGHT IN FEET							τ	RTI	ΔT	CEILING HEIGHT IN FEET						
			4.0	8.0	12.0	16.0	20.0	24.0	28.0				4.0	8.0	12.0	16.0	20.0	24.0	28.0
			FIRE SIZE AT DETECTOR RESPONSE (BTU / SEC)										FIRE SIZE AT DETECTOR RESPONSE (BTU / SEC)						
25	56	40	91	131	183	249	330	424	537	225	503	40	323	415	526	650	787	936	1099
25	56	60	126	187	271	379	513	675	863	225	503	60	432	570	723	901	1101	1324	1570
25	56	80	160	246	367	524	720	957	1234	225	503	80	535	713	914	1150	1417	1718	2056
25	56	100	195	309	471	682	947	1268	1645	225	503	100	639	853	1103	1399	1738	2125	2563
25	56	120	230	377	582	853	1193	1605	2091	225	503	120	734	990	1299	1650	2066	2545	3092
25	56	140	267	448	701	1036	1456	1967	2574	225	503	140	828	1125	1487	1906	2403	2980	3645
50	112	40	131	180	241	313	399	500	616	250	559	40	345	443	560	690	833	989	1159
50	112	60	179	250	344	459	597	762	956	250	559	60	461	608	768	955	1163	1394	1649
50	112	80	224	322	451	615	816	1054	1338	250	559	80	572	759	969	1215	1493	1804	2151
50	112	100	269	395	565	782	1051	1377	1758	250	559	100	682	907	1168	1476	1826	2224	2672
50	112	120	314	470	684	960	1305	1723	2213	250	559	120	783	1052	1374	1738	2166	2656	3214
50	112	140	359	549	810	1150	1575	2092	2700	250	559	140	888	1196	1572	2003	2514	3103	3778
75	168	40	164	222	293	371	465	572	694	275	615	40	366	470	593	729	878	1041	1217
75	168	60	224	306	410	532	679	851	1050	275	615	60	490	644	812	1007	1224	1463	1725
75	168	80	279	389	530	703	909	1155	1444	275	615	80	613	806	1024	1280	1567	1888	2244
75	168	100	333	473	654	881	1158	1485	1874	275	615	100	723	961	1232	1551	1913	2321	2780
75	168	120	386	558	784	1068	1420	1839	2338	275	615	120	831	1113	1439	1823	2264	2766	3334
75	168	140	440	645	918	1266	1697	2220	2833	275	615	140	938	1264	1655	2098	2622	3223	3910
100	224	40	194	260	335	424	526	640	769	300	671	40	387	496	624	767	922	1091	1273
100	224	60	265	357	471	602	757	937	1143	300	671	60	518	680	855	1058	1283	1530	1800
100	224	80	328	451	604	784	1001	1255	1550	300	671	80	647	850	1077	1342	1639	1969	2335
100	224	100	391	545	739	977	1258	1596	1991	300	671	100	763	1012	1294	1624	1997	2416	2885
100	224	120	453	639	879	1175	1534	1959	2464	300	671	120	883	1173	1510	1907	2360	2873	3453
100	224	140	519	735	1022	1380	1821	2344	2967	300	671	140	988	1330	1725	2191	2728	3342	4040
125	280	40	222	296	377	474	583	705	841	325	727	40	407	521	656	804	965	1140	1328
125	280	60	302	404	528	667	832	1020	1234	325	727	60	546	715	897	1107	1340	1595	1873
125	280	80	374	508	673	863	1090	1353	1656	325	727	80	680	892	1128	1403	1710	2050	2425
125	280	100	445	612	820	1065	1360	1706	2108	325	727	100	803	1063	1355	1696	2080	2510	2989
125	280	120	514	716	969	1278	1642	2079	2591	325	727	120	925	1230	1579	1989	2454	2979	3570
125	280	140	587	820	1122	1493	1938	2473	3103	325	727	140	1038	1395	1803	2283	2833	3459	4168
150	335	40	248	329	417	521	637	766	909	350	783	40	427	546	686	840	1007	1188	1382
150	335	60	366	448	577	729	903	1100	1321	350	783	60	572	749	937	1156	1396	1659	1945
150	335	80	417	563	739	939	1176	1448	1759	350	783	80	713	934	1179	1463	1779	2128	2513
150	335	100	495	676	897	1153	1458	1814	2224	350	783	100	842	1113	1415	1766	2161	2602	3091
150	335	120	577	789	1056	1372	1752	2199	2718	350	783	120	967	1287	1647	2069	2546	3083	3685
150	335	140	651	901	1218	1603	2058	2602	3239	350	783	140	1092	1459	1879	2372	2935	3574	4295
175	391	40	273	361	455	566	689	825	975	375	839	40	446	571	716	875	1048	1235	1435
175	391	60	370	490	627	789	972	1177	1407	375	839	60	606	774	977	1203	1451	1722	2016
175	391	80	458	615	796	1012	1259	1541	1860	375	839	80	745	976	1228	1521	1847	2205	2599
175	391	100	544	737	971	1237	1554	1920	2339	375	839	100	886	1162	1473	1835	2241	2692	3192
175	391	120	631	858	1140	1468	1859	2316	2844	375	839	120	1009	1342	1714	2148	2637	3185	3798
175	391	140	712	979	1311	1703	2175	2730	3375	375	839	140	1141	1521	1953	2460	3036	3688	4421
200	447	40	301	386	491	609	739	882	1038	400	894	40	465	595	745	910	1089	1281	1487
200	447	60	401	531	676	846	1038	1251	1489	400	894	60	631	807	1017	1250	1505	1784	2085
200	447	80	497	665	856	1082	1339	1631	1959	400	894	80	776	1016	1277	1579	1913	2281	2684
200	447	100	595	796	1042	1319	1647	2023	2452	400	894	100	920	1209	1530	1903	2319	2781	3291
200	447	120	684	925	1220	1560	1964	2432	2969	400	894	120	1055	1397	1780	2225	2726	3286	3910
200	447	140	771	1053	1400	1806	2290	2856	3510	400	894	140	1190	1582	2027	2547	3136	3800	4544

NOTE: Detector time constant at a reference velocity of 5 ft/sec.

For SI Units: 1 ft = 0.305 m

1000 BTU/sec = 1055 kW

Table B-4.1(i) Installed Spacing of Heat Detector: 10 feet; t_g: 500 Seconds to 1000 Btu/sec; α: 0.004 Btu/sec³

τ	RTI	ΔT	CEILING HEIGHT IN FEET FIRE SIZE AT DETECTOR RESPONSE (BTU / SEC) 4.0	8.0	12.0	16.0	20.0	24.0	28.0
25	56	40	65	98	143	202	274	362	465
25	56	60	92	146	222	322	447	598	775
25	56	80	121	200	312	458	643	867	1133
25	56	100	150	258	410	609	861	1166	1530
25	56	120	181	320	516	773	1098	1493	1963
25	56	140	214	386	630	950	1352	1844	2430
50	112	40	91	130	179	241	316	405	510
50	112	60	127	186	266	369	495	649	828
50	112	80	162	246	361	511	698	924	1191
50	112	100	198	309	464	666	920	1229	1593
50	112	120	235	376	574	835	1162	1561	2031
50	112	140	272	447	692	1015	1422	1917	2506
75	168	40	114	158	212	279	356	449	556
75	168	60	158	223	308	415	545	699	882
75	168	80	199	290	410	564	754	981	1251
75	168	100	241	359	519	725	982	1293	1660
75	168	120	283	431	634	898	1228	1630	2103
75	168	140	326	506	756	1082	1492	1991	2580
100	224	40	134	183	243	312	395	491	602
100	224	60	185	257	348	460	592	751	937
100	224	805	233	330	458	616	810	1040	1313
100	224	100	280	406	572	783	1044	1356	1727
100	224	120	328	484	693	961	1295	1700	2176
100	224	140	376	564	819	1150	1563	2065	2657
125	280	40	153	207	272	345	432	533	647
125	280	60	210	288	386	501	640	803	992
125	280	80	264	369	503	667	863	1099	1375
125	280	100	317	451	624	841	1106	1421	1795
125	280	120	370	534	751	1024	1362	1768	2250
125	280	140	422	619	882	1218	1635	2141	2736
150	335	40	171	229	297	377	468	572	690
150	335	60	234	318	422	542	686	853	1046
150	335	80	293	405	546	717	918	1158	1437
150	335	100	351	493	674	897	1165	1486	1864
150	335	120	409	582	807	1087	1430	1838	2324
150	335	140	466	673	944	1285	1707	2213	2815
175	391	40	187	250	322	407	503	611	733
175	391	60	257	346	456	582	731	903	1100
175	391	80	321	440	588	762	971	1216	1499
175	391	100	384	534	723	953	1225	1551	1933
175	391	120	447	629	861	1148	1494	1909	2398
175	391	140	512	725	1004	1352	1780	2289	2895
200	447	40	204	271	347	435	536	648	774
200	447	60	279	374	489	620	774	951	1152
200	447	80	348	474	629	809	1023	1273	1561
200	447	100	416	574	770	1007	1285	1616	2002
200	447	120	483	674	915	1209	1560	1980	2473
200	447	140	552	775	1062	1419	1848	2365	2975

τ	RTI	ΔT	CEILING HEIGHT IN FEET FIRE SIZE AT DETECTOR RESPONSE (BTU / SEC) 4.0	8.0	12.0	16.0	20.0	24.0	28.0
225	503	40	219	291	370	463	568	685	815
225	503	60	300	400	518	657	817	998	1204
225	503	80	374	507	668	854	1074	1329	1622
225	503	100	446	612	816	1056	1343	1680	2070
225	503	120	518	717	966	1268	1625	2051	2548
225	503	140	591	823	1120	1484	1919	2441	3055
250	559	40	234	310	393	490	599	720	854
250	559	60	320	426	549	693	858	1045	1255
250	559	80	399	538	706	898	1124	1385	1683
250	559	100	476	649	861	1107	1401	1743	2139
250	559	120	555	760	1017	1323	1689	2121	2623
250	559	140	628	870	1176	1548	1990	2517	3135
275	615	40	249	328	415	517	630	755	893
275	615	60	339	451	578	728	898	1090	1305
275	615	80	423	569	743	941	1173	1440	1742
275	615	100	505	685	904	1157	1458	1806	2206
275	615	120	588	801	1066	1379	1753	2190	2697
275	615	140	665	916	1231	1611	2059	2593	3216
300	671	40	266	343	437	542	659	788	930
300	671	60	358	475	607	762	938	1135	1355
300	671	80	447	599	755	983	1221	1493	1801
300	671	100	533	720	947	1207	1513	1868	2273
300	671	120	619	841	1115	1435	1815	2259	2772
300	671	140	701	961	1285	1668	2128	2668	3296
325	727	40	280	361	458	567	688	821	967
325	727	60	377	499	636	796	976	1178	1403
325	727	80	470	628	810	1024	1269	1546	1860
325	727	100	564	755	988	1255	1568	1929	2340
325	727	120	650	880	1162	1489	1877	2327	2845
325	727	140	735	1005	1338	1729	2196	2742	3376
350	783	40	293	378	479	592	717	854	1003
350	783	60	395	522	664	829	1014	1221	1451
350	783	80	493	657	844	1064	1315	1598	1917
350	783	100	591	789	1024	1302	1622	1989	2406
350	783	120	681	919	1209	1543	1938	2394	2919
350	783	140	770	1048	1389	1788	2263	2816	3456
375	839	40	306	394	499	616	744	885	1039
375	839	60	413	544	691	861	1051	1263	1498
375	839	80	515	685	878	1104	1360	1649	1974
375	839	100	616	822	1064	1348	1675	2048	2471
375	839	120	710	957	1254	1596	1998	2461	2991
375	839	140	803	1091	1440	1847	2329	2889	3535
400	894	40	319	410	519	639	772	916	1074
400	894	60	431	567	718	893	1088	1305	1544
400	894	80	537	712	911	1143	1405	1700	2030
400	894	100	642	854	1102	1394	1728	2107	2536
400	894	120	740	994	1299	1648	2057	2527	3063
400	894	140	836	1132	1490	1905	2395	2692	3614

NOTE: Detector time constant at a reference velocity of 5 ft/sec.

For SI Units: 1 ft = 0.305 m

1000 BTU/sec = 1055 kW

Table B-4.1(j)   Installed Spacing of Heat Detector: 10 feet; t_g: 600 Seconds to 1000 Btu/sec; α: 0.003 Btu/sec^3

τ	RTI	ΔT	CEILING HEIGHT IN FEET — FIRE SIZE AT DETECTOR RESPONSE (BTU / SEC)						
			4.0	8.0	12.0	16.0	20.0	24.0	28.0
25	56	40	58	89	133	190	260	346	446
25	56	60	84	136	210	307	429	577	752
25	56	80	110	188	298	442	623	843	1105
25	56	100	139	244	394	591	838	1139	1498
25	56	120	168	305	499	753	1073	1463	1928
25	56	140	200	371	611	927	1325	1811	2391
50	112	40	81	116	163	222	295	381	483
50	112	60	114	170	246	346	469	619	795
50	112	80	146	226	338	485	668	890	1153
50	112	100	179	287	439	637	887	1191	1550
50	112	120	214	352	547	803	1126	1519	1987
50	112	140	249	420	662	981	1382	1871	2454
75	168	40	100	140	191	254	328	417	521
75	168	60	140	201	282	384	510	661	839
75	168	80	178	264	379	528	714	938	1203
75	168	100	216	329	484	685	937	1244	1605
75	168	120	256	398	596	855	1180	1576	2043
75	168	140	295	470	715	1036	1440	1932	2515
100	224	40	118	162	218	283	361	452	559
100	224	60	164	230	316	422	550	704	884
100	224	80	207	298	419	572	760	985	1253
100	224	100	250	369	529	734	988	1297	1660
100	224	120	294	443	645	907	1235	1633	2103
100	224	140	338	519	767	1091	1498	1994	2579
125	280	40	134	183	242	311	393	487	596
125	280	60	186	257	348	459	590	746	930
125	280	80	234	332	458	615	806	1034	1304
125	280	100	282	408	573	782	1040	1349	1716
125	280	120	330	486	694	959	1290	1692	2164
125	280	140	379	566	820	1148	1557	2056	2643
150	335	40	150	202	266	338	423	521	633
150	335	60	207	283	379	493	629	789	975
150	335	80	260	363	495	657	851	1083	1356
150	335	100	312	445	616	829	1091	1403	1773
150	335	120	365	527	741	1012	1346	1747	2225
150	335	140	417	612	872	1204	1617	2118	2708
175	391	40	164	221	286	364	453	554	669
175	391	60	226	308	409	527	667	831	1020
175	391	80	284	394	531	698	896	1131	1407
175	391	100	341	480	657	876	1140	1457	1830
175	391	120	398	568	788	1064	1402	1806	2286
175	391	140	454	657	923	1260	1677	2178	2774
200	447	40	178	239	308	389	482	587	705
200	447	60	245	331	437	559	704	872	1064
200	447	80	308	423	566	736	940	1180	1459
200	447	100	369	514	698	922	1190	1511	1887
200	447	120	429	606	833	1115	1458	1865	2348
200	447	140	489	700	973	1316	1737	2241	2840
225	503	40	192	256	328	413	509	618	739
225	503	60	264	354	465	591	741	912	1108
225	503	80	330	451	600	775	984	1227	1510
225	503	100	395	547	737	968	1240	1564	1944
225	503	120	460	644	878	1165	1510	1924	2410
225	503	140	526	742	1022	1371	1797	2304	2906
250	559	40	205	272	348	437	536	648	773
250	559	60	281	377	492	623	776	952	1152
250	559	80	352	478	633	813	1026	1274	1561
250	559	100	421	579	776	1009	1289	1618	2001
250	559	120	489	681	921	1215	1564	1982	2472
250	559	140	559	783	1070	1425	1853	2367	2972
275	615	40	218	289	367	459	563	678	806
275	615	60	298	398	515	653	811	991	1194
275	615	80	373	505	665	850	1068	1321	1611
275	615	100	446	611	813	1052	1337	1671	2058
275	615	120	518	716	964	1264	1618	2040	2534
275	615	140	591	823	1118	1479	1912	2430	3039
300	671	40	230	304	386	482	588	707	839
300	671	60	315	419	541	683	845	1029	1237
300	671	80	394	531	697	886	1110	1367	1661
300	671	100	471	641	850	1094	1385	1723	2114
300	671	120	549	752	1006	1309	1671	2098	2596
300	671	140	622	862	1164	1532	1970	2493	3106
325	727	40	242	320	405	503	614	736	871
325	727	60	331	440	565	712	879	1067	1278
325	727	80	414	557	728	922	1150	1412	1711
325	727	100	495	671	886	1136	1432	1775	2170
325	727	120	576	786	1047	1356	1724	2156	2658
325	727	140	652	900	1210	1585	2028	2556	3173
350	783	40	257	332	423	525	638	764	902
350	783	60	347	460	589	740	911	1104	1319
350	783	80	434	581	754	957	1190	1457	1760
350	783	100	518	701	922	1177	1478	1826	2226
350	783	120	602	819	1087	1402	1776	2213	2719
350	783	140	682	937	1254	1632	2085	2618	3239
375	839	40	268	346	440	546	662	791	933
375	839	60	363	480	613	768	944	1140	1359
375	839	80	453	606	783	991	1230	1501	1808
375	839	100	545	730	957	1217	1524	1877	2281
375	839	120	628	852	1127	1447	1828	2270	2781
375	839	140	711	974	1299	1683	2142	2680	3306
400	894	40	280	361	458	566	686	818	963
400	894	60	378	499	636	796	975	1176	1399
400	894	80	472	630	812	1025	1268	1544	1856
400	894	100	567	758	991	1257	1569	1927	2336
400	894	120	654	884	1166	1492	1878	2326	2842
400	894	140	740	1010	1342	1733	2198	2742	3372

NOTE: Detector time constant at a reference velocity of 5 ft/sec.

For SI Units: 1 ft = 0.305 m

1000 BTU/sec = 1055 kW

Table B-4.1(k)   Installed Spacing of Heat Detector: 12 feet; t_g: 50 Seconds to 1000 Btu/sec; α: 0.400 Btu/sec³

τ	RTI	ΔT	CEILING HEIGHT IN FEET						
			4.0	8.0	12.0	16.0	20.0	24.0	28.0
			FIRE SIZE AT DETECTOR RESPONSE (BTU / SEC)						
25	56	40	482	585	730	897	1085	1291	1518
25	56	60	593	751	952	1184	1446	1740	2069
25	56	80	722	913	1168	1467	1810	2200	2640
25	56	100	821	1090	1384	1753	2182	2675	3238
25	56	120	927	1243	1598	2043	2565	3169	3864
25	56	140	1007	1389	1836	2339	2959	3683	4518
50	112	40	668	831	1026	1244	1482	1740	2017
50	112	60	878	1078	1338	1633	1958	2314	2702
50	112	80	993	1328	1635	2006	2422	2880	3385
50	112	100	1162	1552	1923	2374	2882	3450	4079
50	112	120	1316	1766	2206	2739	3345	4026	4788
50	112	140	1469	1977	2486	3105	3812	4613	5516
75	168	40	867	1043	1273	1533	1815	2117	2438
75	168	60	1021	1370	1663	2011	2392	2804	3248
75	168	80	1255	1668	2030	2466	2947	3472	4042
75	168	100	1461	1946	2382	2907	3491	4132	4834
75	168	120	1665	2216	2726	3342	4031	4793	5632
75	168	140	1867	2480	3065	3774	4570	5458	6441
100	224	40	957	1257	1497	1790	2111	2452	2814
100	224	60	1226	1625	1954	2349	2781	3245	3741
100	224	80	1486	1971	2383	2878	3420	4006	4638
100	224	100	1742	2303	2795	3388	4042	4753	5525
100	224	120	1993	2624	3195	3888	4654	5494	6410
100	224	140	2242	2937	3587	4380	5263	6234	7300
125	280	40	1047	1409	1695	2026	2382	2761	3160
125	280	60	1399	1852	2220	2661	3139	3651	4195
125	280	80	1708	2253	2709	3258	3857	4501	5190
125	280	100	2011	2636	3176	3833	4552	5330	6168
125	280	120	2309	2933	3625	4393	5233	6147	7137
125	280	140	2605	3283	4067	4943	5907	6960	8106
150	335	40	1199	1584	1886	2246	2636	3049	3483
150	335	60	1566	2067	2470	2953	3475	4032	4622
150	335	80	1923	2520	3015	3615	4267	4965	5710
150	335	100	2272	2874	3530	4250	5031	5872	6774
150	335	120	2618	3277	4032	4868	5778	6763	7824
150	335	140	2961	3668	4522	5473	6514	7646	8870
175	391	40	1313	1736	2064	2455	2876	3321	3789
175	391	60	1729	2273	2707	3229	3793	4392	5026
175	391	80	2133	2699	3300	3953	4656	5406	6204
175	391	100	2529	3161	3869	4647	5487	6388	7350
175	391	120	2922	3605	4419	5319	6297	7350	8479
175	391	140	3317	4035	4954	5977	7092	8299	9599
200	447	40	1425	1882	2234	2654	3105	3581	4081
200	447	60	1889	2471	2934	3493	4097	4737	5412
200	447	80	2339	2933	3577	4277	5028	5828	6676
200	447	100	2783	3436	4194	5026	5923	6882	7903
200	447	120	3226	3919	4789	5752	6793	7911	9107
200	447	140	3538	4378	5367	6460	7646	8925	10298

τ	RTI	ΔT	CEILING HEIGHT IN FEET						
			4.0	8.0	12.0	16.0	20.0	24.0	28.0
			FIRE SIZE AT DETECTOR RESPONSE (BTU / SEC)						
225	503	40	1535	2023	2398	2845	3325	3831	4360
225	503	60	2046	2586	3145	3747	4389	5068	5783
225	503	80	2543	3159	3843	4588	5386	6234	7130
225	503	100	3036	3702	4507	5391	6342	7357	8435
225	503	120	3416	4215	5145	6168	7271	8452	9712
225	503	140	3834	4714	5765	6925	8180	9529	10973
250	559	40	1643	2161	2556	3029	3537	4071	4630
250	559	60	2021	2763	3355	3992	4671	5388	6142
250	559	80	2746	3378	4101	4889	5732	6626	7569
250	559	100	3289	3959	4809	5744	6748	7816	8949
250	559	120	3671	4505	5490	6570	7733	8976	10298
250	559	140	4123	5039	6151	7375	8697	10114	11627
275	615	40	1750	2294	2709	3208	3742	4304	4892
275	615	60	2354	2936	3558	4230	4944	5698	6489
275	615	80	2947	3590	4351	5181	6067	7006	7995
275	615	100	3423	4200	5102	6087	7141	8262	9448
275	615	120	3921	4788	5824	6961	8182	9484	10866
275	615	140	4408	5355	6525	7811	9198	10681	12261
300	671	40	1856	2425	2858	3381	3942	4531	5145
300	671	60	2507	3104	3756	4461	5210	5999	6826
300	671	80	3149	3797	4594	5465	6393	7375	8409
300	671	100	3635	4441	5387	6420	7524	8696	9933
300	671	120	4168	5063	6150	7341	8618	9978	11419
300	671	140	4690	5663	6889	8236	9686	11233	12879
325	727	40	1960	2475	2995	3550	4136	4751	5392
325	727	60	2658	3268	3949	4686	5469	6292	7155
325	727	80	3259	3992	4831	5741	6711	7736	8813
325	727	100	3844	4676	5666	6745	7897	9118	10407
325	727	120	4412	5331	6467	7711	9044	10460	11959
325	727	140	4840	5956	7244	8651	10162	11772	13482
350	783	40	2064	2594	3136	3715	4326	4966	5634
350	783	60	2809	3428	4138	4906	5721	6579	7476
350	783	80	3430	4188	5062	6012	7022	8087	9207
350	783	100	4050	4906	5938	7062	8262	9531	10869
350	783	120	4653	5594	6778	8074	9460	10931	12486
350	783	140	5089	6249	7591	9056	10627	12299	14071
375	839	40	2166	2712	3275	3876	4511	5176	5869
375	839	60	3599	4379	5289	6276	7326	8432	9593
375	839	80	3599	4379	5289	6276	7326	8432	9593
375	839	100	4254	5132	6204	7373	8619	9936	11322
375	839	120	4765	5844	7081	8428	9867	11392	13003
375	839	140	5334	6535	7931	9453	11082	12815	14649
400	894	40	2268	2827	3410	4034	4693	5382	6100
400	894	60	3112	3740	4503	5332	6211	7134	8098
400	894	80	3766	4567	5511	6535	7623	8769	9970
400	894	100	4456	5353	6465	7677	8968	10332	11766
400	894	120	4979	6095	7379	8776	10266	11845	13509
400	894	140	5575	6817	8264	9842	11529	13321	15215

NOTE: Detector time constant at a reference velocity of 5 ft/sec.

For SI Units: 1 ft = 0.305 m

1000 BTU/sec = 1055 kW

Table B-4.1(I)   Installed Spacing of Heat Detector: 12 feet; t_g: 150 Seconds to 1000 Btu/sec; α: 0.044 Btu/sec^3

τ	RTI	ΔT	4.0	8.0	12.0	16.0	20.0	24.0	28.0
			CEILING HEIGHT IN FEET — FIRE SIZE AT DETECTOR RESPONSE (BTU / SEC)						
25	56	40	190	254	327	418	525	648	790
25	56	60	256	345	461	602	766	964	1194
25	56	80	316	437	596	794	1030	1313	1644
25	56	100	375	531	738	999	1318	1693	2136
25	56	120	435	628	888	1217	1623	2100	2665
25	56	140	501	729	1045	1448	1946	2542	3230
50	112	40	275	359	447	556	679	818	972
50	112	60	368	483	613	774	959	1173	1415
50	112	80	452	602	779	997	1253	1550	1893
50	112	100	534	722	954	1228	1561	1953	2407
50	112	120	621	842	1127	1475	1886	2380	2957
50	112	140	700	963	1303	1725	2227	2832	3540
75	168	40	353	439	549	675	814	969	1139
75	168	60	462	590	747	929	1134	1366	1625
75	168	80	567	745	943	1184	1461	1778	2137
75	168	100	675	889	1138	1443	1798	2209	2679
75	168	120	772	1030	1334	1708	2148	2660	3251
75	168	140	878	1173	1543	1980	2510	3132	3854
100	224	40	415	516	641	782	937	1107	1292
100	224	60	545	692	869	1070	1295	1545	1822
100	224	80	678	862	1091	1355	1655	1993	2371
100	224	100	794	1039	1311	1642	2021	2454	2943
100	224	120	914	1202	1531	1932	2397	2932	3542
100	224	140	1017	1361	1752	2227	2783	3427	4168
125	280	40	472	587	726	881	1050	1235	1435
125	280	60	634	788	982	1202	1445	1714	2009
125	280	80	771	978	1230	1516	1837	2196	2594
125	280	100	910	1164	1473	1829	2233	2688	3199
125	280	120	1029	1359	1715	2144	2635	3194	3826
125	280	140	1168	1540	1957	2462	3045	3714	4476
150	335	40	526	654	806	974	1157	1356	1570
150	335	60	704	877	1088	1326	1587	1873	2186
150	335	80	871	1089	1360	1668	2010	2390	2809
150	335	100	1004	1293	1626	2006	2434	2913	3446
150	335	120	1159	1493	1889	2345	2863	3446	4101
150	335	140	1306	1708	2152	2686	3297	3992	4778
175	391	40	578	717	881	1062	1259	1471	1698
175	391	60	771	962	1190	1443	1722	2026	2356
175	391	80	944	1193	1484	1812	2176	2576	3014
175	391	100	1111	1416	1772	2176	2627	3129	3684
175	391	120	1276	1633	2056	2538	3081	3690	4369
175	391	140	1438	1847	2337	2900	3450	4262	5072
200	447	40	640	779	954	1147	1356	1581	1821
200	447	60	836	1043	1286	1556	1852	2172	2519
200	447	80	1017	1293	1604	1951	2334	2754	3213
200	447	100	1209	1534	1912	2339	2813	3338	3915
200	447	120	1389	1768	2215	2723	3292	3926	4629
200	447	140	1566	1998	2515	3107	3775	4523	5359
225	503	40	685	837	1023	1228	1450	1687	1940
225	503	60	905	1122	1380	1665	1976	2313	2676
225	503	80	1104	1390	1718	2085	2488	2927	3405
225	503	100	1304	1647	2047	2496	2993	3540	4139
225	503	120	1499	1898	2369	2902	3497	4155	4882
225	503	140	1691	2143	2688	3307	4002	4778	5639
250	559	40	729	894	1091	1307	1540	1789	2054
250	559	60	958	1197	1470	1771	2097	2450	2829
250	559	80	1182	1483	1829	2214	2636	3095	3592
250	559	100	1396	1757	2178	2648	3167	3736	4357
250	559	120	1606	2024	2519	3076	3695	4378	5129
250	559	140	1813	2284	2855	3501	4223	5025	5913
275	615	40	772	949	1156	1383	1628	1889	2166
275	615	60	1012	1271	1557	1873	2215	2583	2977
275	615	80	1257	1574	1937	2340	2780	3258	3774
275	615	100	1487	1864	2305	2796	3336	3927	4569
275	615	120	1712	2146	2664	3245	3888	4595	5370
275	615	140	1934	2421	3017	3690	4439	5267	6180
300	671	40	814	1002	1220	1458	1713	1985	2274
300	671	60	1079	1343	1643	1972	2329	2712	3121
300	671	80	1332	1662	2042	2463	2921	3417	3951
300	671	100	1576	1968	2428	2940	3501	4113	4776
300	671	120	1816	2265	2805	3409	4076	4807	5605
300	671	140	2027	2554	3175	3874	4649	5503	6442
325	727	40	870	1055	1282	1530	1796	2080	2379
325	727	60	1137	1413	1726	2069	2440	2838	3262
325	727	80	1404	1749	2145	2582	3058	3572	4124
325	727	100	1664	2070	2549	3081	3663	4295	4979
325	727	120	1919	2381	2943	3570	4260	5014	5836
325	727	140	2136	2684	3330	4054	4855	5735	6699
350	783	40	904	1105	1342	1601	1878	2172	2483
350	783	60	1193	1481	1807	2164	2549	2961	3400
350	783	80	1476	1833	2245	2699	3192	3723	4293
350	783	100	1750	2169	2667	3218	3820	4473	5178
350	783	120	2021	2495	3078	3727	4440	5218	6062
350	783	140	2243	2812	3481	4230	5056	5961	6951
375	839	40	938	1155	1401	1670	1957	2262	2583
375	839	60	1249	1548	1887	2257	2656	3082	3535
375	839	80	1547	1916	2343	2814	3324	3872	4459
375	839	100	1836	2267	2783	3353	3975	4648	5373
375	839	120	2088	2606	3210	3881	4617	5417	6284
375	839	140	2348	2937	3629	4402	5253	6184	7198
400	894	40	973	1203	1460	1738	2035	2350	2682
400	894	60	1304	1614	1965	2348	2760	3200	3668
400	894	80	1617	1997	2440	2926	3453	4018	4622
400	894	100	1921	2362	2896	3486	4127	4819	5564
400	894	120	2180	2715	3340	4033	4790	5612	6502
400	894	140	2451	3059	3775	4572	5447	6403	7441

NOTE: Detector time constant at a reference velocity of 5 ft/sec.

For SI Units: 1 ft = 0.305 m

1000 BTU/sec = 1055 kW

Table B-4.1(m)   Installed Spacing of Heat Detector: 12 feet t_g: 300 Seconds to 1000 Btu/sec α: 0.011 Btu/sec^3

τ	RTI	ΔT	4.0	8.0	12.0	16.0	20.0	24.0	28.0
			CEILING HEIGHT IN FEET — FIRE SIZE AT DETECTOR RESPONSE (BTU / SEC)						
25	56	40	115	160	218	290	375	479	599
25	56	60	159	228	322	442	588	761	964
25	56	80	203	302	437	611	825	1082	1380
25	56	100	247	380	562	797	1087	1435	1840
25	56	120	294	463	697	998	1370	1818	2339
25	56	140	341	551	840	1212	1673	2228	2875
50	112	40	164	219	283	363	456	564	688
50	112	60	224	305	408	534	682	861	1069
50	112	80	282	392	536	716	932	1192	1496
50	112	100	339	482	672	911	1206	1554	1967
50	112	120	396	575	815	1121	1498	1946	2476
50	112	140	458	672	966	1343	1809	2369	3022
75	168	40	205	270	342	429	530	645	775
75	168	60	280	372	482	617	776	961	1174
75	168	80	349	473	629	814	1040	1305	1615
75	168	100	418	576	777	1025	1322	1679	2096
75	168	120	487	680	931	1245	1628	2080	2616
75	168	140	558	787	1092	1476	1948	2508	3170
100	224	40	243	312	395	491	600	722	859
100	224	60	330	433	552	697	865	1058	1278
100	224	80	411	548	711	910	1145	1418	1734
100	224	100	490	662	877	1132	1440	1804	2227
100	224	120	573	778	1042	1367	1753	2216	2757
100	224	140	649	895	1213	1608	2083	2653	3321
125	280	40	281	354	445	548	665	795	939
125	280	60	376	485	618	773	950	1151	1379
125	280	80	468	617	791	1001	1246	1528	1852
125	280	100	562	744	966	1237	1556	1928	2359
125	280	120	648	871	1149	1480	1880	2352	2899
125	280	140	734	998	1331	1732	2220	2798	3473
150	335	40	313	393	491	602	727	864	1015
150	335	60	419	537	680	845	1031	1241	1477
150	335	80	521	683	867	1089	1344	1636	1967
150	335	100	624	821	1055	1338	1668	2050	2489
150	335	120	720	958	1245	1594	2005	2486	3042
150	335	140	814	1095	1444	1857	2356	2943	3626
175	391	40	344	430	536	654	786	930	1089
175	391	60	461	587	739	913	1109	1328	1573
175	391	80	579	739	940	1172	1438	1740	2081
175	391	100	684	895	1140	1435	1777	2170	2617
175	391	120	788	1042	1342	1704	2127	2618	3183
175	391	140	895	1189	1546	1979	2490	3087	3778
200	447	40	373	466	578	704	842	994	1160
200	447	60	500	635	796	979	1184	1413	1666
200	447	80	626	798	1010	1253	1530	1842	2192
200	447	100	741	958	1222	1530	1883	2287	2744
200	447	120	861	1115	1435	1811	2247	2749	3323
200	447	140	964	1279	1649	2097	2621	3230	3930
225	503	40	401	500	619	751	897	1056	1229
225	503	60	538	682	851	1043	1257	1494	1756
225	503	80	672	855	1077	1332	1619	1941	2301
225	503	100	796	1025	1301	1621	1987	2401	2869
225	503	120	919	1201	1525	1914	2363	2877	3462
225	503	140	1034	1366	1750	2212	2750	3370	4081
250	559	40	428	534	659	798	950	1116	1295
250	559	60	575	726	904	1105	1328	1574	1844
250	559	80	717	911	1143	1408	1705	2037	2407
250	559	100	849	1090	1378	1710	2087	2513	2991
250	559	120	977	1266	1613	2015	2477	3003	3598
250	559	140	1106	1451	1848	2324	2875	3508	4230
275	615	40	455	566	697	843	1002	1174	1360
275	615	60	617	770	956	1165	1397	1651	1930
275	615	80	761	964	1207	1482	1789	2132	2511
275	615	100	905	1153	1453	1797	2186	2623	3111
275	615	120	1035	1339	1699	2114	2588	3126	3733
275	615	140	1174	1522	1944	2434	2999	3644	4377
300	671	40	481	598	735	886	1052	1231	1423
300	671	60	651	812	1006	1224	1464	1726	2013
300	671	80	804	1017	1269	1554	1872	2224	2613
300	671	100	952	1215	1526	1882	2282	2730	3229
300	671	120	1097	1409	1782	2211	2698	3248	3865
300	671	140	1240	1601	2037	2542	3120	3778	4523
325	727	40	506	628	771	929	1101	1286	1485
325	727	60	685	854	1056	1281	1529	1800	2096
325	727	80	846	1068	1329	1624	1952	2315	2713
325	727	100	999	1275	1598	1965	2376	2836	3346
325	727	120	1155	1478	1864	2305	2805	3367	3996
325	727	140	1305	1679	2128	2647	3239	3910	4667
350	783	40	531	658	807	971	1148	1340	1546
350	783	60	717	894	1104	1337	1593	1873	2176
350	783	80	892	1118	1389	1693	2031	2403	2812
350	783	100	1047	1334	1668	2046	2469	2939	3460
350	783	120	1211	1546	1943	2398	2910	3484	4125
350	783	140	1368	1754	2218	2751	3356	4040	4809
375	839	40	555	688	842	1011	1195	1393	1605
375	839	60	749	934	1151	1392	1656	1944	2255
375	839	80	929	1167	1447	1761	2109	2490	2908
375	839	100	1099	1392	1736	2125	2560	3041	3573
375	839	120	1267	1612	2022	2488	3013	3600	4252
375	839	140	1431	1828	2305	2852	3472	4169	4950
400	894	40	579	717	876	1052	1241	1445	1663
400	894	60	781	972	1197	1446	1718	2013	2333
400	894	80	966	1215	1504	1827	2185	2576	3003
400	894	100	1146	1448	1803	2204	2649	3141	3684
400	894	120	1321	1677	2099	2578	3115	3714	4378
400	894	140	1493	1901	2391	2952	3585	4295	5089

NOTE: Detector time constant at a reference velocity of 5 ft/sec.

For SI Units: 1 ft = 0.305 m
1000 BTU/sec = 1055 kW

Table B-4.1(n)   Installed Spacing of Heat Detector: 12 feet; t_g: 500 Seconds to 1000 Btu/sec; α: 0.004 Btu/sec^3

τ	RTI	ΔT	\multicolumn{7}{c}{CEILING HEIGHT IN FEET — FIRE SIZE AT DETECTOR RESPONSE (BTU / SEC)}
			4.0	8.0	12.0	16.0	20.0	24.0	28.0
25	56	40	82	120	170	235	314	408	519
25	56	60	117	180	266	376	512	676	866
25	56	80	154	246	373	536	738	982	1265
25	56	100	192	318	491	713	989	1321	1710
25	56	120	233	395	618	906	1262	1692	2195
25	56	140	275	478	755	1113	1556	2090	2717
50	112	40	115	158	213	280	361	457	570
50	112	60	160	227	317	430	568	732	926
50	112	80	205	301	431	596	801	1046	1332
50	112	100	251	379	554	779	1057	1392	1783
50	112	120	298	462	687	976	1336	1768	2274
50	112	140	347	549	828	1188	1634	2173	2803
75	168	40	143	191	252	322	407	506	621
75	168	60	198	271	366	483	623	790	987
75	168	80	251	353	488	657	864	1110	1400
75	168	100	304	439	618	846	1127	1462	1858
75	168	120	358	528	756	1048	1411	1846	2355
75	168	140	413	620	902	1265	1714	2256	2889
100	224	40	168	222	286	362	451	555	673
100	224	60	232	312	413	533	678	849	1048
100	224	80	292	402	543	717	926	1176	1469
100	224	100	353	495	680	913	1197	1535	1933
100	224	120	414	591	825	1121	1487	1922	2437
100	224	140	477	690	976	1342	1795	2336	2976
125	280	40	191	251	319	400	494	601	723
125	280	60	263	350	455	582	732	907	1109
125	280	80	331	449	596	774	989	1243	1539
125	280	100	398	549	741	979	1265	1608	2010
125	280	120	466	651	892	1194	1563	2001	2519
125	280	140	535	756	1050	1421	1877	2421	3064
150	335	40	213	275	350	436	535	646	772
150	335	60	293	386	496	629	784	964	1170
150	335	80	368	493	647	830	1051	1309	1609
150	335	100	441	600	800	1042	1335	1682	2087
150	335	120	518	709	958	1266	1637	2081	2602
150	335	140	589	820	1122	1498	1959	2506	3153
175	391	40	237	301	380	471	574	690	819
175	391	60	321	420	536	675	835	1020	1230
175	391	80	402	535	693	885	1111	1374	1678
175	391	100	482	650	857	1105	1403	1755	2164
175	391	120	564	765	1022	1337	1712	2161	2686
175	391	140	641	883	1192	1576	2038	2592	3242
200	447	40	257	325	409	504	612	732	866
200	447	60	348	450	574	719	885	1074	1289
200	447	80	436	575	740	939	1171	1439	1747
200	447	100	522	697	913	1167	1471	1828	2241
200	447	120	609	819	1085	1403	1787	2241	2770
200	447	140	691	943	1261	1652	2119	2678	3332

τ	RTI	ΔT	\multicolumn{7}{c}{CEILING HEIGHT IN FEET — FIRE SIZE AT DETECTOR RESPONSE (BTU / SEC)}
			4.0	8.0	12.0	16.0	20.0	24.0	28.0
225	503	40	276	348	436	536	648	773	911
225	503	60	374	481	611	762	933	1128	1348
225	503	80	468	615	786	991	1229	1503	1816
225	503	100	563	743	962	1227	1538	1900	2318
225	503	120	652	872	1146	1471	1861	2320	2854
225	503	140	740	1001	1328	1723	2200	2764	3422
250	559	40	295	371	463	567	684	813	955
250	559	60	399	512	647	803	980	1180	1405
250	559	80	499	653	830	1042	1286	1565	1884
250	559	100	599	788	1014	1286	1604	1972	2394
275	615	40	313	393	489	598	719	852	998
275	615	60	424	541	682	844	1026	1231	1461
275	615	80	530	685	873	1091	1342	1627	1951
275	615	100	635	832	1064	1344	1668	2042	2470
275	615	120	734	972	1258	1601	2007	2478	3021
275	615	140	833	1113	1459	1868	2359	2934	3602
300	671	40	330	414	515	627	752	890	1041
300	671	60	447	570	716	883	1071	1282	1516
300	671	80	563	721	915	1140	1396	1688	2017
300	671	100	669	874	1114	1400	1732	2112	2546
300	671	120	774	1021	1314	1666	2078	2555	3104
300	671	140	881	1168	1517	1939	2437	3019	3692
325	727	40	348	435	539	656	785	927	1082
325	727	60	471	598	750	922	1115	1331	1570
325	727	80	592	755	956	1187	1450	1747	2082
325	727	100	703	910	1162	1456	1794	2181	2620
325	727	120	813	1068	1369	1729	2148	2632	3187
325	727	140	922	1221	1578	2009	2514	3103	3782
350	783	40	364	455	564	685	818	964	1122
350	783	60	493	626	782	960	1158	1379	1624
350	783	80	619	789	996	1234	1503	1806	2146
350	783	100	736	950	1209	1510	1855	2249	2694
350	783	120	851	1115	1423	1791	2218	2708	3269
350	783	140	963	1272	1638	2077	2591	3186	3871
375	839	40	381	475	587	712	849	999	1162
375	839	60	516	653	814	997	1201	1427	1676
375	839	80	646	823	1036	1280	1555	1864	2210
375	839	100	768	989	1256	1564	1916	2316	2767
375	839	120	891	1154	1476	1852	2286	2784	3350
375	839	140	1005	1323	1697	2145	2667	3269	3960
400	894	40	397	495	611	739	881	1034	1201
400	894	60	538	679	846	1034	1243	1474	1728
400	894	80	673	856	1075	1325	1606	1921	2273
400	894	100	800	1028	1301	1617	1976	2382	2840
400	894	120	926	1198	1528	1912	2354	2858	3431
400	894	140	1047	1373	1756	2212	2741	3351	4048

NOTE: Detector time constant at a reference velocity of 5 ft/sec.

For SI Units: 1 ft = 0.305 m
1000 BTU/sec = 1055 kW

Table B-4.1(o)   Installed Spacing of Heat Detector: 12 feet; t_g: 600 Seconds to 1000 Btu/sec; α: 0.003 Btu/sec³

NOTE: Detector time constant at a reference velocity of 5 ft/sec.

For SI Units: 1 ft = 0.305 m
1000 BTU/sec = 1055 kW

τ	RTI	ΔT	CEILING HEIGHT IN FEET — FIRE SIZE AT DETECTOR RESPONSE (BTU / SEC)						
			4.0	8.0	12.0	16.0	20.0	24.0	28.0
25	56	40	74	110	159	221	299	391	498
25	56	60	106	167	251	359	493	654	841
25	56	80	141	232	356	517	716	955	1235
25	56	100	178	302	472	692	964	1291	1675
25	56	120	217	378	598	882	1234	1658	2156
25	56	140	258	459	734	1087	1525	2054	2675
50	112	40	102	142	194	259	337	430	540
50	112	60	143	207	293	403	538	699	889
50	112	80	185	278	404	566	767	1008	1290
50	112	100	228	353	524	745	1020	1350	1736
50	112	120	273	433	654	940	1294	1721	2222
50	112	140	319	518	793	1149	1590	2121	2745
75	168	40	126	171	227	294	375	471	582
75	168	60	176	245	335	447	585	747	939
75	168	80	224	322	451	616	819	1061	1346
75	168	100	274	403	577	800	1077	1409	1797
75	168	120	324	488	711	999	1356	1786	2288
75	168	140	375	577	854	1212	1655	2190	2816
100	224	40	148	197	258	328	413	511	625
100	224	60	205	280	374	491	630	796	990
100	224	80	261	364	498	666	872	1115	1403
100	224	100	316	451	630	856	1134	1467	1859
100	224	120	372	542	769	1059	1418	1848	2355
100	224	140	429	636	916	1276	1722	2259	2888
125	280	40	168	222	285	361	449	550	667
125	280	60	233	312	412	531	675	844	1040
125	280	80	294	404	543	716	922	1170	1460
125	280	100	355	497	681	911	1193	1527	1922
125	280	120	417	593	825	1119	1481	1913	2423
125	280	140	481	692	977	1340	1789	2326	2961
150	335	40	187	246	312	392	484	589	708
150	335	60	259	344	447	572	719	892	1091
150	335	80	326	442	587	762	974	1225	1517
150	335	100	393	541	731	966	1249	1588	1985
150	335	120	460	643	881	1180	1545	1979	2491
150	335	140	528	747	1038	1405	1857	2396	3033
175	391	40	206	266	338	422	517	626	749
175	391	60	283	373	481	611	73	939	1141
175	391	80	356	478	629	809	1026	1280	1575
175	391	100	429	584	780	1020	1306	1649	2049
175	391	120	503	691	936	1239	1605	2045	2560
175	391	140	573	801	1097	1469	1925	2467	3107
200	447	40	225	287	363	451	550	662	788
200	447	60	307	402	514	649	805	985	1191
200	447	80	385	513	668	855	1076	1334	1633
200	447	100	463	625	827	1070	1364	1710	2113
200	447	120	542	738	989	1298	1668	2111	2629
200	447	140	617	853	1156	1534	1990	2538	3181
225	503	40	242	307	387	478	582	698	827
225	503	60	329	427	547	686	847	1031	1240
225	503	80	414	547	707	900	1126	1388	1690
225	503	100	496	665	874	1122	1420	1770	2177
225	503	120	580	783	1041	1353	1731	2177	2699
225	503	140	659	903	1214	1597	2057	2609	3255
250	559	40	259	327	410	506	613	732	865
250	559	60	351	454	578	722	887	1076	1289
250	559	80	441	581	746	944	1174	1441	1747
250	559	100	532	704	916	1173	1476	1830	2241
250	559	120	616	828	1093	1410	1792	2243	2769
250	559	140	700	952	1270	1660	2125	2680	3330
275	615	40	274	346	433	532	643	766	902
275	615	60	373	479	608	757	927	1120	1337
275	615	80	467	613	783	986	1222	1494	1804
275	615	100	562	742	960	1222	1531	1890	2304
275	615	120	651	871	1143	1466	1854	2309	2838
275	615	140	740	1000	1326	1718	2191	2751	3404
300	671	40	290	365	455	558	672	798	938
300	671	60	393	504	638	791	966	1163	1384
300	671	80	493	644	819	1028	1269	1545	1860
300	671	100	592	779	1002	1271	1585	1949	2368
300	671	120	686	913	1188	1521	1914	2375	2908
300	671	140	779	1047	1380	1779	2258	2822	3479
325	727	40	305	383	477	583	701	831	974
325	727	60	414	529	667	825	1004	1205	1431
325	727	80	518	671	855	1069	1316	1597	1915
325	727	100	622	815	1044	1319	1639	2008	2430
325	727	120	720	954	1235	1575	1974	2440	2977
325	727	140	817	1093	1434	1838	2323	2893	3554
350	783	40	319	401	498	607	729	862	1009
350	783	60	434	553	695	858	1041	1247	1476
350	783	80	543	700	890	1110	1361	1647	1970
350	783	100	650	850	1085	1366	1692	2066	2493
350	783	120	753	994	1282	1628	2034	2504	3046
350	783	140	854	1138	1482	1897	2389	2963	3629
375	839	40	334	418	519	631	756	893	1043
375	839	60	453	576	723	890	1078	1288	1522
375	839	80	571	729	924	1150	1406	1697	2024
375	839	100	679	880	1126	1413	1744	2123	2555
375	839	120	785	1034	1328	1681	2092	2568	3115
375	839	140	894	1183	1533	1956	2453	3033	3703
400	894	40	348	435	539	655	783	924	1077
400	894	60	472	599	750	922	1114	1329	1566
400	894	80	594	758	958	1189	1450	1746	2078
400	894	100	706	914	1165	1458	1795	2180	2617
400	894	120	817	1073	1373	1733	2150	2632	3183
400	894	140	928	1226	1584	2013	2517	3103	3777

Table B-4.1(p)   Installed Spacing of Heat Detector: 15 ft.; t_g: 50 Seconds to 1000 Btu/sec; α: 0.400 Btu/sec³

τ	RTI	ΔT	4.0	8.0	12.0	16.0	20.0	24.0	28.0
			CEILING HEIGHT IN FEET / FIRE SIZE AT DETECTOR RESPONSE (BTU / SEC)						
25	56	40	618	745	903	1085	1287	1509	1753
25	56	60	790	962	1181	1434	1720	2040	2395
25	56	80	935	1169	1452	1781	2157	2583	3061
25	56	100	1046	1393	1722	2131	2605	3146	3759
25	56	120	1215	1600	1992	2488	3066	3731	4490
25	56	140	1357	1799	2266	2852	3541	4340	5255
50	112	40	894	1063	1269	1503	1758	2034	2331
50	112	60	1098	1370	1656	1976	2327	2711	3128
50	112	80	1321	1664	2025	2431	2883	3380	3925
50	112	100	1541	1949	2385	2880	3435	4052	4735
50	112	120	1758	2226	2739	3327	3991	4734	5563
50	112	140	1974	2499	3091	3775	4552	5429	6412
75	168	40	1035	1356	1577	1852	2153	2476	2819
75	168	60	1370	1712	2055	2432	2843	3286	3763
75	168	80	1668	2083	2511	2986	3507	4074	4687
75	168	100	1960	2439	2951	3525	4159	4853	5611
75	168	120	2248	2783	3380	4055	4806	5634	6543
75	168	140	2533	3119	3804	4582	5453	6420	7487
100	224	40	1245	1547	1843	2162	2505	2870	3256
100	224	60	1625	2019	2412	2841	3305	3803	4334
100	224	80	1994	2460	2947	3484	4069	4701	5379
100	224	100	2356	2880	3459	4105	4813	5582	6413
100	224	120	2714	3285	3958	4714	5547	6457	7446
100	224	140	3071	3680	4447	5315	6276	7332	8485
125	280	40	1417	1760	2091	2447	2827	3231	3657
125	280	60	1870	2304	2741	3218	3731	4280	4862
125	280	80	2309	2809	3348	3944	4589	5281	6021
125	280	100	2741	3289	3930	4643	5420	6259	7160
125	280	120	3172	3752	4493	5325	6236	7224	8291
125	280	140	3485	4195	5044	5996	7043	8184	9422
150	335	40	1584	1960	2323	2713	3129	3569	4032
150	335	60	2107	2571	3049	3570	4130	4726	5357
150	335	80	2616	3137	3726	4375	5076	5827	6625
150	335	100	3121	3674	4371	5148	5990	6896	7865
150	335	120	3501	4183	4995	5900	6884	7947	9090
150	335	140	3927	4681	5605	6637	7766	8989	10309
175	391	40	1746	2151	2543	2965	3414	3889	4387
175	391	60	2339	2825	3341	3905	4509	5150	5827
175	391	80	2920	3449	4083	4784	5539	6345	7199
175	391	100	3390	4034	4790	5627	6532	7502	8535
175	391	120	3879	4598	5473	6446	7501	8636	9851
175	391	140	4356	5145	6138	7247	8454	9757	11157
200	447	40	1906	2333	2753	3205	3687	4194	4726
200	447	60	2569	3068	3620	4224	4870	5554	6276
200	447	80	3224	3749	4426	5176	5982	6840	7748
200	447	100	3707	4382	5192	6086	7051	8082	9177
200	447	120	4247	4995	5930	6969	8092	9296	10581
200	447	140	4670	5583	6649	7831	9113	10493	11970
225	503	40	2062	2508	2955	3436	3948	4487	5051
225	503	60	2796	3303	3889	4531	5218	5943	6707
225	503	80	3407	4029	4755	5553	6408	7317	8276
225	503	100	4017	4717	5578	6528	7551	8640	9796
225	503	120	4609	5378	6371	7472	8661	9932	11285
225	503	140	5045	6009	7141	8393	9749	11203	12755
250	559	40	2217	2678	3150	3659	4200	4769	5364
250	559	60	3023	3530	4148	4828	5553	6319	7123
250	559	80	3660	4305	5073	5917	6820	7777	8787
250	559	100	4321	5042	5951	6955	8033	9180	10394
250	559	120	4835	5740	6796	7959	9211	10547	11966
250	559	140	5409	6421	7617	8938	10363	11889	13515
275	615	40	2370	2843	3339	3875	4445	5043	5668
275	615	60	3251	3751	4440	5115	5878	6683	7527
275	615	80	3909	4573	5381	6269	7219	8224	9282
275	615	100	4622	5357	6313	7369	8502	9704	10974
275	615	120	5151	6098	7209	8432	9745	11144	12627
275	615	140	5765	6821	8079	9466	10960	12556	14254
300	671	40	2522	3003	3522	4085	4682	5309	5963
300	671	60	3362	3956	4644	5395	6194	7037	7919
300	671	80	4153	4834	5682	6613	7607	8658	9763
300	671	100	4785	5655	6666	7772	8957	10214	11539
300	671	120	5461	6446	7612	8891	10265	11725	13271
300	671	140	6114	7211	8529	9980	11541	13206	14973
325	727	40	2673	3160	3701	4289	4913	5567	6250
325	727	60	3552	4163	4882	5667	6502	7381	8302
325	727	80	4396	5089	5974	6947	7985	9081	10232
325	727	100	5049	5655	6666	7772	8957	10214	11539
325	727	120	5764	6787	8004	9340	10772	12292	13898
325	727	140	6456	7591	8968	10482	12108	13839	15674
350	783	40	2824	3313	3876	4489	5139	5820	6530
350	783	60	3739	4366	5115	5933	6803	7718	8675
350	783	80	4522	5330	6260	7274	8355	9495	10691
350	783	100	5308	6244	7345	8550	9836	11196	12628
350	783	120	6062	7119	8387	9778	11267	12846	14512
350	783	140	6793	7963	9397	10972	12662	14459	16360
375	839	40	2975	3463	4047	4684	5359	6067	6804
375	839	60	3925	4564	5342	6193	7097	8047	9040
375	839	80	4736	5573	6540	7594	8717	9899	11139
375	839	100	5563	6530	7674	8926	10261	11671	13154
375	839	120	6356	7445	8763	10207	11752	13388	15113
375	839	140	7126	8328	9817	11452	13204	15065	17032
400	894	40	3126	3611	4214	4875	5575	6309	7072
400	894	60	4108	4759	5566	6449	7386	8370	9398
400	894	80	4948	5811	6814	7908	9071	10296	11579
400	894	100	5814	6809	7997	9294	10677	12137	13670
400	894	120	6647	7764	9131	10628	12227	13920	15702
400	894	140	7455	8685	10229	11923	13736	15660	17690

NOTE: Detector time constant at a reference velocity of 5 ft/sec.

For SI Units: 1 ft = 0.305 m

1000 BTU/sec = 1055 kW

Table B-4.1(q)   Installed Spacing of Heat Detector: 15 feet; t_g: 150 Seconds to 1000 Btu/sec; α: 0.044 Btu/sec^3

τ	RTI	ΔT	4.0	8.0	12.0	16.0	20.0	24.0	28.0
			CEILING HEIGHT IN FEET — FIRE SIZE AT DETECTOR RESPONSE (BTU / SEC)						
25	56	40	259	322	407	509	627	762	917
25	56	60	341	447	572	731	918	1137	1390
25	56	80	423	567	749	968	1237	1552	1916
25	56	100	511	691	930	1228	1581	2003	2492
25	56	120	592	819	1120	1498	1950	2488	3112
25	56	140	674	951	1321	1785	2348	3004	3774
50	112	40	371	451	555	675	810	961	1129
50	112	60	488	610	763	942	1147	1381	1645
50	112	80	610	767	972	1216	1500	1829	2204
50	112	100	718	923	1186	1501	1873	2307	2806
50	112	120	826	1089	1404	1798	2266	2815	3449
50	112	140	936	1247	1637	2107	2679	3352	4132
75	168	40	463	560	681	818	970	1138	1322
75	168	60	622	755	929	1129	1354	1607	1889
75	168	80	758	943	1173	1441	1747	2095	2487
75	168	100	900	1129	1419	1759	2154	2606	3121
75	168	120	1021	1314	1667	2086	2575	3142	3791
75	168	140	1161	1499	1918	2421	3014	3703	4497
100	224	40	546	658	794	947	1116	1300	1500
100	224	60	729	886	1079	1300	1545	1818	2118
100	224	80	900	1103	1357	1648	1978	2347	2759
100	224	100	1044	1316	1633	2000	2419	2894	3428
100	224	120	1212	1527	1909	2356	2872	3461	4129
100	224	140	1368	1736	2187	2719	3337	4049	4861
125	280	40	635	750	899	1067	1251	1450	1666
125	280	60	829	1006	1219	1458	1724	2016	2335
125	280	80	1011	1251	1528	1842	2195	2586	3019
125	280	100	1203	1490	1832	2226	2670	3169	3726
125	280	120	1384	1725	2136	2612	3154	3768	4458
125	280	140	1563	1957	2440	3002	3649	4385	5219
150	335	40	703	834	998	1180	1378	1592	1822
150	335	60	927	1120	1350	1608	1893	2203	2541
150	335	80	1138	1391	1689	2026	2400	2814	3268
150	335	100	1345	1654	2021	2440	2910	3433	4012
150	335	120	1549	1912	2351	2855	3425	4064	4779
150	335	140	1750	2166	2679	3272	3948	4712	5570
175	391	40	768	915	1091	1287	1499	1727	1972
175	391	60	1008	1228	1475	1751	2053	2382	2738
175	391	80	1252	1524	1842	2201	2597	3032	3507
175	391	100	1482	1810	2201	2645	3139	3687	4289
175	391	120	1708	2090	2556	3088	3685	4351	5090
175	391	140	1931	2366	2908	3532	4237	5028	5912
200	447	40	832	992	1181	1389	1615	1857	2115
200	447	60	1103	1331	1594	1888	2208	2554	2928
200	447	80	1363	1651	1990	2369	2786	3242	3738
200	447	100	1615	1960	2374	2842	3360	3932	4558
200	447	120	1863	2261	2753	3312	3936	4628	5392
200	447	140	2081	2557	3129	3782	4516	5335	6246
225	503	40	902	1067	1267	1488	1726	1981	2253
225	503	60	1189	1431	1710	2019	2356	2720	3111
225	503	80	1471	1774	2131	2531	2968	3445	3962
225	503	100	1745	2105	2541	3032	3574	4169	4818
225	503	120	2015	2426	2944	3529	4179	4897	5687
225	503	140	2242	2742	3342	4023	4786	5634	6571
250	559	40	953	1138	1350	1583	1834	2101	2386
250	559	60	1273	1527	1821	2147	2500	2881	3288
250	559	80	1576	1893	2269	2687	3145	3642	4179
250	559	100	1873	2244	2703	3216	3782	4399	5072
250	559	120	2131	2586	3128	3739	4415	5159	5974
250	559	140	2398	2920	3548	4258	5050	5925	6890
275	615	40	1006	1208	1431	1675	1938	2218	2515
275	615	60	1354	1621	1929	2270	2640	3037	3461
275	615	80	1680	2008	2402	2840	3317	3834	4390
275	615	100	1999	2380	2859	3395	3983	4624	5319
275	615	120	2266	2741	3307	3943	4645	5414	6254
275	615	140	2550	3094	3748	4487	5306	6209	7201
300	671	40	1071	1276	1509	1765	2040	2332	2641
300	671	60	1434	1712	2034	2391	2776	3188	3629
300	671	80	1782	2121	2532	2988	3484	4020	4597
300	671	100	2089	2512	3012	3570	4180	4843	5560
300	671	120	2398	2892	3482	4142	4869	5664	6528
300	671	140	2699	3264	3944	4709	5556	6487	7506
325	727	40	1128	1342	1586	1853	2139	2443	2764
325	727	60	1513	1801	2137	2508	2908	3337	3793
325	727	80	1883	2230	2658	3133	3648	4203	4798
325	727	100	2202	2641	3161	3740	4372	5057	5796
325	727	120	2528	3040	3652	4337	5088	5907	6797
325	727	140	2846	3429	4135	4927	5801	6759	7804
350	783	40	1184	1407	1661	1938	2235	2551	2884
350	783	60	1591	1888	2238	2623	3038	3482	3953
350	783	80	1983	2338	2782	3274	3808	4381	4995
350	783	100	2312	2767	3307	3906	4560	5266	6027
350	783	120	2655	3184	3819	4527	5303	6146	7060
350	783	140	2989	3591	4322	5140	6041	7026	8098
375	839	40	1239	1470	1734	2022	2330	2657	3001
375	839	60	1667	1973	2336	2735	3165	3623	4111
375	839	80	2049	2442	2903	3413	3964	4556	5188
375	839	100	2421	2891	3450	4070	4744	5472	6254
375	839	120	2780	3325	3982	4713	5513	6380	7318
375	839	140	3131	3749	4505	5349	6276	7287	8385
400	894	40	1293	1532	1806	2104	2423	2761	3116
400	894	60	1743	2056	2432	2845	3289	3763	4265
400	894	80	2139	2545	3022	3549	4118	4727	5378
400	894	100	2528	3012	3590	4230	4925	5673	6477
400	894	120	2904	3464	4143	4896	5719	6610	7572
400	894	140	3271	3905	4684	5554	6507	7544	8669

NOTE: Detector time constant at a reference velocity of 5 ft/sec.

For SI Units: 1 ft = 0.305 m

1000 BTU/sec = 1055 kW

Table B-4.1(r) Installed Spacing of Heat Detector: 15 feet; t_g: 300 Seconds to 1000 Btu/sec; α: 0.011 Btu/sec³

τ	RTI	ΔT	CEILING HEIGHT IN FEET — FIRE SIZE AT DETECTOR RESPONSE (BTU/SEC)						
			4.0	8.0	12.0	16.0	20.0	24.0	28.0
25	56	40	156	207	273	353	451	565	697
25	56	60	215	297	407	543	706	900	1125
25	56	80	275	394	553	754	997	1280	1612
25	56	100	337	498	713	984	1314	1700	2151
25	56	120	403	608	885	1234	1657	2155	2736
25	56	140	469	725	1069	1501	2026	2648	3365
50	112	40	221	279	353	442	545	664	801
50	112	60	300	394	508	650	819	1017	1246
50	112	80	378	508	674	876	1121	1411	1747
50	112	100	456	626	847	1121	1450	1843	2298
50	112	120	538	749	1029	1381	1808	2309	2895
50	112	140	617	877	1222	1657	2186	2808	3535
75	168	40	276	342	426	523	634	760	902
75	168	60	373	475	602	753	930	1134	1368
75	168	80	467	611	784	996	1249	1544	1885
75	168	100	564	745	976	1254	1591	1988	2449
75	168	120	656	882	1172	1526	1957	2466	3057
75	168	140	748	1022	1376	1816	2345	2975	3707
100	224	40	326	399	491	597	716	850	999
100	224	60	439	551	688	850	1036	1248	1489
100	224	80	554	700	888	1112	1373	1676	2023
100	224	100	658	850	1093	1385	1731	2135	2601
100	224	120	763	1006	1303	1671	2109	2624	3221
100	224	140	874	1160	1525	1969	2508	3144	3882
125	280	40	371	452	552	666	794	935	1092
125	280	60	500	621	769	941	1137	1358	1606
125	280	80	627	787	987	1222	1493	1805	2159
125	280	100	746	951	1207	1512	1868	2280	2753
125	280	120	870	1117	1432	1812	2260	2783	3386
125	280	140	980	1289	1663	2123	2671	3314	4059
150	335	40	414	502	610	732	867	1017	1181
150	335	60	557	688	846	1028	1233	1464	1721
150	335	80	697	869	1081	1327	1610	1931	2294
150	335	100	829	1047	1317	1633	2001	2423	2904
150	335	120	958	1226	1556	1949	2408	2941	3552
150	335	140	1088	1405	1800	2273	2832	3484	4236
175	391	40	454	549	665	794	938	1095	1267
175	391	60	617	752	919	1111	1326	1566	1832
175	391	80	764	947	1170	1428	1722	2054	2426
175	391	100	911	1139	1422	1751	2130	2563	3054
175	391	120	1047	1331	1675	2081	2553	3096	3716
175	391	140	1191	1522	1932	2420	2991	3653	4413
200	447	40	493	595	717	854	1005	1170	1349
200	447	60	668	812	990	1191	1415	1665	1940
200	447	80	828	1022	1257	1526	1831	2173	2555
200	447	100	982	1228	1523	1865	2256	2700	3201
200	447	120	1137	1431	1790	2210	2695	3249	3879
200	447	140	1289	1635	2060	2562	3146	3820	4589
225	503	40	530	639	768	912	1070	1242	1429
225	503	60	718	871	1057	1268	1502	1760	2044
225	503	80	894	1095	1340	1621	1937	2289	2681
225	503	100	1058	1313	1621	1976	2379	2835	3346
225	503	120	1222	1529	1902	2335	2833	3399	4040
225	503	140	1385	1744	2184	2701	3299	3984	4764
250	559	40	567	681	817	968	1133	1313	1506
250	559	60	766	928	1123	1343	1586	1854	2147
250	559	80	950	1165	1421	1713	2039	2403	2805
250	559	100	1129	1396	1716	2083	2499	2966	3488
250	559	120	1305	1623	2010	2458	2968	3547	4198
250	559	140	1478	1849	2305	2837	3449	4147	4938
275	615	40	609	722	864	1022	1194	1381	1582
275	615	60	813	983	1187	1416	1668	1945	2247
275	615	80	1006	1233	1500	1802	2140	2514	2926
275	615	100	1199	1476	1808	2188	2616	3095	3628
275	615	120	1385	1715	2116	2577	3101	3692	4355
275	615	140	1569	1952	2423	2970	3596	4306	5109
300	671	40	642	762	910	1075	1254	1448	1655
300	671	60	867	1037	1249	1486	1748	2033	2344
300	671	80	1066	1299	1576	1890	2238	2622	3045
300	671	100	1267	1554	1899	2291	2731	3221	3765
300	671	120	1464	1805	2219	2694	3231	3834	4509
300	671	140	1658	2053	2538	3100	3740	4464	5279
325	727	40	674	801	956	1127	1312	1513	1727
325	727	60	907	1090	1310	1556	1826	2120	2440
325	727	80	1122	1364	1651	1975	2334	2728	3161
325	727	100	1334	1631	1987	2391	2843	3345	3901
325	727	120	1541	1892	2319	2808	3358	3975	4661
325	727	140	1746	2151	2651	3227	3881	4619	5446
350	783	40	706	839	1000	1177	1369	1576	1798
350	783	60	947	1141	1369	1624	1902	2205	2534
350	783	80	1177	1428	1725	2058	2427	2833	3276
350	783	100	1399	1706	2073	2489	2953	3467	4034
350	783	120	1617	1978	2418	2920	3483	4113	4812
350	783	140	1819	2247	2761	3352	4021	4772	5612
375	839	40	738	877	1043	1226	1425	1638	1866
375	839	60	988	1191	1427	1690	1977	2289	2626
375	839	80	1231	1490	1796	2140	2520	2935	3388
375	839	100	1464	1779	2158	2586	3061	3587	4165
375	839	120	1693	2062	2515	3030	3606	4248	4960
375	839	140	1900	2341	2870	3475	4158	4923	5775
400	894	40	769	913	1085	1275	1480	1700	1934
400	894	60	1030	1241	1484	1755	2051	2371	2716
400	894	80	1284	1551	1867	2221	2610	3036	3499
400	894	100	1528	1851	2241	2680	3168	3705	4294
400	894	120	1767	2144	2610	3138	3727	4382	5106
400	894	140	1980	2433	2976	3596	4293	5071	5937

NOTE: Detector time constant at a reference velocity of 5 ft/sec.

For SI Units: 1 ft = 0.305 m
1000 BTU/sec = 1055 kW

Table B-4.1(s)   Installed Spacing of Heat Detector: 15 feet; t_g: 500 Seconds to 1000 Btu/sec; α: 0.004 Btu/sec³

τ	RTI	ΔT	CEILING HEIGHT IN FEET — FIRE SIZE AT DETECTOR RESPONSE (BTU / SEC)						
			4.0	8.0	12.0	16.0	20.0	24.0	28.0
25	56	40	112	156	215	289	379	484	606
25	56	60	160	235	337	464	619	801	1013
25	56	80	210	323	474	664	894	1164	1480
25	56	100	264	419	625	884	1198	1571	2001
25	56	120	321	523	789	1124	1530	2013	2570
25	56	140	381	633	965	1381	1887	2488	3183
50	112	40	155	204	267	342	433	541	665
50	112	60	216	295	399	528	683	868	1082
50	112	80	278	392	544	735	967	1239	1558
50	112	100	341	496	702	962	1279	1651	2087
50	112	120	408	606	872	1208	1617	2099	2663
50	112	140	474	722	1053	1471	1980	2580	3282
75	168	40	192	245	313	393	488	598	724
75	168	60	265	351	460	590	749	936	1152
75	168	80	337	458	614	808	1040	1316	1637
75	168	100	411	571	780	1042	1358	1735	2173
75	168	120	486	688	956	1294	1705	2190	2756
75	168	140	561	811	1143	1563	2074	2677	3382
100	224	40	226	283	356	442	541	654	784
100	224	60	309	400	514	652	814	1004	1223
100	224	80	392	521	682	878	1115	1393	1717
100	224	100	477	642	857	1123	1441	1820	2261
100	224	120	559	768	1041	1381	1792	2281	2851
100	224	140	642	898	1234	1656	2170	2775	3484
125	280	40	257	319	397	487	591	709	842
125	280	60	351	448	568	711	878	1072	1294
125	280	80	443	579	745	948	1189	1471	1798
125	280	100	537	710	931	1199	1524	1906	2350
125	280	120	627	844	1123	1469	1882	2374	2947
125	280	140	718	982	1324	1750	2262	2874	3586
150	335	40	286	353	436	531	640	762	899
150	335	60	390	492	619	768	941	1139	1365
150	335	80	491	632	807	1016	1262	1549	1879
150	335	100	593	776	1000	1277	1606	1992	2439
150	335	120	691	918	1204	1551	1972	2467	3043
150	335	140	790	1063	1412	1840	2359	2973	3689
175	391	40	313	385	472	573	686	813	954
175	391	60	427	535	668	823	1001	1204	1435
175	391	80	541	684	866	1082	1334	1626	1960
175	391	100	647	834	1070	1353	1687	2078	2529
175	391	120	753	989	1279	1636	2061	2561	3140
175	391	140	863	1142	1499	1931	2455	3074	3792
200	447	40	340	416	508	613	731	863	1008
200	447	60	463	577	715	876	1060	1269	1504
200	447	80	584	735	924	1147	1405	1701	2040
200	447	100	698	894	1137	1427	1768	2163	2618
200	447	120	812	1053	1355	1719	2150	2654	3238
200	447	140	926	1218	1579	2022	2552	3174	3896
225	503	40	365	445	542	652	775	911	1061
225	503	60	497	617	761	928	1117	1331	1571
225	503	80	626	784	980	1210	1474	1776	2119
225	503	100	748	951	1202	1500	1847	2248	2707
225	503	120	873	1119	1429	1800	2238	2748	3335
225	503	140	988	1292	1661	2112	2647	3275	4001
250	559	40	390	474	575	690	817	958	1112
250	559	60	530	655	806	978	1173	1393	1638
250	559	80	667	832	1035	1271	1542	1850	2198
250	559	100	796	1007	1266	1571	1925	2332	2796
250	559	120	925	1182	1501	1880	2325	2840	3432
250	559	140	1052	1359	1741	2200	2742	3375	4106
275	615	40	414	502	607	726	858	1004	1162
275	615	60	567	693	849	1027	1228	1453	1703
275	615	80	706	878	1088	1331	1608	1922	2276
275	615	100	843	1062	1328	1641	2001	2415	2885
275	615	120	978	1245	1572	1959	2411	2932	3529
275	615	140	1112	1428	1820	2287	2836	3475	4210
300	671	40	437	529	639	762	899	1048	1212
300	671	60	598	729	891	1075	1282	1512	1767
300	671	80	745	923	1140	1389	1673	1993	2353
300	671	100	893	1115	1389	1709	2077	2496	2972
300	671	120	1030	1305	1641	2036	2495	3023	3626
300	671	140	1172	1496	1897	2372	2929	3574	4315
325	727	40	460	555	669	797	938	1092	1260
325	727	60	628	765	932	1122	1334	1570	1830
325	727	80	783	968	1191	1447	1737	2063	2429
325	727	100	935	1167	1449	1776	2151	2577	3059
325	727	120	1083	1365	1709	2112	2579	3113	3722
325	727	140	1230	1563	1972	2456	3021	3673	4419
350	783	40	482	581	699	831	976	1135	1307
350	783	60	658	800	972	1168	1386	1627	1893
350	783	80	820	1011	1240	1503	1800	2132	2504
350	783	100	977	1218	1507	1842	2224	2657	3145
350	783	120	1134	1423	1776	2187	2661	3203	3818
350	783	140	1287	1628	2046	2539	3112	3770	4523
375	839	40	504	607	729	865	1014	1177	1353
375	839	60	687	834	1012	1213	1436	1683	1954
375	839	80	863	1053	1289	1558	1862	2201	2578
375	839	100	1020	1268	1565	1907	2296	2736	3230
375	839	120	1183	1480	1841	2261	2742	3291	3912
375	839	140	1343	1692	2119	2621	3201	3867	4626
400	894	40	525	632	758	898	1051	1218	1399
400	894	60	715	868	1051	1257	1486	1738	2014
400	894	80	895	1095	1337	1613	1922	2268	2651
400	894	100	1064	1317	1621	1971	2367	2814	3314
400	894	120	1232	1536	1905	2333	2823	3379	4006
400	894	140	1398	1754	2191	2702	3290	3964	4729

NOTE: Detector time constant at a reference velocity of 5 ft/sec.

For SI Units: 1 ft = 0.305 m

1000 BTU/sec = 1055 kW

Table B-4.1(t) Installed Spacing of Heat Detector: 15 feet; t_g: 600 Seconds to 1000 Btu/sec; α: 0.003 Btu/sec³

τ	RTI	ΔT	CEILING HEIGHT IN FEET						
			4.0	8.0	12.0	16.0	20.0	24.0	28.0
			FIRE SIZE AT DETECTOR RESPONSE (BTU / SEC)						
25	56	40	100	143	201	273	360	462	582
25	56	60	145	220	319	444	596	775	983
25	56	80	193	305	454	640	867	1135	1445
25	56	100	245	399	603	858	1168	1536	1961
25	56	120	300	501	765	1095	1496	1973	2525
25	56	140	358	610	938	1350	1850	2445	3134
50	112	40	137	184	244	316	405	509	630
50	112	60	193	270	371	497	650	830	1040
50	112	80	251	363	512	699	927	1195	1509
50	112	100	310	462	666	922	1234	1602	2032
50	112	120	372	569	832	1164	1568	2045	2602
50	112	140	437	683	1011	1424	1927	2525	3216
75	168	40	170	219	283	360	450	557	679
75	168	60	236	317	421	548	703	885	1097
75	168	80	303	418	570	759	987	1258	1574
75	168	100	370	525	730	988	1301	1671	2103
75	168	120	441	638	902	1235	1641	2119	2679
75	168	140	511	756	1085	1499	2004	2601	3299
100	224	40	199	252	320	401	495	604	728
100	224	60	275	361	468	600	757	942	1156
100	224	80	350	472	627	819	1049	1322	1640
100	224	100	426	586	794	1054	1367	1741	2175
100	224	120	504	705	972	1307	1714	2195	2757
100	224	140	581	829	1160	1576	2083	2682	3382
125	280	40	226	283	355	440	538	650	777
125	280	60	311	401	514	650	811	998	1215
125	280	80	394	522	683	876	1110	1386	1706
125	280	100	480	644	857	1121	1436	1811	2248
125	280	120	563	770	1041	1379	1786	2271	2836
125	280	140	646	901	1234	1653	2163	2763	3466
150	335	40	251	313	389	478	579	695	825
150	335	60	345	440	558	699	864	1055	1273
150	335	80	436	570	734	934	1172	1451	1773
150	335	100	529	701	919	1184	1504	1882	2322
150	335	120	618	834	1110	1451	1860	2348	2915
150	335	140	709	970	1309	1731	2239	2845	3551
175	391	40	275	340	421	514	619	738	872
175	391	60	377	477	601	746	915	1110	1332
175	391	80	476	614	785	990	1233	1515	1840
175	391	100	576	755	976	1248	1573	1954	2395
175	391	120	672	895	1177	1520	1935	2425	2995
175	391	140	769	1038	1382	1808	2319	2928	3636
200	447	40	298	367	452	549	658	781	918
200	447	60	408	513	642	792	966	1164	1390
200	447	80	515	658	835	1046	1293	1579	1907
200	447	100	621	803	1034	1311	1640	2025	2470
200	447	120	723	954	1239	1590	2009	2503	3075
200	447	140	826	1104	1454	1881	2399	3011	3722

τ	RTI	ΔT	CEILING HEIGHT IN FEET						
			4.0	8.0	12.0	16.0	20.0	24.0	28.0
			FIRE SIZE AT DETECTOR RESPONSE (BTU / SEC)						
225	503	40	321	393	481	582	696	822	963
225	503	60	438	548	681	837	1015	1218	1447
225	503	80	555	700	883	1100	1351	1642	1974
225	503	100	664	853	1090	1373	1707	2096	2544
225	503	120	773	1012	1303	1659	2083	2580	3156
225	503	140	885	1168	1521	1957	2479	3094	3808
250	559	40	342	418	510	615	732	863	1007
250	559	60	467	581	720	880	1063	1270	1504
250	559	80	590	742	931	1152	1409	1704	2040
250	559	100	706	902	1145	1434	1773	2166	2618
250	559	120	822	1063	1365	1727	2156	2658	3237
250	559	140	937	1230	1590	2032	2559	3177	3895
275	615	40	363	442	538	647	768	902	1051
275	615	60	495	614	757	923	1110	1322	1559
275	615	80	625	782	977	1204	1466	1766	2106
275	615	100	747	949	1199	1494	1839	2236	2692
275	615	120	873	1117	1426	1795	2229	2735	3318
275	615	140	989	1286	1658	2106	2638	3261	3982
300	671	40	383	466	565	678	803	941	1093
300	671	60	523	646	794	964	1157	1373	1614
300	671	80	658	821	1022	1255	1522	1827	2171
300	671	100	787	996	1252	1553	1903	2306	2766
300	671	120	916	1170	1486	1861	2301	2812	3398
300	671	140	1041	1345	1724	2179	2716	3344	4068
325	727	40	403	489	592	708	837	979	1134
325	727	60	554	677	830	1005	1202	1423	1668
325	727	80	691	860	1066	1305	1577	1887	2235
325	727	100	826	1041	1303	1611	1967	2375	2839
325	727	120	959	1222	1544	1926	2372	2888	3479
325	727	140	1092	1403	1789	2251	2794	3427	4155
350	783	40	423	512	618	738	871	1016	1175
350	783	60	580	707	865	1045	1246	1472	1722
350	783	80	723	898	1109	1353	1631	1946	2299
350	783	100	869	1085	1354	1668	2030	2443	2911
350	783	120	1003	1272	1602	1991	2443	2964	3559
350	783	140	1141	1460	1854	2322	2871	3509	4242
375	839	40	442	534	644	767	903	1053	1215
375	839	60	605	737	899	1084	1290	1520	1774
375	839	80	755	934	1151	1401	1685	2004	2362
375	839	100	904	1129	1404	1724	2092	2510	2984
375	839	120	1047	1322	1659	2054	2512	3039	3639
375	839	140	1190	1515	1917	2392	2948	3591	4329
400	894	40	460	555	669	796	936	1088	1255
400	894	60	630	767	933	1122	1333	1567	1826
400	894	80	786	971	1193	1448	1737	2062	2425
400	894	100	939	1171	1453	1779	2153	2577	3055
400	894	120	1089	1371	1715	2117	2581	3113	3718
400	894	140	1237	1570	1979	2462	3024	3673	4415

NOTE: Detector time constant at a reference velocity of 5 ft/sec.

For SI Units: 1 ft = 0.305 m
1000 BTU/sec = 1055 kW

Table B-4.1(u)   Installed Spacing of Heat Detector: 20 feet; t$_g$: 50 Seconds to 1000 Btu/sec; α: 0.400 Btu/sec^3

τ	RTI	ΔT	4.0	8.0	12.0	16.0	20.0	24.0	28.0
25	56	40	906	1047	1221	1424	1650	1898	2169
25	56	60	1122	1352	1602	1890	2214	2575	2974
25	56	80	1351	1649	1975	2354	2786	3271	3813
25	56	100	1578	1940	2347	2825	3373	3994	4693
25	56	120	1804	2229	2722	3305	3979	4747	5616
25	56	140	2029	2519	3103	3797	4605	5532	6583
50	112	40	1236	1478	1711	1969	2251	2556	2883
50	112	60	1603	1922	2239	2596	2989	3417	3881
50	112	80	1960	2340	2744	3202	3710	4270	4881
50	112	100	2311	2744	3236	3800	4431	5129	5898
50	112	120	2657	3139	3723	4397	5156	6002	6940
50	112	140	3002	3527	4207	4996	5891	6893	8010
75	168	40	1543	1841	2119	2425	2756	3110	3487
75	168	60	2035	2402	2775	3192	3648	4139	4666
75	168	80	2514	2926	3396	3925	4508	5141	5824
75	168	100	2986	3429	3995	4641	5355	6135	6983
75	168	120	3362	3912	4582	5347	6197	7131	8153
75	168	140	3768	4387	5162	6050	7041	8136	9340
100	224	40	1831	2165	2483	2830	3205	3605	4028
100	224	60	2445	2833	3255	3726	4239	4789	5374
100	224	80	3048	3454	3980	4576	5227	5929	6682
100	224	100	3518	4039	4678	5400	6192	7051	7977
100	224	120	4022	4608	5358	6208	7145	8166	9273
100	224	140	4514	5164	6026	7007	8094	9282	10577
125	280	40	2109	2465	2816	3202	3618	4059	4525
125	280	60	2845	3231	3696	4219	4784	5388	6029
125	280	80	3451	3934	4520	5177	5892	6660	7478
125	280	100	4060	4606	5309	6103	6989	7903	8904
125	280	120	4651	5254	6076	7007	8027	9131	10321
125	280	140	5095	5877	6826	7897	9074	10354	11739
150	335	40	2379	2746	3129	3550	4004	4484	4990
150	335	60	3243	3607	4110	4680	5295	5950	6643
150	335	80	3886	4389	5026	5742	6517	7346	8228
150	335	100	4583	5139	5902	6763	7700	8705	9778
150	335	120	5111	5853	6751	7758	8857	10042	11312
150	335	140	5716	6553	7580	8736	10000	11368	12841
175	391	40	2646	3014	3424	3880	4369	4887	5430
175	391	60	3469	3952	4502	5117	5780	6483	7226
175	391	80	4309	4821	5507	6277	7110	7999	8940
175	391	100	4948	5637	6465	7391	8394	9468	10610
175	391	120	5642	6428	7392	8473	9648	10910	12257
175	391	140	6313	7194	8296	9533	10882	12335	13893
200	447	40	2910	3270	3706	4194	4718	5271	5850
200	447	60	3821	4289	4877	5535	6243	6993	7783
200	447	80	4603	5227	5966	6790	7678	8622	9621
200	447	100	5395	6120	7004	7991	9059	10198	11408
200	447	120	6155	6978	8006	9157	10405	11741	13164
200	447	140	6891	7809	8983	10298	11727	13263	14903
225	503	40	3174	3517	3978	4496	5052	5639	6254
225	503	60	4140	4613	5237	5937	6688	7483	8318
225	503	80	4969	5623	6408	7282	8223	9223	10277
225	503	100	5828	6585	7523	8569	9699	10902	12176
225	503	120	6654	7508	8598	9816	11134	12542	14038
225	503	140	7454	8402	9644	11034	12541	14157	15879
250	559	40	3323	3744	4239	4788	5375	5995	6643
250	559	60	4454	4927	5586	6325	7118	7956	8836
250	559	80	5325	6006	6835	7759	8751	9803	10911
250	559	100	6251	7034	8024	9128	10317	11582	12919
250	559	120	7142	8021	9170	10453	11839	13317	14885
250	559	140	8006	8976	10283	11746	13329	15023	16823
275	615	40	3543	3973	4492	5070	5688	6339	7019
275	615	60	4628	5221	5923	6701	7534	8414	9337
275	615	80	5672	6377	7250	8221	9262	10366	11526
275	615	100	6665	7471	8511	9670	10918	12242	13640
275	615	120	7621	8519	9725	11072	12524	14070	15706
275	615	140	8373	9525	10904	12438	14095	15863	17740
300	671	40	3759	4195	4738	5344	5992	6674	7385
300	671	60	4901	5516	6251	7066	7939	8860	9824
300	671	80	6013	6739	7653	8670	9760	10913	12124
300	671	100	7071	7896	8984	10198	11502	12885	14342
300	671	120	7931	8996	10265	11674	13190	14802	16506
300	671	140	8867	10066	11508	13111	14839	16682	18634
325	727	40	3972	4411	4978	5612	6288	7000	7742
325	727	60	5169	5803	6571	7423	8334	9294	10299
325	727	80	6348	7092	8046	9108	10245	11446	12707
325	727	100	7472	8311	9446	10712	12071	13511	15026
325	727	120	8362	9468	10792	12261	13840	15517	17286
325	727	140	9350	10593	12097	13767	15566	17481	19506
350	783	40	4182	4623	5213	5873	6577	7318	8089
350	783	60	5431	6083	6883	7771	8719	9718	10763
350	783	80	6678	7437	8431	9536	10719	11967	13277
350	783	100	7706	8707	9897	11215	12628	14123	15695
350	783	120	8784	9929	11307	12835	14475	16215	18049
350	783	140	9824	11109	12674	14409	16277	18261	20359
375	839	40	4392	4830	5442	6128	6860	7629	8430
375	839	60	5690	6358	7189	8112	9097	10133	11216
375	839	80	7003	7775	8807	9955	11182	12477	13834
375	839	100	8067	9103	10339	11707	13173	14722	16349
375	839	120	9199	10381	11811	13396	15097	16899	18796
375	839	140	10289	11615	13238	15038	16972	19026	21194
400	894	40	4470	5021	5666	6378	7137	7933	8763
400	894	60	5945	6627	7489	8445	9466	10540	11661
400	894	80	7326	8107	9175	10365	11637	12977	14380
400	894	100	8423	9491	10772	12190	13707	15309	16991
400	894	120	9606	10824	12306	13947	15707	17569	19528
400	894	140	10747	12111	13791	15654	17654	19776	22013

FIRE SIZE AT DETECTOR RESPONSE (BTU / SEC)

NOTE: Detector time constant at a reference velocity of 5 ft/sec.

For SI Units: 1 ft = 0.305 m
1000 BTU/sec = 1055 kW

Table B-4.1(v)   Installed Spacing of Heat Detector: 20 feet; t_g: 150 Seconds to 1000 Btu/sec; α: 0.044 Btu/sec³

τ	RTI	ΔT	CEILING HEIGHT IN FEET — FIRE SIZE AT DETECTOR RESPONSE (BTU / SEC)						
			4.0	8.0	12.0	16.0	20.0	24.0	28.0
25	56	40	379	456	556	675	812	969	1145
25	56	60	506	629	786	975	1197	1452	1743
25	56	80	635	805	1028	1299	1618	1988	2410
25	56	100	759	994	1285	1646	2074	2571	3139
25	56	120	889	1182	1561	2017	2564	3199	3926
25	56	140	1009	1378	1845	2417	3085	3869	4766
50	112	40	540	635	754	891	1045	1217	1406
50	112	60	723	863	1041	1249	1487	1756	2058
50	112	80	898	1087	1331	1619	1952	2333	2764
50	112	100	1059	1312	1628	2004	2444	2950	3526
50	112	120	1226	1540	1934	2407	2963	3606	4341
50	112	140	1393	1771	2250	2828	3510	4301	5207
75	168	40	684	787	923	1078	1250	1439	1645
75	168	60	906	1063	1263	1493	1751	2040	2359
75	168	80	1111	1331	1600	1911	2266	2666	3114
75	168	100	1319	1597	1939	2339	2800	3324	3914
75	168	120	1525	1862	2283	2780	3355	4014	4762
75	168	140	1729	2128	2634	3234	3933	4739	5656
100	224	40	801	923	1075	1247	1436	1642	1866
100	224	60	1061	1244	1465	1716	1996	2304	2643
100	224	80	1312	1553	1846	2182	2560	2982	3450
100	224	100	1558	1856	2225	2653	3138	3684	4294
100	224	120	1801	2157	2607	3132	3732	4413	5179
100	224	140	2043	2457	2992	3621	4345	5171	6105
125	280	40	915	1049	1216	1404	1609	1832	2072
125	280	60	1211	1412	1652	1923	2224	2553	2911
125	280	80	1501	1759	2075	2435	2837	3282	3772
125	280	100	1784	2098	2493	2947	3459	4029	4662
125	280	120	2064	2432	2911	3465	4093	4798	5587
125	280	140	2304	2763	3330	3989	4742	5592	6548
150	335	40	1007	1167	1348	1551	1772	2011	2266
150	335	60	1355	1570	1828	2119	2440	2789	3167
150	335	80	1681	1953	2291	2675	3101	3569	4080
150	335	100	2002	2326	2747	3227	3765	4361	5017
150	335	120	2278	2691	3199	3782	4438	5171	5984
150	335	140	2570	3053	3651	4342	5124	6002	6982
175	391	40	1114	1280	1474	1691	1927	2181	2451
175	391	60	1492	1720	1996	2306	2646	3014	3412
175	391	80	1856	2138	2497	2904	3352	3843	4377
175	391	100	2178	2542	2988	3495	4059	4680	5361
175	391	120	2504	2939	3475	4086	4771	5531	6369
175	391	140	2824	3330	3959	4680	5493	6399	7406
200	447	40	1211	1387	1594	1825	2075	2343	2629
200	447	60	1626	1864	2156	2485	2843	3231	3647
200	447	80	2028	2315	2695	3123	3594	4108	4664
200	447	100	2367	2750	3220	3753	4342	4988	5694
200	447	120	2722	3176	3739	4379	5092	5880	6744
200	447	140	3070	3596	4254	5007	5850	6785	7819
225	503	40	1305	1490	1710	1954	2218	2500	2800
225	503	60	1757	2002	2311	2657	3034	3440	3875
~~225~~	~~503~~	~~80~~	~~2157~~	~~2485~~	~~2885~~	~~3335~~	~~3828~~	~~4364~~	4942
225	503	100	2551	2951	3444	4001	4616	5287	6017
225	503	120	2934	3405	3995	4663	5404	6218	7110
225	503	140	3309	3852	4540	5323	6196	7161	8224
250	559	40	1396	1590	1821	2078	2356	2652	2966
250	559	60	1886	2136	2461	2824	3218	3642	4095
250	559	80	2308	2650	3070	3540	4055	4612	5212
250	559	100	2730	3145	3661	4242	4881	5578	6332
250	559	120	3140	3627	4242	4937	5706	6548	7466
250	559	140	3542	4100	4817	5630	6533	7528	8619
275	615	40	1486	1687	1930	2199	2490	2799	3126
275	615	60	2013	2267	2606	2985	3397	3839	4310
275	615	80	2454	2810	3249	3740	4275	4853	5474
275	615	100	2905	3334	3872	4477	5140	5860	6639
275	615	120	3342	3843	4483	5205	6000	6869	7814
275	615	140	3729	4341	5086	5929	6861	7886	9005
300	671	40	1574	1782	2035	2317	2620	2942	3283
300	671	60	2100	2392	2747	3143	3571	4030	4518
300	671	80	2598	2966	3423	3934	4489	5088	5730
300	671	100	3076	3517	4077	4705	5392	6136	6939
300	671	120	3541	4053	4717	5465	6288	7184	8155
300	671	140	3942	4577	5349	6220	7182	8236	9383
325	727	40	1661	1874	2138	2431	2747	3082	3435
325	727	60	2213	2516	2885	3297	3741	4217	4722
325	727	80	2739	3119	3593	4123	4699	5318	5981
325	727	100	3244	3697	4277	4928	5638	6406	7232
325	727	120	3690	4257	4946	5720	6568	7491	8488
325	727	140	4150	4806	5605	6505	7496	8578	9755
350	783	40	1746	1964	2239	2543	2871	3218	3584
350	783	60	2324	2637	3020	3447	3908	4400	4922
350	783	80	2877	3268	3760	4308	4904	5543	6226
350	783	100	3411	3872	4473	5146	5879	6670	7520
350	783	120	3871	4458	5170	5969	6843	7792	8815
350	783	140	4354	5031	5856	6784	7803	8914	10119
375	839	40	1831	2052	2337	2653	2992	3352	3730
375	839	60	2432	2755	3152	3594	4071	4578	5117
375	839	80	3014	3413	3923	4490	5105	5763	6466
375	839	100	3530	4043	4665	5360	6115	6929	7802
375	839	120	4050	4654	5390	6213	7113	8087	9136
375	839	140	4554	5251	6102	7057	8105	9244	10477
400	894	40	1916	2139	2433	2761	3111	3483	3873
400	894	60	2539	2871	3282	3738	4230	4754	5308
400	894	80	3149	3557	4082	4668	5301	5980	6702
400	894	100	3683	4212	4854	5569	6347	7183	8079
400	894	120	4225	4847	5605	6453	7378	8377	9452
400	894	140	4751	5467	6343	7325	8401	9568	10829

NOTE: Detector time constant at a reference velocity of 5 ft/sec.

For SI Units: 1 ft = 0.305 m
1000 BTU/sec = 1055 kW

Table B-4.1(w)   Installed Spacing of Heat Detector: 20 feet; t_g: 300 Seconds to 1000 Btu/sec; α: 0.011 Btu/sec^3

τ	RTI	ΔT	4.0	8.0	12.0	16.0	20.0	24.0	28.0
			CEILING HEIGHT IN FEET						
			FIRE SIZE AT DETECTOR RESPONSE (BTU / SEC)						
25	56	40	232	294	375	473	589	723	876
25	56	60	322	429	566	731	928	1157	1419
25	56	80	415	572	774	1022	1311	1649	2037
25	56	100	513	726	1002	1338	1733	2193	2720
25	56	120	613	890	1247	1681	2190	2784	3464
25	56	140	717	1065	1508	2047	2685	3418	4264
50	112	40	326	395	484	588	709	847	1003
50	112	60	445	557	700	870	1071	1303	1567
50	112	80	567	725	929	1177	1471	1812	2203
50	112	100	685	902	1174	1509	1908	2371	2902
50	112	120	805	1082	1436	1865	2378	2976	3660
50	112	140	928	1270	1709	2248	2881	3623	4473
75	168	40	405	483	580	693	821	966	1128
75	168	60	556	673	824	1004	1211	1448	1718
75	168	80	695	864	1077	1333	1632	1977	2372
75	168	100	835	1059	1342	1683	2085	2552	3087
75	168	120	975	1259	1618	2054	2570	3171	3860
75	168	140	1118	1469	1907	2445	3085	3831	4685
100	224	40	477	562	668	790	927	1079	1249
100	224	60	650	778	940	1129	1346	1591	1866
100	224	80	812	992	1217	1482	1790	2142	2542
100	224	100	971	1208	1502	1852	2262	2735	3275
100	224	120	1133	1427	1797	2240	2763	3369	4062
100	224	140	1294	1650	2102	2647	3292	4041	4901
125	280	40	543	636	750	880	1026	1187	1364
125	280	60	737	876	1049	1248	1474	1728	2012
125	280	80	922	1112	1349	1625	1943	2304	2711
125	280	100	1101	1348	1655	2017	2436	2916	3462
125	280	120	1280	1586	1968	2423	2954	3567	4265
125	280	140	1459	1827	2291	2846	3498	4254	5118
150	335	40	611	705	827	966	1120	1289	1474
150	335	60	820	968	1151	1361	1598	1861	2154
150	335	80	1020	1226	1474	1763	2091	2462	2877
150	335	100	1223	1481	1801	2175	2605	3095	3649
150	335	120	1420	1738	2133	2601	3142	3763	4469
150	335	140	1616	1995	2473	3040	3702	4466	5337
175	391	40	668	771	901	1048	1210	1387	1580
175	391	60	903	1057	1250	1470	1716	1989	2291
175	391	80	1122	1334	1595	1895	2235	2616	3041
175	391	100	1339	1609	1942	2328	2770	3271	3834
175	391	120	1554	1883	2293	2773	3326	3958	4672
175	391	140	1768	2157	2650	3230	3904	4677	5555
200	447	40	724	834	971	1126	1296	1482	1682
200	447	60	973	1141	1344	1574	1831	2114	2425
200	447	80	1216	1439	1711	2022	2374	2766	3201
200	447	100	1452	1731	2077	2477	2931	3443	4016
200	447	120	1685	2022	2447	2941	3507	4149	4873
200	447	140	1902	2313	2821	3416	4102	4886	5773

τ	RTI	ΔT	4.0	8.0	12.0	16.0	20.0	24.0	28.0
			CEILING HEIGHT IN FEET						
			FIRE SIZE AT DETECTOR RESPONSE (BTU / SEC)						
225	503	40	777	895	1039	1202	1380	1573	1782
225	503	60	1045	1223	1435	1675	1942	2235	2555
225	503	80	1308	1539	1822	2146	2509	2912	3357
225	503	100	1561	1850	2208	2621	3088	3612	4195
225	503	120	1812	2157	2596	3104	3684	4338	5072
225	503	140	2040	2464	2987	3597	4296	5092	5989
250	559	40	830	954	1105	1275	1460	1661	1878
250	559	60	1120	1302	1523	1773	2049	2352	2682
250	559	80	1397	1637	1931	2266	2640	3055	3511
250	559	100	1668	1964	2336	2762	3242	3777	4372
250	559	120	1919	2288	2741	3264	3857	4523	5269
250	559	140	2173	2611	3149	3774	4487	5295	6204
275	615	40	887	1011	1169	1346	1539	1748	1972
275	615	60	1189	1379	1609	1868	2154	2467	2806
275	615	80	1484	1731	2037	2383	2769	3194	3661
275	615	100	1773	2076	2460	2899	3391	3939	4545
275	615	120	2034	2416	2883	3419	4026	4706	5463
275	615	140	2303	2753	3307	3947	4675	5496	6416
300	671	40	932	1066	1231	1415	1615	1831	2063
300	671	60	1257	1453	1692	1961	2256	2578	2927
300	671	80	1570	1824	2140	2497	2894	3331	3808
300	671	100	1859	2185	2581	3033	3538	4098	4716
300	671	120	2147	2540	3021	3572	4192	4885	5654
300	671	140	2430	2892	3461	4117	4859	5694	6626
325	727	40	977	1120	1292	1483	1690	1913	2152
325	727	60	1324	1526	1774	2052	2357	2688	3046
325	727	80	1654	1914	2240	2609	3017	3465	3953
325	727	100	1956	2291	2700	3164	3682	4254	4884
325	727	120	2258	2662	3156	3721	4355	5061	5843
325	727	140	2554	3029	3612	4283	5040	5889	6834
350	783	40	1023	1173	1351	1549	1763	1994	2240
350	783	60	1389	1598	1854	2140	2454	2795	3163
350	783	80	1737	2002	2339	2718	3138	3596	4095
350	783	100	2050	2395	2815	3292	3823	4407	5049
350	783	120	2366	2780	3288	3867	4515	5235	6029
350	783	140	2676	3162	3760	4446	5218	6081	7039
375	839	40	1073	1225	1409	1613	1835	2072	2325
375	839	60	1453	1668	1932	2227	2550	2900	3277
375	839	80	1803	2088	2435	2826	3256	3725	4235
375	839	100	2142	2497	2929	3419	3961	4558	5212
375	839	120	2472	2897	3418	4011	4673	5406	6213
375	839	140	2795	3292	3906	4607	5394	6270	7242
400	894	40	1119	1276	1466	1677	1905	2149	2409
400	894	60	1517	1736	2008	2313	2645	3003	3389
400	894	80	1880	2173	2530	2931	3372	3852	4373
400	894	100	2233	2597	3041	3542	4097	4707	5372
400	894	120	2576	3011	3546	4152	4828	5574	6394
400	894	140	2913	3421	4048	4764	5566	6457	7442

NOTE: Detector time constant at a reference velocity of 5 ft/sec.

For SI Units: 1 ft = 0.305 m

1000 BTU/sec = 1055 kW

Table B-4.1(x) Installed Spacing of Heat Detector: 20 feet; t_g: 500 Seconds to 1000 Btu/sec; α: 0.004 Btu/sec³

τ	RTI	ΔT	\ 4.0	8.0	12.0	16.0	20.0	24.0	28.0
			\ FIRE SIZE AT DETECTOR RESPONSE (BTU / SEC)						
25	56	40	168	266	300	390	497	622	764
25	56	60	242	343	473	631	816	1033	1280
25	56	80	322	474	668	904	1184	1505	1875
25	56	100	407	617	884	1207	1589	2030	2537
25	56	120	497	772	1118	1537	2032	2603	3260
25	56	140	592	938	1370	1891	2508	3220	4040
50	112	40	230	290	366	458	567	693	836
50	112	60	323	425	553	711	898	1116	1365
50	112	80	418	568	760	996	1273	1598	1970
50	112	100	517	721	985	1308	1688	2132	2642
50	112	120	618	884	1227	1645	2138	2714	3374
50	112	140	723	1057	1485	2006	2625	3339	4162
75	168	40	284	348	429	525	636	764	910
75	168	60	394	499	632	792	981	1201	1452
75	168	80	506	656	850	1087	1367	1693	2067
75	168	100	617	823	1087	1407	1790	2237	2749
75	168	120	730	996	1337	1755	2249	2827	3489
75	168	140	846	1177	1603	2124	2739	3459	4285
100	224	40	332	401	487	587	703	835	983
100	224	60	458	569	707	872	1064	1286	1540
100	224	805	585	741	939	1179	1462	1790	2166
100	224	100	709	918	1185	1510	1894	2343	2857
100	224	120	834	1105	1448	1864	2361	2941	3606
100	224	140	962	1296	1722	2240	2859	3581	4410
125	280	40	377	450	541	647	767	903	1055
125	280	60	522	634	778	948	1145	1371	1627
125	280	80	658	820	1025	1269	1556	1887	2266
125	280	100	795	1011	1283	1611	1999	2450	2966
125	280	120	933	1206	1553	1975	2474	3056	3724
125	280	140	1072	1412	1837	2359	2980	3704	4536
150	335	40	419	497	593	704	829	969	1126
150	335	60	578	697	847	1022	1225	1455	1715
150	335	80	728	896	1107	1357	1649	1984	2365
150	335	100	880	1099	1378	1711	2103	2557	3076
150	335	120	1025	1307	1659	2085	2588	3173	3843
150	335	140	1176	1519	1952	2478	3102	3829	4663
175	391	40	459	541	643	759	889	1034	1194
175	391	60	631	756	912	1094	1302	1537	1801
175	391	80	794	969	1187	1444	1740	2080	2465
175	391	100	955	1185	1470	1810	2206	2664	3186
175	391	120	1116	1403	1763	2194	2701	3289	3962
175	391	140	1277	1626	2066	2597	3224	3954	4790
200	447	40	498	584	690	811	946	1096	1261
200	447	60	683	813	976	1164	1377	1617	1886
200	447	80	863	1040	1265	1527	1830	2175	2564
200	447	100	1030	1267	1560	1906	2309	2771	3297
200	447	120	1202	1497	1864	2302	2814	3406	4083
200	447	140	1374	1730	2177	2714	3346	4079	4919

τ	RTI	ΔT	\ 4.0	8.0	12.0	16.0	20.0	24.0	28.0
			\ FIRE SIZE AT DETECTOR RESPONSE (BTU / SEC)						
225	503	40	535	625	736	862	1002	1157	1326
225	503	60	732	869	1037	1231	1450	1696	1970
225	503	80	921	1108	1340	1609	1918	2268	2663
225	503	100	1103	1347	1647	2001	2409	2877	3407
225	503	120	1286	1587	1963	2407	2925	3522	4203
225	503	140	1468	1831	2286	2830	3467	4204	5048
250	559	40	575	665	781	911	1056	1216	1390
250	559	60	780	922	1097	1297	1522	1773	2052
250	559	80	978	1174	1413	1689	2004	2361	2760
250	559	100	1174	1424	1733	2093	2509	2982	3517
250	559	120	1367	1676	2059	2511	3036	3638	4323
250	559	140	1559	1929	2393	2944	3587	4330	5177
275	615	40	609	704	824	959	1109	1273	1452
275	615	60	827	974	1155	1361	1592	1849	2133
275	615	80	1036	1238	1484	1767	2089	2451	2857
275	615	100	1243	1500	1816	2184	2606	3086	3626
275	615	120	1446	1762	2154	2614	3145	3753	4443
275	615	140	1649	2026	2498	3057	3707	4454	5306
300	671	40	643	742	866	1006	1161	1330	1513
300	671	60	877	1025	1211	1423	1660	1923	2213
300	671	80	1094	1301	1553	1844	2172	2541	2952
300	671	100	1310	1574	1898	2273	2703	3188	3735
300	671	120	1524	1846	2246	2714	3253	3867	4562
300	671	140	1737	2120	2601	3168	3825	4579	5435
325	727	40	675	779	907	1052	1211	1385	1573
325	727	60	919	1075	1266	1484	1727	1996	2291
325	727	80	1150	1362	1621	1919	2254	2629	3046
325	727	100	1376	1646	1977	2361	2797	3290	3842
325	727	120	1601	1928	2337	2813	3359	3980	4681
325	727	140	1814	2212	2702	3277	3942	4702	5564
350	783	40	707	815	947	1097	1260	1439	1632
350	783	60	960	1123	1320	1544	1793	2068	2368
350	783	80	1204	1422	1688	1993	2335	2716	3139
350	783	100	1441	1716	2056	2447	2891	3390	3948
350	783	120	1676	2009	2426	2911	3465	4092	4799
350	783	140	1896	2302	2801	3385	4058	4824	5692
375	839	40	738	850	987	1140	1309	1492	1689
375	839	60	1002	1171	1374	1603	1858	2138	2444
375	839	80	1257	1481	1754	2065	2414	2802	3230
375	839	100	1505	1786	2133	2532	2983	3489	4054
375	839	120	1750	2088	2514	3007	3568	4203	4917
375	839	140	1977	2390	2899	3492	4172	4946	5820
400	894	40	769	884	1025	1183	1356	1544	1746
400	894	60	1045	1217	1426	1661	1921	2207	2519
400	894	80	1310	1538	1818	2136	2492	2886	3321
400	894	100	1568	1854	2208	2615	3074	3587	4158
400	894	120	1811	2166	2600	3101	3671	4313	5033
400	894	140	2056	2478	2995	3597	4286	5067	5947

NOTE: Detector time constant at a reference velocity of 5 ft/sec.

For SI Units: 1 ft = 0.305 m

1000 BTU/sec = 1055 kW

Table B-4.1(y)   Installed Spacing of Heat Detector: 20 feet; t_g: 600 Seconds to 1000 Btu/sec; α: 0.003 Btu/sec^3

τ	RTI	ΔT	CEILING HEIGHT IN FEET							τ	RTI	ΔT	CEILING HEIGHT IN FEET						
			4.0	8.0	12.0	16.0	20.0	24.0	28.0				4.0	8.0	12.0	16.0	20.0	24.0	28.0
			FIRE SIZE AT DETECTOR RESPONSE (BTU / SEC)										FIRE SIZE AT DETECTOR RESPONSE (BTU / SEC)						
25	56	40	151	208	280	369	474	595	735	225	503	40	470	553	654	771	901	1046	1206
25	56	60	222	321	449	605	787	1000	1244	225	503	60	647	773	930	1112	1319	1554	1816
25	56	80	298	449	641	874	1149	1466	1831	225	503	80	815	991	1210	1466	1762	2100	2483
25	56	100	380	590	854	1173	1551	1986	2487	225	503	100	980	1211	1497	1836	2232	2687	3206
25	56	120	467	743	1085	1499	1989	2554	3205	225	503	120	1146	1434	1794	2224	2729	3313	3982
25	56	140	560	906	1334	1850	2461	3167	3979	225	503	140	1311	1661	2100	2629	3253	3979	4810
50	112	40	205	262	336	425	531	654	794	250	559	40	501	587	693	814	948	1097	1260
50	112	60	290	390	517	671	855	1069	1314	250	559	60	688	819	982	1169	1381	1620	1886
50	112	80	379	527	717	949	1223	1543	1910	250	559	80	871	1049	1273	1535	1836	2178	2565
50	112	100	472	675	936	1255	1632	2071	2574	250	559	100	1042	1278	1571	1916	2316	2775	3297
50	112	120	569	834	1174	1588	2079	2646	3299	250	559	120	1216	1510	1877	2312	2822	3410	4082
50	112	140	669	1003	1428	1945	2558	3265	4080	250	559	140	1390	1746	2192	2726	3354	4083	4917
75	168	40	251	312	389	481	588	713	854	275	615	40	531	621	731	855	994	1146	1314
75	168	60	352	453	582	738	923	1138	1385	275	615	60	729	865	1032	1224	1441	1684	1955
75	168	80	455	603	793	1024	1300	1622	1990	275	615	80	918	1105	1335	1602	1908	2255	2646
75	168	100	558	760	1021	1340	1716	2157	2662	275	615	100	1102	1344	1643	1994	2399	2863	3389
75	168	120	664	927	1265	1679	2167	2739	3394	275	615	120	1285	1585	1958	2400	2914	3506	4182
75	168	140	774	1103	1525	2042	2652	3365	4182	275	615	140	1467	1829	2282	2822	3455	4187	5024
100	224	40	293	358	439	534	645	771	915	300	671	40	565	654	767	896	1038	1195	1366
100	224	60	408	513	645	804	992	1209	1457	300	671	60	769	909	1081	1278	1500	1748	2023
100	224	80	524	674	866	1101	1378	1701	2071	300	671	80	966	1159	1395	1668	1980	2332	2727
100	224	100	638	844	1105	1422	1802	2244	2751	300	671	100	1160	1408	1713	2070	2481	2949	3479
100	224	120	754	1019	1357	1768	2260	2833	3491	300	671	120	1352	1658	2038	2486	3006	3602	4281
100	224	140	876	1202	1623	2140	2750	3465	4285	300	671	140	1543	1911	2370	2916	3554	4291	5131
125	280	40	332	400	486	585	699	829	975	325	727	40	594	686	803	936	1082	1243	1418
125	280	60	460	569	706	869	1060	1279	1529	325	727	60	808	952	1129	1331	1558	1811	2091
125	280	80	587	742	939	1176	1456	1781	2153	325	727	80	1014	1212	1454	1733	2050	2407	2807
125	280	100	713	921	1185	1507	1888	2332	2841	325	727	100	1217	1471	1782	2146	2562	3036	3570
125	280	120	839	1108	1448	1860	2353	2928	3587	325	727	120	1418	1730	2116	2571	3096	3698	4381
125	280	140	968	1300	1722	2235	2850	3566	4389	325	727	140	1617	1990	2457	3010	3653	4394	5239
150	335	40	369	441	530	634	752	885	1035	350	783	40	621	717	838	975	1125	1289	1469
150	335	60	509	623	765	933	1127	1349	1602	350	783	60	846	994	1176	1383	1615	1873	2157
150	335	80	648	808	1010	1251	1534	1861	2235	350	783	80	1062	1265	1512	1797	2119	2482	2886
150	335	100	784	997	1266	1591	1974	2420	2931	350	783	100	1274	1532	1850	2220	2642	3121	3660
150	335	120	921	1192	1535	1952	2447	3024	3685	350	783	120	1482	1800	2194	2655	3186	3793	4480
150	335	140	1059	1395	1817	2334	2950	3668	4493	350	783	140	1691	2069	2543	3103	3752	4497	5346
175	391	40	404	480	573	681	803	940	1093	375	839	40	648	748	873	1013	1167	1335	1518
175	391	60	559	675	822	994	1192	1418	1674	375	839	60	886	1036	1222	1434	1671	1933	2222
175	391	80	705	871	1078	1324	1611	1941	2318	375	839	80	1108	1316	1569	1860	2188	2555	2964
175	391	100	852	1071	1345	1674	2061	2509	3022	375	839	100	1329	1593	1917	2293	2721	3205	3749
175	391	120	998	1275	1623	2044	2541	3120	3784	375	839	120	1546	1869	2270	2737	3275	3887	4579
175	391	140	1145	1484	1913	2433	3051	3771	4598	375	839	140	1763	2146	2628	3194	3849	4600	5453
200	447	40	437	517	615	727	853	994	1150	400	894	40	675	778	906	1050	1208	1381	1567
200	447	60	604	725	877	1054	1257	1486	1746	400	894	60	921	1077	1268	1485	1726	1993	2287
200	447	80	761	932	1145	1396	1687	2021	2401	400	894	80	1154	1366	1625	1921	2255	2628	3042
200	447	100	918	1142	1422	1756	2146	2598	3114	400	894	100	1383	1652	1983	2365	2799	3289	3838
200	447	120	1073	1355	1709	2134	2635	3217	3883	400	894	120	1609	1937	2344	2819	3363	3980	4677
200	447	140	1229	1573	2007	2531	3152	3875	4704	400	894	140	1824	2222	2711	3284	3946	4702	5559

NOTE: Detector time constant at a reference velocity of 5 ft/sec.

For SI Units: 1 ft = 0.305 m

1000 BTU/sec = 1055 kW

Table B-4.1(z)   Installed Spacing of Heat Detector: 25 ft.; t_g: 50 Seconds to 1000 Btu/sec; α: 0.400 Btu/sec³

τ	RTI	ΔT	4.0	8.0	12.0	16.0	20.0	24.0	28.0
			\multicolumn{7}{CEILING HEIGHT IN FEET — FIRE SIZE AT DETECTOR RESPONSE (BTU / SEC)}						
25	56	40	1187	1381	1575	1797	2046	2319	2617
25	56	60	1529	1795	2072	2393	2756	3158	3601
25	56	80	1864	2194	2560	2990	3477	4023	4630
25	56	100	2194	2586	3050	3596	4220	4924	5711
25	56	120	2520	2977	3545	4217	4989	5865	6847
25	56	140	2846	3368	4048	4855	5786	6846	8038
50	112	40	1687	1953	2201	2479	2786	3117	3472
50	112	60	2226	2545	2885	3276	3708	4179	4688
50	112	80	2752	3103	3541	4048	4613	5233	5908
50	112	100	3272	3642	4183	4813	5518	6298	7152
50	112	120	3655	4164	4818	5578	6432	7382	8429
50	112	140	4098	4682	5452	6348	7360	8490	9742
75	168	40	2138	2432	2722	3049	3407	3791	4199
75	168	60	2869	3178	3570	4023	4520	5056	5631
75	168	80	3467	3868	4373	4954	5595	6291	7041
75	168	100	4071	4532	5151	5866	6656	7519	8455
75	168	120	4657	5177	5915	6767	7713	8752	9884
75	168	140	5114	5806	6670	7665	8774	9997	11336
100	224	40	2567	2861	3187	3557	3961	4392	4849
100	224	60	3380	3738	4183	4692	5249	5846	6483
100	224	80	4147	4555	5120	5770	6482	7250	8073
100	224	100	4770	5330	6024	6817	7688	8633	9650
100	224	120	5433	6084	6906	7845	8882	10010	11230
100	224	140	6076	6820	7773	8864	10071	11390	12823
125	280	40	2988	3258	3614	4024	4470	4945	5447
125	280	60	3894	4257	4747	5309	5921	6576	7270
125	280	80	4675	5183	5810	6523	7302	8139	9031
125	280	100	5471	6070	6830	7697	8646	9670	10765
125	280	120	6236	6926	7823	8846	9969	11185	12491
125	280	140	6978	7760	8796	9979	11280	12694	14220
150	335	40	3296	3620	4012	4460	4946	5462	6006
150	335	60	4390	4745	5276	5887	6551	7260	8010
150	335	80	5239	5777	6457	7231	8073	8975	9933
150	335	100	6139	6765	7588	8525	9548	10646	11818
150	335	120	7003	7718	8686	9788	10994	12293	13685
150	335	140	7841	8645	9759	11029	12422	13928	15546
175	391	40	3641	3968	4390	4873	5397	5952	6536
175	391	60	4728	5197	5776	6435	7149	7909	8711
175	391	80	5781	6340	7071	7902	8805	9769	10791
175	391	100	6781	7426	8308	9312	10405	11575	12820
175	391	120	7744	8471	9506	10684	11969	13350	14823
175	391	140	8515	9479	10675	12030	13510	15106	16813
200	447	40	3977	4301	4750	5267	5827	6420	7042
200	447	60	5149	5637	6256	6959	7721	8530	9382
200	447	80	6305	6880	7659	8545	9506	10529	11612
200	447	100	7405	8059	8997	10065	11225	12465	13780
200	447	120	8293	9185	10291	11542	12903	14363	15915
200	447	140	9269	10283	11552	12988	14553	16235	18031

τ	RTI	ΔT	4.0	8.0	12.0	16.0	20.0	24.0	28.0
			\multicolumn{7}{CEILING HEIGHT IN FEET — FIRE SIZE AT DETECTOR RESPONSE (BTU / SEC)}						
225	503	40	4308	4621	5097	5646	6240	6868	7527
225	503	60	5557	6060	6716	7463	8270	9127	10027
225	503	80	6816	7399	8224	9163	10179	11261	12402
225	503	100	7852	8660	9660	10790	12016	13323	14706
225	503	120	8944	9877	11047	12368	13804	15339	16968
225	503	140	9998	11057	12397	13911	15559	17325	19205
250	559	40	4498	4918	5430	6011	6638	7301	7995
250	559	60	5955	6469	7161	7949	8801	9703	10650
250	559	80	7316	7902	8770	9760	10830	11968	13165
250	559	100	8404	9247	10301	11491	12779	14151	15601
250	559	120	9576	10547	11778	13168	14675	16283	17988
250	559	140	10706	11807	13215	14805	16532	18380	20343
275	615	40	4783	5216	5754	6365	7025	7721	8449
275	615	60	6345	6866	7593	8421	9315	10261	11254
275	615	80	7635	8379	9300	10339	11462	12653	13906
275	615	100	8942	9818	10923	12171	13521	14956	16470
275	615	120	10192	11198	12487	13943	15520	17200	18978
275	615	140	11397	12534	14008	15672	17477	19404	21449
300	671	40	5061	5505	6069	6709	7400	8128	8890
300	671	60	6728	7252	8012	8880	9815	10804	11841
300	671	80	8080	8852	9815	10903	12076	13320	14627
300	671	100	9468	10373	11528	12833	14242	15738	17316
300	671	120	10794	11831	13178	14698	16343	18093	19942
300	671	140	12074	13242	14780	16516	18397	20402	22526
325	727	40	5335	5787	6375	7044	7765	8525	9319
325	727	60	7107	7628	8421	9327	10302	11333	12413
325	727	80	8516	9312	10317	11452	12675	13970	15330
325	727	100	9982	10914	12119	13478	14945	16502	18140
325	727	120	11385	12449	13852	15435	17145	18963	20882
325	727	140	12737	13933	15534	17340	19294	21376	23577
350	783	40	5604	6063	6675	7372	8122	8913	9738
350	783	60	7308	7985	8820	9763	10778	11850	12972
350	783	80	8943	9763	10809	11989	13260	14605	16016
350	783	100	10488	11443	12696	14109	15633	17248	18946
350	783	120	11965	13053	14510	16155	17930	19815	21801
350	783	140	13391	14609	16271	18146	20171	22327	24605
375	839	40	5869	6333	6968	7692	8471	9292	10148
375	839	60	7647	8344	9211	10190	11244	12356	13518
375	839	80	9363	10204	11289	12514	13833	15227	16688
375	839	100	10985	11961	13260	14726	16305	17978	19735
375	839	120	12537	13645	15155	16859	18698	20648	22702
375	839	140	14034	15271	16992	18934	21031	23260	25611
400	894	40	6130	6597	7255	8005	8813	9663	10549
400	894	60	7979	8695	9594	10609	11700	12851	14054
400	894	80	9776	10636	11760	13029	14394	15836	17346
400	894	100	11475	12470	13814	15331	16965	18694	20509
400	894	120	13101	14225	15787	17551	19451	21465	23584
400	894	140	14670	15921	17700	19707	21873	24174	26598

NOTE: Detector time constant at a reference velocity of 5 ft/sec.

For SI Units: 1 ft = 0.305 m

1000 BTU/sec = 1055 kW

Table B-4.1(aa)   Installed Spacing of Heat Detector: 25 feet; t_g: 150 Seconds to 1000 Btu/sec; α: 0.044 Btu/sec³

τ	RTI	ΔT	4.0	8.0	12.0	16.0	20.0	24.0	28.0
			CEILING HEIGHT IN FEET — FIRE SIZE AT DETECTOR RESPONSE (BTU / SEC)						
25	56	40	520	609	724	861	1018	1195	1394
25	56	60	701	843	1029	1250	1507	1801	2131
25	56	80	881	1084	1352	1672	2045	2472	2954
25	56	100	1045	1335	1695	2126	2629	3205	3855
25	56	120	1231	1596	2058	2611	3256	3994	4828
25	56	140	1414	1875	2449	3126	3924	4836	5867
50	112	40	742	843	976	1130	1303	1495	1707
50	112	60	981	1148	1352	1590	1861	2166	2507
50	112	80	1223	1450	1734	2068	2452	2887	3376
50	112	100	1460	1755	2127	2568	3078	3659	4316
50	112	120	1696	2065	2534	3092	3740	4483	5322
50	112	140	1933	2380	2955	3640	4439	5355	6392
75	168	40	925	1042	1192	1364	1555	1765	1993
75	168	60	1228	1410	1635	1895	2186	2510	2867
75	168	80	1526	1770	2076	2433	2837	3290	3794
75	168	100	1820	2127	2522	2985	3513	4110	4779
75	168	120	2092	2484	2976	3554	4219	4974	5824
75	168	140	2371	2845	3440	4143	4955	5881	6927
100	224	40	1084	1220	1386	1576	1785	2012	2258
100	224	60	1451	1647	1893	2174	2487	2831	3208
100	224	80	1806	2060	2390	2771	3198	3673	4197
100	224	100	2129	2466	2887	3375	3928	4546	5234
100	224	120	2456	2869	3388	3993	4681	5456	6322
100	224	140	2778	3273	3896	4624	5458	6403	7463
125	280	40	1237	1385	1566	1772	1998	2243	2506
125	280	60	1661	1867	2132	2434	2768	3134	3531
125	280	80	2047	2329	2683	3088	3539	4037	4583
125	280	100	2426	2781	3229	3744	4322	4965	5675
125	280	120	2796	3228	3776	4409	5123	5923	6811
125	280	140	3160	3673	4326	5084	5945	6913	7993
150	335	40	1382	1540	1735	1957	2199	2460	2740
150	335	60	1863	2074	2357	2679	3034	3421	3838
150	335	80	2285	2583	2959	3388	3864	4385	4954
150	335	100	2707	3079	3552	4094	4699	5368	6101
150	335	120	3119	3568	4143	4805	5549	6374	7287
150	335	140	3524	4052	4735	5524	6414	7409	8513
175	391	40	1522	1687	1896	2132	2390	2667	2963
175	391	60	2031	2270	2571	2913	3288	3695	4133
175	391	80	2513	2825	3222	3674	4174	4719	5311
175	391	100	2978	3363	3861	4430	5062	5756	6514
175	391	120	3431	3891	4495	5186	5958	6812	7750
175	391	140	3839	4413	5128	5948	6868	7891	9021
200	447	40	1658	1828	2049	2300	2573	2866	3177
200	447	60	2207	2458	2776	3137	3532	3959	4416
200	447	80	2733	3056	3475	3950	4473	5041	5656
200	447	100	3239	3635	4158	4752	5411	6131	6915
200	447	120	3695	4202	4833	5553	6354	7236	8201
200	447	140	4164	4761	5505	6356	7308	8360	9518
225	503	40	1791	1963	2197	2462	2749	3057	3383
225	503	60	2378	2639	2974	3353	3767	4213	4690
225	503	80	2946	3279	3718	4215	4761	5353	5991
225	503	100	3494	3898	4444	5064	5748	6495	7304
225	503	120	3975	4501	5159	5908	6738	7648	8640
225	503	140	4477	5096	5870	6752	7734	8817	10004
250	559	40	1922	2095	2340	2618	2919	3241	3583
250	559	60	2544	2815	3166	3562	3995	4459	4956
250	559	80	3154	3495	3954	4473	5041	5655	6316
250	559	100	3697	4151	4721	5366	6076	6848	7683
250	559	120	4246	4792	5476	6252	7111	8049	9069
250	559	140	4781	5421	6223	7136	8150	9263	10479
275	615	40	2018	2221	2478	2770	3085	3421	3776
275	615	60	2706	2985	3351	3765	4216	4699	5214
275	615	80	3359	3705	4183	4722	5312	5949	6632
275	615	100	3927	4398	4990	5660	6395	7193	8053
275	615	120	4510	5074	5783	6587	7474	8440	9488
275	615	140	5077	5737	6567	7511	8555	9698	10944
300	671	40	2134	2344	2613	2917	3245	3595	3965
300	671	60	2865	3151	3532	3963	4431	4932	5466
300	671	80	3514	3909	4405	4966	5577	6236	6941
300	671	100	4152	4638	5252	5946	6706	7529	8414
300	671	120	4767	5348	6083	6914	7828	8823	9898
300	671	140	5366	6044	6903	7876	8951	10124	11399
325	727	40	2248	2465	2745	3061	3402	3765	4148
325	727	60	3022	3312	3709	4156	4641	5160	5711
325	727	80	3700	4108	4623	5204	5836	6516	7242
325	727	100	4371	4873	5508	6226	7010	7857	8767
325	727	120	5019	5617	6375	7233	8175	9197	10299
325	727	140	5650	6345	7230	8233	9338	10541	11846
350	783	40	2359	2583	2873	3201	3555	3932	4328
350	783	60	3176	3470	3881	4344	4846	5383	5952
350	783	80	3882	4303	4836	5436	6089	6790	7538
350	783	100	4586	5102	5759	6499	7308	8179	9114
350	783	120	5266	5879	6662	7546	8514	9563	10693
350	783	140	5928	6639	7551	8583	9717	10951	12285
375	839	40	2469	2698	2999	3339	3705	4094	4504
375	839	60	3288	3624	4050	4529	5048	5601	6187
375	839	80	4061	4494	5044	5664	6337	7059	7827
375	839	100	4798	5327	6004	6767	7599	8495	9453
375	839	120	5509	6136	6942	7852	8847	9923	11080
375	839	140	6201	6927	7865	8925	10089	11352	12716
400	894	40	2577	2811	3123	3474	3852	4254	4676
400	894	60	3429	3776	4215	4710	5245	5815	6418
400	894	80	4236	4681	5248	5887	6580	7322	8111
400	894	100	5006	5548	6245	7030	7885	8805	9787
400	894	120	5748	6388	7218	8153	9174	10277	11460
400	894	140	6471	7209	8174	9262	10455	11747	13140

NOTE: Detector time constant at a reference velocity of 5 ft/sec.

For SI Units: 1 ft = 0.305 m
1000 BTU/sec = 1055 kW

Table B-4.1(bb)   Installed Spacing of Heat Detector: 25 feet; t_g: 300 Seconds to 1000 Btu/sec; α: 0.011 Btu/sec³

τ	RTI	ΔT	4.0	8.0	12.0	16.0	20.0	24.0	28.0	τ	RTI	ΔT	4.0	8.0	12.0	16.0	20.0	24.0	28.0
			CEILING HEIGHT IN FEET — FIRE SIZE AT DETECTOR RESPONSE (BTU / SEC)										CEILING HEIGHT IN FEET — FIRE SIZE AT DETECTOR RESPONSE (BTU / SEC)						
25	56	40	321	396	493	609	744	899	1074	225	503	40	1054	1181	1339	1519	1715	1929	2159
25	56	60	451	580	746	946	1178	1444	1744	225	503	60	1430	1617	1853	2122	2420	2748	3104
25	56	80	582	781	1030	1324	1669	2063	2508	225	503	80	1794	2039	2359	2725	3134	3588	4087
25	56	100	719	994	1337	1740	2210	2747	3354	225	503	100	2124	2453	2863	3335	3866	4459	5116
25	56	120	866	1223	1667	2194	2797	3491	4275	225	503	120	2460	2866	3371	3956	4619	5364	6195
25	56	140	1013	1467	2021	2675	3426	4289	5265	225	503	140	2791	3278	3885	4591	5396	6305	7324
50	112	40	447	527	630	751	890	1048	1225	250	559	40	1126	1258	1423	1610	1815	2037	2275
50	112	60	616	747	916	1118	1351	1619	1921	250	559	60	1529	1720	1966	2245	2553	2890	3257
50	112	80	781	976	1223	1518	1863	2258	2706	250	559	80	1900	2166	2497	2875	3297	3761	4271
50	112	100	948	1214	1550	1952	2422	2961	3571	250	559	100	2265	2603	3026	3511	4055	4659	5328
50	112	120	1119	1468	1898	2418	3025	3721	4509	250	559	120	2621	3037	3557	4155	4832	5589	6430
50	112	140	1294	1729	2271	2914	3670	4535	5514	250	559	140	2972	3469	4092	4812	5631	6552	7581
75	168	40	560	642	753	882	1028	1192	1374	275	615	40	1196	1333	1505	1699	1912	2142	2388
75	168	60	757	897	1075	1283	1522	1794	2100	275	615	60	1625	1821	2075	2364	2683	3030	3406
75	168	80	954	1156	1410	1710	2059	2457	2907	275	615	80	2016	2290	2632	3022	3455	3931	4452
75	168	100	1151	1422	1761	2166	2638	3178	3791	275	615	100	2401	2749	3184	3682	4239	4856	5536
75	168	120	1350	1695	2130	2650	3258	3956	4746	275	615	120	2778	3204	3737	4350	5040	5810	6663
75	168	140	1550	1978	2517	3163	3918	4785	5767	275	615	140	3149	3656	4293	5028	5861	6795	7835
100	224	40	654	746	865	1003	1157	1329	1518	300	671	40	1264	1405	1584	1786	2007	2244	2498
100	224	60	890	1035	1222	1439	1687	1966	2278	300	671	60	1720	1919	2182	2481	2809	3166	3552
100	224	80	1113	1323	1587	1896	2251	2655	3110	300	671	80	2129	2411	2764	3165	3610	4097	4629
100	224	100	1338	1616	1964	2376	2853	3398	4014	300	671	100	2535	2891	3339	3850	4420	5049	5741
100	224	120	1563	1914	2356	2881	3492	4193	4987	300	671	120	2932	3366	3913	4541	5245	6027	6893
100	224	140	1789	2218	2762	3411	4169	5038	6024	300	671	140	3323	3837	4490	5241	6088	7035	8087
125	280	40	742	842	969	1116	1279	1459	1656	325	727	40	1331	1476	1662	1871	2099	2344	2606
125	280	60	1002	1163	1360	1588	1845	2133	2452	325	727	60	1797	2014	2286	2594	2932	3299	3695
125	280	80	1260	1480	1755	2074	2439	2851	3312	325	727	80	2239	2528	2892	3305	3761	4260	4803
125	280	100	1513	1799	2158	2580	3065	3617	4238	325	727	100	2666	3030	3490	4014	4597	5239	5943
125	280	120	1765	2120	2573	3107	3725	4431	5230	325	727	120	3083	3525	4086	4727	5446	6242	7120
125	280	140	2004	2447	3001	3657	4420	5294	6283	325	727	140	3493	4015	4682	5449	6310	7271	8336
150	335	40	825	932	1068	1223	1395	1584	1789	350	783	40	1397	1546	1738	1954	2189	2442	2711
150	335	60	1118	1284	1491	1729	1997	2294	2622	350	783	60	1884	2107	2388	2706	3053	3430	3835
150	335	80	1401	1629	1914	2245	2621	3042	3511	350	783	80	2347	2644	3018	3442	3910	4420	4975
150	335	100	1680	1972	2344	2777	3273	3833	4462	350	783	100	2794	3166	3638	4175	4771	5426	6142
150	335	120	1944	2318	2783	3328	3955	4669	5473	350	783	120	3231	3680	4255	4911	5643	6453	7343
150	335	140	2213	2667	3232	3898	4669	5549	6544	350	783	140	3638	4189	4871	5653	6530	7505	8583
175	391	40	908	1019	1162	1325	1506	1703	1917	375	839	40	1461	1614	1812	2035	2277	2537	2814
175	391	60	1226	1399	1616	1865	2143	2450	2787	375	839	60	1970	2199	2488	2815	3172	3558	3973
175	391	80	1536	1771	2068	2411	2797	3228	3707	375	839	80	2453	2757	3141	3576	4055	4577	5143
175	391	100	1842	2139	2523	2969	3475	4045	4683	375	839	100	2920	3299	3783	4333	4942	5609	6337
175	391	120	2122	2507	2985	3543	4181	4904	5715	375	839	120	3377	3833	4421	5091	5837	6660	7564
175	391	140	2412	2877	3456	4135	4915	5803	6805	375	839	140	3797	4360	5057	5854	6746	7735	8827
200	447	40	979	1101	1252	1424	1613	1818	2040	400	894	40	1525	1680	1885	2115	2364	2631	2915
200	447	60	1330	1510	1737	1996	2284	2601	2948	400	894	60	2053	2288	2586	2922	3288	3684	4108
200	447	80	1667	1907	2216	2570	2968	3410	3899	400	894	80	2558	2868	3262	3708	4199	4732	5308
200	447	100	1979	2299	2696	3154	3673	4254	4901	400	894	100	3045	3430	3926	4488	5110	5790	6531
200	447	120	2294	2689	3181	3752	4402	5135	5956	400	894	120	3497	3982	4584	5268	6028	6865	7782
200	447	140	2605	3081	3674	4365	5158	6056	7065	400	894	140	3953	4528	5239	6051	6958	7962	9067

NOTE: Detector time constant at a reference velocity of 5 ft/sec.

For SI Units: 1 ft = 0.305 m

    1000 BTU/sec = 1055 kW

Table B-4.1(cc)   Installed Spacing of Heat Detector: 25 feet; t_g: 300 Seconds to 1000 Btu/sec; α: 0.011 Btu/sec^3

| τ | RTI | ΔT | CEILING HEIGHT IN FEET — FIRE SIZE AT DETECTOR RESPONSE (BTU / SEC) | | | | | | | τ | RTI | ΔT | CEILING HEIGHT IN FEET — FIRE SIZE AT DETECTOR RESPONSE (BTU / SEC) | | | | | | |
			4.0	8.0	12.0	16.0	20.0	24.0	28.0				4.0	8.0	12.0	16.0	20.0	24.0	28.0
25	56	40	234	307	397	505	632	776	939	225	503	40	729	827	951	1093	1250	1423	1612
25	56	60	341	470	631	820	1041	1293	1578	225	503	60	996	1152	1344	1566	1815	2093	2402
25	56	80	456	652	895	1181	1510	1887	2314	225	503	80	1257	1473	1741	2053	2407	2807	3254
25	56	100	578	853	1186	1579	2030	2548	3134	225	503	100	1513	1795	2147	2559	3031	3568	4172
25	56	120	708	1069	1503	2013	2599	3270	4030	225	503	120	1769	2120	2564	3086	3689	4377	5154
25	56	140	847	1301	1843	2479	3210	4048	4996	225	503	140	2012	2450	2993	3634	4379	5232	6198
50	112	40	317	389	481	589	716	861	1026	250	559	40	778	879	1008	1155	1317	1495	1689
50	112	60	450	573	730	919	1140	1393	1680	250	559	60	1063	1222	1420	1648	1903	2187	2501
50	112	80	583	772	1009	1291	1621	2000	2428	250	559	80	1338	1559	1835	2153	2514	2919	3371
50	112	100	722	984	1312	1699	2153	2672	3259	250	559	100	1610	1896	2256	2674	3153	3695	4303
50	112	120	869	1211	1638	2146	2731	3404	4166	250	559	120	1871	2235	2686	3215	3824	4516	5298
50	112	140	1017	1452	1987	2621	3352	4191	5142	250	559	140	2134	2578	3128	3776	4526	5383	6352
75	168	40	389	465	559	671	801	948	1113	275	615	40	824	930	1063	1215	1382	1565	1764
75	168	60	545	670	829	1020	1241	1495	1783	275	615	60	1126	1290	1494	1728	1989	2279	2598
75	168	80	698	885	1121	1404	1735	2114	2544	275	615	80	1418	1643	1925	2250	2618	3029	3486
75	168	100	855	1113	1436	1824	2277	2797	3386	275	615	100	1705	1995	2362	2788	3273	3821	4434
75	168	120	1016	1354	1776	2277	2865	3540	4304	275	615	120	1976	2347	2807	3343	3957	4655	5441
75	168	140	1182	1606	2134	2761	3495	4336	5289	275	615	140	2253	2703	3261	3916	4672	5534	6507
100	224	40	455	534	633	750	882	1032	1201	300	671	40	875	980	1117	1273	1446	1633	1837
100	224	60	631	761	924	1118	1342	1598	1887	300	671	60	1188	1357	1567	1806	2074	2369	2694
100	224	80	804	994	1233	1518	1849	2230	2661	300	671	80	1495	1725	2014	2346	2720	3138	3601
100	224	100	978	1237	1562	1950	2403	2925	3515	300	671	100	1788	2091	2466	2899	3391	3945	4564
100	224	120	1156	1491	1910	2413	3001	3677	4443	300	671	120	2080	2457	2924	3468	4090	4793	5584
100	224	140	1337	1758	2278	2906	3640	4483	5438	300	671	140	2369	2826	3392	4055	4817	5684	6661
125	280	40	515	598	703	824	961	1115	1287	325	727	40	916	1028	1170	1331	1508	1701	1909
125	280	60	712	846	1015	1213	1441	1700	1992	325	727	60	1249	1422	1637	1883	2156	2458	2788
125	280	80	905	1098	1341	1630	1964	2347	2780	325	727	80	1571	1805	2101	2440	2821	3245	3714
125	280	100	1095	1357	1685	2075	2530	3053	3645	325	727	100	1876	2185	2567	3008	3508	4068	4692
125	280	120	1289	1625	2046	2550	3139	3816	4583	325	727	120	2181	2565	3040	3591	4220	4930	5726
125	280	140	1484	1902	2425	3053	3787	4631	5588	325	727	140	2482	2946	3521	4191	4960	5833	6815
150	335	40	576	659	769	895	1037	1196	1371	350	783	40	958	1075	1221	1387	1569	1766	1980
150	335	60	788	927	1102	1305	1538	1801	2096	350	783	60	1308	1485	1706	1958	2238	2545	2881
150	335	80	996	1197	1446	1739	2077	2463	2899	350	783	80	1647	1884	2186	2532	2920	3350	3825
150	335	100	1206	1472	1805	2199	2657	3182	3777	350	783	100	1962	2277	2667	3116	3623	4189	4820
150	335	120	1415	1755	2180	2686	3277	3955	4725	350	783	120	2279	2670	3153	3713	4349	5066	5868
150	335	140	1626	2044	2571	3200	3935	4780	5739	350	783	140	2593	3064	3647	4326	5102	5981	6968
175	391	40	629	718	832	963	1110	1274	1454	375	839	40	1000	1121	1272	1442	1628	1831	2049
175	391	60	865	1005	1186	1395	1633	1900	2199	375	839	60	1367	1547	1774	2032	2317	2630	2972
175	391	80	1087	1292	1548	1846	2189	2579	3018	375	839	80	1721	1960	2269	2623	3017	3454	3935
175	391	100	1312	1583	1922	2321	2783	3311	3908	375	839	100	2047	2368	2765	3222	3736	4310	4946
175	391	120	1536	1880	2310	2821	3415	4096	4867	375	839	120	2376	2774	3265	3833	4477	5200	6008
175	391	140	1763	2183	2714	3346	4083	4930	5891	375	839	140	2702	3179	3771	4459	5243	6128	7121
200	447	40	680	773	893	1029	1181	1349	1534	400	894	40	1042	1166	1321	1496	1687	1894	2117
200	447	60	931	1080	1266	1482	1725	1998	2301	400	894	60	1424	1608	1841	2104	2396	2715	3062
200	447	80	1173	1384	1646	1951	2299	2694	3136	400	894	80	1781	2036	2352	2712	3113	3557	4044
200	447	100	1414	1691	2036	2441	2908	3440	4040	400	894	100	2130	2457	2862	3326	3848	4428	5072
200	447	120	1654	2002	2438	2954	3552	4236	5011	400	894	120	2472	2875	3375	3951	4603	5334	6148
200	447	140	1887	2318	2855	3491	4231	5081	6044	400	894	140	2809	3293	3894	4590	5382	6274	7273

NOTE: Detector time constant at a reference velocity of 5 ft/sec.

For SI Units: 1 ft = 0.305 m

1000 BTU/sec = 1055 kW

Table B-4.1(dd) Installed Spacing of Heat Detector: 25 feet; t_g: 600 Seconds to 1000 Btu/sec; α: 0.003 Btu/sec³

τ	RTI	ΔT	\multicolumn{7}{c}{CEILING HEIGHT IN FEET}
			4.0	8.0	12.0	16.0	20.0	24.0	28.0
			\multicolumn{7}{c}{FIRE SIZE AT DETECTOR RESPONSE (BTU / SEC)}
25	56	40	211	283	373	479	603	744	904
25	56	60	313	441	601	789	1005	1254	1534
25	56	80	423	620	860	1143	1468	1841	2262
25	56	100	542	817	1148	1536	1983	2495	3074
25	56	120	669	1030	1460	1965	2546	3211	3964
25	56	140	804	1259	1797	2427	3153	3983	4923
50	112	40	283	353	442	548	672	814	975
50	112	60	404	529	683	869	1087	1336	1618
50	112	80	531	719	954	1233	1559	1933	2356
50	112	100	663	925	1250	1638	2084	2597	3178
50	112	120	801	1146	1571	2075	2655	3322	4076
50	112	140	946	1383	1915	2544	3269	4102	5044
75	168	40	345	417	509	617	742	885	1047
75	168	60	488	610	766	952	1170	1419	1702
75	168	80	630	814	1047	1326	1653	2027	2451
75	168	100	776	1034	1356	1738	2186	2701	3283
75	168	120	928	1265	1685	2183	2766	3434	4190
75	168	140	1085	1509	2036	2663	3387	4221	5166
100	224	40	402	477	572	683	811	956	1119
100	224	60	563	688	846	1034	1253	1504	1788
100	224	80	721	907	1141	1421	1747	2122	2548
100	224	100	885	1138	1458	1841	2290	2805	3389
100	224	120	1048	1383	1796	2295	2878	3546	4304
100	224	140	1218	1636	2158	2779	3507	4342	5288
125	280	40	454	533	632	747	878	1026	1191
125	280	60	633	761	923	1115	1336	1589	1874
125	280	80	807	996	1233	1514	1842	2219	2645
125	280	100	983	1240	1561	1946	2395	2911	3496
125	280	120	1162	1494	1910	2408	2991	3661	4420
125	280	140	1344	1762	2278	2900	3628	4464	5412
150	335	40	504	586	689	808	942	1094	1262
150	335	60	699	832	998	1193	1418	1673	1961
150	335	80	891	1082	1322	1607	1937	2315	2743
150	335	100	1079	1339	1663	2049	2500	3017	3604
150	335	120	1271	1605	2022	2521	3105	3776	4537
150	335	140	1466	1880	2399	3021	3750	4587	5536
175	391	40	555	637	743	866	1005	1160	1332
175	391	60	762	899	1070	1270	1498	1757	2047
175	391	80	966	1164	1409	1698	2031	2412	2842
175	391	100	1171	1435	1763	2152	2605	3124	3712
175	391	120	1376	1713	2133	2634	3219	3891	4654
175	391	140	1583	1999	2520	3143	3872	4710	5662
200	447	40	599	685	796	923	1066	1225	1400
200	447	60	822	964	1140	1344	1577	1839	2133
200	447	80	1042	1243	1494	1787	2124	2508	2941
200	447	100	1260	1527	1860	2254	2710	3231	3822
200	447	120	1478	1817	2242	2746	3334	4008	4772
200	447	140	1697	2114	2639	3265	3995	4835	5788

τ	RTI	ΔT	\multicolumn{7}{c}{CEILING HEIGHT IN FEET}
			4.0	8.0	12.0	16.0	20.0	24.0	28.0
			\multicolumn{7}{c}{FIRE SIZE AT DETECTOR RESPONSE (BTU / SEC)}
225	503	40	642	732	847	978	1126	1288	1467
225	503	60	884	1027	1208	1417	1654	1921	2218
225	503	80	1114	1320	1576	1874	2216	2603	3039
225	503	100	1345	1617	1956	2354	2814	3338	3931
225	503	120	1576	1919	2349	2857	3448	4124	4891
225	503	140	1809	2228	2757	3385	4118	4960	5915
250	559	40	684	777	896	1032	1183	1350	1533
250	559	60	938	1088	1273	1488	1730	2001	2302
250	559	80	1185	1395	1656	1960	2306	2698	3137
250	559	100	1429	1705	2049	2452	2916	3445	4040
250	559	120	1673	2019	2454	2967	3561	4241	5010
250	559	140	1908	2338	2872	3505	4241	5085	6042
275	615	40	724	822	944	1084	1240	1411	1597
275	615	60	992	1147	1338	1557	1804	2079	2384
275	615	80	1253	1468	1735	2044	2395	2791	3234
275	615	100	1510	1791	2141	2549	3018	3550	4150
275	615	120	1767	2116	2557	3076	3674	4357	5129
275	615	140	2011	2447	2986	3624	4363	5210	6170
300	671	40	764	865	991	1135	1295	1470	1661
300	671	60	1047	1205	1400	1625	1877	2157	2466
300	671	80	1320	1540	1812	2126	2483	2884	3331
300	671	100	1590	1874	2230	2645	3119	3656	4258
300	671	120	1850	2212	2659	3183	3786	4473	5248
300	671	140	2112	2553	3099	3742	4485	5336	6298
325	727	40	803	907	1037	1185	1349	1528	1723
325	727	60	1100	1261	1462	1691	1948	2233	2547
325	727	80	1386	1609	1887	2207	2570	2975	3427
325	727	100	1669	1956	2318	2739	3219	3760	4367
325	727	120	1938	2305	2759	3289	3897	4588	5367
325	727	140	2210	2657	3209	3858	4606	5461	6426
350	783	40	841	948	1082	1234	1402	1585	1783
350	783	60	1151	1316	1522	1756	2018	2308	2626
350	783	80	1451	1677	1961	2287	2655	3065	3522
350	783	100	1746	2036	2405	2832	3317	3863	4474
350	783	120	2024	2396	2857	3393	4007	4703	5485
350	783	140	2307	2759	3318	3973	4727	5585	6554
375	839	40	882	988	1126	1282	1453	1641	1843
375	839	60	1202	1371	1581	1820	2087	2382	2705
375	839	80	1514	1744	2034	2365	2739	3155	3616
375	839	100	1812	2115	2490	2923	3414	3966	4581
375	839	120	2108	2486	2953	3496	4116	4817	5603
375	839	140	2402	2859	3425	4087	4846	5709	6682
400	894	40	916	1027	1169	1329	1504	1696	1902
400	894	60	1251	1424	1639	1883	2155	2455	2782
400	894	80	1577	1810	2105	2443	2822	3243	3709
400	894	100	1884	2193	2574	3013	3510	4067	4688
400	894	120	2191	2575	3049	3598	4224	4930	5721
400	894	140	2495	2958	3531	4199	4965	5833	6809

NOTE: Detector time constant at a reference velocity of 5 ft/sec.

For SI Units: 1 ft = 0.305 m

1000 BTU/sec = 1055 kW

Table B-4.1(ee) Installed Spacing of Heat Detector: 30 ft.; t_g: 50 Seconds to 1000 Btu/sec; α: 0.400 Btu/sec^3

τ	RTI	ΔT	CEILING HEIGHT IN FEET							τ	RTI	ΔT	CEILING HEIGHT IN FEET						
			4.0	8.0	12.0	16.0	20.0	24.0	28.0				4.0	8.0	12.0	16.0	20.0	24.0	28.0
			FIRE SIZE AT DETECTOR RESPONSE (BTU / SEC)										FIRE SIZE AT DETECTOR RESPONSE (BTU / SEC)						
25	56	40	1541	1757	1963	2204	2475	2773	3096	225	503	40	5480	5821	6309	6882	7509	8174	8873
25	56	60	2013	2288	2589	2944	3344	3788	4276	225	503	60	7117	7642	8320	9106	9963	10875	11835
25	56	80	2472	2801	3207	3687	4231	4839	5512	225	503	80	8688	9332	10193	11189	12273	13430	14652
25	56	100	2925	3307	3828	4445	5148	5936	6813	225	503	100	10171	10936	11980	13185	14498	15902	17388
25	56	120	3311	3811	4458	5223	6098	7084	8183	225	503	120	11590	12477	13707	15122	16667	18321	20077
25	56	140	3718	4318	5100	6024	7084	8283	9621	225	503	140	12958	13973	15390	17018	18797	20707	22739
50	112	40	2231	2480	2735	3032	3361	3718	4101	250	559	40	5868	6207	6721	7327	7988	8689	9425
50	112	60	2986	3235	3592	4014	4484	4997	5550	250	559	60	7610	8154	8869	9698	10600	11560	12569
50	112	80	3587	3941	4414	4970	5590	6270	7009	250	559	80	9298	9962	10868	11915	13056	14271	15552
50	112	100	4208	4627	5222	5918	6698	7558	8499	250	559	100	10891	11674	12772	14038	15416	16888	18443
50	112	120	4810	5300	6023	6868	7819	8873	10031	250	559	120	12415	13320	14610	16095	17713	19444	21279
50	112	140	5293	5963	6823	7827	8959	10219	11610	250	559	140	13884	14914	16399	18105	19967	21961	24080
75	168	40	2867	3087	3379	3725	4106	4517	4955	275	615	40	6248	6582	7121	7758	8452	9188	9959
75	168	60	3727	4024	4437	4922	5458	6038	6659	275	615	60	8091	8651	9402	10272	11218	12224	13280
75	168	80	4572	4905	5441	6070	6768	7525	8340	275	615	80	9892	10572	11522	12620	13815	15086	16425
75	168	100	5231	5745	6416	7196	8061	9006	10029	275	615	100	11594	12391	13540	14865	16307	17844	19467
75	168	120	5958	6567	7375	8311	9354	10496	11739	275	615	120	13221	14137	15486	17039	18729	20535	22446
75	168	140	6664	7375	8324	9424	10652	12003	13479	275	615	140	14789	15828	17379	19161	21102	23180	25383
100	224	40	3365	3621	3952	4342	4771	5232	5719	300	671	40	6623	6947	7510	8176	8902	9672	10478
100	224	60	4462	4734	5193	5736	6333	6976	7661	300	671	60	8560	9135	9919	10830	11819	12870	13972
100	224	80	5310	5761	6363	7062	7832	8664	9554	300	671	80	10474	11166	12158	13306	14553	15879	17274
100	224	100	6213	6751	7492	8352	9300	10329	11434	300	671	100	12283	13089	14287	15670	17174	18775	20464
100	224	120	7081	7711	8596	9621	10755	11989	13321	300	671	120	14013	14933	16339	17957	19718	21596	23582
100	224	140	7923	8650	9684	10880	12207	13655	15225	300	671	140	15680	16719	18333	20188	22208	24366	26652
125	280	40	3870	4115	4479	4909	5383	5888	6423	325	727	40	6827	7290	7888	8584	9342	10144	10984
125	280	60	4990	5377	5889	6486	7141	7843	8588	325	727	60	9019	9606	10424	11373	12405	13499	14646
125	280	80	6091	6557	7214	7978	8817	9720	10682	325	727	80	11045	11746	12779	13974	15273	16652	18102
125	280	100	7137	7682	8488	9423	10450	11560	12747	325	727	100	12961	13769	15016	16455	18019	19683	21436
125	280	120	8143	8770	9728	10838	12060	13383	14805	325	727	120	14793	15710	17170	18853	20683	22632	24690
125	280	140	8954	9826	10946	12235	13658	15203	16869	325	727	140	16294	17578	19263	21190	23287	25524	27891
150	335	40	4248	4569	4970	5440	5954	6503	7082	350	783	40	7161	7635	8257	8982	9770	10605	11478
150	335	60	5589	5987	6541	7189	7898	8856	9459	350	783	60	9470	10067	10917	11905	12977	14114	15305
150	335	80	6837	7304	8013	8838	9743	10713	11744	350	783	80	11607	12312	13385	14627	15976	17408	18912
150	335	100	7873	8550	9423	10430	11533	12720	13986	350	783	100	13629	14435	15728	17223	18845	20570	22386
150	335	120	8963	9758	10793	11984	13291	14701	16210	350	783	120	15303	16459	17983	19729	21626	23645	25774
150	335	140	10016	10935	12134	13513	15029	16669	18430	350	783	140	17103	18426	20172	22170	24341	26656	29103
175	391	40	4673	5004	5436	5942	6496	7085	7706	375	839	40	7488	7973	8619	9371	10190	11055	11961
175	391	60	6166	6565	7160	7856	8616	9428	10285	375	839	60	9914	10518	11400	12424	13537	14715	15949
175	391	80	7411	8005	8771	9655	10622	11657	12754	375	839	80	12161	12867	13978	15266	16664	18147	19703
175	391	100	8667	9379	10311	11387	12563	13825	15166	375	839	100	14054	15076	16425	17974	19654	21439	23316
175	391	120	9870	10702	11805	13073	14462	15956	17550	375	839	120	16012	17201	18779	20587	22549	24636	26835
175	391	140	11031	11989	13264	14729	16336	18068	19921	375	839	140	17897	19256	21063	23129	25374	27765	30289
200	447	40	5082	5421	5881	6422	7012	7641	8301	400	894	40	7810	8304	8973	9753	10600	11497	12433
200	447	60	6726	7119	7752	8493	9303	10165	11075	400	894	60	10351	10960	11873	12934	14085	15304	16580
200	447	80	8060	8681	9496	10437	11464	12561	13721	400	894	80	12709	13411	14560	15893	17339	18872	20480
200	447	100	9432	10172	11162	12303	13549	14883	16298	400	894	100	14664	15714	17108	18710	20447	22291	24228
200	447	120	10744	11607	12774	14117	15585	17160	18837	400	894	120	16709	17929	19559	21427	23454	25608	27876
200	447	140	12010	12999	14347	15895	17589	19411	21355	400	894	140	18677	20070	21936	24070	26387	28853	31453

NOTE: Detector time constant at a reference velocity of 5 ft/sec.

For SI Units: 1 ft = 0.305 m

1000 BTU/sec = 1055 kW

Table B-4.1(ff)   Installed Spacing of Heat Detector: 30 feet; t_g: 150 Seconds to 1000 Btu/sec; α: 0.044 Btu/sec³

τ	RTI	ΔT	CEILING HEIGHT IN FEET								τ	RTI	ΔT	CEILING HEIGHT IN FEET						
			4.0	8.0	12.0	16.0	20.0	24.0	28.0					4.0	8.0	12.0	16.0	20.0	24.0	28.0
			FIRE SIZE AT DETECTOR RESPONSE (BTU / SEC)											FIRE SIZE AT DETECTOR RESPONSE (BTU / SEC)						
25	56	40	684	780	911	1066	1243	1443	1665		225	503	40	2301	2481	2727	3010	3319	3650	4002
25	56	60	919	1084	1300	1556	1850	2183	2555		225	503	60	3092	3339	3697	4107	4556	5040	5558
25	56	80	1150	1399	1715	2088	2518	3005	3550		225	503	80	3788	4151	4626	5169	5766	6413	7110
25	56	100	1387	1728	2157	2663	3245	3904	4641		225	503	100	4480	4938	5535	6217	6971	7791	8680
25	56	120	1629	2072	2626	3278	4026	4872	5819		225	503	120	5150	5709	6433	7262	8180	9185	10279
25	56	140	1877	2432	3122	3931	4858	5906	7077		225	503	140	5805	6468	7326	8308	9400	10601	11914
50	112	40	958	1073	1220	1391	1583	1796	2030		250	559	40	2461	2645	2903	3200	3524	3870	4237
50	112	60	1284	1465	1696	1965	2270	2613	2992		250	559	60	3312	3559	3933	4361	4829	5333	5871
50	112	80	1603	1856	2182	2563	3000	3492	4042		250	559	80	4047	4423	4917	5482	6102	6772	7492
50	112	100	1919	2251	2683	3192	3775	4436	5177		250	559	100	4786	5257	5877	6585	7364	8211	9126
50	112	120	2217	2654	3203	3852	4598	5445	6394		250	559	120	5501	6073	6823	7680	8627	9662	10783
50	112	140	2526	3066	3743	4544	5467	6514	7688		250	559	140	6198	6876	7762	8775	9898	11130	12472
75	168	40	1194	1323	1486	1675	1885	2116	2366		275	615	40	2616	2805	3075	3385	3722	4083	4465
75	168	60	1604	1795	2044	2334	2659	3019	3414		275	615	60	3479	3772	4162	4608	5095	5618	6175
75	168	80	2004	2257	2602	3004	3459	3967	4529		275	615	80	4299	4686	5200	5786	6428	7122	7865
75	168	100	2361	2716	3168	3694	4294	4967	5717		275	615	100	5083	5567	6209	6942	7747	8620	956
75	168	120	2732	3178	3745	4408	5166	6022	6978		275	615	120	5842	6427	7203	8087	9063	10126	11276
75	168	140	3100	3645	4337	5148	6078	7131	8311		275	615	140	6582	7272	8186	9230	10385	11647	13019
100	224	40	1409	1547	1726	1932	2161	2409	2678		300	671	40	2769	2960	3241	3564	3916	4291	4687
100	224	60	1901	2093	2362	2672	3019	3399	3814		300	671	60	3677	3980	4385	4848	5354	5895	6472
100	224	80	2335	2620	2988	3414	3891	4420	5001		300	671	80	4545	4942	5475	6082	6747	7463	8228
100	224	100	2773	3141	3616	4167	4789	5481	6248		300	671	100	5374	5869	6533	7290	8121	9020	9986
100	224	120	3203	3660	4251	4938	5716	6589	7560		300	671	120	6177	6772	7572	8485	9489	10581	11759
100	224	140	3628	4180	4895	5728	6677	7745	8937		300	671	140	6959	7659	8600	9674	10859	12153	13555
125	280	40	1611	1755	1948	2171	2417	2683	2969		325	727	40	2919	3112	3403	3739	4104	4493	4904
125	280	60	2148	2368	2657	2988	3356	3758	4193		325	727	60	3871	4183	4603	5083	5606	6166	6761
125	280	80	2662	2958	3349	3798	4300	4851	5454		325	727	80	4786	5193	5743	6371	7058	7796	8584
125	280	100	3159	3537	4037	4614	5261	5977	6765		325	727	100	5659	6164	6849	7630	8486	9411	10402
125	280	120	3647	4110	4728	5441	6245	7141	8131		325	727	120	6505	7109	7933	8873	9905	11025	12232
125	280	140	4092	4682	5424	6284	7257	8346	9556		325	727	140	7329	8036	9004	10108	11324	12649	14081
150	335	40	1806	1950	2157	2396	2658	2941	3244		350	783	40	3066	3260	3562	3910	4288	4691	5116
150	335	60	2398	2628	2934	3287	3676	4098	4555		350	783	60	4061	4382	4816	5312	5853	6431	7044
150	335	80	2972	3277	3689	4163	4689	5264	5890		350	783	80	5022	5437	6006	6654	7362	8122	8932
150	335	100	3528	3911	4436	5039	5712	6454	7266		350	783	100	5940	6452	7158	7963	8843	9793	10810
150	335	120	4029	4536	5181	5923	6754	7675	8689		350	783	120	6828	7439	8287	9253	10313	11461	12696
150	335	140	4548	5157	5928	6818	7819	8933	10164		350	783	140	7693	8405	9400	10533	11780	13135	14597
175	391	40	1996	2135	2355	2609	2888	3187	3507		375	839	40	3213	3405	3717	4077	4468	4884	5323
175	391	60	2637	2875	3199	3571	3981	4424	4901		375	839	60	4248	4576	5024	5537	6095	6691	7322
175	391	80	3271	3581	4014	4511	5061	5661	6310		375	839	80	5254	5677	6263	6931	7660	8441	9273
175	391	100	3839	4267	4816	5447	6146	6914	7751		375	839	100	6215	6734	7461	8289	9194	10168	11211
175	391	120	4417	4942	5614	6385	7245	8193	9233		375	839	120	7147	7762	8633	9625	10713	11889	13151
175	391	140	4982	5611	6411	7331	8362	9504	10761		375	839	140	7974	8766	9788	10950	12227	13612	15105
200	447	40	2137	2311	2544	2813	3108	3423	3759		400	894	40	3308	3546	3870	4242	4645	5074	5526
200	447	60	2867	3111	3452	3844	4273	4738	5235		400	894	60	4431	4766	5229	5757	6332	6945	7594
200	447	80	3521	3871	4326	4846	5420	6043	6716		400	894	80	5482	5912	6515	7203	7952	8755	9608
200	447	100	4165	4609	5182	5839	6565	7359	8222		400	894	100	6488	7011	7758	8609	9538	10537	11604
200	447	120	4790	5332	6030	6830	7720	8696	9762		400	894	120	7389	8077	8973	9991	11106	12309	13599
200	447	140	5400	6047	6876	7827	8888	10060	11344		400	894	140	8311	9121	10168	11359	12666	14081	15604

NOTE: Detector time constant at a reference velocity of 5 ft/sec.

For SI Units: 1 ft = 0.305 m

1000 BTU/sec = 1055 kW

Table B-4.1(gg)   Installed Spacing of Heat Detector: 30 feet; t_g: 300 Seconds to 1000 Btu/sec; α: 0.011 Btu/sec^3

τ	RTI	ΔT	CEILING HEIGHT IN FEET FIRE SIZE AT DETECTOR RESPONSE (BTU / SEC) 4.0	8.0	12.0	16.0	20.0	24.0	28.0
25	56	40	422	511	625	761	916	1093	1290
25	56	60	594	753	952	1187	1456	1761	2101
25	56	80	772	1018	1316	1666	2068	2520	3026
25	56	100	958	1302	1717	2194	2743	3361	4052
25	56	120	1154	1606	2147	2766	3476	4275	5169
25	56	140	1359	1930	2605	3378	4262	5257	6369
50	112	40	587	675	793	931	1089	1267	1466
50	112	60	805	961	1159	1392	1661	1965	2307
50	112	80	1023	1259	1552	1898	2297	2749	3256
50	112	100	1249	1572	1974	2447	2993	3611	4304
50	112	120	1478	1901	2424	3038	3745	4545	5441
50	112	140	1714	2251	2901	3668	4549	5545	6660
75	168	40	723	818	943	1089	1253	1437	1640
75	168	60	985	1148	1352	1591	1864	2171	2515
75	168	80	1248	1484	1780	2128	2528	2982	3491
75	168	100	1510	1830	2230	2703	3248	3867	4561
75	168	120	1775	2187	2705	3315	4020	4820	5717
75	168	140	2036	2558	3203	3964	4842	5838	6955
100	224	40	846	949	1081	1235	1408	1599	1809
100	224	60	1152	1320	1533	1780	2060	2374	2723
100	224	80	1453	1693	1997	2352	2757	3215	3728
100	224	100	1753	2072	2478	2955	3503	4124	4821
100	224	120	2041	2459	2979	3592	4297	5098	5998
100	224	140	2337	2856	3501	4261	5138	6135	7253
125	280	40	959	1069	1210	1373	1554	1753	1971
125	280	60	1307	1481	1703	1960	2249	2571	2927
125	280	80	1646	1889	2203	2567	2981	3446	3963
125	280	100	1968	2300	2716	3201	3755	4381	5082
125	280	120	2294	2717	3245	3864	4574	5378	6280
125	280	140	2619	3142	3792	4556	5436	6434	7555
150	335	40	1068	1183	1332	1503	1693	1901	2126
150	335	60	1454	1633	1864	2131	2430	2761	3126
150	335	80	1831	2075	2399	2774	3198	3671	4196
150	335	100	2178	2518	2944	3439	4003	4636	5342
150	335	120	2534	2964	3502	4130	4847	5657	6564
150	335	140	2888	3416	4075	4847	5732	6734	7859
175	391	40	1171	1292	1448	1627	1826	2042	2276
175	391	60	1595	1778	2019	2296	2605	2946	3319
175	391	80	1989	2253	2588	2974	3409	3892	4425
175	391	100	2379	2726	3164	3671	4245	4887	5601
175	391	120	2763	3201	3751	4389	5116	5934	6847
175	391	140	3145	3679	4350	5131	6025	7034	8163
200	447	40	1270	1395	1559	1747	1954	2179	2421
200	447	60	1733	1917	2167	2455	2774	3125	3508
200	447	80	2152	2424	2770	3168	3614	4107	4650
200	447	100	2572	2927	3377	3896	4481	5133	5856
200	447	120	2985	3429	3991	4642	5380	6207	7129
200	447	140	3394	3934	4617	5410	6313	7331	8467

τ	RTI	ΔT	CEILING HEIGHT IN FEET FIRE SIZE AT DETECTOR RESPONSE (BTU / SEC) 4.0	8.0	12.0	16.0	20.0	24.0	28.0
225	503	40	1367	1495	1666	1862	2077	2311	2562
225	503	60	1849	2051	2311	2608	2938	3299	3692
225	503	80	2310	2589	2946	3356	3813	4318	4871
225	503	100	2759	3121	3583	4115	4712	5375	6108
225	503	120	3200	3651	4226	4889	5639	6477	7407
225	503	140	3636	4181	4877	5683	6598	7625	8770
250	559	40	1460	1592	1770	1974	2197	2439	2698
250	559	60	1973	2181	2450	2758	3098	3469	3872
250	559	80	2464	2749	3117	3539	4007	4523	5088
250	559	100	2941	3309	3783	4328	4939	5613	6357
250	559	120	3409	3865	4453	5131	5893	6743	7684
250	559	140	3872	4421	5131	5951	6878	7916	9070
275	615	40	1552	1686	1871	2082	2314	2564	2831
275	615	60	2093	2307	2585	2903	3253	3635	4048
275	615	80	2613	2905	3284	3717	4197	4725	5300
275	615	100	3118	3492	3979	4536	5159	5846	6601
275	615	120	3614	4074	4676	5367	6143	7005	7956
275	615	140	4074	4654	5378	6213	7153	8204	9368
300	671	40	1642	1778	1969	2188	2427	2685	2961
300	671	60	2211	2430	2717	3044	3405	3796	4219
300	671	80	2760	3057	3446	3890	4383	4922	5509
300	671	100	3293	3671	4169	4740	5375	6075	6842
300	671	120	3789	4278	4893	5598	6388	7263	8226
300	671	140	4292	4882	5621	6470	7424	8487	9663
325	727	40	1732	1867	2065	2291	2538	2804	3088
325	727	60	2325	2550	2845	3182	3553	3955	4388
325	727	80	2903	3205	3604	4061	4565	5116	5713
325	727	100	3464	3845	4356	4939	5588	6300	7079
325	727	120	3979	4477	5105	5825	6628	7516	8492
325	727	140	4505	5105	5858	6722	7691	8767	9955
350	783	40	1801	1954	2158	2392	2646	2920	3212
350	783	60	2438	2667	2971	3318	3698	4110	4553
350	783	80	3044	3350	3760	4227	4743	5305	5915
350	783	100	3605	4016	4538	5135	5796	6522	7313
350	783	120	4165	4672	5314	6047	6865	7767	8755
350	783	140	4714	5324	6091	6970	7954	9044	10245
375	839	40	1882	2039	2250	2490	2753	3034	3333
375	839	60	2549	2782	3094	3450	3840	4262	4715
375	839	80	3183	3492	3912	4390	4918	5492	6113
375	839	100	3764	4183	4717	5326	6001	6740	7543
375	839	120	4347	4864	5518	6266	7098	8013	9014
375	839	140	4918	5538	6319	7213	8212	9317	10531
400	894	40	1962	2123	2340	2587	2857	3146	3453
400	894	60	2657	2894	3215	3580	3980	4412	4875
400	894	80	3320	3631	4061	4551	5090	5676	6308
400	894	100	3920	4347	4893	5515	6203	6954	7770
400	894	120	4526	5051	5719	6481	7327	8256	9271
400	894	140	5119	5748	6544	7453	8467	9586	10814

NOTE: Detector time constant at a reference velocity of 5 ft/sec.

For SI Units: 1 ft = 0.305 m
1000 BTU/sec = 1055 kW

Table B-4.1(hh)   Installed Spacing of Heat Detector: 30 feet; t_g: 500 Seconds to 1000 Btu/sec; α: 0.004 Btu/sec^3

τ	RTI	ΔT	CEILING HEIGHT IN FEET						
			4.0	8.0	12.0	16.0	20.0	24.0	28.0
			FIRE SIZE AT DETECTOR RESPONSE (BTU / SEC)						
25	56	40	310	397	506	634	781	947	1132
25	56	60	455	615	810	1034	1292	1582	1906
25	56	80	609	858	1153	1490	1877	2312	2798
25	56	100	777	1124	1531	1997	2527	3125	3793
25	56	120	955	1413	1943	2552	3237	4013	4880
25	56	140	1145	1722	2385	3145	4001	4970	6052
50	112	40	416	501	608	735	881	1047	1232
50	112	60	591	742	930	1153	1409	1699	2024
50	112	80	770	1002	1288	1623	2008	2444	2931
50	112	100	958	1285	1683	2142	2672	3270	3939
50	112	120	1154	1586	2106	2705	3393	4170	5039
50	112	140	1359	1906	2559	3309	4169	5138	6222
75	168	40	512	595	705	834	981	1148	1334
75	168	60	712	862	1050	1272	1528	1818	2144
75	168	80	918	1144	1426	1759	2142	2578	3066
75	168	100	1127	1443	1832	2291	2819	3417	4087
75	168	120	1344	1762	2268	2866	3552	4329	5199
75	168	140	1567	2095	2736	3480	4338	5308	6394
100	224	40	594	682	795	928	1078	1248	1436
100	224	60	823	975	1166	1390	1647	1939	2266
100	224	805	1053	1280	1562	1894	2278	2714	3202
100	224	100	1285	1598	1984	2441	2968	3566	4237
100	224	120	1522	1930	2434	3028	3713	4490	5361
100	224	140	1765	2277	2910	3653	4509	5480	6567
125	280	40	671	763	881	1018	1173	1345	1537
125	280	60	927	1082	1277	1505	1765	2059	2388
125	280	80	1180	1409	1694	2029	2414	2850	3340
125	280	100	1435	1747	2134	2591	3118	3716	4389
125	280	120	1693	2096	2599	3191	3875	4652	5524
125	280	140	1956	2459	3087	3828	4683	5653	6742
150	335	40	744	839	962	1104	1263	1441	1636
150	335	60	1023	1184	1384	1616	1881	2178	2509
150	335	80	1301	1533	1822	2161	2549	2987	3479
150	335	100	1578	1890	2281	2740	3268	3868	4541
150	335	120	1857	2258	2761	3354	4038	4816	5689
150	335	140	2128	2637	3264	4004	4858	5828	6918
175	391	40	813	912	1040	1187	1351	1533	1733
175	391	60	1118	1281	1487	1725	1994	2295	2630
175	391	80	1417	1652	1947	2290	2682	3124	3618
175	391	100	1715	2029	2424	2887	3418	4020	4695
175	391	120	2004	2414	2921	3516	4202	4981	5855
175	391	140	2300	2809	3438	4179	5033	6004	7095
200	447	40	883	982	1115	1267	1436	1623	1827
200	447	60	1208	1375	1586	1830	2104	2410	2749
200	447	80	1529	1767	2067	2416	2813	3259	3757
200	447	100	1840	2163	2564	3031	3566	4171	4849
200	447	120	2153	2566	3077	3677	4365	5146	6022
200	447	140	2466	2977	3610	4353	5209	6181	7274

τ	RTI	ΔT	CEILING HEIGHT IN FEET						
			4.0	8.0	12.0	16.0	20.0	24.0	28.0
			FIRE SIZE AT DETECTOR RESPONSE (BTU / SEC)						
225	503	40	943	1050	1187	1344	1518	1710	1919
225	503	60	1295	1466	1682	1932	2212	2523	2867
225	503	80	1638	1878	2185	2540	2942	3393	3894
225	503	100	1966	2293	2700	3173	3713	4322	5003
225	503	120	2296	2714	3231	3835	4527	5311	6190
225	503	140	2626	3142	3779	4525	5384	6359	7453
250	559	40	1004	1116	1257	1419	1599	1795	2009
250	559	60	1379	1554	1776	2031	2317	2634	2983
250	559	80	1745	1986	2299	2661	3069	3525	4031
250	559	100	2088	2420	2833	3313	3858	4471	5156
250	559	120	2436	2858	3381	3991	4688	5475	6357
250	559	140	2783	3302	3945	4696	5559	6536	7633
275	615	40	1066	1179	1325	1492	1677	1878	2097
275	615	60	1462	1639	1867	2129	2421	2743	3097
275	615	80	1837	2092	2411	2779	3193	3655	4166
275	615	100	2206	2543	2964	3450	4001	4620	5309
275	615	120	2572	2998	3529	4145	4847	5639	6525
275	615	140	2935	3458	4108	4865	5732	6713	7813
300	671	40	1125	1241	1392	1564	1753	1960	2183
300	671	60	1543	1723	1956	2224	2522	2850	3210
300	671	80	1935	2194	2520	2895	3316	3784	4300
300	671	100	2322	2664	3091	3585	4142	4767	5460
300	671	120	2704	3136	3673	4296	5005	5802	6691
300	671	140	3084	3612	4268	5032	5904	6890	7993
325	727	40	1183	1302	1457	1633	1828	2039	2268
325	727	60	1623	1804	2043	2317	2621	2955	3320
325	727	80	2031	2295	2627	3009	3436	3910	4433
325	727	100	2435	2782	3217	3717	4282	4912	5611
325	727	120	2834	3271	3815	4446	5161	5963	6858
325	727	140	3230	3762	4426	5197	6075	7065	8172
350	783	40	1239	1361	1520	1701	1901	2117	2351
350	783	60	1701	1884	2128	2408	2718	3059	3430
350	783	80	2125	2393	2732	3121	3555	4035	4564
350	783	100	2546	2898	3340	3848	4419	5056	5761
350	783	120	2962	3403	3955	4593	5315	6124	7023
350	783	140	3374	3910	4581	5360	6244	7240	8352
375	839	40	1295	1419	1582	1768	1973	2194	2433
375	839	60	1766	1962	2212	2498	2814	3160	3527
375	839	80	2217	2490	2835	3230	3672	4159	4693
375	839	100	2655	3011	3460	3976	4555	5198	5909
375	839	120	3087	3532	4092	4738	5467	6283	7188
375	839	140	3515	4054	4734	5520	6412	7414	8531
400	894	40	1349	1476	1643	1834	2043	2270	2513
400	894	60	1839	2038	2294	2586	2908	3260	3643
400	894	80	2307	2584	2936	3339	3787	4280	4821
400	894	100	2762	3123	3579	4102	4688	5339	6056
400	894	120	3210	3660	4228	4881	5618	6440	7351
400	894	140	3654	4197	4885	5679	6578	7586	8709

NOTE: Detector time constant at a reference velocity of 5 ft/sec.

For SI Units: 1 ft = 0.305 m

1000 BTU/sec = 1055 kW

Table B-4.1(ii) Installed Spacing of Heat Detector: 30 feet; t_g: 600 Seconds to 1000 Btu/sec; α: 0.003 Btu/sec^3

τ	RTI	ΔT	4.0	8.0	12.0	16.0	20.0	24.0	28.0		τ	RTI	ΔT	4.0	8.0	12.0	16.0	20.0	24.0	28.0
			CEILING HEIGHT IN FEET											CEILING HEIGHT IN FEET						
			FIRE SIZE AT DETECTOR RESPONSE (BTU / SEC)											FIRE SIZE AT DETECTOR RESPONSE (BTU / SEC)						
25	56	40	281	369	476	602	747	909	1091		225	503	40	830	930	1058	1205	1369	1550	1749
25	56	60	419	579	773	995	1249	1535	1855		225	503	60	1144	1308	1514	1751	2019	2319	2651
25	56	80	568	817	1110	1445	1826	2257	2737		225	503	80	1452	1686	1981	2323	2713	3152	3643
25	56	100	730	1079	1484	1948	2470	3062	3723		225	503	100	1758	2070	2465	2925	3453	4051	4721
25	56	120	906	1364	1890	2494	3174	3943	4802		225	503	120	2053	2462	2967	3559	4240	5013	5882
25	56	140	1091	1669	2328	3082	3932	4893	5967		225	503	140	2356	2863	3489	4225	5073	6038	7122
50	112	40	372	455	561	685	828	991	1173		250	559	40	887	987	1119	1270	1438	1623	1826
50	112	60	534	685	872	1092	1345	1631	1952		250	559	60	1217	1384	1595	1837	2110	2413	2750
50	112	80	703	938	1221	1554	1935	2365	2847		250	559	80	1543	1780	2080	2427	2821	3263	3757
50	112	100	883	1211	1608	2065	2589	3181	3844		250	559	100	1858	2180	2579	3044	3575	4176	4848
50	112	120	1070	1506	2024	2620	3303	4072	4933		250	559	120	2175	2587	3096	3691	4375	5150	6020
50	112	140	1268	1820	2471	3217	4070	5032	6107		250	559	140	2492	3001	3631	4369	5219	6185	7270
75	168	40	451	536	642	767	911	1074	1257		275	615	40	937	1042	1178	1333	1506	1695	1901
75	168	60	639	787	972	1191	1443	1729	2050		275	615	60	1289	1459	1674	1921	2198	2507	2847
75	168	80	829	1056	1335	1665	2045	2475	2957		275	615	80	1632	1872	2176	2528	2927	3374	3871
75	168	100	1026	1345	1731	2187	2710	3302	3966		275	615	100	1961	2287	2691	3161	3697	4301	4976
75	168	120	1230	1651	2161	2752	3433	4203	5066		275	615	120	2293	2708	3222	3822	4509	5287	6159
75	168	140	1443	1975	2616	3358	4210	5172	6249		275	615	140	2624	3137	3770	4512	5364	6332	7419
100	224	40	527	610	719	847	993	1157	1341		300	671	40	987	1096	1236	1395	1572	1765	1976
100	224	60	735	884	1070	1289	1542	1828	2150		300	671	60	1358	1531	1751	2003	2285	2598	2943
100	224	80	946	1171	1449	1777	2156	2587	3070		300	671	80	1720	1961	2271	2628	3032	3483	3984
100	224	100	1162	1473	1857	2310	2833	3425	4089		300	671	100	2062	2392	2802	3277	3817	4424	5103
100	224	120	1383	1792	2295	2886	3565	4336	5199		300	671	120	2408	2827	3347	3951	4642	5423	6298
100	224	140	1611	2131	2760	3500	4351	5314	6392		300	671	140	2753	3269	3907	4653	5509	6479	7568
125	280	40	594	680	793	924	1072	1239	1425		325	727	40	1037	1149	1292	1456	1637	1834	2049
125	280	60	825	976	1165	1386	1640	1928	2250		325	727	60	1427	1602	1827	2084	2371	2688	3037
125	280	80	1056	1281	1561	1890	2269	2700	3183		325	727	80	1796	2048	2364	2726	3135	3591	4096
125	280	100	1290	1600	1983	2435	2956	3549	4214		325	727	100	2160	2495	2910	3391	3936	4548	5229
125	280	120	1529	1933	2432	3020	3699	4469	5334		325	727	120	2521	2944	3469	4079	4775	5559	6437
125	280	140	1773	2281	2907	3644	4494	5456	6536		325	727	140	2879	3399	4043	4794	5653	6626	7717
150	335	40	657	747	863	998	1150	1320	1508		350	783	40	1087	1201	1348	1515	1700	1902	2120
150	335	60	911	1064	1256	1481	1737	2027	2351		350	783	60	1494	1671	1901	2163	2455	2777	3131
150	335	80	1161	1388	1670	2001	2381	2813	3298		350	783	80	1878	2134	2454	2823	3237	3698	4207
150	335	100	1413	1723	2107	2559	3081	3674	4339		350	783	100	2257	2595	3016	3503	4053	4670	5355
150	335	120	1669	2070	2568	3156	3834	4604	5469		350	783	120	2631	3059	3590	4205	4906	5694	6576
150	335	140	1930	2431	3054	3789	4637	5600	6681		350	783	140	3003	3527	4177	4932	5796	6773	7867
175	391	40	717	810	930	1069	1225	1398	1590		375	839	40	1135	1251	1402	1573	1762	1968	2191
175	391	60	990	1148	1345	1573	1833	2126	2452		375	839	60	1560	1740	1973	2241	2538	2865	3223
175	391	80	1261	1491	1776	2110	2493	2926	3413		375	839	80	1958	2218	2544	2918	3338	3804	4318
175	391	100	1532	1842	2229	2683	3205	3799	4466		375	839	100	2351	2694	3121	3613	4169	4791	5481
175	391	120	1805	2204	2703	3291	3969	4740	5606		375	839	120	2740	3171	3708	4330	5036	5829	6714
175	391	140	2073	2578	3201	3935	4782	5745	6827		375	839	140	3125	3653	4308	5070	5939	6919	8016
200	447	40	774	871	995	1138	1298	1475	1670		400	894	40	1182	1301	1455	1630	1824	2034	2260
200	447	60	1069	1229	1430	1663	1927	2223	2552		400	894	60	1625	1806	2045	2317	2619	2951	3314
200	447	80	1358	1590	1880	2217	2603	3039	3528		400	894	80	2037	2300	2632	3012	3437	3908	4427
200	447	100	1646	1958	2348	2805	3330	3925	4593		400	894	100	2444	2791	3224	3722	4284	4911	5605
200	447	120	1929	2335	2836	3426	4105	4877	5744		400	894	120	2846	3282	3825	4453	5164	5962	6851
200	447	140	2217	2723	3346	4080	4928	5891	6974		400	894	140	3245	3777	4438	5206	6080	7065	8165

NOTE: Detector time constant at a reference velocity of 5 ft/sec.

For SI Units: 1 ft = 0.305 m

1000 BTU/sec = 1055 kW

Table B-4.1(jj) Installed Spacing of Heat Detector: 50 ft.; t_g: 50 Seconds to 1000 Btu/sec; α: 0.400 Btu/sec^3

τ	RTI	ΔT	CEILING HEIGHT IN FEET — FIRE SIZE AT DETECTOR RESPONSE (BTU / SEC) 4.0	8.0	12.0	16.0	20.0	24.0	28.0
25	56	40	3499	3611	3846	4155	4512	4907	5336
25	56	60	4555	4723	5115	5604	6162	6780	7454
25	56	80	5425	5806	6382	7078	7867	8740	9695
25	56	100	6348	6890	7670	8597	9645	10804	12073
25	56	120	7255	7981	8987	10169	11502	12976	14589
25	56	140	8154	9087	10339	11799	13441	15255	17239
50	112	40	4909	5030	5295	5650	6060	6511	6996
50	112	60	6378	6566	6990	7530	8145	8819	9549
50	112	80	7744	8031	8633	9376	10217	11141	12143
50	112	100	9044	9459	10257	11224	12313	13511	14814
50	112	120	10300	10869	11881	13089	14449	15946	17578
50	112	140	11383	12270	13513	14982	16634	18454	20442
75	168	40	6159	6216	6507	6906	7368	7874	8417
75	168	60	7858	8112	8577	9173	9851	10593	11390
75	168	80	9536	9909	10558	11365	12277	13273	14347
75	168	100	11130	11646	12493	13527	14690	15963	17337
75	168	120	12669	13346	14407	15683	17116	18685	20385
75	168	140	14167	15023	16312	17847	19567	21454	23502
100	224	40	7107	7258	7585	8027	8537	9096	9692
100	224	60	9264	9497	9999	10649	11389	12196	13060
100	224	80	11254	11597	12290	13161	14144	15216	16366
100	224	100	13136	13616	14514	15619	16861	18215	19671
100	224	120	14944	15582	16698	18051	19567	21222	23005
100	224	140	16700	17513	18859	20474	22281	24255	26387
125	280	40	8093	8224	8576	9057	9613	10220	10868
125	280	60	10578	10773	11309	12011	12809	13679	14607
125	280	80	12863	13156	13891	14823	15876	17021	18247
125	280	100	15016	15438	16384	17559	18880	20317	21856
125	280	120	17078	17654	18823	20254	21856	23600	25472
125	280	140	19076	19822	21227	22925	24823	26891	29115
150	335	40	9028	9128	9503	10022	10620	11273	11969
150	335	60	11831	11970	12538	13288	14142	15072	16063
150	335	80	14401	14620	15394	16385	17505	18722	20021
150	335	100	16567	17143	18143	19387	20784	22302	23924
150	335	120	18854	19594	20824	22331	24018	25851	27814
150	335	140	21069	21988	23459	25240	27230	29393	31714
175	391	40	9927	9984	10381	10935	11574	12271	13013
175	391	60	12795	13093	13702	14500	15407	16394	17445
175	391	80	15570	15997	16820	17869	19053	20339	21709
175	391	100	18190	18766	19814	21124	22596	24192	25896
175	391	120	20701	21442	22727	24308	26078	27999	30052
175	391	140	23131	24052	25583	27445	29526	31784	34201
200	447	40	10597	10789	11218	11806	12484	13223	14009
200	447	60	13882	14175	14815	15658	16617	17659	18767
200	447	80	16902	17322	18185	19288	20535	21888	23327
200	447	100	19748	20319	21414	22788	24332	26005	27789
200	447	120	22474	23212	24549	26202	28054	30060	32202
200	447	140	25110	26028	27618	29560	31730	34082	36596

τ	RTI	ΔT	CEILING HEIGHT IN FEET — FIRE SIZE AT DETECTOR RESPONSE (BTU / SEC) 4.0	8.0	12.0	16.0	20.0	24.0	28.0
225	503	40	11385	11571	12023	12643	13359	14138	14966
225	503	60	14931	15215	15886	16771	17780	18875	20039
225	503	80	18187	18598	19498	20654	21962	23379	24886
225	503	100	21253	21815	22954	24389	26004	27752	29613
225	503	120	24186	24916	26304	28028	29959	32049	34277
225	503	140	27020	27932	29579	31599	33857	36301	38909
250	559	40	12148	12327	12799	13451	14203	15021	15890
250	559	60	15948	16220	16919	17847	18904	20050	21268
250	559	80	19433	19830	20766	21974	23341	24820	26393
250	559	100	27713	23260	24443	25938	27621	29442	31379
250	559	120	25848	26564	28002	29794	31802	33974	36287
250	559	140	28875	29774	31477	33573	35915	38450	41151
275	615	40	12890	13059	13552	14234	15020	15876	16785
275	615	60	16939	17194	17922	18890	19993	21189	22460
275	615	80	20648	21026	21997	23255	24678	26219	27855
275	615	100	24135	24663	25887	27442	29191	31083	33094
275	615	120	27466	28163	29649	31508	33591	35844	38240
275	615	140	30681	31562	33318	35489	37915	40539	43332
300	671	40	13614	13770	14283	14994	15815	16707	17655
300	671	60	17906	18141	18896	19903	21052	22297	23618
300	671	80	21835	22189	23194	24501	25979	27579	29278
300	671	100	25525	26028	27293	28904	30719	32680	34763
300	671	120	29048	29720	31253	33176	35334	37665	40143
300	671	140	32446	33302	35111	37354	39863	42573	45456
325	727	40	14322	14463	14995	15735	16589	17517	18502
325	727	60	18854	19065	19846	20892	22084	23377	24748
325	727	80	22998	23323	24361	25715	27248	28906	30665
325	727	100	26887	27359	28663	30331	32208	34238	36392
325	727	120	30598	31238	32817	34804	37034	39442	42000
325	727	140	34176	35000	36861	39175	41763	44559	47531
350	783	40	15017	15141	15691	16459	17344	18307	19329
350	783	60	19786	19968	20774	21857	23093	24432	25851
350	783	80	24142	24431	25501	26902	28487	30203	32021
350	783	100	28226	28660	30003	31725	33665	35761	37984
350	783	120	32121	32723	34346	36395	38695	41179	43815
350	783	140	35876	36660	38571	40954	43622	46502	49561
375	839	40	15700	15803	16371	17166	18083	19080	20137
375	839	60	20703	20851	21681	22801	24079	25463	26930
375	839	80	25267	25516	26617	28063	29701	31471	33348
375	839	100	29544	29934	31314	33089	35090	37251	39543
375	839	120	33621	34176	35842	37953	40322	42880	45594
375	839	140	37550	38285	40245	42697	45442	48404	51549
400	894	40	16373	16452	17038	17859	18807	19837	20929
400	894	60	21607	21717	22571	23727	25046	26475	27988
400	894	80	26378	26580	27711	29201	30890	32715	34649
400	894	100	30845	31184	32600	34427	36488	38713	41071
400	894	120	35101	35602	37310	39480	41918	44549	47338
400	894	140	39201	39880	41887	44406	47227	50270	53499

NOTE: Detector time constant at a reference velocity of 5 ft/sec.

For SI Units: 1 ft = 0.305 m
1000 BTU/sec = 1055 kW

Table B-4.1(kk) Installed Spacing of Heat Detector: 50 feet; t_g: 150 Seconds to 1000 Btu/sec; α: 0.044 Btu/sec³

τ	RTI	ΔT	4.0	8.0	12.0	16.0	20.0	24.0	28.0
			\multicolumn{7}{c}{CEILING HEIGHT IN FEET — FIRE SIZE AT DETECTOR RESPONSE (BTU / SEC)}
25	56	40	1494	1637	1837	2075	2342	2639	2963
25	56	60	2041	2305	2663	3076	3539	4049	4605
25	56	80	2568	3009	3555	4177	4868	5625	6449
25	56	100	3120	3754	4516	5373	6319	7353	8475
25	56	120	3687	4541	5542	6658	7883	9219	10667
25	56	140	4272	5369	6630	8025	9552	11213	13012
50	112	40	2075	2207	2409	2651	2924	3225	3553
50	112	60	2794	3039	3386	3792	4248	4753	5304
50	112	80	3496	3880	4398	4998	5672	6417	7232
50	112	100	4193	4741	5456	6279	7200	8217	9328
50	112	120	4893	5628	6564	7635	8831	10147	11583
50	112	140	5571	6544	7724	9066	10559	12199	13986
75	168	40	2572	2696	2907	3163	3450	3765	4105
75	168	60	3460	3681	4034	4450	4917	5431	5990
75	168	80	4274	4656	5174	5778	6455	7202	8018
75	168	100	5097	5637	6343	7160	8076	9087	10194
75	168	120	5914	6633	7549	8603	9784	11089	12517
75	168	140	6730	7647	8794	10109	11580	13204	14980
100	224	40	3026	3136	3358	3628	3932	4263	4621
100	224	60	4020	4262	4627	5059	5543	6072	6647
100	224	80	4991	5364	5892	6510	7201	7960	8787
100	224	100	5937	6462	7174	8000	8924	9942	11054
100	224	120	6869	7564	8480	9538	10721	12026	13453
100	224	140	7793	8678	9817	11127	12593	14211	15983
125	280	40	3409	3542	3775	4060	4381	4730	5105
125	280	60	4563	4802	5179	5629	6132	6681	7274
125	280	80	5658	6024	6566	7201	7911	8688	9531
125	280	100	6721	7233	7957	8800	9740	10773	11898
125	280	120	7762	8439	9364	10434	11630	12945	14381
125	280	140	8790	9650	10794	12112	13585	15208	16983
150	335	40	3791	3924	4167	4468	4805	5172	5564
150	335	60	5076	5310	5701	6169	6692	7261	7874
150	335	80	6292	6648	7204	7859	8590	9388	10251
150	335	100	7466	7962	8701	9564	10525	11577	12720
150	335	120	8612	9268	10207	11296	12510	13842	15293
150	335	140	9741	10574	11729	13063	14551	16188	17973
175	391	40	4156	4286	4540	4855	5209	5593	6003
175	391	60	5567	5794	6198	6684	7227	7817	8451
175	391	80	6900	7241	7812	8489	9241	10062	10947
175	391	100	8183	8658	9414	10298	11282	12357	13520
175	391	120	9367	10060	11016	12126	13362	14716	16185
175	391	140	10583	11457	12628	13983	15491	17146	18948
200	447	40	4507	4633	4897	5226	5596	5996	6424
200	447	60	6041	6257	6674	7179	7741	8352	9007
200	447	80	7427	7810	8397	9094	9870	10713	11621
200	447	100	8795	9326	10099	11005	12014	13113	14300
200	447	120	10128	10821	11795	12929	14189	15567	17059
200	447	140	11435	12307	13496	14874	16406	18084	19906
225	503	40	4846	4967	5241	5584	5969	6386	6830
225	503	60	6504	6703	7133	7655	8238	8869	9545
225	503	80	7974	8358	8961	9679	10478	11345	12276
225	503	100	9439	9970	10761	11690	12723	13848	15060
225	503	120	10862	11556	12549	13707	14993	16396	17913
225	503	140	12256	13128	14337	15739	17297	19000	20846
250	559	40	5177	5289	5573	5929	6329	6762	7224
250	559	60	6877	7133	7577	8117	8719	9371	10068
250	559	80	8503	8889	9508	10246	11068	11958	12914
250	559	100	10062	10594	11402	12355	13413	14563	15801
250	559	120	11574	12268	13280	14463	15776	17206	18749
250	559	140	13053	13924	15153	16582	18167	19897	21768
275	615	40	5500	5602	5895	6264	6679	7128	7606
275	615	60	7295	7552	8009	8565	9187	9859	10577
275	615	80	9018	9404	10039	10798	11642	12556	13535
275	615	100	10668	11200	12026	13002	14086	15262	16525
275	615	120	12266	12960	13992	15200	16540	17997	19567
275	615	140	13828	14698	15948	17403	19017	20775	22674
300	671	40	5818	5907	6209	6590	7020	7484	7977
300	671	60	7702	7959	8428	9002	9642	10334	11072
300	671	80	9521	9906	10556	11335	12201	13139	14143
300	671	100	11259	11790	12634	13633	14742	15944	17234
300	671	120	12942	13635	14685	15919	17286	18772	20370
300	671	140	14584	15453	16724	18205	19848	21635	23563
325	727	40	6055	6202	6514	6908	7352	7831	8340
325	727	60	8100	8356	8838	9428	10087	10799	11557
325	727	80	10011	10396	11061	11860	12748	13710	14737
325	727	100	11837	12367	13228	14250	15384	16612	17928
325	727	120	13602	14293	15363	16622	18017	19531	21157
325	727	140	15323	16190	17482	18991	20663	22480	24437
350	783	40	6344	6491	6813	7219	7676	8170	8695
350	783	60	8489	8745	9238	9845	10522	11253	12031
350	783	80	10492	10875	11554	12373	13284	14268	15319
350	783	100	12403	12930	13808	14853	16012	17267	18610
350	783	120	14250	14937	16027	17310	18733	20276	21930
350	783	140	16048	16911	18224	19760	21462	23309	25296
375	839	40	6628	6775	7105	7523	7994	8503	9042
375	839	60	8871	9125	9631	10253	10948	11698	12496
375	839	80	10964	11344	12038	12876	13808	14815	15890
375	839	100	12960	13482	14377	15445	16629	17909	19278
375	839	120	14886	15568	16677	17986	19436	21007	22690
375	839	140	16760	17618	18951	20515	22246	24124	26141
400	894	40	6907	7053	7391	7822	8306	8828	9383
400	894	60	9246	9498	10015	10653	11365	12134	12952
400	894	80	11429	11804	12512	13369	14323	15353	16450
400	894	100	13507	14024	14936	16025	17234	18540	19936
400	894	120	15512	16187	17315	18649	20126	21725	23438
400	894	140	17461	18311	19666	21256	23017	24926	26973

NOTE: Detector time constant at a reference velocity of 5 ft/sec.

For SI Units: 1 ft = 0.305 m

1000 BTU/sec = 1055 kW

Table B-4.1(II)   Installed Spacing of Heat Detector: 50 feet; t_g: 300 Seconds to 1000 Btu/sec; α: 0.011 Btu/sec³

τ	RTI	ΔT	CEILING HEIGHT IN FEET — FIRE SIZE AT DETECTOR RESPONSE (BTU / SEC)							τ	RTI	ΔT	CEILING HEIGHT IN FEET — FIRE SIZE AT DETECTOR RESPONSE (BTU / SEC)						
			4.0	8.0	12.0	16.0	20.0	24.0	28.0				4.0	8.0	12.0	16.0	20.0	24.0	28.0
25	56	40	942	1100	1296	1520	1769	2043	2340	225	503	40	2892	3014	3230	3488	3775	4087	4422
25	56	60	1344	1649	2008	2409	2849	3328	3848	225	503	60	3910	4155	4510	4924	5385	5888	6432
25	56	80	1772	2256	2807	3414	4075	4794	5572	225	503	80	4898	5273	5786	6380	7041	7765	8551
25	56	100	2227	2923	3686	4523	5433	6420	7487	225	503	100	5862	6386	7077	7873	8759	9733	10794
25	56	120	2703	3634	4636	5726	6908	8189	9573	225	503	120	6813	7503	8392	9411	10547	11798	13165
25	56	140	3209	4392	5653	7016	8493	10091	11818	225	503	140	7757	8631	9736	10999	12408	13963	15663
50	112	40	1272	1412	1598	1815	2058	2326	2620	250	559	40	3087	3203	3425	3690	3985	4306	4649
50	112	60	1775	2039	2374	2758	3187	3660	4175	250	559	60	4164	4409	4770	5194	5664	6177	6730
50	112	80	2275	2704	3221	3806	4454	5165	5939	250	559	80	5211	5585	6106	6709	7380	8113	8908
50	112	100	2795	3413	4138	4951	5847	6826	7889	250	559	100	6231	6752	7451	8255	9150	10132	11201
50	112	120	3333	4165	5123	6187	7355	8628	10008	250	559	120	7235	7920	8815	9842	10985	12242	13614
50	112	140	3890	4959	6171	7508	8970	10560	12283	250	559	140	8229	9096	10205	11474	12888	14446	16150
75	168	40	1559	1690	1876	2094	2333	2607	2901	275	615	40	3251	3387	3614	3887	4190	4519	4871
75	168	60	2143	2398	2726	3104	3528	3996	4508	275	615	60	4410	4655	5024	5456	5936	6459	7021
75	168	80	2728	3131	3630	4201	4840	5544	6313	275	615	80	5516	5889	6417	7030	7711	8454	9259
75	168	100	3320	3895	4593	5386	6269	7240	8298	275	615	100	6591	7109	7815	8629	9534	10525	11602
75	168	120	3922	4695	5617	6657	7810	9074	10449	275	615	120	7646	8327	9229	10264	11416	12681	14060
75	168	140	4537	5531	6699	8008	9454	11036	12754	275	615	140	8690	9549	10664	11940	13362	14926	16634
100	224	40	1811	1944	2133	2356	2605	2878	3176	300	671	40	3429	3566	3799	4078	4390	4727	5088
100	224	60	2484	2731	3058	3438	3862	4330	4842	300	671	60	4651	4896	5271	5712	6202	6734	7307
100	224	80	3145	3532	4025	4591	5225	5925	6691	300	671	80	5814	6185	6721	7344	8035	8789	9604
100	224	100	3807	4356	5040	5822	6695	7659	8713	300	671	100	6942	7457	8171	8996	9911	10912	11999
100	224	120	4474	5208	6107	7130	8271	9526	10896	300	671	120	8049	8724	9633	10679	11841	13115	14502
100	224	140	5150	6091	7228	8515	9946	11518	13231	300	671	140	9141	9993	11114	12399	13829	15401	17115
125	280	40	2048	2180	2374	2603	2858	3138	3442	325	727	40	3603	3741	3979	4265	4585	4930	5300
125	280	60	2802	3043	3374	3758	4186	4658	5173	325	727	60	4886	5130	5512	5963	6462	7005	7587
125	280	80	3538	3911	4404	4971	5605	6305	7070	325	727	80	6105	6474	7018	7651	8353	9118	9943
125	280	100	4269	4795	5474	6251	7120	8080	9131	325	727	100	7287	7797	8520	9355	10282	11294	12390
125	280	120	4977	5701	6589	7602	8733	9982	11348	325	727	120	8443	9112	10030	11086	12259	13543	14939
125	280	140	5706	6632	7750	9023	10441	12005	13712	325	727	140	9583	10426	11556	12851	14291	15871	17593
150	335	40	2272	2403	2602	2638	3101	3388	3699	350	783	40	3774	3911	4154	4448	4775	5129	5507
150	335	60	3104	3339	3675	4065	4500	4978	5498	350	783	60	5116	5359	5748	6208	6717	7270	7863
150	335	80	3892	4272	4768	5339	5977	6680	7447	350	783	80	6391	6757	7309	7952	8665	9441	10277
150	335	100	4680	5215	5893	6671	7541	8500	9550	350	783	100	7625	8130	8861	9707	10646	11669	12776
150	335	120	5464	6175	7058	8067	9194	10439	11802	350	783	120	8831	9492	10419	11486	12670	13965	15371
150	335	140	6249	7156	8264	9527	10938	12495	14198	350	783	140	10018	10851	11990	13295	14746	16336	18067
175	391	40	2486	2615	2820	3063	3334	3629	3948	375	839	40	3941	4079	4327	4628	4962	5324	5710
175	391	60	3395	3622	3963	4361	4804	5289	5816	375	839	60	5342	5584	5980	6448	6967	7530	8133
175	391	80	4241	4618	5118	5696	6340	7048	7821	375	839	80	6672	7034	7594	8248	8972	9759	10607
175	391	100	5089	5619	6300	7082	7954	8916	9968	375	839	100	7958	8456	9196	10054	11004	12039	13158
175	391	120	5930	6632	7514	8524	9651	10896	12258	375	839	120	9214	9865	10801	11880	13076	14382	15799
175	391	140	6770	7663	8765	10025	11432	12985	14686	375	839	140	10397	11268	12416	13733	15195	16796	18537
200	447	40	2692	2818	3028	3279	3558	3861	4188	400	894	40	4105	4242	4496	4803	5146	5516	5910
200	447	60	3648	3893	4241	4647	5099	5593	6128	400	894	60	5564	5805	6207	6684	7213	7786	8399
200	447	80	4675	4951	5457	6043	6695	7410	8189	400	894	80	6949	7306	7874	8538	9274	10072	10931
200	447	100	5482	6009	6694	7482	8361	9327	10383	400	894	100	8287	8777	9526	10395	11357	12404	13534
200	447	120	6379	7075	7959	8972	10102	11349	12712	400	894	120	9536	10231	11177	12267	13475	14794	16222
200	447	140	7271	8154	9256	10516	11923	13475	15175	400	894	140	10806	11678	12835	14164	15638	17251	19002

NOTE: Detector time constant at a reference velocity of 5 ft/sec.

For SI Units: 1 ft = 0.305 m

1000 BTU/sec = 1055 kW

Table B-4.1(mm) Installed Spacing of Heat Detector: 50 feet; t_g: 500 Seconds to 1000 Btu/sec; α: 0.004 Btu/sec^3

			CEILING HEIGHT IN FEET						
τ	RTI	ΔT	4.0	8.0	12.0	16.0	20.0	24.0	28.0
			FIRE SIZE AT DETECTOR RESPONSE (BTU / SEC)						
25	56	40	706	875	1073	1292	1532	1793	2076
25	56	60	1053	1381	1738	2130	2557	3020	3520
25	56	80	1435	1948	2496	3091	3735	4432	5186
25	56	100	1851	2574	3337	4159	5047	6008	7047
25	56	120	2299	3253	4258	5323	6481	7732	9083
25	56	140	2777	3981	5241	6577	8025	9590	11280
50	112	40	920	1068	1252	1462	1697	1955	2236
50	112	60	1322	1610	1950	2331	2751	3210	3708
50	112	80	1748	2208	2734	3316	3953	4647	5399
50	112	100	2200	2859	3597	4406	5287	6245	7281
50	112	120	2672	3564	4532	5590	6740	7988	9337
50	112	140	3173	4311	5535	6863	8303	9864	11552
75	168	40	1107	1247	1427	1633	1864	2119	2398
75	168	60	1568	1836	2163	2534	2948	3403	3899
75	168	80	2039	2469	2975	3545	4175	4864	5614
75	168	100	2529	3148	3862	4657	5531	6484	7518
75	168	120	3040	3874	4818	5862	7003	8246	9594
75	168	140	3572	4644	5840	7152	8584	10141	11827
100	224	40	1278	1415	1593	1799	2029	2284	2562
100	224	60	1798	2051	2372	2738	3147	3599	4093
100	224	80	2312	2724	3217	3777	4400	5085	5832
100	224	100	2845	3435	4130	4912	5777	6726	7757
100	224	120	3393	4188	5108	6136	7270	8508	9853
100	224	140	3959	4981	6149	7446	8869	10420	12104
125	280	40	1439	1572	1751	1958	2190	2446	2725
125	280	60	2004	2257	2574	2939	3346	3796	4288
125	280	80	2571	2969	3455	4009	4626	5308	6052
125	280	100	3146	3715	4396	5168	6026	6970	7999
125	280	120	3732	4498	5399	6414	7539	8772	10114
125	280	140	4333	5317	6461	7742	9156	10702	12383
150	335	40	1592	1721	1902	2112	2347	2605	2886
150	335	60	2204	2453	2771	3135	3542	3992	4484
150	335	80	2817	3206	3688	4238	4853	5532	6274
150	335	100	3433	3988	4660	5425	6277	7216	8243
150	335	120	4058	4801	5689	6693	7810	9038	10377
150	335	140	4694	5648	6773	8041	9445	10986	12664
175	391	40	1740	1863	2047	2260	2499	2760	3044
175	391	60	2396	2642	2960	3327	3736	4186	4679
175	391	80	3053	3436	3916	4465	5078	5755	6497
175	391	100	3710	4253	4920	5680	6528	7464	8488
175	391	120	4373	5098	5976	6972	8082	9306	10642
175	391	140	5045	5973	7084	8340	9737	11273	12948
200	447	40	1867	2000	2187	2404	2646	2911	3199
200	447	60	2580	2824	3145	3514	3925	4378	4873
200	447	80	3280	3658	4138	4687	5301	5979	6720
200	447	100	3978	4511	5175	5932	6778	7712	8735
200	447	120	4679	5388	6259	7250	8356	9575	10909
200	447	140	5369	6292	7392	8640	10030	11561	13232

			CEILING HEIGHT IN FEET						
τ	RTI	ΔT	4.0	8.0	12.0	16.0	20.0	24.0	28.0
			FIRE SIZE AT DETECTOR RESPONSE (BTU / SEC)						
225	503	40	2000	2132	2322	2544	2790	3059	3351
225	503	60	2759	3000	3324	3696	4112	4568	5065
225	503	80	3501	3873	4355	4906	5522	6200	6943
225	503	100	4240	4762	5425	6182	7027	7960	8982
225	503	120	4958	5671	6538	7526	8628	9845	11177
225	503	140	5695	6605	7697	8939	10323	11850	13519
250	559	40	2128	2260	2453	2679	2930	3204	3500
250	559	60	2932	3172	3498	3875	4294	4754	5255
250	559	80	3717	4083	4567	5121	5740	6420	7164
250	559	100	4474	5006	5670	6428	7274	8208	9229
250	559	120	5241	5948	6813	7799	8900	10115	11445
250	559	140	6011	6911	7999	9236	10617	12140	13806
275	615	40	2253	2385	2581	2811	3067	3345	3646
275	615	60	3101	3338	3668	4050	4474	4938	5442
275	615	80	3910	4287	4774	5332	5954	6638	7384
275	615	100	4716	5245	5911	6671	7519	8453	9476
275	615	120	5517	6219	7084	8070	9170	10385	11715
275	615	140	6319	7212	8297	9531	10909	12430	14095
300	671	40	2374	2506	2706	2940	3200	3484	3789
300	671	60	3267	3501	3835	4221	4650	5119	5627
300	671	80	4111	4487	4977	5540	6166	6853	7603
300	671	100	4952	5479	6147	6910	7761	8698	9722
300	671	120	5787	6485	7350	8337	9439	10654	11984
300	671	140	6621	7507	8590	9824	11201	12720	14384
325	727	40	2493	2624	2828	3066	3331	3619	3929
325	727	60	3430	3660	3997	4388	4822	5296	5810
325	727	80	4307	4683	5176	5744	6375	7066	7820
325	727	100	5182	5709	6379	7146	8001	8940	9967
325	727	120	6050	6745	7612	8602	9705	10922	12252
325	727	140	6916	7797	8880	10114	11491	13010	14673
350	783	40	2610	2740	2947	3190	3460	3753	4068
350	783	60	3570	3815	4157	4553	4992	5472	5990
350	783	80	4498	4874	5372	5944	6580	7277	8035
350	783	100	5408	5933	6607	7379	8237	9181	10211
350	783	120	6309	7001	7871	8864	9970	11188	12521
350	783	140	7205	8082	9166	10402	11779	13299	14963
375	839	40	2724	2854	3063	3311	3586	3884	4204
375	839	60	3722	3968	4313	4715	5160	5644	6168
375	839	80	4686	5062	5564	6142	6783	7485	8247
375	839	100	5630	6154	6832	7609	8472	9420	10453
375	839	120	6562	7252	8125	9122	10232	11453	12788
375	839	140	7489	8362	9448	10686	12066	13587	15252
400	894	40	2837	2965	3178	3430	3710	4012	4337
400	894	60	3871	4117	4467	4874	5324	5815	6344
400	894	80	4871	5247	5753	6336	6983	7691	8458
400	894	100	5848	6371	7053	7835	8703	9656	10694
400	894	120	6812	7500	8376	9377	10491	11716	13054
400	894	140	7768	8638	9727	10968	12351	13874	15540

NOTE: Detector time constant at a reference velocity of 5 ft/sec.

For SI Units: 1 ft = 0.305 m
1000 BTU/sec = 1055 kW

Table B-4.1(nn) Installed Spacing of Heat Detector: 50 feet; t_g: 600 Seconds to 1000 Btu/sec; α: 0.003 Btu/sec^3

τ	RTI	ΔT	4.0	8.0	12.0	16.0	20.0	24.0	28.0
			FIRE SIZE AT DETECTOR RESPONSE (BTU / SEC)						
25	56	40	645	818	1016	1233	1470	1728	2007
25	56	60	979	1311	1669	2058	2480	2938	3432
25	56	80	1350	1868	2416	3006	3645	4336	5082
25	56	100	1757	2484	3250	4063	4945	5898	6928
25	56	120	2195	3154	4156	5217	6366	7608	8950
25	56	140	2665	3874	5130	6460	7899	9454	11133
50	112	40	826	978	1163	1373	1605	1860	2137
50	112	60	1204	1500	1842	2222	2639	3094	3587
50	112	80	1610	2081	2611	3191	3824	4513	5258
50	112	100	2044	2721	3460	4266	5142	6093	7122
50	112	120	2503	3409	4382	5437	6581	7820	9160
50	112	140	2991	4145	5372	6696	8129	9681	11359
75	168	40	987	1131	1309	1514	1743	1995	2271
75	168	60	1414	1689	2018	2390	2801	3253	3744
75	168	80	1860	2298	2809	3379	4007	4692	5435
75	168	100	2322	2958	3677	4473	5343	6290	7317
75	168	120	2811	3665	4617	5660	6798	8034	9372
75	168	140	3323	4423	5623	6935	8362	9910	11587
100	224	40	1136	1274	1450	1653	1880	2131	2406
100	224	60	1610	1872	2193	2558	2966	3414	3903
100	224	80	2092	2512	3010	3570	4192	4873	5614
100	224	100	2591	3197	3898	4682	5546	6489	7514
100	224	120	3110	3926	4856	5886	7017	8249	9586
100	224	140	3649	4698	5878	7176	8596	10141	11815
125	280	40	1274	1410	1586	1789	2016	2266	2541
125	280	60	1797	2048	2364	2726	3130	3576	4064
125	280	80	2314	2721	3209	3762	4378	5056	5795
125	280	100	2849	3433	4120	4894	5751	6690	7713
125	280	120	3399	4185	5096	6115	7238	8466	9801
125	280	140	3967	4978	6135	7420	8832	10373	12046
150	335	40	1406	1538	1715	1920	2148	2400	2675
150	335	60	1965	2217	2531	2892	3294	3739	4226
150	335	80	2526	2924	3405	3954	4566	5240	5978
150	335	100	3096	3664	4340	5106	5957	6893	7914
150	335	120	3678	4441	5337	6345	7461	8685	10018
150	335	140	4275	5256	6393	7666	9071	10607	12277
175	391	40	1532	1662	1840	2047	2278	2532	2808
175	391	60	2131	2380	2694	3054	3457	3901	4388
175	391	80	2730	3120	3598	4144	4753	5426	6162
175	391	100	3335	3890	4559	5319	6165	7097	8116
175	391	120	3948	4693	5577	6576	7686	8906	10236
175	391	140	4574	5530	6652	7914	9311	10843	12510
200	447	40	1654	1781	1961	2171	2404	2660	2939
200	447	60	2291	2537	2852	3214	3618	4063	4549
200	447	80	2927	3311	3788	4332	4940	5611	6346
200	447	100	3565	4111	4775	5530	6373	7302	8319
200	447	120	4211	4941	5816	6807	7912	9128	10456
200	447	140	4866	5801	6910	8162	9552	11080	12745

τ	RTI	ΔT	4.0	8.0	12.0	16.0	20.0	24.0	28.0
			FIRE SIZE AT DETECTOR RESPONSE (BTU / SEC)						
225	503	40	1762	1895	2078	2291	2528	2787	3068
225	503	60	2445	2690	3007	3370	3776	4222	4710
225	503	80	3117	3497	3974	4518	5126	5797	6531
225	503	100	3790	4327	4988	5740	6581	7508	8523
225	503	120	4467	5183	6052	7038	8138	9351	10676
225	503	140	5152	6068	7167	8411	9795	11318	12981
250	559	40	1874	2006	2192	2408	2648	2911	3195
250	559	60	2595	2838	3157	3523	3932	4380	4870
250	559	80	3303	3679	4156	4701	5310	5981	6717
250	559	100	4009	4538	5197	5949	6788	7714	8728
250	559	120	4718	5421	6285	7268	8365	9574	10898
250	559	140	5414	6330	7421	8659	10038	11557	13218
275	615	40	1982	2115	2303	2522	2766	3032	3320
275	615	60	2741	2983	3304	3674	4085	4536	5029
275	615	80	3484	3856	4334	4882	5492	6164	6900
275	615	100	4223	4745	5403	6155	6994	7919	8933
275	615	120	4944	5654	6516	7496	8591	9798	11120
275	615	140	5681	6588	7674	8907	10281	11798	13456
300	671	40	2087	2220	2411	2634	2882	3151	3443
300	671	60	2884	3124	3448	3821	4236	4691	5186
300	671	80	3662	4029	4509	5059	5672	6346	7083
300	671	100	4415	4947	5606	6359	7198	8124	9138
300	671	120	5178	5883	6743	7723	8816	10023	11343
300	671	140	5943	6842	7924	9153	10525	12038	13694
325	727	40	2190	2323	2517	2744	2995	3269	3564
325	727	60	3024	3262	3589	3966	4384	4843	5341
325	727	80	3837	4199	4682	5234	5850	6527	7266
325	727	100	4616	5145	5806	6560	7401	8328	9342
325	727	120	5407	6108	6968	7947	9040	10246	11567
325	727	140	6199	7091	8171	9398	10768	12279	13934
350	783	40	2291	2424	2621	2851	3106	3384	3683
350	783	60	3161	3397	3728	4108	4531	4993	5495
350	783	80	3989	4365	4851	5407	6026	6706	7448
350	783	100	4812	5340	6003	6759	7602	8531	9546
350	783	120	5631	6330	7190	8170	9263	10470	11790
350	783	140	6450	7337	8415	9641	11010	12520	14174
375	839	40	2390	2523	2722	2956	3215	3497	3800
375	839	60	3296	3530	3863	4248	4675	5141	5647
375	839	80	4152	4528	5017	5577	6200	6883	7628
375	839	100	5005	5532	6197	6956	7802	8733	9750
375	839	120	5852	6547	7409	8391	9485	10692	12013
375	839	140	6697	7579	8657	9883	11251	12761	14414
400	894	40	2487	2620	2822	3059	3323	3608	3916
400	894	60	3429	3660	3997	4386	4817	5288	5797
400	894	80	4314	4689	5181	5745	6372	7059	7807
400	894	100	5195	5720	6388	7151	8000	8933	9953
400	894	120	6068	6762	7625	8609	9706	10914	12236
400	894	140	6939	7818	8896	10123	11491	13001	14654

NOTE: Detector time constant at a reference velocity of 5 ft/sec.

For SI Units: 1 ft = 0.305 m

1000 BTU/sec = 1055 kW

B-4.1.1 Example.

Input:

Ceiling height: 8 ft (2.4 m)

Detector type: Fixed temperature 135°F (57°C)

UL listed spacing: 30 ft (9.1 m)

Installed spacing: 15 ft (4.6 m)

Fire:

Fire growth rate: Slow

t_g: 600 sec

α: 0.003 Btu/sec^3

Environmental conditions:

T_o: 55°F (12.8°C)

Threshold fire size (Q_d) at detector response:

From Table B-3.2.2 the detector time constant (τ_o) is 80 seconds.

$$\Delta T = T_s - T_o = 135 - 55 = 80°F$$

From Table B-4.1(t)

For $\tau_o = 75$ sec; $Q_d = 418$ Btu/sec

For $\tau_o = 100$ sec; $Q_d = 472$ Btu/sec

By interpolation:

$$Q_d = 418 - [(75 - 80)(418 - 472)/(75 - 100)]$$

$$Q_d = 429 \text{ Btu/sec}$$

B-5 Smoke Detector Spacing for Flaming Fires.

B-5.1 Ideally, the placement of smoke detectors should be based on a knowledge of fire plume and ceiling jet flows, smoke production rates, particulate changes due to aging, and the unique operating characteristics of the detector being used. Knowledge of plume and jet flows enabled the heat detector spacing information presented in Section B-3 to be developed. Unfortunately, that knowledge does not apply to smoke originating from smoldering fires. Understanding of smoke production and aging lags considerably behind that of heat production. The operating characteristics of smoke detectors in specific fire environments are not often measured or made generally available for other than a very few combustible materials. Hence, the existing data base precludes the development of complete engineering design information for smoke detector location and spacing.

B-5.2 In a flaming fire, smoke detector response is affected by ceiling height, and size and rate of growth of the fire in much the same way as heat detector response. The thermal energy of the flaming fire transports smoke particles to the smoke sensor just as it does heat to a heat sensor. While the relationship between the amount of smoke and the amount of heat produced by a fire is highly dependent upon the fuel and the way it is burning, research has shown that the relationship between temperature and optical density of smoke remains essentially constant within the fire plume and on the ceiling in the proximity of the plume.

B-5.3 In smoldering fires, thermal energy also provides a force for transporting smoke particles to the smoke sensor. However, because the rate of energy release is usually small and the rate of growth of the fire is slow, other factors such as airflow may have a stronger influence on the transport of smoke particles to the smoke sensor. Additionally, for smoldering fires, the relationship between temperature and the optical density of smoke is not constant and, therefore, not useful.

B-5.4 Smoke detectors, regardless of whether they detect by sensing scattered light, loss of light transmission (light extinction), or reduction of ion current, are particle detectors. Particle concentration, size, color, and size distribution affect each sensing technology differently. It is generally accepted that the concentration of sub-micron-diameter particles produced by a flaming fire is greater than that produced by a smoldering fire. Conversely, the concentration of larger particles is greater from a smoldering fire. It is also known that the smaller particles agglomerate and form larger ones as they age and are carried away from the fire source. More research is required to provide sufficient data to first predict particle concentration and behavior and, secondly, to predict the response of a particular detector.

B-5.5 Unlike heat detectors, listed smoke detectors are not given a listed spacing. It has become general practice to install smoke detectors on 30-ft (9.1-m) centers on smooth ceilings with reductions made empirically to that spacing for beamed or joisted ceilings and for areas having high rates of air movement. Spacing adjustments for ceiling height are also necessary as discussed herein.

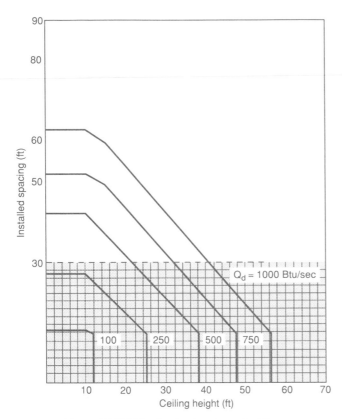

Figure B-5.5.1(a)   Smoke detector — fast fire.

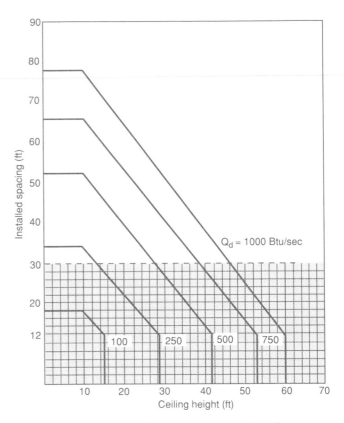

Figure B-5.5.1(b)   Smoke detector — medium fire.

B-5.5.1   Figures B-5.5.1(a), (b), and (c) are based on the assumption that smoke transport to the detector is entirely from fire plume dynamics. It assumes that the ratio of the gas temperature rise to the optical density of the smoke is a constant and that the detector will actuate at a constant value of temperature rise equal to 20°F (–6.7C), which is considered indicative of concentrations of smoke from a number of common fuels that would cause detection by a relatively sensitive detector. It is cautioned that many fuel/detector combinations may cause operation at a higher temperature rise. In addition, it is assumed that the detector design does not significantly impair smoke entry. The data presented in Figures B-5.5.1(a), (b), and (c) clearly indicate that spacings considerably greater than 30 ft (9.1 m) are acceptable for detecting geometrically growing flaming fires when Q_d is 1000 Btu/sec or more.

B-5.5.2   In the early stages of development of a growing fire, where the heat release rate is approximately 250 Btu/sec or less, the environmental effects in spaces having high ceilings may dominate the transportation of smoke. Examples of such envi-

ronmental effects are heating, cooling, humidity, and ventilation. Greater thermal energy release from the fire may be required to overcome such environmental effects. Until the growing fire reaches a sufficiently high level of heat release, closer spacing of smoke detectors on the ceiling may not significantly improve the response of the detectors to the fire. Therefore, where considering ceiling height alone, smoke detector spacing closer than 30 ft (9.1 m) may not be warranted except in instances where an engineering analysis indicates additional benefit will result. Other construction characteristics must also be considered; see the appropriate sections of Chapter 5 dealing with smoke detectors and smoke detectors for the control of smoke spread.

B-5.6   The method used to determine the spacing of smoke detectors is similar to that used for heat detectors and is based on fire size, fire growth rate, and ceiling height.

B-5.6.1   In order to use Figures B-5.5.1(a), (b), or (c) to determine the installed spacing of a smoke detector, the designer must

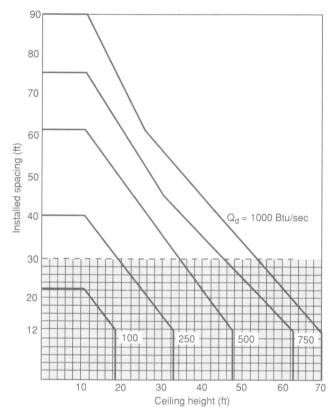

Figure B-5.5.1(c) *Smoke detector — slow fire.*

first select Q_d, the threshold flaming fire size at which detection is desired.

B-5.6.2 In addition to threshold flaming fire size, Q_d, the designer must consider the expected fire growth rate. Figures B-5.5.1(a), (b), and (c) are used for fast-, medium-, and slow-growing flaming fires, respectively. See section B-2.2.2 for heat release rate and fire growth rate information.

B-5.6.2.1 Figures B-5.5.1(a), (b), and (c) use the following values for t_g:

Fast fire growth rate, t_g = 150 seconds
Medium fire growth rate, t_g = 300 seconds
Slow fire growth rate, t_g = 600 seconds.

B-5.6.3 **Example 1.** Determine the installed spacing of a smoke detector on a 30-ft (9.1-m) ceiling required to detect a 750 Btu/sec fire that is growing at a medium rate, use Figure B-5.5.1(b).

Input:

Ceiling height: 30 ft (9.1 m)

Q_d: 750 Btu/sec (791 kW)

Fire growth rate: Medium, t_g = 300 seconds

Required installed spacing:

From Figure B-5.5.1(b), using the 750 Btu/sec (791 kW) curve, installed spacing is 41 ft (12.8 m).

B-5.6.4 **Example 2.** Consider a 20-ft (6.1-m) ceiling with a threshold fire size of 250 Btu/sec growing at a medium rate.

Input:

Ceiling height: 20 ft (6.1 m)

Q_d: 250 Btu/Sec (264 kW)

Fire growth rate: Medium, t_g = 300 seconds

Required installed spacing:

From Figure B-5.5.1(b), using the 250 Btu/sec (264 kW) curve, installed spacing for the smoke detector is 30 ft (9.1 m), since the intersection of a vertical line at 20 ft (6.1 m) and the Q_d = 250 curve falls within the shaded area below 30 ft (9.1 m) spacing. (*See B-5.2.1.2.*)

NOTE: Both a slow and fast rate-of-growth fire would result in the same 30-ft (9.1-m) spacing using Figures B-5.5.1(a) and (c). Engineering judgment and consideration of all factors affecting smoke movement and transport should be exercised in deciding whether a reduced spacing is warranted.

B-5.6.5 Smoke detector spacings of less than 30 ft (9.1 m) may be used for detection of flaming fires where no other detector type is suitable and where environmental conditions allow the use of a smoke detector.

B-6   Theoretical Considerations.

B-6.1 **Introduction.** The design methods of this appendix are the joint result of extensive experimental work and of mathematical modeling of the heat and mass transfer processes involved. This section outlines models and data correlations used to generate the design data presented in this appendix. Only the general principles are described. More detailed information may be obtained from references 4, 9, 10, and 14 in Appendix C.

B-6.2 **Temperature and Velocity Correlations.** In order to predict the operation of any detector, it is necessary to characterize the local environment created by the fire at the detector location. For a heat detector, the important variables are temperature and velocity of the gases at the detector. Through a program of full-scale tests and using mathematical modeling techniques, general expressions for temperature and velocity at a detector location have been developed. (*See references 4, 9, 10, and 14 in Appendix C.*) The expressions are valid for fires that grow according to:

$$Q = \alpha t^2$$

Where:

Q = the theoretical fire heat release rate

α = the fire intensity coefficient characteristic of a particular fuel and configuration

t = time.

The calculations used to produce the spacing curves assume that the ratio of the actual convective heat release to the theoretical heat release for all fuels is equal to that ratio for wood crib fires.

B-6.3 **Heat Detector Model.** The heating of a heat detector is given by the equation:

$$\Delta T_s/dt = (T_g - T_d)/\tau$$

Where:

T_s = the temperature rating or set point of the detector

T_g = the gas temperature at the detector

τ = detector time constant.

The time constant is a measure of the detector's sensitivity and is given by:

$$\tau = (mc)/(hA)$$

Where:

m = the detector element's mass

c = the detector element's specific heat

h = the convection heat transfer coefficient for the detector

A = the surface area of the detector's element.

h varies approximately as the square root of the gas velocity, u. It is customary to speak of the time constant at a reference velocity of u_o = 5 ft/sec.

$$\tau = \tau_o(u_o/u)^{1/2}$$

t_o can be measured most easily by a plunge test. It can also be related to the listed spacing of a detector through a calculation; Table B-3.2.2 results from these calculations. This model uses the temperature and velocity of the gases at the detector to predict the temperature rise of the detector element. Detector operation occurs when the preset conditions are reached.

The detector's sensitivity can also be expressed in units that are independent of the air velocity used in the test to determine the time constant. This is known as the response time index (RTI).

$$RTI = \tau\sqrt{u}$$

The RTI value can therefore be obtained by multiplying t_o values by $\sqrt{u_o}$; for example, where u_o = 5 ft/sec, a τ_o of 30 sec corresponds to an RTI of 67 $sec^{1/2}$ $ft^{1/2}$ or 36 $sec^{1/2}$ $m^{1/2}$

A detector having a RTI of 67 $sec^{1/2}$ $ft^{1/2}$ would have a τ_o of 23.7 sec, if measured in an air velocity of 8 ft/sec.

B-6.4 **Ambient Temperature Considerations.** (*See also B-3.2.3.*) The maximum ambient temperature expected to occur at the ceiling dictates the choice of temperature rating for a fixed temperature heat detector application. But the minimum ambient temperature likely at the ceiling constitutes the worst-case condition for response by that detector to a fire.

The mass, specific heat, heat transfer coefficient, and surface area of a detector's sensing element characterize that detector's time constant. The response time of a given detector to a given fire depends only on the detector's time constant and the difference between the detector's temperature rating and the ambient temperature at the detector where the fire starts. When ambient temperature at the ceiling decreases, more heat from a fire will be needed to bring the air surrounding the detector's sensing element up to its rated (operating) temperature; this translates into slower response and, in the case of a growing fire, a larger fire size at the time of detection. In a room or work area that has central heating, the minimum ambient temperature would usually be about 68°F (20°C). Certain warehouse occupancies may only be heated enough to prevent water pipe freeze-up; in this case, the minimum ambient should be considered 35°F (2°C) even though during many months of the year the actual ambient is much higher. An unheated building in northern states and in Canada should be presumed to have a minimum ambient of −40°F (−40°C) or lower.

B-6.5 Heat and Smoke Analogy — Smoke Detector Model.

For smoke detectors, the temperature of the gases at the detector is not directly relevant to detection, but the mass concentration and size distribution of particulates is relevant. For many types of smoke, the mass concentration of particles is directly proportional to the optical density of the smoke, D_o. A general correlation for flaming fires has been shown to exist between the temperature rise of the fire gases at a given location and the optical densiy.

If the optical density at which a detector responds, D_o, was known and was independent of particle size distribution, the response of the detector could be approximated as a function of heat release rate of the burning fuel, rate of fire growth, and ceiling height, assuming that the above correlation held.

However, the more popular ionization and light-scattering detectors exhibit widely different D_o when particle size distribution is changed; hence, where D_o for these detectors is measured in order to predict response, the test aerosol used must be very carefully controlled so that particle size distribution is constant.

C

Referenced
Publications

The following documents or portions thereof are referenced within this code for informational purposes only and thus are not considered part of the requirements of this document. The edition indicated for each reference is the current edition as of the date of the NFPA issuance of this document.

C-1.1   **NFPA Publications.** National Fire Protection Association, 1 Batterymarch Park, P.O. Box 9101, Quincy, MA 02269-9101.

NFPA 11, *Standard for Low Expansion Foam and Combined Agent Systems*, 1988 edition.

NFPA 11A, *Standard for Medium- and High-Expansion Foam Systems*, 1988 edition.

NFPA 12, *Standard on Carbon Dioxide Extinguishing Systems*, 1993 edition.

NFPA 12B, *Standard on Halon 1211 Fire Extinguishing Systems*, 1990 edition.

NFPA 13, *Standard for the Installation of Sprinkler Systems*, 1991 edition.

NFPA 14, *Standard for the Installation of Standpipe and Hose Systems*, 1993 edition.

NFPA 15, *Standard for Water Spray Fixed Systems for Fire Protection*, 1990 edition.

NFPA 17, *Standard for Dry Chemical Extinguishing Systems*, 1990 edition.

NFPA 70, *National Electrical Code,* 1993 edition.

NFPA 80, *Standard for Fire Doors and Fire Windows*, 1992 edition.

NFPA 90A, *Standard for the Installation of Air Conditioning and Ventilating Systems,* 1993 edition.

NFPA 90B, *Standard for the Installation of Warm Air Heating and Air Conditioning Systems*, 1993 edition.

NFPA 92A, *Recommended Practice for Smoke-Control Systems,* 1993 edition.

NFPA 92B, *Guide for Smoke Management Systems in Malls, Atria, and Large Areas,* 1991 edition.

NFPA *101, Life Safety Code,* 1991 edition.

NFPA 170, *Standard for Firesafety Symbols,* 1991 edition.

NFPA 231C, *Standard for Rack Storage of Materials*, 1991 edition.

NFPA 1221, *Standard on the Installation, Maintenance, and Use of Public Fire Service Communication Systems*, 1991 edition.

C-2 Bibliography.

This part of the appendix lists other publications pertinent to the subject of this NFPA document and which may or may not be referenced.

1. Alpert, R. "Ceiling Jets." *Fire Technology*, August 1972.

2. Alpert and Ward, "Hazard Analysis of Unsprinklered Risk." *SFPE Technology Report*, 1984.

3. Babrauskas, V.; Lawson, J. R.; Walton, W. D.; and Twilley, W. H. "Upholstered Furniture Heat Release Rates Measured with a Furniture Calorimeter," (NBSIR 82-2604) (Dec. 1982). National Institute of Standards and Technology (formerly National Bureau of Standards), Center for Fire Research, Gaithersburg, MD 20889.

4. Beyler, C. "A Design Method for Flaming Fire Detection." *Fire Technology*, Volume 20, Number 4, November 1984.

5. DiNenno, P., ed. Chapter 3-1 of "SFPE Handbook of Fire Protection Engineering," by R. Schifiliti, September 1988.

6. Evans, D. D., and Stroup, D. W. "Methods to Calculate Response Time of Heat and Smoke Detectors Installed Below Large Unobstructed Ceilings," (NBSIR 85-3167) (February 1985, Issued July 1986). National Institute of Standards and Technology (formerly National Bureau of Standards), Center for Fire Research, Gaithersburg, MD 20889.

7. Heskestad, G. "Characterization of Smoke Entry and Response for Products-of-Combustion Detectors." Proceedings, 7th International Conference on Problems of Automatic Fire Detection, Rheinish-Westfalischen Technischen Hochschule Aachen (March 1975).

8. Heskestad, G. "Investigation of a New Sprinkler Sensitivity Approval Test: The Plunge Test." FMCR Tech. Report 22485, Factory Mutual Research Corporation, 1151 Providence Turnpike, Norwood, MA 02062.

9. Heskestad, G. "The Initial Convective Flow in Fire: Seventeenth Symposium on Combustion." The Combustion Institute, Pittsburgh, PA (1979).

10. Heskestad, G., and Delichatsios, M. A. "Environments of Fire Detectors — Phase 1: Effect of Fire Size, Ceiling Height and Material." Measurements vol. I (NBS-GCR-77-86), Analysis vol. II (NBS-GCR-77-95). National Technical Information Service (NTIS), Springfield, VA 22151.

11. Heskestad, G., and Delichatsios, M. A. "Update: The Initial Convective Flow in Fire." *Fire Safety Journal*, Volume 15, Number 5, 1989.

12. International Organization for Standardization, *Audible Emergency Evacuation Signal*, ISO 8201:1987.

13. Lawson, J. R.; Walton, W. D.; and Twilley, W. H. "Fire Performance of Furnishings as Measured in the NBS Furniture Calorimeter, Part 1," (NBSIR 83-2787) (August 1983). National Institute of Standards and Technology (formerly National Bureau of Standards), Center for Fire Research, Gaithersburg, MD 20889.

14. Schifiliti, R. "Use of Fire Plume Theory in the Design and Analysis of Fire Detector and Sprinkler Response." Master's thesis, Worcester Polytechnic Institute, Center for Firesafety Studies, Worcester, MA, 1986.

15. Title 47, Code of Federal Regulations, Communications Act of 1934 Amended.

Cross-References to Previous Editions

The 1993 edition of NFPA 72 is a consolidation of four standards and two guides. In the code, cross-references to the previous editions of these signaling documents were provided at the end of each code section and in a list at the end of the code. In this handbook, the internal cross-references have been deleted for ease of reading, but the following list has been retained for the reader's convenience.

Cross-References to Previous Editions

Former Reference	New Reference
NFPA 71-1989	
NFPA 71	3-8.7.7
NFPA 71	3-8.7.8
NFPA 71, Title	4-3
NFPA 71, 1-1.1	4-3.1
NFPA 71, 1-1.2	4-3.2.1
NFPA 71, 1-1.3	1-3, 1-3.1, and 1-3.2
NFPA 71, 1-2.2	4-3.2.2
NFPA 71, 1-2.3	1-7.2.3
NFPA 71, 1-2.3.1	1-7.2.3.1 and 1-7.2.3.1.2
NFPA 71, 1-2.3.1.1	1-7.2.3.1.1
NFPA 71, 1-2.3.2	1-7.2.3.2, 4-3.2.3, 4-3.2.3.1, 4-3.2.3.1.1, 4-3.2.3.1.2, 4-3.2.3.2, 4-3.2.3.2.1, and 4-3.2.3.2.2
NFPA 71, 1-2.3.2.1	1-7.2.3.2.1
NFPA 71, 1-2.3.2.2	1-7.2.3.2.2
NFPA 71, 1-2.4	4-3.2.4
NFPA 71, 1-2.5	4-3 Note
NFPA 71, 1-3	1-4
NFPA 71, 1-4.1	1-7.1.1 and 4-3.2.5
NFPA 71, 1-4.1(b)	7-1.5, 7-1.5.1, 7-1.5.2, 7-1.5.3, 7-1.6, 7-2, and 7-7.1
NFPA 71, 1-4.2	1-5.3, 1-5.3.1, and 7-2.1.1
NFPA 71, 1-4.3	1-7.2.1 and Figure 1-7.2.1
NFPA 71, 1-4.5	1-7.3, 4-3.6.2.1, and 7-5.3
NFPA 71, 1-5	1-5.5.2
NFPA 71, 1-5.1	1-5.5.2.1
NFPA 71, 1-5.2	1-5.5.2.2
NFPA 71, 1-5.3	1-5.5.5
NFPA 71, 1-5.4	1-5.5.2.3 and 7-4.3
NFPA 71, 1-5.5	1-5.5.3
NFPA 71, 1-6.2	4-3.3.1
NFPA 71, 1-7	4-3.3
NFPA 71, 1-7.1	4-3.4.1
NFPA 71, 1-7.2	4-3.4.4
NFPA 71, 1-7.2.1	4-3.4.4.1
NFPA 71, 1-7.2.2	4-3.4.4.2
NFPA 71, 1-7.2.3	4-3.4.4.3
NFPA 71, 1-7.2.4	4-3.4.4.4 and 7-4.4
NFPA 71, 1-9.1	4-3.5.1
NFPA 71, 1-9.1.2	4-3.5.1.1
NFPA 71, 1-9.5	A-7-2.1
NFPA 71, 1-9.5(a), (h), (i)	7-1.3
NFPA 71, 1-10	7-5

Former Reference	New Reference
NFPA 71, 1-10.1	4-3.6.2.2
NFPA 71, 1-10.2	4-3.6.1
NFPA 71, 1-10.2.1	4-3.6.1.1
NFPA 71, 1-10.2.2	4-3.6.1.2
NFPA 71, 1-10.2.2.1	4-3.6.1.2.1
NFPA 71, 1-10.2.3	4-3.6.1.3
NFPA 71, 1-10.2.4	4-3.6.1.4
NFPA 71, 1-10.2.5	4-3.6.1.5
NFPA 71, 2-1	1-5.5.4
NFPA 71, 2-2.1.1	1-5.2.1
NFPA 71, 2-2.1.1 3rd paragraph	1-5.2.3 2nd paragraph
NFPA 71, 2-2.1.1 2nd paragraph	1-5.2.3 Exception No. 1, Exception No. 2, and Note
NFPA 71, 2-2.1.3	1-5.2.2
NFPA 71, 2-2.1.4	1-5.2.4
NFPA 71, 2-2.1.5	1-5.2.5
NFPA 71, 2-2.1.6	1-5.2.6
NFPA 71, 2-2.1.8	1-5.2.6.1
NFPA 71, 2-2.2	1-5.2.5
NFPA 71, 2-2.3.1	1-5.2.9.1
NFPA 71, 2-2.3.2	1-5.2.9.1
NFPA 71, 2-2.3.3	1-5.2.9.1
NFPA 71, 2-2.3.4	1-5.2.9.2
NFPA 71, 2-2.3.5	1-5.2.9.3
NFPA 71, 2-2.3.6	1-5.2.9.4 and 1-5.2.9.5
NFPA 71, 2-2.4	1-5.2.8
NFPA 71, 2-2.4.1	1-5.2.8.2
NFPA 71, 2-2.4.2	1-5.2.8.3
NFPA 71, 2-2.5.1	1-5.2.11 and 1-5.2.11.1
NFPA 71, 2-2.5.2	1-5.2.11.2
NFPA 71, 2-2.5.4	1-5.2.11.3
NFPA 71, 2-2.6	1-5.2.10
NFPA 71, 2-2.6.1 excluding Exception	1-5.2.10.1
NFPA 71, 2-2.6.2	1-5.2.10.2
NFPA 71, 2-2.6.3	1-5.2.10.4
NFPA 71, 2-2.6.4	1-5.2.10.3
NFPA 71, 2-2.6.5	1-5.2.10.6
NFPA 71, 2-4	1-5.8
NFPA 71, 2-4.1	1-5.8.1
NFPA 71, 2-4.1 Exception No. 5	1-5.8.6.1 Exception No. 2
NFPA 71, 2-4.1 Exception No. 8	1-5.8.4 Exception No. 3

Former Reference	New Reference	Former Reference	New Reference
NFPA 71, 2-4.1 Exception No. 9	1-5.8.6.1 Exception No. 3	NFPA 71, 5-1.3	4-2.2.2.3
NFPA 71, 2-4.1 Exception No. 11	1-5.8.4 Exception No. 1	NFPA 71, 5-2	4-2.3.2.1
		NFPA 71, 5-2.1	4-2.3.2.1.1
NFPA 71, 2-4.1 Exception No. 12	1-5.8.1 Exception No. 11	NFPA 71, 5-2.2	4-2.3.2.1.2
		NFPA 71, 5-2.3	4-2.3.2.1.3
NFPA 71, 2-4.1 Exception No. 13	1-5.8.1 Exception No. 7 and Note	NFPA 71, 5-2.4	4-2.3.2.1.4
		NFPA 71, 5-2.5	4-2.3.2.1.5
NFPA 71, 2-4.2	1-5.4.6	NFPA 71, 5-2.6	4-2.3.2.1.6 and 4-2.3.2.1.6.1
NFPA 71, 2-4.2.1	1-5.4.6.1 and 1-5.4.6.2	NFPA 71, 5-2.7	4-2.3.2.1.7
NFPA 71, 3-1.1.1	5-9 and 5-9.1	NFPA 71, 5-2.8	4-2.3.2.1.8
NFPA 71, 3-2.2	3-8.1.1	NFPA 71, 5-2.8.1	4-2.3.2.1.9
NFPA 71, 3-2.2.1	3-8.13.1	NFPA 71, 5-2.9	4-2.3.2.1.10
NFPA 71, 3-2.2.2	3-8.13.2	NFPA 71, 5-3.1	4-2.3.2.2.1
NFPA 71, 3-2.2.3	3-8.13.3	NFPA 71, 5-3.1.1	4-2.3.2.2.1.1
NFPA 71, 3-2.2.4	3-8.13.4	NFPA 71, 5-3.1.2	4-2.3.2.2.1.2
NFPA 71, 3-2.2.6	3-8.13.5	NFPA 71, 5-3.2.1	4-2.4.1
NFPA 71, 3-2.2.7	3-8.13.6	NFPA 71, 5-3.2.2	4-2.4.2
NFPA 71, 3-3.2.1	3-8.1.2	NFPA 71, 5-3.3	4-2.3.2.2.2
NFPA 71, 3-4.1.1	5-9.1.1	NFPA 71, 5-3.3.1	4-2.3.2.2.2.1
NFPA 71, 3-4.2.1	5-7.3	NFPA 71, 5-3.3.2	4-2.3.2.2.2.2
NFPA 71, 3-4.3	3-8.9	NFPA 71, 5-3.3.3	4-2.3.2.2.2.3
NFPA 71, 3-4.3.1	3-8.9.1 and 5-8.1	NFPA 71, Table 5-3.3.3	4-2.3.2.2.2.3(a) and (b)
NFPA 71, 3-4.4	5-10	NFPA 71, 5-3.3.4	4-2.3.2.2.2.4
NFPA 71, 3-4.4.1	3-8.9.2	NFPA 71, 5-3.3.5	4-2.3.2.2.2.5
NFPA 71, 3-4.4.2	5-10.1	NFPA 71, 5-4.1	1-5.8.6.2
NFPA 71, 3-4.4.3	5-10.2	NFPA 71, 5-4.2	1-5.8.6.3
NFPA 71, 3-4.4.4	5-10.3	NFPA 71, 5-5	4-2.3.2.3, 7-4.4.2, and 7-4.4.3
NFPA 71, 3-4.4.5	5-10.4	NFPA 71, 5-5.1	4-2.3.2.3.1
NFPA 71, 3-4.4.6	3-8.10.1	NFPA 71, 5-5.2	4-2.3.2.3.2
NFPA 71, 3-4.4.7	5-10.5	NFPA 71, 5-5.3	4-2.3.2.3.3
NFPA 71, 4-1	4-2.2	NFPA 71, 5-5.4	4-2.3.2.3.4
NFPA 71, 4-1.1	4-2.2.1	NFPA 71, 5-5.5	4-2.3.2.3.5
NFPA 71, 4-1.2	4-2.2.2	NFPA 71, 5-6	4-2.3.2.4
NFPA 71, 4-1.2.1	4-2.2.2.1	NFPA 71, 5-7	4-2.3.2.5 and 7-4.4.3.1
NFPA 71, 4-1.2.2	4-2.2.2.2	NFPA 71, 5-7.1	4-2.3.2.5.1
NFPA 71, 4-1.2.3	4-3.4.5	NFPA 71, 5-7.1.1	4-2.3.2.5.1.1
NFPA 71, 4-2.1.1	4-2.4.1	NFPA 71, 5-7.1.2	4-2.3.2.5.1.2
NFPA 71, 4-2.1.2	4-2.4.2	NFPA 71, 5-7.2.1	4-2.4.1
NFPA 71, 4-2.1.3	4-2.3.1.2.3	NFPA 71, 5-7.2.2	4-2.4.2
NFPA 71, 4-3.1	4-2.3.1.1	NFPA 71, 6-1	7-4.4.3.2
NFPA 71, 4-3.1.1	4-2.3.1.2	NFPA 71, 6-1.1	4-2.2.1
NFPA 71, 4-3.1.1.1	4-2.3.1.2.1	NFPA 71, 6-2.1	4-2.3.3.1
NFPA 71, 4-3.1.1.2	4-2.3.1.2.2	NFPA 71, 6-2.1.1	1-5.4.2.1 and 4-2.3.3.1.1
NFPA 71, 4-3.1.2	4-2.2.3, 4-2.2.3.1, and 4-2.2.3.3	NFPA 71, 6-2.1.2	1-5.4.2 and 4-2.3.3.1.2
		NFPA 71, 6-2.2	4-2.3.3.1.3
NFPA 71, 4-3.1.3	4-2.2.4, 4-2.2.4.1, and 4-2.2.4.2	NFPA 71, 6-2.3	4-2.3.3.1.4
		NFPA 71, 6-3	4-2.3.3.2
NFPA 71, 4-3.1.4	4-2.3.1.3	NFPA 71, 6-3.1	4-2.3.3.2.1
NFPA 71, 4-4.1	4-2.3.1.4	NFPA 71, 6-3.2	4-2.3.3.2.2
NFPA 71, Table 4-4.1	Table 4-2.3.1.4	NFPA 71, 6-3.3	4-2.3.3.2.3
NFPA 71, 4-4.2	4-2.3.1.5	NFPA 71, 6-3.4	4-2.3.3.2.4
NFPA 71, 5-1	7-4.4.1(a)	NFPA 71, 6-3.4.1	4-2.3.3.2.5
NFPA 71, 5-1.1	4-2.2.1	NFPA 71, 6-3.4.2	4-2.3.3.2.6
		NFPA 71, 6-4	4-2.3.3.3

Former Reference	New Reference
NFPA 71, 6-4.1.1	4-2.3.3.3.1
NFPA 71, 6-4.1.2	3-8.6.3 and 4-2.3.3.3.2
NFPA 71, 6-4.1.3	4-2.3.3.3.3
NFPA 71, 6-4.2	3-8.7.1.2 and 4-2.3.3.3.4
NFPA 71, 6-4.3	4-2.3.3.3.5
NFPA 71, 6-4.4	4-2.3.3.3.6
NFPA 71, 6-4.5	4-2.3.3.3.7
NFPA 71, 6-4.6	4-2.3.3.3.8
NFPA 71, 7-1.1	4-2.2.1
NFPA 71, 7-1.2.1	4-2.2.2.1
NFPA 71, 7-1.2.2	4-2.2.2.4 and 4-2.2.2.5
NFPA 71, 7-1.2.3	4-3.4.5
NFPA 71, 7-2.1.1	4-2.4.1
NFPA 71, 7-2.1.2	4-2.4.2
NFPA 71, 7-2.1.3	4-2.3.4.1
NFPA 71, 7-2.1.4	4-2.3.4.2
NFPA 71, 7-3	4-2.3.4.3
NFPA 71, 7-3.1	4-2.3.4.3.1
NFPA 71, 7-3.1.1	4-2.3.4.3.2
NFPA 71, 7-3.1.2	4-2.2.3, 4-2.2.3.1, and 4-2.2.3.3
NFPA 71, 7-3.1.3	4-2.2.4, 4-2.2.4.1, and 4-2.2.4.2
NFPA 71, 7-3.1.4	4-2.3.4.4
NFPA 71, 7-4	4-2.3.4.5
NFPA 71, 7-4.1	4-2.3.4.5.1
NFPA 71, 8-1.1	4-2.2.1 and 4-2.2.2.3
NFPA 71, 8-1.2.1	4-2.2.2.1
NFPA 71, 8-1.2.2	4-2.2.2.4 and 4-2.2.2.5
NFPA 71, 8-1.2.3	4-2.3.5.1 and 4-3.4.5
NFPA 71, 8-2.1.1	4-2.4.1
NFPA 71, 8-2.1.2	4-2.4.2
NFPA 71, 8-2.1.3	4-2.3.5.2
NFPA 71, 8-2.1.4	4-2.3.5.3 and 4-2.3.5.3.1
NFPA 71, 8-3	4-2.3.5.4
NFPA 71, 8-3.1	4-2.3.5.4.1
NFPA 71, 8-3.1.1	4-2.3.5.4.2
NFPA 71, 8-3.1.2	4-2.2.3, 4-2.2.3.1, 4-2.2.3.2, and 4-2.2.3.3
NFPA 71, 8-3.1.3	4-2.3.5.5
NFPA 71, 8-4.1	4-2.3.5.6
NFPA 71, Table 8-4.1	Table 4-2.3.5.6
NFPA 71, 8-4.2	4-2.3.5.7
NFPA 71, A-1-2.2	A-4-3.2.2
NFPA 71, A-1-2.4	A-4-3.2.4
NFPA 71, A-1-4.1	A-4-3.2.5
NFPA 71, A-1-7.1	A-4-2.4.2
NFPA 71, A-1-7.2.2	A-4-3.4.4.2
NFPA 71, A-1-10.2.3	A-4-3.6.1.3
NFPA 71, A-1-10.2.5(b)	A-4-3.6.1.5(b)
NFPA 71, A-2-2.1.6(c)	A-1-5.2.6(c)
NFPA 71, A-2-2.1.8	A-1-5.2.6.1
NFPA 71, A-2-2.3	A-1-5.2.9
NFPA 71, A-2-2.3.4(d)	A-1-5.2.9.2(d)

Former Reference	New Reference
NFPA 71, A-2-2.5	A-1-5.2.11
NFPA 71, A-3-4.2.1	A-5-7.2
NFPA 71, A-3-4.3	A-5-8.1
NFPA 71, A-4-1.2.3	A-4-3.4.5
NFPA 71, A-4-2.1.3(b)	A-4-2.3.1.2.3(b)
NFPA 71, A-4-3.1	A-4-2.3.1.2
NFPA 71, A-5-2.3	A-4-2.3.2.1.3
NFPA 71, A-5-2.5	A-4-2.3.2.1.5
NFPA 71, A-5-2.7	A-4-2.3.2.1.7
NFPA 71, A-5-3.2.1	A-4-2.4.1
NFPA 71, A-5-3.3.1	A-4-2.3.2.2.2.1
NFPA 71, A-5-3.3.3	A-4-2.3.2.2.2.3
NFPA 71, A-5-3.3.4	A-4-2.3.2.2.2.4
NFPA 71, A-6-2.1.2	A-4-2.3.3.1.2
NFPA 71, A-6-3.4.1(c)	A-4-2.3.3.2.5(c)
NFPA 71, A-6-3.4.2(d)(3)	A-4-2.3.3.2.6(d)(3)
NFPA 71, A-6-4.3.1	A-4-2.3.3.3.5
NFPA 71, A-7-1.2.3	A-4-3.4.5
NFPA 71, A-7-3.1.4	A-4-2.3.4.4
NFPA 71, A-8-1.2.3	A-4-2.3.5.2

NFPA 72–1990

Former Reference	New Reference
NFPA 72, 1-1	1-1
NFPA 72, 1-2.1	1-2 and 1-2.1
NFPA 72, 1-2.2	1-2.2
NFPA 72, 1-3.2	1-3, 1-3.1, 1-3.2, and 7-1.1.2
NFPA 72, 1-3.3	1-3.3
NFPA 72, 1-4	1-4
NFPA 72, 2-1	1-5.5.2
NFPA 72, 2-1.1	1-5.5.2.1
NFPA 72, 2-1.2	1-5.3 and 1-5.3.1
NFPA 72, 2-1.3	1-5.5.2.2
NFPA 72, 2-1.4	1-5.5.4
NFPA 72, 2-1.5	1-5.5.5
NFPA 72, 2-1.6	1-5.6
NFPA 72, 2-2	1-7
NFPA 72, 2-2.1	1-7.1.1 and 7-1.4
NFPA 72, 2-2.2	1-7.2.1, Figure 1-7.2.1, and 7-1.4
NFPA 72, 2-3	1-5.5
NFPA 72, 2-3.1	1-5.5.1
NFPA 72, 2-3.2	1-5.5.6
NFPA 72, 2-3.2.1	1-5.5.6.1
NFPA 72, 2-3.2.2	1-5.5.6.2
NFPA 72, 2-4.1	1-5.4.10
NFPA 72, 2-4.2	1-5.7.1.2
NFPA 72, 2-4.3	1-5.4.2, 1-5.4.2.1, and 5-9.1.3
NFPA 72, 2-4.4	1-5.4.3 and 1-5.4.3.1
NFPA 72, 2-4.5	1-5.4.3.2
NFPA 72, 2-4.6	1-5.7.1
NFPA 72, 2-4.6.1	1-5.7.1.1
NFPA 72, 2-4.6.2	1-5.4.4
NFPA 72, 2-4.6.3	1-5.4.5
NFPA 72, 2-4.7	1-5.4.6

Former Reference	New Reference
NFPA 72, 2-4.7.1	1-5.4.6.1 and 1-5.4.6.2
NFPA 72, 2-4.7.2	1-5.4.6.4
NFPA 72, 2-4.7.2.1	1-5.4.6.4.1
NFPA 72, 2-4.7.2.2	1-5.4.6.4.2
NFPA 72, 2-4.8.3	6-2.2.2 and 6-2.3
NFPA 72, 2-4.10	1-5.4.7(a) - (c)
NFPA 72, 2-4.11	1-5.4.8
NFPA 72, 2-4.12	1-5.4.9
NFPA 72, 2-5.4	7-1.2 and 7-1.2.1
NFPA 72, 2-5.6	1-5.5.2.3
NFPA 72, 2-5.7	1-7.3
NFPA 72, 2-6	3-4.2 Exceptions No. 2 and No. 3, and 3-5
NFPA 72, 2-6.2	Table 3-5.1
NFPA 72, 2-6.3	3-5.3
NFPA 72, 2-7	3-6
NFPA 72, 2-7.2	Table 3-6.1
NFPA 72, 3-1	3-3
NFPA 72, 3-2.1	5-9 and 5-9.1
NFPA 72, 3-2.2	3-8.1.1 and 5-9.1.1
NFPA 72, 3-2.3	5-9.1.2
NFPA 72, 3-2.4	3-8.1.2 and 5-9.1.2
NFPA 72, 3-2.5	3-8.1.3
NFPA 72, 3-3	3-8.2
NFPA 72, 3-3.1	3-8.2.1
NFPA 72, 3-3.2	3-8.2.2
NFPA 72, 3-3.3	3-8.2.3
NFPA 72, 3-3.4	3-8.2.4
NFPA 72, 3-3.5	3-8.2.5
NFPA 72, 3-3.6	3-8.3 and 3-8.3.1
NFPA 72, 3-3.6.1	3-8.3.1.1
NFPA 72, 3-3.6.2	3-8.3.1.2
NFPA 72, 3-3.6.3	3-8.3.2
NFPA 72, 3-3.6.4	3-8.3.3
NFPA 72, 3-3.6.5	3-8.3.4
NFPA 72, 3-3.7	3-8.4
NFPA 72, 3-4	5-10
NFPA 72, 3-4.1	3-8.6
NFPA 72, 3-4.1.1	3-8.6.1, 5-7, and 5-7.1
NFPA 72, 3-4.1.2	5-7.2
NFPA 72, 3-4.1.4	3-8.6.2
NFPA 72, 3-4.2	3-8.7
NFPA 72, 3-4.2.1	3-8.7.1
NFPA 72, 3-4.2.2	3-8.7.2
NFPA 72, 3-4.2.3	3-8.7.3
NFPA 72, 3-4.2.4	3-8.7.4
NFPA 72, 3-4.2.5	3-8.7.3 Note, 3-8.7.5, and 5-10.1
NFPA 72, 3-4.2.6	5-10.2
NFPA 72, 3-4.2.7	5-10.3
NFPA 72, 3-4.2.8	5-10.4
NFPA 72, 3-4.2.9	3-8.10 and 3-8.10.2
NFPA 72, 3-4.3	3-8.11
NFPA 72, 3-4.3.1	3-8.11.1

Former Reference	New Reference
NFPA 72, 3-4.3.2	3-8.11.2
NFPA 72, 3-5	3-8.12
NFPA 72, 3-5.1.1	3-8.12.2
NFPA 72, 3-5.1.3	3-8.12.3
NFPA 72, 3-5.2	3-8.13
NFPA 72, 3-5.2.1	3-8.13.1
NFPA 72, 3-5.2.2	3-8.13.2
NFPA 72, 3-5.2.3	3-8.13.3
NFPA 72, 3-5.2.4	3-8.13.4
NFPA 72, 3-5.2.5	3-8.13.5
NFPA 72, 3-5.2.6	3-8.13.6
NFPA 72, 3-5.2.7	3-8.12.3
NFPA 72, 3-6	3-8.14
NFPA 72, 3-6.1	3-8.14.1
NFPA 72, 3-6.2	3-8.14.2
NFPA 72, 3-6.3	3-8.14.3
NFPA 72, 3-6.4	3-8.14.4 Exception
NFPA 72, 3-6.5	3-8.14.5
NFPA 72, 3-7.1	1-5.4.1
NFPA 72, 3-7.2.1	3-8.9.3
NFPA 72, 3-7.3	3-8.15
NFPA 72, 3-7.3.1	3-8.15.1
NFPA 72, 3-7.3.2	3-8.15.2
NFPA 72, 3-7.3.3	3-8.15.3
NFPA 72, 3-7.3.5	3-8.15.4 and (a)
NFPA 72, 4-2	1-5.8
NFPA 72, 4-2.1	1-5.8.1
NFPA 72, 4-2.1 Exception No. 1	1-5.8.1 Exception No. 1
NFPA 72, 4-2.1 Exception No. 2	1-5.8.1 Exception No. 2
NFPA 72, 4-2.1 Exception No. 5	1-5.8.1 Exception No. 4
NFPA 72, 4-2.1 Exception No. 6	1-5.8.6.1 Exception No. 2
NFPA 72, 4-2.1 Exception No. 7	1-5.8.1 Exception No. 3
NFPA 72, 4-2.1 Exception No. 8	1-5.8.1 Exception No. 5 and 1-5.8.4 Exception No. 2
NFPA 72, 4-2.1 Exception No. 9	1-5.8.1 Exception No. 6
NFPA 72, 4-2.1 Exception No. 10	1-5.8.1 Exception No. 7 and Note
NFPA 72, 4-2.1 Exception No. 11	1-5.8.1 Exception No. 8
NFPA 72, 4-2.1 Exception No. 12	1-5.8.1 Exception No. 9
NFPA 72, 4-2.2	1-5.8.2 and 4-2.3.6.3
NFPA 72, 4-2.3	1-5.8.3
NFPA 72, 4-3	1-5.8.1
NFPA 72, 4-4	1-5.8.5.1
NFPA 72, 4-5	1-5.8.6
NFPA 72, 4-5.1	1-5.8.6.1
NFPA 72, 4-5.2	1-5.8.6.1
NFPA 72, 5-1	1-5.2.1

Former Reference	New Reference	Former Reference	New Reference
NFPA 72, 5-2	1-5.2.2	NFPA 72, 7-3(c)	4-7.4.1(c) and Note
NFPA 72, 5-3	1-5.2.3 Exception No. 1, Exception No. 2, and Note	NFPA 72, 7-4.2	4-7.4.4
		NFPA 72, 7-4.2.1	4-7.4.4.1
NFPA 72, 5-3.1	1-5.2.3 Exception No. 1, Exception No. 2, and Note	NFPA 72, 7-4.2.2	4-7.4.4.2
		NFPA 72, 7-4.2.3	4-7.4.4.3
NFPA 72, 5-3.2	1-5.2.4	NFPA 72, 7-4.2.4	4-7.4.4.4
NFPA 72, 5-3.3	1-5.2.5	NFPA 72, 7-4.7	4-7.4.4.5
NFPA 72, 5-3.5	1-5.2.7	NFPA 72, 7-6.1.1	4-7.4.1(b)1
NFPA 72, 5-4	1-5.2.8	NFPA 72, 7-6.1.2	4-7.4.1(b)2
NFPA 72, 5-4.1	1-5.2.8.1	NFPA 72, 7-6.1.3	4-7.4.1(b)3
NFPA 72, 5-4.2	1-5.2.8.2	NFPA 72, 7-6.1.4	4-7.4.1(b)4
NFPA 72, 5-4.3	1-5.2.8.4	NFPA 72, 7-6.1.5	4-7.4.1(b)5
NFPA 72, 5-5	1-5.2.9	NFPA 72, 7-6.1.6	4-7.4.1(b)6
NFPA 72, 5-5.1	1-5.2.9.2	NFPA 72, 7-6.1.7	4-7.4.1(b)7
NFPA 72, 5-5.2	1-5.2.9.1	NFPA 72, 7-6.1.8	4-7.4.1(b)8
NFPA 72, 5-5.3	1-5.2.9.4 and 1-5.2.9.5	NFPA 72, 7-6.2.1	4-7.4.1(a)1
NFPA 72, 5-5.4	1-5.2.9.3	NFPA 72, 7-6.2.2	4-7.4.1(a)2
NFPA 72, 5-6	1-5.2.10	NFPA 72, 7-6.3.1	4-7.4.1(c)1
NFPA 72, 5-6.1	1-5.2.10.1	NFPA 72, 7-6.3.2	4-7.4.1(c)2
NFPA 72, 5-6.2	1-5.2.10.2	NFPA 72, 7-6.3.3	4-7.4.1(c)3
NFPA 72, 5-6.3	1-5.2.10.4	NFPA 72, Chap. 8	4-5
NFPA 72, 5-6.4	1-5.2.10.5	NFPA 72, 8-1.1	4-5.1
NFPA 72, 5-6.5	1-5.2.10.6	NFPA 72, 8-2	4-5.2
NFPA 72, 6-1	3-2 Exception	NFPA 72, 8-2.1	4-5.2.1
NFPA 72, 6-2.2	3-8.6.3	NFPA 72, 8-2.2	4-5.3.1
NFPA 72, 6-2.3	3-8.7.1.2	NFPA 72, 8-2.3	4-5.3.2
NFPA 72, 6-3	1-5.8.1	NFPA 72, 8-2.4	4-5.3.3
NFPA 72, 6-4	Table 3-7.1	NFPA 72, 8-2.5	4-5.5
NFPA 72, 6-5	3-13	NFPA 72, 8-2.6	4-5.2.2
NFPA 72, 6-5.1	3-13.1	NFPA 72, 8-3.1	4-5.6.1
NFPA 72, 6-5.2	2-3.1.2 Exception No. 2 and 3-13.2	NFPA 72, 8-3.2	4-5.4.4
		NFPA 72, 8-3.3	4-5.6.2
NFPA 72, 6-5.3	3-13.3	NFPA 72, 8-3.4	4-5.4.1.2 and 4-5.4.1.5
NFPA 72, 6-5.3.1	3-13.3.1	NFPA 72, 8-3.5	4-5.4.1.2
NFPA 72, 6-5.3.2	3-13.3.2	NFPA 72, 8-3.6	4-5.4.1.5
NFPA 72, 6-5.3.3	3-13.3.3	NFPA 72, 8-4	4-5.6.4
NFPA 72, 6-5.3.4	3-13.3.4	NFPA 72, 8-5	4-2.3.6.5
NFPA 72, 6-5.3.5	3-13.3.5	NFPA 72, 8-5.1	4-2.3.6.5.1
NFPA 72, 6-5.4	3-13.4	NFPA 72, 8-5.4	4-2.3.6.5.2
NFPA 72, 6-5.4.1	3-13.4.1	NFPA 72, 8-5.9	4-2.3.6.5.2 Note
NFPA 72, 6-5.4.2	3-13.4.2	NFPA 72, 8-6.1	4-2.3.6.1
NFPA 72, 6-5.4.3	3-13.4.3	NFPA 72, 8-6.2	4-2.3.6.2
NFPA 72, 6-5.4.4	3-13.4.4	NFPA 72, 8-6.3	4-2.3.6.4
NFPA 72, 6-5.4.5	3-13.4.5	NFPA 72, 8-7	4-2.3.2
NFPA 72, 6-5.4.6	3-13.4.6	NFPA 72, 8-7.1.3	4-2.2.2.3
NFPA 72, Chap. 7	4-7	NFPA 72, 8-7.2	4-2.3.2.1
NFPA 72, 7-1	4-7.1	NFPA 72, 8-7.2.1	4-2.3.2.1.1
NFPA 72, 7-2	4-7.2	NFPA 72, 8-7.2.2	4-2.3.2.1.2
NFPA 72, 7-2.3	4-7.2.1	NFPA 72, 8-7.2.3	4-2.3.2.1.3
NFPA 72, 7-2.4	4-7.2.2	NFPA 72, 8-7.2.4	4-2.3.2.1.4
NFPA 72, 7-2.5	4-7.2.3	NFPA 72, 8-7.2.5	4-2.3.2.1.5
NFPA 72, 7-2.6	4-7.2.4	NFPA 72, 8-7.2.6	4-2.3.2.1.6 and 4-2.3.2.1.6.1
NFPA 72, 7-3	4-7.4.1	NFPA 72, 8-7.2.7	4-2.3.2.1.7
NFPA 72, 7-3(a)	4-7.4.1(a)	NFPA 72, 8-7.2.8	4-2.3.2.1.8
NFPA 72, 7-3(b)	4-7.4.1(b)	NFPA 72, 8-7.2.9	4-2.3.2.1.10

Former Reference	New Reference
NFPA 72, 8-7.3.1	4-2.3.2.2.1
NFPA 72, 8-7.3.1.1	4-2.3.2.2.1.1
NFPA 72, 8-7.3.1.2	4-2.3.2.2.1.2
NFPA 72, 8-7.3.2.1	4-2.4.1
NFPA 72, 8-7.3.2.2	4-2.4.2
NFPA 72, 8-7.3.3	4-2.3.2.2.2
NFPA 72, 8-7.3.3.1	4-2.3.2.2.2.1
NFPA 72, 8-7.3.3.2	4-2.3.2.2.2.2
NFPA 72, 8-7.3.3.3	4-2.3.2.2.2.3
NFPA 72, 8-7.3.3.4	4-2.3.2.2.2.4
NFPA 72, 8-7.3.3.5	4-2.3.2.2.2.5
NFPA 72, Chap 9	4-4
NFPA 72, 9-1	4-4.2.3
NFPA 72, 9-2	4-4.2
NFPA 72, 9-2.1	4-4.3.1
NFPA 72, 9-2.2	4-4.3.2
NFPA 72, 9-2.3	4-4.5.1 and 4-4.5.2
NFPA 72, 9-2.4	4-4.5.3
NFPA 72, 9-3.1	4-4.6.4
NFPA 72, 9-3.2	4-4.6.5
NFPA 72, 9-3.3	4-4.6.6
NFPA 72, 9-3.4	4-4.6.7.4
NFPA 72, 9-4	4-4.6.2
NFPA 72, 9-5	4-2.3.7
NFPA 72, 9-5.1	4-2.3.7.1
NFPA 72, 9-5.2	4-2.3.7.2
NFPA 72, 9-5.3	4-2.3.7.3
NFPA 72, 9-6	4-4.3.5
NFPA 72, 9-7.1	4-4.4 and 4-4.4.1
NFPA 72, 9-7.3	4-4.4.2
NFPA 72, 9-8.1.1	4-4.4.3
NFPA 72, 9-8.1.2	4-4.4.4
NFPA 72, 9-8.1.3	4-4.4.5
NFPA 72, 9-8.2	4-4.4.6
NFPA 72, 9-8.3.1	4-4.4.7
NFPA 72, 9-8.3.2	4-4.4.8
NFPA 72, 9-8.3.3	1-5.4.6.3 and 4-4.4.9
NFPA 72, 9-9	4-4.4.10
NFPA 72, 9-10	4-4.6.7
NFPA 72, 9-10.1	4-4.6.8.1 and 4-4.6.8.2
NFPA 72, 9-10.2	4-4.6.7.1
NFPA 72, 9-10.3	4-4.6.7.2
NFPA 72, 9-10.4	4-4.6.7.3
NFPA 72, 9-10.5	4-4.6.7.4
NFPA 72, Chap. 10	3-12
NFPA 72, 10-2.2	1-5.8.5.2
NFPA 72, 10-2.4	7-5.2
NFPA 72, 10-3.1	3-2.4
NFPA 72, 10-3.2	3-12.1, 3-12.2, 3-12.3, and 3-12.3.1
NFPA 72, 10-3.3	3-12.3.2
NFPA 72, 10-3.4	3-12.3.3
NFPA 72, 10-3.5	3-12.3.4
NFPA 72, 10-4	3-12.4
NFPA 72, 10-4.1	3-12.4.1

Former Reference	New Reference
NFPA 72, 10-4.2	3-12.4.2
NFPA 72, 10-4.3	3-12.4.3
NFPA 72, 10-4.3.1	3-12.4.3.1
NFPA 72, 10-4.3.2	3-12.4.3.2
NFPA 72, 10-4.3.3	3-12.4.3.3
NFPA 72, 10-4.4.1	3-12.4.4
NFPA 72, 10-4.5	3-12.4.5
NFPA 72, 10-4.5.1	3-12.4.5.1
NFPA 72, 10-4.5.2	3-12.4.5.2
NFPA 72, 10-4.6	3-12.4.6
NFPA 72, 10-4.6.1	3-12.4.6.1
NFPA 72, 10-4.6.2	3-12.4.6.2
NFPA 72, 10-4.6.3	3-12.4.6.3
NFPA 72, 10-5	3-12.6
NFPA 72, 10-5.1	3-12.6.1
NFPA 72, 10-5.2	3-12.6.2
NFPA 72, 10-5.3	3-12.6.3
NFPA 72, 10-5.5	3-12.6.4
NFPA 72, 10-5.6	3-12.6.5
NFPA 72, 10-5.7	3-12.6.6
NFPA 72, A-1-2.1	A-1-2.1
NFPA 72, A-2-1.4	A-1-5.5.4
NFPA 72, A-2-2.2	A-1-5.8.5.1 and A-1-7.2.1
NFPA 72, A-2-2.3(a)	1-7.2.2, 1-7.2.2.1, 1-7.2.2.2, and 1-7.2.2.2.1
NFPA 72, A-2-3.1(a)	A-1-5.5.1(a)
NFPA 72, Figure A-2-4.10(a)(1)	Figure A-2-2.2.2(a) and Figure A-3-7.2(a)(1)
NFPA 72, Figure A-2-4.10(a)(2)	Figure A-2-2.2.2(b) and Figure A-3-7.2(a)(2)
NFPA 72, Figure A-2-4.10(a)(3)	Figure A-2-2.2.2(c) and Figure A-3-7.2(a)(3)
NFPA 72, A-2-4.10(b)	A-1-5.4.7(b)
NFPA 72, A-2-4.3	A-1-5.4.2.1 and A-4-2.3.3.1.2
NFPA 72, A-2-6.2	A-3-4.1, A-3-4.2, A-3-5.1, and A-3-6.1
NFPA 72, A-2-7.2	A-3-4.1, A-3-4.2, A-3-5.1, and A-3-6.1
NFPA 72, A-3-3.6.5	A-3-8.2.3 and A-3-8.3.4
NFPA 72, A-3-3.7	A-3-8.4
NFPA 72, A-3-4.1.2	A-5-7.2
NFPA 72, A-3-4.3.2	A-3-8.11.2 and A-3-8.14.1
NFPA 72, A-3-7.3.5(a) and (b)	A-3-8.15.4
NFPA 72, Figure A-3-7.3.5(a)	Figure A-3-8.15.4(a)
NFPA 72, Figure A-3-7.3.5(b)	Figure A-3-8.15.4(b)
NFPA 72, Figure A-7-3(a)(1)	Figure A-4-7.4.1(a)(1)
NFPA 72, Figure A-7-3(a)(2)	Figure A-4-7.4.1(a)(2)
NFPA 72, Figure A-7-3(b)(1)	Figure A-4-7.4.1(b)(1)
NFPA 72, Figure A-7-3(b)(2)	Figure A-4-7.4.1(b)(2)
NFPA 72, Figure A-7-3(c)	Figure A-4-7.4.1(c)

Former Reference	New Reference	Former Reference	New Reference
NFPA 72, A-8-7.2.3	A-4-2.3.2.1.3	NFPA 72E, 2-7.7	5-1.4.1
NFPA 72, A-8-7.2.5	A-4-2.3.2.1.5	NFPA 72E, Chap. 3	5-2
NFPA 72, A-8-7.2.7	A-4-2.3.2.1.7	NFPA 72E, 3-1	5-2.1
NFPA 72, A-8-7.3.2.1	A-4-2.4.1	NFPA 72E, 3-1.1.1	5-2.1
NFPA 72, A-8-7.3.3.1	A-4-2.3.2.2.2.1	NFPA 72E, 3-1.1.2	5-2.2
NFPA 72, A-8-7.3.3.3	A-4-2.3.2.2.2.3	NFPA 72E, 3-2	5-2.3
NFPA 72, A-8-7.3.3.4	A-4-2.3.2.2.2.4	NFPA 72E, 3-2.1	5-2.3.1
NFPA 72, A-9-1	A-4-4.2.3	NFPA 72E, 3-2.1.2	5-2.3.1.2
NFPA 72, A-9-3.2	A-4-4.6.5	NFPA 72E, 3-2.1.3	5-2.3.1.3
NFPA 72, A-9-3.3	A-4-4.6.6	NFPA 72E, 3-2.2	5-2.3.2
NFPA 72, A-9-5.2(b)	A-4-2.3.7.2(b)	NFPA 72E, 3-2.2.1	5-2.3.2.1
NFPA 72, A-10-4.1	A-3-12.4.1	NFPA 72E, 3-2.2.2	5-2.3.2.2
NFPA 72, A-10-4.6.2	6-3.1.3, 6-3.1.4, and 6-3.1.5	NFPA 72E, 3-2.3	5-2.3.3
		NFPA 72E, 3-2.3.1	5-2.3.3.1
NFPA 72E–1990		NFPA 72E, 3-2.3.2	5-2.3.3.2
NFPA 72E, 1-1	5-1.1	NFPA 72E, 3-3	5-2.4
NFPA 72E, 1-1.2	5-1.2.1	NFPA 72E, 3-3.1	5-2.4.1 and Table 5-2.4.1
NFPA 72E, 1-2.1	5-1.1	NFPA 72E, 3-3.1.1	5-2.4.1.1
NFPA 72E, 1-2.2	5-1.2.2	NFPA 72E, 3-4	5-2.5
NFPA 72E, 1-2.3	5-1.2.2	NFPA 72E, 3-4.1	5-2.5.1
NFPA 72E, 2-1	1-4	NFPA 72E, 3-4.2	5-2.5.2
NFPA 72E, 2-1.1	1-4	NFPA 72E, 3-4.3	5-2.6
NFPA 72E, 2-2.1	1-4	NFPA 72E, 3-5	5-2.7
NFPA 72E, 2-2.1.1	5-2.3.1.1	NFPA 72E, 3-5.1	5-2.7.1
NFPA 72E, 2-2.1.3	1-4	NFPA 72E, 3-5.1.1	5-2.7.1.1
NFPA 72E, 2-2.1.3.1	5-1.2.3	NFPA 72E, 3-5.1.2	5-2.7.1.2
NFPA 72E, 2-2.1.3.2	5-1.2.4	NFPA 72E, Table 3-5.1.2	Table 5-2.7.1.2
NFPA 72E, 2-2.1.4	1-4	NFPA 72E, 3-5.2	5-2.7.2
NFPA 72E, 2-2.1.5	1-4	NFPA 72E, 3-5.3	5-2.7.3
NFPA 72E, 2-2.2.1	1-4	NFPA 72E, 3-5.4	5-2.7.4
NFPA 72E, 2-2.2.2	1-4	NFPA 72E, 3-5.4.1	5-2.7.4.1
NFPA 72E, 2-2.2.3	1-4	NFPA 72E, 3-5.4.2	5-2.7.4.2
NFPA 72E, 2-2.3.1	1-4	NFPA 72E, 3-5.4.3	5-2.7.4.3
NFPA 72E, 2-2.3.2	1-4	NFPA 72E, Chap. 4	5-3
NFPA 72E, 2-3	1-4	NFPA 72E, 4-1	5-3.1
NFPA 72E, 2-3.1	1-4	NFPA 72E, 4-1.1	5-3.1.1
NFPA 72E, 2-3.1.1	1-4	NFPA 72E, 4-1.2	5-3.1.2
NFPA 72E, 2-3.1.2	1-4	NFPA 72E, 4-1.2.1	5-3.1.3
NFPA 72E, 2-4	1-4	NFPA 72E, 4-1.2.2	5-3.1.4
NFPA 72E, 2-4.1	1-4	NFPA 72E, 4-1.3	5-3.2
NFPA 72E, 2-4.1.1	1-4	NFPA 72E, 4-2	5-3.3 and 5-3.4
NFPA 72E, 2-4.1.2	1-4	NFPA 72E, 4-2.1	5-3.3.1
NFPA 72E, 2-4.1.3	1-4	NFPA 72E, 4-2.1.1	5-3.3.1.1
NFPA 72E, 2-4.1.4	1-4	NFPA 72E, 4-2.1.2	5-3.3.1.2
NFPA 72E, 2-5.1	1-5.3 and 1-5.3.1	NFPA 72E, 4-2.2	5-3.3.2
NFPA 72E, 2-5.1.1	1-5.3.2	NFPA 72E, 4-2.2.1	5-3.3.2.1
NFPA 72E, 2-5.1.2	1-7.1.1	NFPA 72E, 4-2.2.2	5-3.3.2.2
NFPA 72E, 2-5.1.3	1-7.1.2	NFPA 72E, 4-2.3	5-3.3.3
NFPA 72E, 2-7	5-1.3	NFPA 72E, 4-2.3.1	5-3.3.3.1
NFPA 72E, 2-7.1	5-1.3.1	NFPA 72E, 4-2.3.2	5-3.3.3.2
NFPA 72E, 2-7.2	5-1.3.2	NFPA 72E, 4-2.4	5-3.3.4
NFPA 72E, 2-7.3	5-1.3.3	NFPA 72E, 4-3.1	5-3.4.1
NFPA 72E, 2-7.4	5-1.3.4	NFPA 72E, 4-3.1.1	5-3.4.2
NFPA 72E, 2-7.5	5-1.3.5 and A-5-1.3.5	NFPA 72E, 4-4	5-3.5
NFPA 72E, 2-7.6	5-1.3.6	NFPA 72E, 4-4.1	5-3.5.1

Former Reference	New Reference	Former Reference	New Reference
NFPA 72E, 4-4.1.1	5-3.5.1.1	NFPA 72E, 5-2.1.7	5-4.2.1
NFPA 72E, 4-4.1.2	5-3.5.1.2	NFPA 72E, 5-2.1.8	5-4.2.1
NFPA 72E, 4-4.2	5-3.5.2	NFPA 72E, 5-2.1.9	5-4.2.1
NFPA 72E, 4-4.2.1	5-3.5.2.1	NFPA 72E, 5-2.2	5-4.2.2
NFPA 72E, 4-4.3.1	5-3.5.3.1	NFPA 72E, 5-2.2.1	5-4.2.2.1
NFPA 72E, 4-4.3.1.1	5-3.5.3.1.1	NFPA 72E, 5-2.2.2	5-4.2.2.2
NFPA 72E, 4-4.4	5-3.5.4	NFPA 72E, 5-2.2.3	5-4.2.2.3
NFPA 72E, 4-4.5	5-3.5.5	NFPA 72E, 5-2.2.4	5-4.2.2.4
NFPA 72E, 4-4.5.1	5-3.5.5.1	NFPA 72E, 5-2.3	5-4.2.3
NFPA 72E, 4-4.5.1.1	5-3.5.5.1.1	NFPA 72E, 5-2.3.1	5-4.2.3
NFPA 72E, 4-4.5.2	5-3.5.5.2	NFPA 72E, 5-3	5-4.3
NFPA 72E, 4-4.6	5-3.5.6	NFPA 72E, 5-3.1	5-4.3.1
NFPA 72E, 4-4.6.1	5-3.5.6.1	NFPA 72E, 5-4	5-4.4
NFPA 72E, 4-4.6.2	5-3.5.6.2	NFPA 72E, 5-4.1	5-4.4.1
NFPA 72E, 4-4.7	5-3.5.7	NFPA 72E, 5-4.1.1	5-4.4.1.1
NFPA 72E, 4-4.7.1	5-3.5.7.1	NFPA 72E, 5-4.1.2	5-4.4.1.2
NFPA 72E, 4-4.7.2	5-3.5.7.2	NFPA 72E, 5-4.2	5-4.4.2
NFPA 72E, 4-4.7.3	5-3.5.7.3	NFPA 72E, 5-4.2.1	5-4.4.2.1
NFPA 72E, 4-4.7.4	5-3.5.7.4	NFPA 72E, 5-4.2.2	5-4.4.2.2
NFPA 72E, 4-4.8	5-3.5.8	NFPA 72E, 5-4.2.3	5-4.4.2.3
NFPA 72E, 4-4.8.1	5-3.5.8.1	NFPA 72E, 5-4.2.4	5-4.4.2.4
NFPA 72E, 4-4.8.2	5-3.5.8.2	NFPA 72E, 5-4.2.5	5-4.4.2.5
NFPA 72E, 4-4.9	5-3.5.9	NFPA 72E, 5-4.2.6	5-4.4.2.6
NFPA 72E, 4-4.10	5-3.5.10	NFPA 72E, 5-4.3	5-4.5
NFPA 72E, 4-5	5-3.6	NFPA 72E, 5-4.3.1	5-4.5.1
NFPA 72E, 4-5.1	5-3.6.1	NFPA 72E, 5-4.3.2	5-4.5.2
NFPA 72E, 4-5.2	5-3.6.2	NFPA 72E, 5-4.3.3	5-4.5.3
NFPA 72E, 4-5.2.1	5-3.6.2.1	NFPA 72E, 5-4.3.4	5-4.5.4
NFPA 72E, 4-6	5-3.7	NFPA 72E, 5-4.3.5	5-4.5.5
NFPA 72E, 4-6.1	5-3.7.1	NFPA 72E, 5-4.3.6	5-4.5.6
NFPA 72E, 4-6.1.1	5-3.7.1.1	NFPA 72E, 5-5.1	5-4.4.2.5
NFPA 72E, 4-6.1.2	5-3.7.1.2	NFPA 72E, 5-6	5-4.6
NFPA 72E, 4-6.1.3	5-3.7.1.3	NFPA 72E, 5-6.1	5-4.3.2
NFPA 72E, 4-6.2.1	5-3.7.2.1	NFPA 72E, 5-6.2	5-4.6.1
NFPA 72E, 4-6.2.2	5-3.7.2.2	NFPA 72E, 5-6.3	5-4.6.2
NFPA 72E, 4-6.3	5-3.7.3	NFPA 72E, 5-6.4	5-4.6.3
NFPA 72E, 4-6.3.1	5-3.7.3.1	NFPA 72E, Chap. 6	5-5
NFPA 72E, 4-6.3.2	5-3.7.3.2	NFPA 72E, 6-1.1.3	5-5.2
NFPA 72E, 4-6.4	5-3.7.5	NFPA 72E, 6-1.1.4	5-5.3
NFPA 72E, 4-6.5	5-3.7.6	NFPA 72E, 6-2	5-5.5
NFPA 72E, 4-6.5.1	5-3.7.6.1	NFPA 72E, 6-2.1	5-5.5.1
NFPA 72E, 4-6.5.3	5-3.7.6.2	NFPA 72E, 6-2.2	5-5.5.2
NFPA 72E, 4-6.5.4	5-3.7.6.3	NFPA 72E, 6-3	5-5.6
NFPA 72E, 4-6.5.4(a)	Figure 5-3.7.6.3	NFPA 72E, 6-3.1	5-5.6.1
NFPA 72E, 4-6.5.4(b)	Table 5-3.7.6.3	NFPA 72E, 6-3.1.1	5-5.6.1.1
NFPA 72E, 5-1.1	5-4.1.2	NFPA 72E, 6-3.1.2	5-5.6.1.2
NFPA 72E, 5-1.2	5-4, 5-4.1, and 5-4.1.1	NFPA 72E, 6-3.2	5-5.6.2
NFPA 72E, 5-2	5-4.2	NFPA 72E, 6-3.3	5-5.6.3
NFPA 72E, 5-2.1	5-4.2.1	NFPA 72E, 6-3.4	5-5.6.4
NFPA 72E, 5-2.1.1	5-4.2.1	NFPA 72E, 6-3.4.1	5-5.6.4.1
NFPA 72E, 5-2.1.2	5-4.2.1	NFPA 72E, 6-3.5	5-5.6.5
NFPA 72E, 5-2.1.3	5-4.2.1	NFPA 72E, 6-3.5.1	5-5.6.5.1
NFPA 72E, 5-2.1.4	5-4.2.1	NFPA 72E, 6-3.5.2	5-5.6.5.2
NFPA 72E, 5-2.1.5	5-4.2.1	NFPA 72E, 6-3.6	5-5.6.6
NFPA 72E, 5-2.1.6	5-4.2.1	NFPA 72E, 6-3.6.1	5-5.6.6.1

Former Reference	New Reference	Former Reference	New Reference
NFPA 72E, 6-3.6.2	5-5.6.6.2	NFPA 72E, 9-4.6	5-11.6.6
NFPA 72E, 6-3.6.3	5-5.6.6.3	NFPA 72E, 9-4.7	5-11.6.7
NFPA 72E, 6-3.7	5-5.6.7	NFPA 72E, 9-5	5-11.7
NFPA 72E, 6-3.7.1	5-5.6.7.1	NFPA 72E, 9-5.1	5-11.7.1
NFPA 72E, 6-3.7.2	5-5.6.7.2	NFPA 72E, 9-5.2	5-11.7.2
NFPA 72E, 6-3.8	5-5.6.8	NFPA 72E, 9-5.3	5-11.7.3
NFPA 72E, 6-3.9	5-5.6.9	NFPA 72E, 9-5.4	5-11.7.4
NFPA 72E, 6-4	5-5.7	NFPA 72E, 9-5.4.1	5-11.7.4.1
NFPA 72E, 6-4.1	5-5.7.1	NFPA 72E, 9-5.4.1.1	5-11.7.4.1.1 and Figure 5-11.7.4.1.1
NFPA 72E, 6-4.2	5-5.7.2		
NFPA 72E, 6-4.2.1	5-5.7.2.1	NFPA 72E, 9-5.4.1.2	5-11.7.4.1.2
NFPA 72E, 6-5	5-5.8	NFPA 72E, 9-5.4.1.3	5-11.7.4.1.3
NFPA 72E, 6-5.1	5-5.8.1	NFPA 72E, 9-5.4.1.4	5-11.7.4.1.4
NFPA 72E, 6-5.1.1	5-5.8.1.1	NFPA 72E, 9-5.4.2	5-11.7.4.2
NFPA 72E, 6-5.1.2	5-5.8.1.2	NFPA 72E, 9-5.4.3	5-11.7.4.3
NFPA 72E, 6-5.1.3	5-5.8.1.3	NFPA 72E, 9-5.4.3.1	5-11.7.4.3.1 and Figure 5-11.7.4.3.1
NFPA 72E, Chap. 7	5-6		
NFPA 72E, 7-1	5-6.1	NFPA 72E, 9-5.4.3.2	5-11.7.4.3.2
NFPA 72E, 7-1.1.2	5-6.2	NFPA 72E, 9-5.4.3.3	5-11.7.4.3.3
NFPA 72E, 7-2.1	5-6.3	NFPA 72E, 9-5.4.4	5-11.7.4.4
NFPA 72E, 7-2.2	5-6.4	NFPA 72E, 9-5.4.4.1	5-11.7.4.4.1
NFPA 72E, 7-3	5-6.5	NFPA 72E, 9-5.5	5-11.7.5
NFPA 72E, 7-3.1	5-6.5.1	NFPA 72E, 9-5.5.1	5-11.7.5.1 and 5-11.7.5.2
NFPA 72E, 7-3.2	5-6.5.2	NFPA 72E, A-2-7.7	A-5-1.4
NFPA 72E, 7-3.3	5-6.5.3	NFPA 72E, Figure A-2-7.7(a)	Figure A-5-1.4(a)
NFPA 72E, 7-3.4	5-6.5.4		
NFPA 72E, 8-1.2	7-1.1.1	NFPA 72E, Figure A-2-7.7(b)	Figure A-5-1.4(b)
NFPA 72E, 8-1.3	7-1.2		
NFPA 72E, 8-1.3.1	7-1.2	NFPA 72E, Figure A-3-4.1	Figure A-5-2.5.1
NFPA 72E, 8-1.3.2	7-1.2.2	NFPA 72E, Figure A-3-5.1	Figure A-5-2.7.1
NFPA 72E, 8-1.4	7-1.3	NFPA 72E, A-3-5.1(a)	Figure A-5-2.7.1(a)
NFPA 72E, 8-3.2	7-3, 7-3.1, 7-3.1.1, and 7-3.1.2	NFPA 72E, A-3-5.1(b)	Figure A-5-2.7.1(b)
		NFPA 72E, Figure A-3-5.1(c)	Figure A-5-2.7.1(c)
NFPA 72E, 8-4.1	7-4.1 and 7-4.2	NFPA 72E, Figure A-3-5.1(d)	Figure A-5-2.7.1(d)
NFPA 72E, Chap. 9	5-11		
NFPA 72E, 9-1	5-11 Note	NFPA 72E, Figure A-3-5.1.1	Figure A-5-2.7.1.1
NFPA 72E, 9-1.1	5-11.1	NFPA 72E, A-3-5.1.2	A-5-2.7.1.2
NFPA 72E, 9-1.2	5-11.2	NFPA 72E, A-3-5.2	Figure A-5-2.7.2
NFPA 72E, 9-1.3	5-11.3	NFPA 72E, A-3-5.3	A-5-2.7.3(a) and (b)
NFPA 72E, 9-2	5-11.4	NFPA 72E, A-3-5.4.1	Figure A-5-2.7.4.1
NFPA 72E, 9-2.1	5-11.4.1	NFPA 72E, A-3-5.4.2	Figure A-5-2.7.4.2
NFPA 72E, 9-2.2	5-11.4.2	NFPA 72E, A-4-1.1	A-5-3.1.1
NFPA 72E, 9-2.3	5-11.4.3	NFPA 72E, A-4-1.3	A-5-3.2
NFPA 72E, 9-2.4	5-11.4.4	NFPA 72E, A-4-2.2	A-5-3.3.2
NFPA 72E, 9-3	5-11.5	NFPA 72E, A-4-2.3	A-5-3.3.3
NFPA 72E, 9-3.1	5-11.5.1	NFPA 72E, A-4-4.1	A-5-3.5.1
NFPA 72E, 9-3.2	5-11.5.2	NFPA 72E, A-4-4.1.2	A-5-3.5.1.2
NFPA 72E, 9-3.2.1	5-11.5.2.1	NFPA 72E, Figure A-4-4.1.2	Figure A-5-3.5.1.2
NFPA 72E, 9-3.2.2	5-11.5.2.2	NFPA 72E, A-4-4.2	A-5-3.5.2
NFPA 72E, 9-4	5-11.6	NFPA 72E, Figure A-4-4.3.1.1	Figure A-5-3.5.2.1
NFPA 72E, 9-4.1	5-11.6.1		
NFPA 72E, 9-4.2	5-11.6.2	NFPA 72E, A-4-4.5.2	A-5-3.5.5.2
NFPA 72E, 9-4.3	5-11.6.3	NFPA 72E, Figure A-4-4.5.2	Figure A-5-3.5.5.2
NFPA 72E, 9-4.4	5-11.6.4	NFPA 72E, A-4-4.6	A-5-3.5.6
NFPA 72E, 9-4.5	5-11.6.5		

Former Reference	New Reference
NFPA 72E, A-4-4.7.4	A-5-3.5.7.4
NFPA 72E, A-4-5.1	A-5-3.6.1
NFPA 72E, A-4-6.1.1	A-5-3.7.1.1
NFPA 72E, A-4-6.1.2	A-5-3.7.1.2
NFPA 72E, Table A-4-6.1.4	Table A-5-3.7.1.1
NFPA 72E, Table A-4-6.1.5(a)	Table A-5-3.7.1.2(a)
NFPA 72E, Table A-4-6.1.5(b)	Table A-5-3.7.1.2(b)
NFPA 72E, Figure A-4-6.1.8(a)	Figure A-5-3.7.5(a)
NFPA 72E, Figure A-4-6.1.8(b)	Figure A-5-3.7.5(b)
NFPA 72E, A-4-6.2.2	A-5-3.7.2.2
NFPA 72E, A-4-6.4	A-5-3.7.5
NFPA 72E, A-5-2.1.1	A-5-4.2.1
NFPA 72E, A-5-2.1.6	A-5-4.2.1
NFPA 72E, A-5-2.1.9	A-5-4.2.1
NFPA 72E, A-5-3.1	A-5-4.3.1
NFPA 72E, A-5-4.1.1	A-5-4.4.1.1 and Figure A-5-4.4.1.1
NFPA 72E, A-5-4.2.1	A-5-4.4.2.1
NFPA 72E, A-5-4.2.3	A-5-4.4.2.3 and Figure A-5-4.4.2.3
NFPA 72E, A-5-4.2.4	A-5-4.4.2.4
NFPA 72E, A-5-4.2.6	A-5-4.4.2.6
NFPA 72E, A-5-4.3.1	A-5-4.5.1
NFPA 72E, A-5-4.3.2	A-5-4.5.2
NFPA 72E, A-5-4.3.5	A-5-4.5.5
NFPA 72E, A-5-4.3.6	A-5-4.5.6
NFPA 72E, A-6-1.1.1	A-5-5.1
NFPA 72E, A-6-3.1	A-5-5.6.1
NFPA 72E, A-6-3.3	A-5-5.6.3
NFPA 72E, A-6-3.6.3	A-5-5.6.6.3
NFPA 72E, A-6-4.1	A-5-5.7.1
NFPA 72E, A-6-5.1.3	A-5-5.8.1.3
NFPA 72E, A-8-1.3.2	7-1.2.2
NFPA 72E, A-9-1.1	A-5-11.1
NFPA 72E, A-9-1.2(a) and (b)	A-5-11.2
NFPA 72E, A-9-3.2.2	A-5-11.5.2.2 and Figure A-5-11.5.2.2(b)
NFPA 72E, A-9-3.2.2(a)	Figure A-5-11.5.2.2(a)
NFPA 72E, A-9-3.2.2(b)	Figure A-5-11.5.2.2(b)
NFPA 72E, A-9-3.2.2(c)	Figure A-5-11.5.2.2(c)
NFPA 72E, A-9-4.8(a)	Figure A-5-11.6.2(a)
NFPA 72E, A-9-4.8(b)	Figure A-5-11.6.2(b)
NFPA 72E, A-9-4.8(c)	Figure A-5-11.6.2(c)
NFPA 72E, A-9-5.4.3.2	Figure A-5-11.7.4.3.2
NFPA 72E, A-9-5.4.3.3	Figure A-5-11.7.4.3.3
NFPA 72E, B-1-1	B-2.1.1
NFPA 72E, B-1-2	B-2.1.2
NFPA 72E, B-1-3	B-2.1.3
NFPA 72E, B-1-4	B-2.1.4
NFPA 72E, B-1-5	B-2.1.5

Former Reference	New Reference
NFPA 72E, B-1-6	B-2.1.6
NFPA 72E, B-1-7	B-2.1.7
NFPA 72E, Appendix C	Appendix B
NFPA 72E, C-2-2.2.3	B-2.2.2.4
NFPA 72E, C-3-4.1.3	B-3.4.1.2 and B-3.4.2
NFPA 72E, C-5-2	B-1.2.2
NFPA 72E, C-5-2.2.1	Table B-3.2.2
NFPA 72E, Table C-5-2.2.2	Table B-2.2.2.3
NFPA 72E, C-5-2.3	B-1.2.4
NFPA 72E, C-5-3.4	B-3.3, B-3.3.1, B-3.3.2, and Table B-3.3.2

NFPA 72G–1989

Former Reference	New Reference
NFPA 72G, 1-2.1	6-1.1
NFPA 72G, 1-2.2	6-1.2
NFPA 72G, 1-2.3	6-1.5
NFPA 72G, 2-1.1	1-4
NFPA 72G, 2-1.1.1	1-4
NFPA 72G, 2-1.1.1.1	1-4
NFPA 72G, 2-1.1.1.2	1-4
NFPA 72G, 2-1.1.1.3	1-4
NFPA 72G, 2-1.1.2	1-4
NFPA 72G, 2-1.1.3	1-4
NFPA 72G, 2-2	1-4
NFPA 72G, 2-2.1	1-4
NFPA 72G, 2-2.2	1-4
NFPA 72G, 2-2.3	1-4
NFPA 72G, 2-3.1	1-4
NFPA 72G, 2-3.2	1-4
NFPA 72G, 2-4	6-2.2
NFPA 72G, 2-4.1	6-2.2.1
NFPA 72G, 3-1	6-3
NFPA 72G, 3-1.1.1	6-3.1.1
NFPA 72G, 3-1.1.2	6-3.1.5 Note and 6-3.2
NFPA 72G, 3-1.1.3	6-3.3
NFPA 72G, 3-1.1.4	6-3.4 and A-6-3.1
NFPA 72G, 3-2.1.1	6-4 and 6-4.1
NFPA 72G, 3-2.1.2	6-4 and 6-4.1
NFPA 72G, 3-2.2.1	6-6
NFPA 72G, 3-2.2.2	6-6.1
NFPA 72G, 3-2.3.1	6-4.2
NFPA 72G, 3-2.3.2	6-4.2.1
NFPA 72G, 3-2.3.3	6-4.2.2
NFPA 72G, 3-2.4.1	6-4.3
NFPA 72G, 3-3.1	6-5
NFPA 72G, 3-4.2	6-1.3
NFPA 72G, 3-5.1	6-2.4
NFPA 72G, 3-5.2	6-2.5
NFPA 72G, 4-1.1	6-3.5, 6-3.5.1, 6-3.5.2, and 6-3.6
NFPA 72G, 4-2.1	6-3.1.2
NFPA 72G, 4-3.1	6-2.2.2 and 6-2.3
NFPA 72G, 4-4.1	6-3.7
NFPA 72G, 5-2.1.1	6-4.4

Former Reference	New Reference	Former Reference	New Reference
NFPA 72G, 5-2.1.4	6-4.4.1, Figure 6-4.4.1, Tables 6-4.4.1(a) and (b), 6-4.4.1.1, 6-4.4.1.2, 6-4.4.2, Table 6-4.4.2, 6-4.4.2.1, 6-4.4.2.2, 6-4.4.3, Table 6-4.4.3, 6-4.4.3.1, 6-4.4.3.2, and 6-4.4.4	NFPA 74, 1-3	2-1.3
		NFPA 74, 1-3.1.1	2-1.3.1
		NFPA 74, 1-3.1.2	2-1.3.2
		NFPA 74, 1-4	1-4
		NFPA 74, Chap. 2	2-2
		NFPA 74, 2-1	2-2.1
		NFPA 74, 2-1.1	2-2.1.1
		NFPA 74, 2-1.1.1	2-2.1.1.1
NFPA 72G, 5-2.1.5	6-6.2	NFPA 74, 2-1.1.2	2-2.1.1.2
NFPA 72G, 5-3.1	6-2.2.2 and 6-2.3	NFPA 74, 2-2	2-2.2
NFPA 72G, 6-1.1	6-7	NFPA 74, 2-2.1	2-2.2.1
NFPA 72G, 7-2.1.1	6-8, 6-8.1, and 6-8.2	NFPA 74, Chap. 3	2-3 and 2-3.1
NFPA 72G, 7-3.1	6-2.2.2 and 6-2.3	NFPA 74, 3-1.1	2-3.1.1
NFPA 72G, 7-4.1	6-8.3	NFPA 74, 3-1.1.1	2-3.1.2 Exception No. 3
NFPA 72G, 7-4.2	6-8.4 and 6-8.5	NFPA 74, 3-2	2-3.2
NFPA 72G, 8-2.1	6-9.1	NFPA 74, 3-2.1	2-3.2.1
NFPA 72G, 8-2.1.1	6-9.1.1	NFPA 74, 3-2.2	2-3.2.2
NFPA 72G, 8-2.1.2	6-9.1.2	NFPA 74, 3-2.3	2-3.2.3
NFPA 72G, 8-2.2	6-9.1.3	NFPA 74, 3-2.4	2-3.2.4
NFPA 72G, 8-4.1	6-9.2	NFPA 74, 3-2.5	2-3.2.5
NFPA 72G, 8-4.1 Exception	6-9.2 Exception	NFPA 74, 3-2.7	2-3.2.6
NFPA 72G, 9-4.1	7-4.1 and 7-4.2	NFPA 74, 3-3	2-3.3
		NFPA 74, 3-3.1	2-3.3.1
NFPA 72H–1988		NFPA 74, 3-3.1(a)	2-3.3.1(a)
NFPA 72H, 1-1.3	7-1.1.2	NFPA 74, 3-3.1(b)	2-3.3.1(b)
NFPA 72H, 2-1	7-5.2	NFPA 74, 3-3.1(c)	2-3.3.1(c)
NFPA 72H, 2-2.1	7-1.3	NFPA 74, 3-3.1(d)	2-3.3.1(d)
NFPA 72H, 4-1	7-1.3, 7-2.2, Table 7-2.2, Table 7-3.1, 7-3.2, Table 7-3.2, 7-3.4	NFPA 74, 3-3.1(e)	2-3.3.1(e)
		NFPA 74, 3-3.1(f)	2-3.3.1(f)
		NFPA 74, 3-3.1(g)	2-3.3.1(g)
NFPA 72H, Figure 7-2	Figure A-7-2.2(a)	NFPA 74, 3-4	2-3.5
NFPA 72H, Figure 7-3	Figure A-7-2.2(b)	NFPA 74, Chap. 4	2-4
NFPA 72H, Figure 7-4	Figure A-7-2.2(c)	NFPA 74, 4-1	2-4.1
NFPA 72H, Figure 7-5	Figure A-7-2.2(d)	NFPA 74, 4-2	2-4.2
NFPA 72H, Figure 7-6	Figure A-7-2.2(e)	NFPA 74, 4-2.1	2-4.2.1
NFPA 72H, Figure 7-7	Figure A-7-2.2(f)	NFPA 74, 4-3	2-4.3 and 5-2.4.2
NFPA 72H, Figure 7-8	Figure A-7-2.2(g)	NFPA 74, 4-3.1	2-4.3.1
NFPA 72H, Figure 7-9	Figure A-7-2.2(h)	NFPA 74, 4-4	2-4.4
NFPA 72H, Figure 7-10.2	Figure A-7-2.2(j)	NFPA 74, 4-4.1	2-4.4.1
NFPA 72H, Figure 7-11	Figure A-7-2.2(k)	NFPA 74, 4-5	2-4.5
NFPA 72H, Figure 7-12	Figure A-7-2.2(l)	NFPA 74, 4-5.1	2-4.5.1
NFPA 72H, Figure 7-13	Figure A-7-2.2(i) and Figure A-7-2.2(m)	NFPA 74, 4-5.2	2-4.5.2
		NFPA 74, 4-5.3	2-4.5.3
NFPA 72H, Figure 7-14	Figure A-7-2.2(n)	NFPA 74, 4-5.4	2-4.5.4
NFPA 72H, Figure 7-15	Figure A-7-2.2(o)	NFPA 74, 4-5.5	2-4.5.5
NFPA 72H, Figure 7-17	Figure A-7-2.2(p)	NFPA 74, 4-7	2-4.7
		NFPA 74, 4-7.1	2-4.7.1
NFPA 74–1989		NFPA 74, 4-7.2	2-4.7.2
NFPA 74, 1-1	2-1.1	NFPA 74, Chap. 5	2-5
NFPA 74, 1-2	2-1.2	NFPA 74, 5-1	2-5.1
NFPA 74, 1-2.1	2-1.2.1	NFPA 74, 5-1.1	2-5.1.1
NFPA 74, 1-2.2	2-1.2.2	NFPA 74, 5-1.1.1	2-5.1.1.1
NFPA 74, 1-2.3	2-1.2.3	NFPA 74, 5-1.1.2	2-5.1.1.2
NFPA 74, 1-2.4	2-1.2.4	NFPA 74, 5-1.1.3	2-5.1.1.3
NFPA 74, 1-2.5	2-1.2.5	NFPA 74, 5-1.1.4	2-5.1.1.4

Former Reference	New Reference
NFPA 74, 5-1.1.5	2-5.1.1.5
NFPA 74, 5-1.2	2-5.1.2
NFPA 74, 5-1.2.1	2-5.1.2.1
NFPA 74, 5-2	2-5.2
NFPA 74, 5-2.1	2-5.2.1
NFPA 74, 5-2.1.1	2-5.2.1.1
NFPA 74, 5-2.1.2	2-5.2.1.2
NFPA 74, 5-2.1.3	2-5.2.1.3
NFPA 74, 5-2.1.4	2-5.2.1.4
NFPA 74, 5-2.1.5	2-5.2.1.5
NFPA 74, 5-2.2	2-5.2.2
NFPA 74, 5-2.2.1	2-5.2.2.1
NFPA 74, 5-2.2.2	2-5.2.2.2
NFPA 74, 5-2.2.4	2-5.2.2.4
NFPA 74, 5-2.2.5	2-5.2.2.5
NFPA 74, 5-3	2-5.3
NFPA 74, Chap. 6	2-6
NFPA 74, 6-1	2-6.1
NFPA 74, 6-2	2-6.2.1
NFPA 74, 7-1.1.1	2-7
NFPA 74, 7-1.2.1	2-7(a)
NFPA 74, 7-1.2.2	2-7(b)
NFPA 74, 7-1.2.3	2-7(c)
NFPA 74, 7-1.2.4	2-7(d)
NFPA 74, 7-1.2.5	2-7(e)
NFPA 74, 7-1.2.6	2-7(f)
NFPA 74, 7-1.2.7	2-7(g)
NFPA 74, 7-1.2.8	2-7(h)
NFPA 74, 7-1.2.9	2-7(i) and Exception
NFPA 74, A-1-1	A-2-1.1
NFPA 74, A-2-1.1	A-2-2.1.1 and Figure A-2-2.1.1.2
NFPA 74, A-2-2	A-2-2.2
NFPA 74, A-4-3	A-2-4.3 and A-5-2.4.2
NFPA 74, A-4-3.1	A-2-4.3.1 and A-5-2.6
NFPA 74, A-5-1.2.1	A-2-5.1.2.1
NFPA 74, A-5-2.1.6	A-2-5.2.1.6 2nd, 3rd, 4th paragraphs
NFPA 74, A-5-2.2.3	A-2-5.2.2.3
NFPA 74, A-5-2.2.5	A-2-5.2.2.5 and A-5-2.7
NFPA 74, A-6-1	A-2-6.1
NFPA 74, A-6-2	A-2-6.2
NFPA 74, B-1.1	A-2-5.2
NFPA 74, B-2	A-2-5.2.1 and A-2-5.2.1.6 1st paragraph
NFPA 74, Figure B-2.1.1	Figure A-2-5.2.1(a)
NFPA 74, Figure B-2.1.2	Figure A-2-5.2.1(b)
NFPA 74, Figure B-2.1.3	Figure A-2-5.2.1(c)
NFPA 74, B-3	A-2-5.2.2
NFPA 74, Figure B-3.2.1	Figure A-2-5.2.2(b)
NFPA 74, Figure B-3.4.2	Figure A-2-5.2.2(d)
NFPA 74, Appendix C	A-2
NFPA 1221–1991	
NFPA 1221, 1-3	1-4

Former Reference	New Reference
NFPA 1221, 2-1.10.2.2	A-7-2.2 and Note; A-7-3.2, Note, and (a) - (e)
NFPA 1221, 2-1.11.2	7-1.2.1
NFPA 1221, 3-1.5.3.2	A-7-2.2 and Note; A-7-3.2, Note, and (a) - (e)
NFPA 1221, Chap. 4	4-6
NFPA 1221, 4-1.1	4-6.2
NFPA 1221, 4-1.1.1	4-6.2.1
NFPA 1221, 4-1.1.2	1-5.3 and 1-5.3.1
NFPA 1221, 4-1.1.3	4-6.2.2
NFPA 1221, 4-1.1.4	4-6.2.3
NFPA 1221, 4-1.1.5	4-6.2.4
NFPA 1221, 4-1.2	4-6.3
NFPA 1221, 4-1.3	4-6.4
NFPA 1221, 4-1.3.1	1-5.3 and 1-5.3.1
NFPA 1221, 4-1.3.3	1-5.5.1
NFPA 1221, 4-1.3.4	4-6.4.1
NFPA 1221, 4-1.3.6	1-5.5.2.3
NFPA 1221, 4-1.3.9	1-5.5.1
NFPA 1221, 4-1.3.10	1-5.5.5
NFPA 1221, 4-1.3.11	1-5.5.4
NFPA 1221, 4-1.4	5-9.2
NFPA 1221, 4-1.4.1.1	4-6.4.2, 4-6.4.3, and 5-9.2.6
NFPA 1221, 4-1.4.1.2	4-6.4.4
NFPA 1221, 4-1.4.1.3	4-6.4.5 and 5-9.2.8
NFPA 1221, 4-1.4.1.4	4-6.4.6 and 5-9.2.8
NFPA 1221, 4-1.4.1.5	4-6.4.7 and 5-9.2.9
NFPA 1221, 4-1.4.1.6	5-9.2.10
NFPA 1221, 4-1.4.1.7	1-5.5.4 and 5-9.2.10
NFPA 1221, 4-1.4.1.8	4-6.4.8 and 5-9.2.7
NFPA 1221, 4-1.3.10	1-5.5.5
NFPA 1221, 4-1.3.11	1-5.5.4
NFPA 1221, 4-1.4.2	4-6.5
NFPA 1221, 4-1.4.2.1	5-9.2.1
NFPA 1221, 4-1.4.2.2	5-9.2.2
NFPA 1221, 4-1.4.2.3	5-9.2.3
NFPA 1221, 4-1.4.2.4	4-6.4.9
NFPA 1221, 4-1.4.2.5	5-9.2.4
NFPA 1221, 4-1.4.2.6	5-9.2.5
NFPA 1221, 4-1.4.3	4-6.6
NFPA 1221, 4-1.4.3.1	4-6.6
NFPA 1221, 4-1.4.3.2	4-6.6 and A-4-6.6
NFPA 1221, 4-1.4.3.3	4-6.6
NFPA 1221, 4-1.5.1.1	1-5.2.2
NFPA 1221, 4-1.5.1.2	1-5.2.8.2
NFPA 1221, 4-1.5.1.4	1-5.2.8.2
NFPA 1221, 4-1.5.2	1-5.2.3 Exception No. 1, Exception No. 2, and Note
NFPA 1221, 4-1.5.2(a)	1-5.2.4
NFPA 1221, 4-1.5.2(c)	1-5.2.5
NFPA 1221, 4-1.5.3	4-6.7
NFPA 1221, 4-1.5.3.1	4-6.7.1.6
NFPA 1221, 4-1.5.3.1.1	4-6.7.1.7

Former Reference	New Reference
NFPA 1221, 4-1.5.3.1.1.1	4-6.7.1.1
NFPA 1221, 4-1.5.3.1.1.2	4-6.7.1.2
NFPA 1221, 4-1.5.3.1.1.3	4-6.7.1.3
NFPA 1221, 4-1.5.3.1.2	4-6.7.1.8
NFPA 1221, 4-1.5.3.1.3	4-6.7.1.9
NFPA 1221, 4-1.5.3.2	4-6.7.1.4
NFPA 1221, 4-1.5.3.3	4-6.7.1.5
NFPA 1221, 4-1.5.4	4-6.7.2
NFPA 1221, 4-1.5.4.1	4-6.7.2.1
NFPA 1221, 4-1.5.4.2	4-6.7.2.2
NFPA 1221, 4-1.5.4.3	4-6.7.2.3
NFPA 1221, 4-1.5.4.4	4-6.7.2.4
NFPA 1221, 4-1.5.5	1-5.2.10 and 4-6.7.3
NFPA 1221, 4-1.5.5.1	4-6.7.3.1
NFPA 1221, 4-1.5.5.2	1-5.2.10.2 and 4-6.7.3.2
NFPA 1221, 4-1.5.5.3	4-6.7.3.3
NFPA 1221, 4-1.5.5.4	1-5.2.10.4, 1-5.2.10.5, and 4-6.7.3.4
NFPA 1221, 4-1.5.5.5	1-5.2.10.5 and 4-6.7.3.5
NFPA 1221, 4-1.5.5.6	1-5.2.10.3 and 4-6.7.3.6
NFPA 1221, 4-1.5.5.7	1-5.2.10.6 and 4-6.7.3.7
NFPA 1221, 4-1.5.5.8	4-6.7.3.8
NFPA 1221, 4-1.5.6	4-6.7.4
NFPA 1221, 4-1.5.6.1	1-5.2.9.1 and 4-6.7.4.1
NFPA 1221, 4-1.5.6.2	1-5.2.9.1 and 4-6.7.4.2
NFPA 1221, 4-1.5.6.3	1-5.2.9.1 and 4-6.7.4.3
NFPA 1221, 4-1.6.2.3	A-7-2.2 and Note; A-7-3.2, Note, and (a) - (e)
NFPA 1221, 4-1.8	4-6.8
NFPA 1221, 4-1.8.1	4-6.8.1
NFPA 1221, 4-1.8.1.1	4-6.8.1.1
NFPA 1221, 4-1.8.1.2	4-6.8.1.2
NFPA 1221, 4-1.8.1.3	4-6.8.1.3
NFPA 1221, 4-1.8.1.4	4-6.8.1.4
NFPA 1221, 4-1.8.2	4-6.8.2
NFPA 1221, 4-1.8.2.1	4-6.8.2.1
NFPA 1221, 4-1.8.2.1(a)	4-6.8.2.1.1
NFPA 1221, 4-1.8.2.1(b)	4-6.8.2.1.2
NFPA 1221, 4-1.8.2.1(c)	4-6.8.2.1.3
NFPA 1221, 4-1.8.2.1(d)	4-6.8.2.1.4
NFPA 1221, 4-1.8.2.1(e)	4-6.8.2.1.5
NFPA 1221, 4-1.8.2.1(f)	4-6.8.2.1.6
NFPA 1221, 4-1.8.2.2	4-6.8.2.2
NFPA 1221, 4-1.8.2.2(a)	4-6.8.2.2.1
NFPA 1221, 4-1.8.2.2(b)	4-6.8.2.2.2
NFPA 1221, 4-1.8.2.2(c)	4-6.8.2.2.3
NFPA 1221, 4-1.8.2.2(d)	4-6.8.2.2.4
NFPA 1221, 4-1.8.2.2(e)	4-6.8.2.2.5
NFPA 1221, 4-1.8.2.2(f)	4-6.8.2.2.6
NFPA 1221, 4-1.8.3	4-6.8.2.3
NFPA 1221, 4-1.8.3.1	4-6.8.2.3.1
NFPA 1221, 4-1.8.3.2	4-6.8.2.3.2
NFPA 1221, 4-1.8.3.3	4-6.8.2.3.3
NFPA 1221, 4-1.8.3.4	4-6.8.2.3.4
NFPA 1221, 4-1.8.3.5	4-6.8.2.3.5

Former Reference	New Reference
NFPA 1221, 4-1.8.4	4-6.8.2.4
NFPA 1221, 4-1.8.4.1	4-6.8.2.4.1
NFPA 1221, 4-1.8.4.2	4-6.8.2.4.2
NFPA 1221, 4-1.8.5	4-6.8.2.5
NFPA 1221, 4-1.8.5.1	4-6.8.2.5.1
NFPA 1221, 4-1.8.5.2	4-6.8.2.5.2
NFPA 1221, 4-1.8.5.3	4-6.8.2.5.3
NFPA 1221, 4-1.8.5.4	4-6.8.2.5.4
NFPA 1221, 4-1.8.5.5	4-6.8.2.5.5
NFPA 1221, 4-1.8.5.6	4-6.8.2.5.6
NFPA 1221, 4-1.8.5.7	4-6.8.2.5.7
NFPA 1221, 4-1.8.5.8	4-6.8.2.5.8
NFPA 1221, 4-2.1	4-6.9.1
NFPA 1221, 4-2.1.1	4-6.9.1.1
NFPA 1221, 4-2.1.1.1	4-6.9.1.1.1
NFPA 1221, 4-2.1.1.2	4-6.9.1.1.2
NFPA 1221, 4-2.1.1.3	4-6.9.1.1.3
NFPA 1221, 4-2.1.1.4	4-6.9.1.1.4
NFPA 1221, 4-2.1.1.5	4-6.9.1.1.5
NFPA 1221, 4-2.1.2	4-6.9.1.2
NFPA 1221, 4-2.1.2.1	4-6.9.1.2.1 and 5-9.2.11
NFPA 1221, 4-2.1.2.2	4-6.9.1.2.2
NFPA 1221, 4-2.1.2.3	4-6.13 and 4-6.13.1
NFPA 1221, 4-2.1.2.4	4-6.13.2
NFPA 1221, 4-2.1.2.5	4-6.13.3
NFPA 1221, 4-2.1.2.6	4-6.13.4
NFPA 1221, 4-2.1.3	4-6.9.1.3
NFPA 1221, 4-2.1.3.1	4-6.9.1.3.1
NFPA 1221, 4-2.1.3.2	4-6.9.1.3.3
NFPA 1221, 4-2.2	4-6.9.1.4
NFPA 1221, 4-2.2.1	4-6.9.1.4.1
NFPA 1221, 4-2.2.1.1	4-6.9.1.4.1.1
NFPA 1221, 4-2.2.1.2	4-6.9.1.4.1.2
NFPA 1221, 4-2.2.1.3	4-6.9.1.4.1.3
NFPA 1221, 4-2.2.1.4	4-6.9.1.4.1.4
NFPA 1221, 4-2.2.1.5	4-6.9.1.4.1.5
NFPA 1221, 4-2.2.1.6	4-6.9.1.4.1.6
NFPA 1221, 4-2.2.2	4-6.9.1.4.2
NFPA 1221, 4-2.2.2.1	4-6.9.1.4.2.1
NFPA 1221, 4-2.2.2.2	4-6.9.1.4.2.2
NFPA 1221, 4-2.2.2.3	4-6.9.1.4.3
NFPA 1221, 4-2.2.3.1	4-6.9.1.4.3.1
NFPA 1221, 4-2.2.3.2	4-6.9.1.4.3.2
NFPA 1221, 4-2.2.3.3	4-6.9.1.4.3.3
NFPA 1221, 4-2.2.3.4	4-6.9.1.4.3.4
NFPA 1221, 4-2.3	4-6.10
NFPA 1221, 4-2.3.1	4-6.10.1
NFPA 1221, 4-2.3.1.1	4-6.10.1.1
NFPA 1221, 4-2.3.1.2	4-6.10.1.2
NFPA 1221, 4-2.3.1.3	4-6.10.1.3
NFPA 1221, 4-2.3.1.4	4-6.10.1.4
NFPA 1221, 4-2.4	4-6.11
NFPA 1221, 4-2.4.1	4-6.11.2
NFPA 1221, 4-2.4.1.1	4-6.11.2.1
NFPA 1221, 4-2.4.1.2	4-6.11.2.2

Former Reference	New Reference	Former Reference	New Reference
NFPA 1221, 4-2.4.2.1	4-6.13.5	NFPA 1221, 4-4.1.2.3	4-6.15.8
NFPA 1221, 4-2.4.3	4-6.11.1	NFPA 1221, 4-4.1.3	4-6.9.1.3
NFPA 1221, 4-2.4.3.1	4-6.11.1.1	NFPA 1221, 4-4.1.3.1	4-6.9.1.3.1
NFPA 1221, 4-2.4.3.2	4-6.11.1.2	NFPA 1221, 4-4.1.3.2	4-6.9.1.3.2
NFPA 1221, 4-2.4.3.3	4-6.11.1.3	NFPA 1221, 4-4.2	4-6.9.1.4
NFPA 1221, 4-2.5	4-6.12	NFPA 1221, 4-4.2.1	4-6.9.1.4.1
NFPA 1221, 4-2.5.1	4-6.12.1	NFPA 1221, 4-4.2.1.1	4-6.9.1.4.1.1
NFPA 1221, 4-2.5.2	4-6.12.2	NFPA 1221, 4-4.2.1.2	4-6.9.1.4.1.2
NFPA 1221, 4-2.5.3	4-6.12.3	NFPA 1221, 4-4.2.1.3	4-6.9.1.4.1.3
NFPA 1221, 4-2.5.4	4-6.12.4	NFPA 1221, 4-4.2.1.4	4-6.9.1.4.1.4
NFPA 1221, 4-2.5.5	4-6.12.5	NFPA 1221, 4-4.2.1.5	4-6.9.1.4.1.5
NFPA 1221, 4-3	4-6.14	NFPA 1221, 4-4.2.1.6	4-6.9.1.4.1.6
NFPA 1221, 4-3.1	4-6.14.1	NFPA 1221, 4-4.2.2	4-6.9.1.4.2
NFPA 1221, 4-3.1.1	4-6.14.1.1	NFPA 1221, 4-4.2.2.1	4-6.9.1.4.2.1
NFPA 1221, 4-3.1.2	4-6.14.1.2	NFPA 1221, 4-4.2.2.2	4-6.9.1.4.2.2
NFPA 1221, 4-3.1.3	4-6.14.1.3	NFPA 1221, 4-4.2.2.3	4-6.9.1.4.3
NFPA 1221, 4-3.2.1	4-6.9.1.1.1 and 4-6.9.1.1.2	NFPA 1221, 4-4.2.3.1	4-6.9.1.4.3.1
NFPA 1221, 4-3.2.2	4-6.14.2	NFPA 1221, 4-4.2.3.2	4-6.9.1.4.3.2
NFPA 1221, 4-3.3	5-9.2.12	NFPA 1221, 4-4.2.3.3	4-6.9.1.4.3.3
NFPA 1221, 4-3.3.1	5-9.2.12.1	NFPA 1221, 4-4.2.3.4	4-6.9.1.4.3.4
NFPA 1221, 4-3.3.2.1	5-9.2.12.2	NFPA 1221, 4-4.3	4-6.10
NFPA 1221, 4-3.3.2.2	5-9.2.12.3	NFPA 1221, 4-4.3.1	4-6.10.1
NFPA 1221, 4-3.3.2.3	5-9.2.12.4	NFPA 1221, 4-4.3.1.1	4-6.10.1.1
NFPA 1221, 4-3.3.2.4	5-9.2.12.5	NFPA 1221, 4-4.3.1.2	4-6.10.1.2
NFPA 1221, 4-3.3.2.5	5-9.2.12.6	NFPA 1221, 4-4.3.1.3	4-6.10.1.3
NFPA 1221, 4-3.3.3	5-9.2.13	NFPA 1221, 4-4.3.1.4	4-6.10.1.4
NFPA 1221, 4-3.3.3.1	5-9.2.13.1	NFPA 1221, 4-4.4	5-9.2.14
NFPA 1221, 4-3.3.3.2	5-9.2.13.2	NFPA 1221, 4-4.4.1	5-9.2.14.1
NFPA 1221, 4-3.3.3.3	5-9.2.13.3	NFPA 1221, 4-4.4.2	5-9.2.14.2
NFPA 1221, 4-3.3.3.4	5-9.2.13.4	NFPA 1221, 4-4.5.1	4-6.15.1
NFPA 1221, 4-3.3.3.5	5-9.2.13.5	NFPA 1221, 4-4.5.2	4-6.15.2
NFPA 1221, 4-3.4	4-6.14.3	NFPA 1221, 4-4.5.3	4-6.15.3
NFPA 1221, 4-3.4.2	4-6.14.3.1	NFPA 1221, 4-4.5.4	4-6.11.1.1
NFPA 1221, 4-3.4.2.1	4-6.14.3.1.1	NFPA 1221, 4-4.5.5	4-6.11.1.2
NFPA 1221, 4-3.4.2.2	4-6.14.3.1.2	NFPA 1221, 4-4.5.6	4-6.15.4
NFPA 1221, 4-3.4.2.3	4-6.14.3.1.3	NFPA 1221, 4-4.5.7	4-6.15.5
NFPA 1221, 4-3.4.3	4-6.14.3.2	NFPA 1221, 4-4.5.8	4-6.11.1.3
NFPA 1221, 4-3.4.3.1	4-6.14.3.2.1	NFPA 1221, 4-4.7	4-6.12
NFPA 1221, 4-3.4.3.2	4-6.14.3.2.2	NFPA 1221, 4-4.7.1	4-6.12.1
NFPA 1221, 4-3.5	4-6.14.4	NFPA 1221, 4-4.7.2	4-6.12.2
NFPA 1221, 4-3.6.1	4-6.14.1.4	NFPA 1221, 4-4.7.3	4-6.12.3
NFPA 1221, 4-3.7	4-6.12	NFPA 1221, 4-4.7.4	4-6.12.4
NFPA 1221, 4-3.7.1	4-6.14.6	NFPA 1221, 4-4.7.5	4-6.12.5
NFPA 1221, 4-3.7.2	4-6.12.2	NFPA 1221, 4-5	4-6.16
NFPA 1221, 4-3.7.3	4-6.12.3	NFPA 1221, 4-5.1.1.2	4-6.9.1.1.2
NFPA 1221, 4-3.7.4	4-6.12.4	NFPA 1221, 4-5.1.1.3	4-6.9.1.1.3
NFPA 1221, 4-3.7.5	4-6.12.5	NFPA 1221, 4-5.1.1.4	4-6.9.1.1.4
NFPA 1221, 4-4	4-6.15	NFPA 1221, 4-5.1.1.5	4-6.9.1.1.5
NFPA 1221, 4-4.1.1.1	4-6.9.1.1.1	NFPA 1221, 4-5.1.2	4-6.16.1
NFPA 1221, 4-4.1.1.2	4-6.9.1.1.2	NFPA 1221, 4-5.1.2.1	4-6.16.1.1
NFPA 1221, 4-4.1.1.3	4-6.9.1.1.3	NFPA 1221, 4-5.1.2.2	4-6.16.1.2
NFPA 1221, 4-4.1.1.4	4-6.9.1.1.4	NFPA 1221, 4-5.1.2.3	4-6.16.1.3
NFPA 1221, 4-4.1.1.5	4-6.9.1.1.5	NFPA 1221, 4-5.1.3	4-6.16.1.4
NFPA 1221, 4-4.1.2.1	4-6.15.6	NFPA 1221, 4-5.2	4-6.9.1.4
NFPA 1221, 4-4.1.2.2	4-6.15.7	NFPA 1221, 4-5.2.1	4-6.9.1.4.1

Former Reference	New Reference
NFPA 1221, 4-5.2.1.1	4-6.9.1.4.1.1
NFPA 1221, 4-5.2.1.2	4-6.9.1.4.1.2
NFPA 1221, 4-5.2.1.3	4-6.9.1.4.1.3
NFPA 1221, 4-5.2.1.4	4-6.9.1.4.1.4
NFPA 1221, 4-5.2.1.5	4-6.9.1.4.1.5
NFPA 1221, 4-5.2.1.6	4-6.9.1.4.1.6
NFPA 1221, 4-5.2.2	4-6.9.1.4.2.1
NFPA 1221, 4-5.2.2.3	4-6.9.1.4.3
NFPA 1221, 4-5.2.3.1	4-6.9.1.4.3.1
NFPA 1221, 4-5.2.3.2	4-6.9.1.4.3.2
NFPA 1221, 4-5.2.3.3	4-6.9.1.4.3.3
NFPA 1221, 4-5.2.3.4	4-6.9.1.4.3.4
NFPA 1221, 4-5.3	4-6.16.1.5
NFPA 1221, 4-5.4	5-9.2.14
NFPA 1221, 4-5.4.1	5-9.2.14.2
NFPA 1221, 4-5.4.2	5-9.2.14.1
NFPA 1221, 4-5.5	4-6.16.2
NFPA 1221, 4-5.5.1	4-6.16.2.1
NFPA 1221, 4-5.5.6	4-6.16.2.2
NFPA 1221, 4-5.5.7	4-6.16.2.3
NFPA 1221, 4-5.5.8	4-6.11.1.3
NFPA 1221, 4-5.7	4-6.12
NFPA 1221, 4-5.7.1	4-6.12.2
NFPA 1221, 4-5.7.2	4-6.12.3
NFPA 1221, 4-5.7.3	4-6.12.4
NFPA 1221, 4-5.7.4	4-6.12.5
NFPA 1221, 4-5.7.5	4-6.16.2.4

Former Reference	New Reference
NFPA 1221, A-2-1.10.2.2(b)	A-7-2.2 Note
NFPA 1221, A-4-1.4.1.5	A-4-6.4.7 and A-5-9.2.9
NFPA 1221, A-4-3.3.2.2	A-5-9.2.12.3
NFPA 1221, Figure B-4.1.5.3.1.1(a)	Figure A-4-6.7.1.7(a)
NFPA 1221, Figure B-4.1.5.3.1.1(b)(1)	Figure A-4-6.7.1.7(b)(1)
NFPA 1221, Figure B-4.1.5.3.1.1(b)(2)	Figure A-4-6.7.1.7(b)(2)
NFPA 1221, Figure B-4.1.5.3.1.1(c)	Figure A-4-6.7.1.7(c)
NFPA 1221, Figure B-4.1.5.3.1.2(a)	Figure A-4-6.7.1.8(a)
NFPA 1221, Figure B-4.1.5.3.1.2(b)(1)	Figure A-4-6.7.1.8(b)(1)
NFPA 1221, Figure B-4.1.5.3.1.2(b)(2)	Figure A-4-6.7.1.8(b)(2)
NFPA 1221, Figure B-4.1.5.3.1.3(a)	Figure A-4-6.7.1.9(a)
NFPA 1221, Figure B-4.1.5.3.1.3(b)(1)	Figure A-4-6.7.1.9(b)(1)
NFPA 1221, Figure B-4.1.5.3.1.3(b)(2)	Figure A-4-6.7.1.9(b)(2)
NFPA 1221, Figure B-4.1.5.3.1.3(c)	Figure A-4-6.7.1.9(c)
NFPA 1221, Figure B-4.3.4.2.1	Figure A-4-6.14.3.1.1

Index

© 1993 National Fire Protection Association, All Rights Reserved.

The copyright in this index is separate and distinct from the copyright in the document which it indexes. The licensing provisions set forth for the document are not applicable to this index. This index may not be reproduced in whole or in part by any means without the express written permission of the National Fire Protection Association, Inc.

A

Active multiplex systems, 4-2.3.1
   Definition, 1-4
Active signaling element (definition), 1-4
Addressable device (definition), 1-4
Adverse condition (definition), 1-4
Air duct systems, smoke detectors for, 5-11.5.2
Air sampling-type detectors, 5-3.7.4, A-5-3.7.4.1
   Definition, 1-4
Alarm (definition), 1-4
Alarm notification appliances, 2-2.2
   Hearing impaired and, 2-2.3
   Testing, Table 7-2.2
Alarm service (definition), 1-4
Alarm signals, 1-5.4.2
   Deactivation, 1-5.4.8
   Intensity, 2-4.4
Alarm-silencing switch, 2-4.5.4
Alarm verification feature (definition), 1-4
Alert tone (definition), 1-4
Ambient temperature, B-6.4
Analog initiating (sensor) device (definition), 1-4
Annunciation (visible initiation), 1-5.7
Annunciators
   Definition, 1-4
   Testing, Table 7-2.2
Approved (definition), 1-4
Audible trouble-silencing switch, 1-5.4.6.4, 2-4.5.4
Authority having jurisdiction (definition), 1-4
Automatic drift compensation, 3-8.5
Automatic extinguishing system operation detector (definition), 1-4
Automatic extinguishing system supervision (definition), 1-4
Automatic fire detectors
   Definition, 1-4
   Spacing, B-1

Auxiliary box (definition), 1-4
Auxiliary fire alarm systems, 4-7
   Definition, 1-4
   Equipment, 4-7.4

B

Batteries
   Charging, A-7-3.2
   Primary, 1-5.2.11
   Storage, 1-5.2.9, A-1-5.2.9
   Testing, Table 7-2.2
Beam construction, 5-2.7.3, 5-3.5.7, 5-5.6.6
   Definition, 1-4
Box battery (definition), 1-4
Bridging point (definition), 1-4

C

Carrier (definition), 1-4
Carrier system (definition), 1-4
Ceiling (definition), 1-4
Ceiling height, B-2.3
   Definition, 1-4
   Fire development and, B-2
Ceiling surface (definition), 1-4
Central station (definition), 1-4
Central station fire alarm systems, 1-7.2.3, 4-3
   Definition, 1-4
Central station service (definition), 1-4
Certificate of completion, 1-7.2, Fig. 1-7.2.1
   Definition, 1-4
Certification (definition), 1-4
Certification of personnel (definition), 1-4
Channel (definition), 1-4
Circuit breakers, 1-5.2.8.4
Circuit interface (definition), 1-4
Circuit styles, A-7-2.2
Coded alarm signals, 1-5.4.2.1
   Definition, 6-2.1
   Designations, A-1-5.4.2.1, A-5-9.1.3
Coded supervisory signals, 1-5.4.3.1

Combination detector (definition), 1-4
Combination fire alarm (definition), 1-4
Combination system, 2-4.7
    Definition, 1-4
Communication channel (definition), 1-4
Compatibility listed (definition), 1-4
Compatibility of equipment (definition), 1-4
Concealed detectors, 3-8.4
Control equipment, protection of, 1-5.6
Control unit (definition), 1-4

D

DACR, *see* Digital alarm communicator receiver (DACR)
DACS, *see* Digital alarm communicator system (DACS)
DACT, *see* Digital alarm communicator transmitter (DACT)
DARR, *see* Digital alarm radio receiver (DARR)
DARS, *see* Digital alarm radio system (DARS)
DART, *see* Digital alarm radio transmitter (DART)
Dead air space, A-2-5.2.1.6, A-2-5.2.2
Definitions, 1-4, 5-4.2.1, 6-2.1
Delinquency signal (definition), 1-4
Derived channel, 4-2.3.2.6
    Definition, 1-4
Digital alarm communicator receiver (DACR)
    Definition, 1-4
    Supervising station systems, 4-2.3.2.2
    Testing, 7-4.4.2
Digital alarm communicator system (DACS) (definition), 1-4
Digital alarm communicator transmitter (DACT)
    Definition, 1-4
    Supervising station systems, 4-2.3.2.1
    Testing, 7-4.4.1
Digital alarm communicators, 2-4.9
Digital alarm radio receiver (DARR)
    Definition, 1-4
    Supervising station systems, 4-2.3.2.5
    Testing, 7-4.4.3.1
Digital alarm radio system (DARS)
    Definition, 1-4
    Supervising station systems, 4-2.3.2.3
    Testing, 7-4.4.3
Digital alarm radio transmitter (DART)
    Definition, 1-4
    Supervising station systems, 4-2.3.2.4
Directly-connected noncoded systems, 4-2.3.6
Display (definition), 1-4
Distinctive signals, 1-5.4.7, 3-7.2
Documentation, 1-7
Door release service, 3-9.4, 5-11.7
Door unlocking devices, 3-9.5

Dry cell battery (definition), 1-4
Dual control (definition), 1-4

E

Elevator recall for fire fighters' service, 3-8.15
Elevator shutdown, 3-8.16
Ember (definition), 5-4.2.1
Ember detector sensitivity (definition), 5-4.2.1
Ember detectors, 5-4.2.3, 5-4.5
    Definition, 5-1.2.4, 5-4.2.1
Emergency voice/alarm communications, 3-12
Engine-driven generator, 1-5.2.10
    Testing, Table 7-2.2
Evacuation (definition), 1-4
Evacuation signals, 3-7.2
    Definition, 1-4
    Tone generator, defined, 1-4
    Zoning, 3-12.5
Exit plan (definition), 1-4

F

Family living unit (definition), 1-4
Field of view (definition), 5-4.2.1
Fire alarm control (panel) unit (definition), 1-4
Fire alarm signal
    Definition, 1-4
    Tone generator, defined, 1-4
Fire alarm signals
    Initiation, 3-8.2
Fire alarm systems, 2-6.2.2; *see also* Inspections; Maintenance; Protected premises fire alarm systems; Testing
    Central station service, 4-3
    Common fundamentals, 1-5.1
    Compatibility, 1-5.3
    Definition, 1-4
    Documentation, 1-7
    Equipment, 1-5.1.2
    Installation/monitoring, 1-5.8
    Interfaces, 1-6
    Notification appliances for, Chap. 6
    Performance/limitations, 1-5.5
    Power supply, 1-5.2
    Protection of control equipment, 1-5.6
    Signaling channels, monitoring, 1-5.8
    System functions, 1-5.4
    Visible initiation (annunciation), 1-5.7
Fire command center (definition), 1-4
Fire command station, 3-12.4.5
Fire development, ceiling height and, B-2
Fire-gas detector (definition), 1-4

Fire growth, B-2.2.3
Fire rating (definition), 1-4
Fire safety function control device (definition), 1-4
Fire safety functions, 3-9
    Definition, 1-4
Fire size vs. distance, Fig. A-5-4.4.1.1
Fire supervisory control units (definition), 1-4
Fire warden (definition), 1-4
Fixed temperature detectors, 5-2.3.1
    Spacing, B-3.2
Flame (definition), 5-4.2.1
Flame detector sensitivity (definition), 5-4.2.1
Flame detectors, 5-4.2.2, 5-4.4.2
    Definition, 5-1.2.3, 5-4.2.1

G

Gas-sensing fire detectors, 5-5
General alarm signals, 6-2.1
Girders (definition), 1-4
Grounding, 1-5.5.5
Guard signal (definition), 1-4
Guard's tour
    Delinquency, 4-4.6.7.2
    Equipment testing, Table 7-2.2
    Supervision, 1-4, 3-8.12
Guard's tour box (definition), 1-4

H

Hearing impaired, alarm notification, 2-2.3
Heat detectors, 2-4.3, 5-2
    Definition, 1-4
    Heating of, calculation, B-6.3
    Installation, A-2-5.2.2
    Location, 2-5.2.2, A-2-5.2.2
    Spacing, 2-5.2.2, B-1, B-3, B-5
    System analysis, B-4
Heating, ventilating, and air conditioning (HVAC), 5-3.6,
    5-5.7
High air movement areas, 5-3.7.6, Fig. 5-3.7.6.3
High ceilings, 5-2.7.1.2
High rack storage, 5-3.7.5, A-5-3.7.5
Hold-up alarms, 1-5.4.7
Household (definition), 1-4
Household fire alarm system (definition), 1-4
Household fire warning equipment, Chap. 2, A-2
    Basic requirements, 2-2
    Combination system, 2-4.7
    Detector location/spacing, 2-5.2
    Equivalency, 2-1.3.3
    Heat detectors, 2-4.3
    Installation, 2-5
    Maintenance/testing, 2-6.2
    Marking/instructions, 2-7
    Performance, 2-4
    Power supplies, 2-3
    Required protection, 2-2.1
    Smoke detectors, 2-4.2
Hunt group (definition), 1-4

I

Indicating lights, A-4-6.4.7
Initiating device circuits
    Definition, 1-4
    Performance/capacities, 3-5, Table 3-5.1
Initiating devices, 1-5.5.6, Chap. 5
    Automatic extinguishing systems, detection of, 5-8
    Definition, 1-4
    Gas-sensing fire detectors, 5-5
    Heat detectors, 5-2
    Location/installation, 5-1.3
    Manually actuated, 5-9
    Other fire detectors, 5-6
    Publicly accessible fire service (street) boxes, 5-9.2
    Radiant energy-sensing detectors, 5-4
    Smoke detectors, 5-3, 5-11
    Sprinkler waterflow alarm, 5-7
    Supervisory signal, 5-10
    Testing, Table 7-2.2
Inspections, Chap. 7, Fig. 7-5.1
    Frequency, 7-3
    Records, 7-5
Integrated system (definition), 1-4
Interconnected fire alarm control units, 3-11
Intermediate fire alarm control unit (definition), 1-4
Ionization smoke detection, 5-3.3.1
Irregular areas, 5-2.7.1.1

L

Labeled (definition), 1-4
Leg facility (definition), 1-4
Level ceilings (definition), 1-4
Light pulse characteristics, 6-4.2
Line-type detector (definition), 1-4
Listed (definition), 1-4
Loading capacity (definition), 1-4
Local control (panel) unit (definition), 1-4
Local fire alarm system (definition), 1-4
Local fire safety functions, 1-5.4.1

Local supervisory system (definition), 1-4
Local system (definition), 1-4
Loss of power (definition), 1-4
Loudspeakers, 3-12.4.6, 6-8.2
Low power wireless systems, 2-4.8, 3-13
   Definition, 1-4
   Special requirements, A-3-13

M

Maintenance, 1-4, Chap. 7, 7-4
Manual fire alarm box (definition), 1-4
Manual fire alarm signal initiation, 3-8.1
Master box (definition), 1-4
Master control (panel) unit (definition), 1-4
McCulloh systems, 4-2.3.3, 4-2.3.3.3, 7-4.4.3.2
Metering, 1-5.2.9.4
Monitoring
   Emergency voice/alarm communication systems, 1-5.8.5
   Installation conductors, 1-5.8, 2-4.6
   Power supplies, 1-5.8.6
   Signaling channels, 1-5.8
Multiple-station alarm device (definition), 1-4
Multiple-station detector interconnection, 2-5.1.2
Multiple-station smoke detectors, 2-6.2.1
Multiplexing (definition), 1-4
Municipal fire alarm (street) box (definition), 1-4
Municipal fire alarm system (definition), 1-4
Municipal transmitter (definition), 1-4

N

Noncoded alarm signals, 6-2.1, 6-3.6
Nonrestorable initiating device (definition), 1-4
Notification appliance circuits, 1-4, 3-7, Table 3-7.1
Notification appliance (definition), 1-4
Notification appliances for fire alarm systems, Chap. 6
   Audible characteristics, 6-3
   Coded appliance characteristics, 6-7
   Definitions, 6-2.1
   Signal classification, 6-2.1
   Sleeping areas, 6-3.5
   Supplementary visible signaling method, 6-6
   Textual audible appliances, 6-8
   Textual visible appliances, 6-9
   Visible characteristics, private mode, 6-5
   Visible characteristics, public mode, 6-4

O

Off-hook (definition), 1-4
Off-premises fire alarm systems, 4-2
One-way private radio alarm systems, 4-2.3.4
On-hook (definition), 1-4
Operating mode, private alarm signals, 6-2.1
Other fire detectors (definition), 1-4
Overcurrent protection
   Light and power service, 1-5.2.8.3
   Storage batteries, 1-5.2.9.3
Owner's manuals, fire alarm systems, 1-7.2.2.1.1
Ownership (definition), 1-4

P

Paging system (definition), 1-4
Parallel telephone system (definition), 1-4
Partitions, 5-3.5.10, 5-5.6.9
Peaked ceiling types, 5-2.7.4.1, 5-3.5.8.1, 5-5.6.7.1
   Definition, 1-4
Permanent visual record (recording) (definition), 1-4
Photoelectric light obscuration smoke detection, 5-3.3.3
Photoelectric light-scattering smoke detection, 5-3.3.2
Plant (definition), 1-4
Positive alarm sequence, 3-8.3
   Definition, 1-4
Power supply
   Continuity, 1-5.2.6
   Definition, 1-4
   Remotely located control equipment, 1-5.2.7
Presignal feature, 1-5.4.10
Pressure supervision, 3-8.7.7
Pressure supervisory signal-initiating devices, 5-10.2
Primary battery, 1-5.2.11
   Definition, 1-4
Primary power supply, 1-5.2.4, 2-3.2, 2-3.3, 2-3.5
Primary trunk facility (definition), 1-4
Prime contractor (definition), 1-4
Private microwave radio systems, 4-2.3.7
Private operating mode alarm signals, 6-2.1, 6-3.2
Private radio signaling (definition), 1-4
Projected beam-type detectors, 5-3.5.3, 5-3.5.5.2, 5-3.7.3
Proprietary fire alarm system (definition), 1-4
Proprietary supervising station
   Definition, 1-4
   Systems, 4-4
Protected premises, 4-2.3.5.3.2
   Definition, 1-4
Protected premises fire alarm systems, Chap. 3
   Applications, 3-3

Emergency voice/alarm communications, 3-12
Fire safety control functions, 3-9
General, 3-2
Initiating device, performance, 3-4
Initiating device circuits, performance, 3-5
Interconnected fire alarm control units, 3-11
Low power radio system requirements, 3-13
Notification appliance, performance, 3-4
Notification appliance circuits, 3-7
Signaling line circuits, performance, 3-4, 3-6
Suppression system actuation, 3-10
System requirements, 3-8
Public fire alarm reporting systems, 4-6, 5-9.2
Coded radio reporting systems, 4-6.14
Coded wired reporting systems, 4-6.13
Design, series/parallel, 5-9.2.14
Design of boxes, 4-6.5
Equipment/installation, 4-6.4
Interconnections, requirements, 4-6.8
Location of boxes, 4-6.6
Power supply, 4-6.7, 4-6.10
Radio coded, 5-9.2.12
Receiving equipment, 4-6.11
Signal transmission facilities, 4-6.9
Supervision, 4-6.12
Telephone (parallel) reporting systems, 4-6.16
Telephone (series) reporting systems, 4-6.15
Testing, Table 7-2.2
Public fire service communication center (definition), 1-4
Public operating mode alarm signals, 6-2.1, 6-3.1
Public switched telephone network (definition), 1-4
Pump supervision, 3-8.10
Purpose of standard, 1-2

R

RACSR, *see* Radio alarm central station receiver (RACSR)
Radiant energy-sensing fire detectors, 5-4
Definition, 1-4
Radio alarm central station receiver (RACSR) (definition), 1-4
Radio alarm satellite station receiver (RASSR) (definition), 1-4
Radio alarm system (RAS) (definition), 1-4
Radio alarm transmitter (RAT) (definition), 1-4
Radio channel, 1-4
Raised floors, 5-3.5.9
RAS, *see* Radio alarm system (RAS)
RASSR, *see* Radio alarm satellite station receiver (RASSR)
RAT, *see* Radio alarm transmitter (RAT)
Rate compensation detectors, 5-2.3.2
Rate-of-rise detector, 5-2.3.3
Record drawings (definition), 1-4

Records, 1-7.3, 7-5
Referenced publications, Chap. 8, C-1
Relocation (definition), 1-4
Remote station fire alarm system (definition), 1-4
Remotely located control equipment
Power supply, 1-5.2.7
Repeater facility (definition), 1-4
Repeater station (definition), 1-4
Residential sprinkler systems, 2-2.1.1.3
Restorable initiating device (definition), 1-4
Return air system, 5-11.5.2.2
Room temperature supervisory signal-initiating devices, 5-10.5
Runner (definition), 1-4
Runner service (definition), 1-4

S

Sampling-type smoke detector, 5-3.5.4
Satellite trunk (definition), 1-4
Scanner (definition), 1-4
Scope of standard, 1-1
Secondary (standby) power supply, 2-3.4
Capacity/sources, 1-5.2.5
Testing, Table 7-2.2
Secondary trunk facility (definition), 1-4
Separate sleeping area (definition), 1-4
Shall (definition), 1-4
Shapes of ceilings (definition), 1-4
Shed ceiling types, 5-2.7.4.2, 5-3.5.8.2, 5-5.6.7.2
Definition, 1-4
Should (definition), 1-4
Signal annunciation, 3-8.8
Signal (definition), 1-4
Signal transmission sequence (definition), 1-4
Signaling line circuit interface (definition), 1-4
Signaling line circuits
Definition, 1-4
Performance/capabilities of, 3-6, Table 3-6.1
Signaling systems, *see* Fire alarm systems
Single-station alarm device (definition), 1-4
Single-station smoke detectors, 2-6.2.1
Sloped ceilings, 5-2.7.4, 5-3.5.8, 5-5.6.7
Definition, 1-4
Smoke detectors, 2-4.2
Air duct systems, 5-11.5.1, 5-11.6
Control of smoke spread, 5-11
Definition, 1-4
Design, 5-3
Door release service, 5-11.7
Environmental conditions and, Table A-5-3.7.1.1, Table A-5-3.7.1.2

Installation, 5-3, A-2-5.2.1.6
  Location, 2-5.2.1, 5-3.5, A-2-5.2.1
  Spacing, 2-5.2.1, 5-3.5
Smoke spread, detectors for control of, 5-11
Smooth ceiling (definition), 1-4
Smooth ceiling spacing, 5-2.7.1, 5-3.5.5, 5-5.6.4
Solid joist construction, 5-2.7.2, 5-3.5.6, 5-5.6.5
  Definition, 1-4
Spacing (definition), 1-4
Spark (definition), 5-4.2.1
Spark detector sensitivity (definition), 5-4.2.1
Spark detectors, 5-4.2.3, 5-4.5
  Definition, 5-1.2.4, 5-4.2.1
Special hazard equipment
  Testing, Table 7-2.2
Spot-type detectors, 5-3.5.2, 5-3.5.5.1, 5-3.7.2, 5-5.6.4.1
  Definition, 1-4
Sprinkler waterflow alarm-initiating devices, 5-7
Standard signal, 2-2.2.2
Storage batteries, 1-5.2.9
Story (definition), 1-4
Street fire alarm box (definition), 1-4
Subscriber (definition), 1-4
Subsidiary station (definition), 1-4
Supervising station fire alarm systems, Chap. 4
  Adverse conditions, 4-2.2.3
  Auxiliary, 4-7
  Central station service, 4-3
  Communication methods, 4-2.3
  Communication methods, off-premises, 4-2
  Definition, 1-4
  Digital alarm communicator systems, 4-2.3.2
  Equipment, 4-2.2.2
  Facilities, 4-3.3, 4-5.3
  Personnel, 4-4.5
  Proprietary, 4-4
  Public alarm reporting systems, 4-6
  Remote, 4-5
  System classification, 4-2.3.1.3
  System loading capacities, 4-2.3.1.4, Table 4-2.3.1.4
  Transmitters, 4-2.3.3.1
Supervisory service (definition), 1-4
Supervisory signals, 1-5.4.3, 4-4.6.7.3
  Definition, 1-4
  Initiating devices, 5-10
  Initiation, 3-8.7
  Silencing, 1-5.4.9
Supplementary (definition), 1-4
Supply air system, 5-11.5.2.1
Suppression system actuation, 3-10
Suspended ceilings, 5-3.5.9, 5-5.6.8
Switched telephone network (definition), 1-4

System interfaces, 1-6
System reacceptance testing, 7-1.6
System unit (definition), 1-4

T

Tampering, 3-8.11
Testing, Chap. 7, Fig. 7-5.1
  Frequency, 7-3
  Methods, A-7-2.2, Table 7-2.2
  Records, 7-5
Textual alarm signals, 6-2.1
Textual audible appliances, 6-8
Textual visible appliances, 6-9
Thermal lag, 5-2.3.1.2
Three-wire power supply, 1-5.2.8
Transmission channel (definition), 1-4
Transmitters, 4-2.3.3.1
  Definition, 1-4
Transponder (definition), 1-4
Trouble signals, 1-5.4.6, 4-4.6.7.4
  Definition, 1-4
  Testing, Table 7-2.2
Troubleshooting, 1-7.2.2.1.1
Trunk facility
  Definition, 1-4
  Primary, 1-4
  Secondary, 1-4
Two-way RF multiplex systems, 4-2.3.4
Two-way telephone communications service, 3-12.6
Two-wire power supply, 1-5.2.8

U

Under-voltage detection, 1-5.2.9.5
Uninterrupted power supply (UPS)
  Testing, Table 7-2.2
Uninterruptible power system bypass, 1-5.2.6.1

V

Visible characteristics, public mode, 6-4, Fig. 6-4.4.1, Table 6-4.4.1
Visible zone alarm indication, 1-5.7.1
Voice/alarm signaling service, 3-12.4
Voice and tone devices, 3-12.4.4

W

Water level supervisory signal-initiating devices, 5-10.3
Water temperature supervision, 3-8.7.8
    Signal-initiating devices, 5-10.4
Waterflow alarm signal initiation, 3-8.6
WATS (Wide Area Telephone Service) (definition), 1-4
Wavelength (definition), 5-4.2.1

Wiring, 1-5.5.4
    Diagrams, A-7-2.2

Z

Zone (definition), 1-4
Zone of origin, 1-5.7.1.2

Automatic Sprinkler Systems Handbook
helps you apply 1994 standards.

Put today's technology to work...

The *Automatic Sprinkler Systems Handbook* is the key to interpreting and applying the sprinkler standards. It includes the complete and current texts of NFPA 13, 13D and 13R, which deals with residential occupancy sprinkler systems. Plus, the handbook provides commentary and explanations of standard provisions that help you implement requirements the right way . . . the first time.

This 1994 edition includes:
   * Obstructions to sprinkler discharge
   * Use of specially listed non-metallic pipe and fittings
   * Use of flexible pipe couplings for systems subject to seismic activity
   * And much more!

☐ **Yes! Send me** _____ copies of the *Automatic Sprinkler Systems Handbook* (Item No. 2H-13HB94). **$74.50 NFPA Members $67.00. Plus $4.15 handling fee. NFPA pays all shipping charges.**

Total amount enclosed $ _____

Name _____

Company _____

Address _____

City, State, Zip _____

Prices subject to change

NFPA Member No._____

☐ I enclose a check (payable to NFPA).

☐ Please bill me.

California residents add 7.25% sales tax

| For easy ordering, call toll-free |
| **1-800-344-3555!** |
| Monday-Friday, 8:30 AM-8:00 PM, ET |

The *FIRE PROTECTION HANDBOOK*
has the answers!

When you've got tough fire protection questions...

This handbook is revised to cover the latest technology and fire facts. Each chapter, written by leading experts, gives in-depth information on the full range of fire-related topics. It has been completely reorganized from 22 sections to 10, so it is easier to use. Plus, it contains 16 new chapters, hundreds of rewritten pages, hundreds of helpful photographs, charts, illustrations, and a time-saving subject index. The *Fire Protection Handbook* covers:
   * Environmental issues in fire protection
   * Fire modeling
   * NFPA 1500 fire fighter OSHA requirements
   * Pre planning warehouse fires
   * Emergency power supply
   * Fire pre-planning
   * And much more!

☐ **Yes! Send me**_____ copies of the *Fire Protection Handbook* (Item No. 2H-FPHI791). **$109.25 NFPA Members $98.25. Plus $4.15 handling fee. NFPA pays all shipping charges.**

Total amount enclosed $ _____

Name _____

Company _____

Address _____

City, State, Zip _____

Prices subject to change

NFPA Member No._____

☐ I enclose a check (payable to NFPA).

☐ Please bill me.

California residents add 7.25% sales tax

| For easy ordering, call toll-free |
| **1-800-344-3555!** |
| Monday-Friday, 8:30 AM-8:00 PM, ET |

The 1994 *Life Safety Code* ® *Handbook*
helps you do your job right!

Apply this vital information to improve safety and health...

The NFPA *101: Life Safety Code Handbook,* just revised and updated, helps you stay up-to-date with the latest building requirements. This handbook contains the complete text of NFPA *101* plus explanatory commentary, charts, diagrams, illustrations, and photographs that help you understand and apply code requirements. It contains:
* New criteria for accessible means of egress for persons with disabilities for coordination with *Americans with Disabilities Act Accessibility Guidelines* (ADAAG)
* Requirements for new day care centers rewritten
* And more!

☐ **Yes! Send me**_____ copies of the *Life Safety Code Handbook* (Item No. 2H-10HB94). **$74.50 NFPA Members $71.75. Plus $4.15 handling fee. NFPA pays all shipping charges.**

Total amount enclosed $ _____

Name_____

Company _____

Address_____

City, State, Zip _____

Prices subject to change

NFPA Member No._____

☐ I enclose a check (payable to NFPA).

☐ Please bill me.

California residents add 7.25% sales tax

| For easy ordering, call toll-free |
| **1-800-344-3555!** |
| Monday-Friday, 8:30 AM-8:00 PM, ET |

	NO POSTAGE NECESSARY IF MAILED IN THE UNITED STATES

BUSINESS REPLY MAIL

FIRST CLASS MAIL   PERMIT NO. 3376   BOSTON, MA

POSTAGE WILL BE PAID BY ADDRESSEE

NATIONAL FIRE PROTECTION ASSOCIATION
1 BATTERYMARCH PARK
PO BOX 9101
QUINCY MA 02269-9904

	NO POSTAGE NECESSARY IF MAILED IN THE UNITED STATES

BUSINESS REPLY MAIL

FIRST CLASS MAIL   PERMIT NO. 3376   BOSTON, MA

POSTAGE WILL BE PAID BY ADDRESSEE

NATIONAL FIRE PROTECTION ASSOCIATION
1 BATTERYMARCH PARK
PO BOX 9101
QUINCY MA 02269-9904

	NO POSTAGE NECESSARY IF MAILED IN THE UNITED STATES

BUSINESS REPLY MAIL

FIRST CLASS MAIL   PERMIT NO. 3376   BOSTON, MA

POSTAGE WILL BE PAID BY ADDRESSEE

NATIONAL FIRE PROTECTION ASSOCIATION
1 BATTERYMARCH PARK
PO BOX 9101
QUINCY MA 02269-9904